Lecture Notes in Computer Science 3536

Commenced Publication in 1973
Founding and Former Series Editors:
Gerhard Goos, Juris Hartmanis, and Jan van Leeuwen

Gianfranco Ciardo   Philippe Darondeau (Eds.)

Applications and Theory of Petri Nets 2005

26th International Conference, ICATPN 2005
Miami, USA, June 20-25, 2005
Proceedings

 Springer

Volume Editors

Gianfranco Ciardo
University of California at Riverside
Department of Computer Science and Engineering
Riverside, CA 92521, USA
E-mail: ciardo@cs.ucr.edu

Philippe Darondeau
INRIA, IRISA, Campus de Beaulieu
35042 Rennes Cedex, France
E-mail: darondeau@irisa.fr

Library of Congress Control Number: 2005927320

CR Subject Classification (1998): F.1-3, C.1-2, G.2.2, D.2, D.4, J.4

ISSN   0302-9743
ISBN-10   3-540-26301-2 Springer Berlin Heidelberg New York
ISBN-13   978-3-540-26301-2 Springer Berlin Heidelberg New York

Springer is a part of Springer Science+Business Media

springeronline.com

© Springer-Verlag Berlin Heidelberg 2005
Printed in Germany

Typesetting: Camera-ready by author, data conversion by Scientific Publishing Services, Chennai, India
Printed on acid-free paper   SPIN: 11494744   06/3142   5 4 3 2 1 0

Preface

This volume contains the proceedings of the 26th International Conference on Application and Theory of Petri Nets and Other Models of Concurrency (ICATPN 2005). The Petri net conferences serve to discuss yearly progress in the field of Petri nets and related models of concurrency, and to foster new advances in the application and theory of Petri nets. The conferences typically have 100–150 participants, one third from industry and the others from universities and research institutions, and they always take place in the last week of June. Successive editions of the conference are coordinated by the Steering Committee, whose members are listed on the next page, which also supervises several other activities—see the Petri Nets World at the URL www.daimi.au.dk/PetriNets.

The 2005 conference was organized in Miami by the School of Computer Science at Florida International University (USA). We would like to express our deep thanks to the Organizing Committee, chaired by Xudong He, for the time and effort invested to the benefit of the community in making the event successful. Several tutorials and workshops were organized within the conference, covering introductory and advanced aspects related to Petri nets. Detailed information can be found at the conference URL www.cs.fiu.edu/atpn2005.

We received altogether 71 submissions from authors in 22 countries. Two submissions were not in the scope of the conference. The Program Committee selected 23 contributions from the remaining 69 submissions, classified into three categories: application papers (6 accepted, 25 submitted), theory papers (14 accepted, 40 submitted), and tool presentations (3 accepted, 4 submitted). We thank all authors who submitted papers. We would like to express our gratitude to the members of the Program Committee and the other reviewers for the extensive, careful evaluation efforts they performed before the Program Committee meeting in Rennes; their names are listed on the next two pages. We also gratefully acknowledge Martin Karusseit, of the University of Dortmund, for his technical support with the Online Conference Service, which relieved many administrative tasks in the management of the review process.

This volume contains the papers that were presented at the conference after selection by the Program Committee, plus a few other papers that summarize invited lectures given at the conference. The invited papers included in this volume reflect the lectures given by Giuliana Franceschinis, Ken McMillan, Manuel Silva, and Jeannette Wing. Two more lectures were delivered at the conference: *New Dimensions for Nets* by Carl Adam Petri, and *Processes for a Service Oriented World* by Francisco Curbera. We are much indebted to Springer for smoothing all difficulties in the preparation of this volume.

April 2005   Gianfranco Ciardo and Philippe Darondeau

Organization

Steering Committee

Wil van der Aalst, The Netherlands
Jonathan Billington, Australia
Jörg Desel, Germany
Susanna Donatelli, Italy
Serge Haddad, France
Kurt Jensen, Denmark (chair)
Jetty Kleijn, The Netherlands
Maciej Koutny, UK

Sadatoshi Kumagai, Japan
Tadao Murata, USA
Carl Adam Petri, Germany (honorary
    member)
Lucia Pomello, Italy
Wolfgang Reisig, Germany
Grzegorz Rozenberg, The Netherlands
Manuel Silva, Spain

Organizing Committee

Alicia Bullard
Shu-Ching Chen
Peter Clarke
Yi Deng

Xudong He (chair)
Steven Luis
Tadao Murata (advisor)
Sol Shatz

Tool Demonstration

Shu-Ching Chen (chair)

Program Committee

Luca Bernardinello, Italy
Soren Christensen, Denmark
Gianfranco Ciardo, USA
    (co-chair, applications)
José Manuel Colom, Spain
Philippe Darondeau, France
    (co-chair, theory)
Luis Gomes, Portugal
Roberto Gorrieri, Italy
Xudong He, USA
Kees M. van Hee, The Netherlands
Petr Jančar, Czech Republic
Guy Juanole, France
Gabriel Juhás, Germany

Hanna Klaudel, France
Jetty Kleijn, The Netherlands
Maciej Koutny, UK
Sadatoshi Kumagai, Japan
Charles Lakos, Australia
Johan Lilius, Finland
Daniel Moldt, Germany
Madhavan Mukund, India
William H. Sanders, USA
P.S. Thiagarajan, Singapore
Enrico Vicario, Italy
Hagen Voelzer, Germany
Alex Yakovlev, UK

Referees

Bharat Adsul
Alessandra Agostini
Marcus Alanen
Joao Paulo Barros
Marek Bednarczyk
Marco Bernardo
Bernard Berthomieu
Eike Best
Jerker Björkqvist
Andrea Bobbio
Frank de Boer
Tommaso Bolognesi
Marcello Bonsangue
Roland Bouroulet
Cyril Briand
Didier Buchs
Cécile Bui Thanh
Frank Burns
Nadia Busi
Lawrence Cabac
Antonella Carbonaro
Josep Carmona
Antonio Cerone
Michel Combacau
Jordi Cortadella
Anikó Costa
Deepak D'Souza
Zhengfan Dai
Pierpaolo Degano
Salem Derisavi
Jörg Desel
Raymond Devillers
Michel Diaz
Roxana Dietze
Junhua Ding
Susanna Donatelli
Zhijiang Dong
Michael Duvigneau
Joost Engelfriet
Rob Esser
Alessandro Fantechi
Berndt Farwer
Carlo Ferigato
Gianluigi Ferrari

Joern Freiheit
Yujian Fu
Guy Gallasch
Mauro Gaspari
Edgar de Graaf
Mark Griffith
Stefan Haar
Serge Haddad
Bing Han
Keijo Heljanko
David Hemer
Kunihiko Hiraishi
Hendrik Jan Hoogeboom
Ying Huang
Guillaume Hutzler
Dorothea Iglezakis
Andreas Jakoby
Ryszard Janicki
Kurt Jensen
Jens Bæk Jørgensen
Kaustubh Joshi
Tommi Junttila
Mohamed Kaaniche
Seiichi Kawata
Victor Khomenko
Ekkart Kindler
Nicolas Knaak
Geoffrey Koh
Michael Köhler
Walter Kosters
Martin Kot
Lars Kristensen
Vinh Lam
Kristian Bisgaard Lassen
Kolja Lehmann
Kamal Lodaya
Robert Lorenz
Roberto Lucchi
Olga Marroquin Alonso
Axel Martens
Cecilia Mascolo
Vesna Milijic
Toshiyuki Miyamoto
Lian Mo

K. Narayan Kumar
Christian Neumair
Apostolos Niaouris
Katsuaki Onogi
Jan Ortmann
Elisabeth Pelz
Laure Petrucci
Sibylle Peuker
Giovanni Michele Pinna
Lucia Pomello
Franck Pommereau
Ivan Porres
Christine Reese
Heiko Rölke
Olivier Roux
Eric Rozier
Luigi Sassoli
Zdenek Sawa
Karsten Schmidt
Carla Seatzu
Yang Shaofa
Christian Shelton
Tianjun Shi
Natalia Sidorova
Jason Steggles
S.P. Suresh
Tatsuya Suzuki
Koji Takahashi
Shigemasa Takai
Mikko Tiusanen
Fernando Tricas
Kohkichi Tsuji
Dietmar Tutsch
Naoshi Uchihira
Robert Valette
Somsak Vanit-Anunchai
Daniele Varacca
Marc Voorhoeve
Lisa Wells
Michael Westergaard
Luke Wildman
Dianxiang Xu
Gianluigi Zavattaro

Table of Contents

Invited Papers

Full Papers

Tool Papers

Expressiveness and Efficient Analysis of Stochastic Well-Formed Nets

Giuliana Franceschinis

Dip.Informatica, Univ. del Piemonte Orientale, Alessandria, Italy
giuliana@mfn.unipmn.it

Abstract. This paper is a survey of the Stochastic Well-formed Net (SWN) formalism evolution, in particular it discusses the expressiveness of the formalism in terms of ease of use from the modeler point of view, and briefly presents the main results that can be found in the literature about efficient (state space based) analysis of SWN models. Software tools supporting SWN design and analysis are also mentioned in the paper. The goal of the paper is not to present in details the formalism nor the analysis algorithms, but rather to recall the achieved results and to highlight open problems and possible directions for new developments in this research area.

1   Introduction

Well-formed Nets (SWN) and their stochastic extension (SWN) were first presented at the beginning of '90s [15, 16], as an evolution of (Stochastic) Regular Nets [34, 29]. SWN belong to the class of High Level Petri Nets (HLPN), and includes (stochastic) timed transitions as well as immediate transitions, with a semantics inherited from Generalized Stochastic Petri Nets (GSPN) [1]. A peculiar feature of the SWN formalism is the ability to capture in its structured color syntax the behavioral symmetries of the model which can be automatically exploited for building a smaller state space, where markings are aggregated according to an equivalence relation that preserves the interesting model properties (both qualitative and quantitative).

This paper is a survey of the SWN formalism evolution, in particular it discusses the expressiveness of the formalism in terms of ease of use from the modeler point of view, proposing some useful extensions, and briefly presents the main results that can be found in the literature about efficient (state space based) analysis of SWN models. Software tools supporting SWN design and analysis are also mentioned in the paper. The ideas presented in this paper and the cited bibliography mainly refer to the work developed in this field by researchers at the University of Piemonte Orientale (Dip. di Informatica), Alessandria, and at the University of Torino (Dip. di Informatica), Torino in Italy, at the University of Paris VI (LIP6), the University of Paris Dauphine (LAMSADE) and at the University of Reims Champagne-Ardenne in France: these research groups are actively cooperating in developing both new theoretical results, application examples and software tools for SWN.

G. Ciardo and P. Darondeau (Eds.): ICATPN 2005, LNCS 3536, pp. 1–14, 2005.

The paper is organized as follows: Sections 2 and 3 contain the discussion of the SWN expressiveness, in particular the first section considers possible extensions to the original syntax that could ease the process of building SWN models, without sacrificing the efficiency in the SWN model analysis, while the second section discusses the issue of compositional SWN model construction methods. Sections 4.1 and 4.2 surveys both consolidated and more recent methods for efficient (state space based) analysis of SWN models. Section 5 presents some existing tools that support design and analysis of SWN models. Section 6 outlines some promising directions for future development.

2   On the SWN Formalism Syntax and Possible Extensions

The Stochastic Well-formed Net (SWN) formalism has been introduced in 1990 [15] as an extension of the Regular Net (RN) formalism: it is a high level Petri net formalism similar to Coloured Petri Nets (CPN) [41, 42], but featuring a constrained colour stucture syntax which allows to automatically discover and exploit the behavioural symmetries of the model for efficient state space based analysis. The Symbolic Marking and Symbolic Firing concepts defined for SWNs lead to a Symbolic Reachability Graph construction algorithm [17] whose size can be orders of magnitude smaller than the complete RG (examples have been presented in [16, 2, 18]).

The constrained color syntax of SWN is based on the definition of a finite set of basic types, called *basic colour classes*, and a limited set of basic functions and basic predicates defined on such classes: the places and transition colour domains, and the arc functions are then constructed upon these basic objects. In fact the colours associated with places and transitions are tuples of typed elements, where the element types are the basic colour classes, and the arc functions are (sums of) tuples whose elements are basic functions defined on basic colour classes. Basic predicates can be used to restrict the possible colours of a transition or to obtain arc functions whose structure may change when applied to different color instances of a given transition.

Basic color classes are finite sets of colors, can be ordered (circular order defined through a successor function) and can be partitioned into *static subclasses* (the partition into static subclasses is called the *static partition* of a basic color class). Intuitively, colors in a basic color class are homogeneous objects (processors, processes, resources, message types, hosts in a network, etc.) and each static subclass groups objects of the same nature with similar behaviour (so if all objects in a class behave in the same way, the static partition will contain a unique static subclass).

The definition of basic color classes static partition is a delicate step in the model definition, because this is the part of the model specification that defines the possible symmetries that can be exploited by the SRG. If the modeler fails to define the coarsest partition allowing to correctly express the system behaviour, the state space aggregation automatically achieved by the SRG algorithm might be less effective.

Observe that the static partition may need refinement if the model is built in a compositional way: in fact the static partition of a given class defined for one submodel might not be the same needed in another submodel, and when the submodels are composed a new static partition taking into account those of the two component submodels must be defined.

The above considerations suggest that the modeler should be supported in the color structure definition phase by some automatic tool able to check the model color inscriptions (initial marking, arc functions, transition guards) and decide whether the static partition definition is adequate, and propose a change if need be. A work in this direction is that presented in [49, 48], where the proposal is to impose less constraints on the syntax, and let the modeler directly refer to (subsets of) color elements within basic color classes when specifying the model: an algorithm is provided that is able to automatically derive the optimal static partition based on the analysis of the model color structure. In the same work some new useful extensions to the formalism are also proposed (e.g. the possibility of considering an ordering among elements in a class, not the circular one allowed in SWN, and as a consequence the introduction of new comparison operators in basic predicates: this is actually only syntactic sugar because it can be expressed using static subclasses, but it is often very convenient for the modeler).

Up to now the "stochastic" part of the formalism has not been mentioned: in SWN transitions can either be timed or immediate (fire in zero time after enabling), the timing semantics of SWN reflects that of Generalized Stochastic Petri Nets (GSPN), and in fact the net obtained by unfolding a SWN is a GSPN. The complete specification of a GSPN model includes the definition of average firing times of timed transitions, and the priority and weights of immediate transitions: the latter task is needed to give a deterministic or probabilistic criteria for immediate transition conflict resolution (required to construct the underlying stochastic process, or to simulate the model behavior in time, for performance evaluation purposes). This task requires a clear knowledge of the potential (direct and indirect) conflicts between immediate transitions: in [47] a technique is proposed that can support a modeler in this complex task. The transition times, priority and probability definition in SWN must take into account the color structure of the model, and may influence the color classes static partition (in fact different instances of a given transition may have different average firing times or different priority/weight). Moreover the priority and weight definition method proposed in [47] can be directly applied only at the level of the net unfolding: to overcome this problem a way of expressing in a parametric and symbolic way the potential conflict relations between different instances of SWN (immediate) transitions is required, together with a symbolic calculus allowing to compute such symbolic expressions using the net color structure information. In [22, 11, 12] a syntax for expressing such structural conflict relations, and a calculus for their computation has been proposed: the chosen syntax is very similar to the SWN arc function syntax, and hence it should not be difficult for the modeler to interpret them: the extension to SWN of the priority/weight definition method developed for GSPN exploiting the above mentioned symbolic expressions is still under development.

3   Composing SWN Models

When modeling complex systems, possibly designing several models of a given system, a compositional model construction method is more convenient than a *flat* and monolithic modeling approach. In the CPN formalism [41] this idea has led to the definition of hierarchical CPNs, with the possibility of defining submodels (pages) with well defined interface nodes (ports), of including *substitution transitions* in a model, that are an abstraction for a submodel and are connected to socket nodes (to be related with port nodes in the submodel represented by the transition), and of using fusion places, which are a mechanism to merge places within or across submodels. In the Stochastic Activity Network (SAN) formalism [44], replicate and join constructs can be used to create instances of submodels (possibly several replicas of the same submodel) and compose them by merging common places. Other approaches to high level Petri nets composition have been defined, for example the Box Calculus[8] or the Cooperative Nets[45, 46], or the compositional WNs (cWN)[43]. In the context of SWN, the compositionality issue has emerged as a natural consequence of the application of the formalism and related tools to non toy case studies: the problem has been tackled both from a pragmatic point of view and from a theoretical point of view. Composition operators working by *superposition of nodes with matching labels* have been defined first for GSPNs [27] and then extended to SWNs: a composition tool, called *algebra*[7] has been implemented and integrated into the *GreatSPN* software package to support such operators.

A relevant aspect to be considered when composing colored submodels is the treatment of the color structure: one possibility that highly simplifies composition is to assume that the submodels to be composed have already a "compatible" color structure (same definition of common color classes), that the identifiers of the color elements coming from the component submodels that are intended to represent the same object match, while identifiers that represent *local* and independent objects in the two submodels are disjoint. In *algebra*, compatible color definition is assumed, shared identifiers are preceded by a special symbol, and must match in the submodels to be composed, while non shared identifiers are automatically renamed in case of a clash. This pragmatic approach however is not flexible enough: in [4] a more parametric method has been proposed, where the submodels to be composed may have *parametric color classes*, which can be instantiated when a parametric submodel is instantiated and composed with another submodel; the original idea proposed in [4] is that color definitions can *flow* from one submodel to another one, by means of color import/export specification: colors exported by one model can be imported by another model when they are composed.

Trying to generalize the idea of parametric submodels, to allow the flexible definition of submodel interface and composition operators, in the OsMoSys framework [52] the concept of *model class* has been defined. Model classes are parametric models represented using a given formalism: they have a precise interface, and they can be instantiated and connected to elements of a larger model to build a complete (solvable) model or a more complex model class. When a model class becomes a part of a larger model class, each parameter can either be istantiated or become a parameter of the new model class, and each interface can either be hidden or become an interface of the new model class (possibly with different name).

Another important issue when defining model composition, is the definition of composition operators: on one hand a minimal set of composition operators allowing to express all interesting composition schema should be defined, on the other hand if the "final user" is not an expert in the use of formal languages but rather an expert in a given application domain which wants to easily build models of different scenarios by composing submodels taken from a library, a rich and application oriented set of composition operators might be desirable. In OsMoSys, composition operators can be part of the formalism used to express the model classes to be composed, or can be an extended version of that formalism, or can be a completely new formalism, defined for the purpose of composition (this is easily realizable in OsMoSys because it is a multi-formalism framework featuring a meta-formalism, that is a language for defining formalisms supporting inheritance between formalisms). In [31] and [30] two examples of application oriented SWN composition formalism defined within the OsMoSys framework are presented.

Last but not least, (de)composition of models can be used for analysis efficiency purposes (specifically, in case of SWNs decomposition can be combined with state space aggregation in a very effective way, see [37, 38, 23], this aspect shall be further discussed in Sec. 4.2): also in this case a flexible language for describing the (de)composition structure of a model to be used for analysis purposes is needed, the OsMoSys framework can be applied also to this case.

4   Efficient Analysis

In this section analysis methods for SWNs are discussed; we focus on state space based analysis methods for both qualitative and quantitative evaluation. In Subsection 4.1 the Symbolic Reachability Graph (SRG), automatically exploiting SWN models behavioral symmetries, and the Extended Symbolic Reachability Graph (ESRG), able to deal with partially symmetric systems, are presented: the aggregate state spaces generated by these algorithms can be used both for qualitative properties analysis, through model checking, and for quantitative properties analysis, by generating and solving a *lumped* Continuous Time Markov Chain (CTMC). Some recent results on the possibility of using very efficient data structures (Data Decision Diagrams [21] and Set Decision Diagrams [50]) for storing the Symbolic Markings and the SRG to increase the state space size that can be generated for model checking purposes are also mentioned in this subsection. In Subsection 4.2 algorithms combining SRG aggregation methods and decomposition methods (based on tensor algebra representation of the CTMC) are presented: the combination of these two methods allow to increase the size of models that can be solved for performance evaluation purposes.

4.1   SRG, ESRG, and Further Evolution of ESRG for Model Checking and Performance Analysis

The main motivation for introducing the SWN formalism has been the possibility of automatically exploiting behavioral symmetries to build an aggregate state space, which can be orders of magnitude smaller than the complete state space (the maximum achievable reduction is the product of the factorial of all static subclasses cardinalities for non ordered classes, multiplied by the product of the cardinalities of ordered classes with only one static subclass): this reduction is achieved *automatically* through the definition of the Symbolic Marking and Symbolic Firing concepts. Each Symbolic Marking represents in a compact way an equivalence class of ordinary markings. Symbolic transition instances can be checkled for enabling directly on the Symbolic Marking representation and a Symbolic Firing Rule can be used to compute the new reached Symbolic Marking without ever explicitly representing the underlying ordinary markings and transition firings. A canonical representation is defined for Symbolic Markings, so that it is easy to check whether a reached Symbolic Marking corresponds to an already reached equivalent class. Hence a Symbolic Reachability graph generation algorithm can be defined [16, 17] and the analysis can be performed on this graph instead of the complete RG. When the model has to be used for performance evaluation purposes, a stochastic process must be derived and analysed: in the case of SWN the underlying stochastic process is a CTMC isomorphic to the RG. It has been proven that, due to the constraints imposed by the formalism on the color dependence of transition firing times and weights, the CTMC of a SWN satisfies both the strong and exact lumpability conditions[9] with respect to the aggregation induced by the SRG, so that performance analysis can be performed on a CTMC which is isomorphic to the SRG, and its transition rates can be directly computed from the Symbolic Marking and Symbolic Firing information without need to ever expanding the complete RG.

It is also interesting to observe that the Symbolic Marking and Symbolic Firing can be successfully exploited also in discrete event stochastic simulation of SWN [32]: the gain in this case is due to the event list management, in fact the event list contains the set of enabled transition instances in the current state, and since one symbolic transition instance may group several ordinary transition instances, fewer enabling tests are required at each state change, furthermore the resulting "symbolic" event list can be much smaller than the ordinary one thus saving both time and space.

Unfortunately the effectiveness of the SRG method vanishes if the behaviour of a model is not completely symmetric: as already discussed in Sec. 2 this is related to the number and cardinality of static subclasses within each basic color class. In the worst case where all non ordered basic color classes are partitioned in cardinality one static subclasses and all ordered basic color classes are partitioned into two or more static subclasses, the SRG coincides with the RG. Often however, it happens that the asymmetries in the model behavior involve only a small part of the state space, but they influence the level of aggregation of the whole RG. This consideration led to the evolution of the SRG called Extended SRG (ESRG) [36]: it exploits symmetries in a more flexible way, and takes into account the partition into static subclasses only when actually needed. The ESRG structure uses a two level representation of the aggregate states: the Extended Symbolic Markings in fact have a Symbolic Representation (SR) (first level) that groups all the states that *would be equivalent* if all color classes had only one static subclass, and a set of Eventualities (Ev) (second level) that are the SRG Symbolic Markings included in the ESM SR. The gain that can be obtained from the ESRG is due to the fact that eventualities are explicitly represented only when actually needed, that is only when asymmetries actually arise in the behavior (typically because some transition which depend on the static subclasses partition is enabled). Hence the effectiveness of ESRG aggregation depends on how often asymmetries influence the system evolution. A weak point of the ESRG, with respect to the SRG, is that not all properties can be checked on the ESRG, and in some case eventualities that were not created by the ESRG algorithm, need to be generated in a second phase when properties are checked on the ESRG or when a lumped CTMC has to be generated for quantitative evaluation [13, 14, 5]. In particular when a CTMC has to be generated from the ESRG, the lumpability condition must be checked (while in the SRG it is ensured by construction) on some markings: depending on the type of performance index to be computed, one can choose whether to check for strong or exact lumpability, achieving different degrees of aggregation. The lumpability check algorithms working on the ESRG [14, 5] are in general more efficient than those that may be applied directly on the SRG (without a clue on the possible additional aggregations not captured by the SRG).

A further evolution of the ESRG consists in allowing a *partial refinement* of a SR into its eventualities, which in some sense means adding a third level in the ESM representation laying between SR and Ev: in fact subsets of eventualities could remain aggregated, and this may happen when in a given state only the static partition of a few classes is involved, while those of the other classes do not influence the local behavior of the system (see [10, 35, 3, 39]). The E^{2}SRG first introduced in [10] and applied in [6] is an example of such three level representation of a ESM: it has not yet been implemented, and still needs some work to be completely worked out. This is one possible direction of further evolution of these SRG-ESRG methods.

Combining two types of symmetries: SRG and Decision Diagrams The type of symmetries exploited by the SRG are based on a notion of equivalence of similar markings, however another type of symmetry can be exploited based on the decomposition of the state into sub-states and factorisation of substates: this type of symmetry in the states structure can be efficiently captured and exploited by using specific data structures, the Decision Diagrams. The SRG technique and DD based techniques however are not easily combined because of the mutual dependence of sub-states in Symbolic Markings: this dependence is due to presence of Dynamic Subclasses in its representation, whose definition and evolution depend on the whole state (so that the representation of a given sub-marking after a symbolic firing may change even if the fired transition had no connections with such sub-marking). Recently the problem has been faced by means of a new type of Decision Diagram called Data Decision Diagrams (DDD) and Set Decision Diagrams (SDD)[21, 48, 50]) that have shown to be very effective for storing the states of huge SRG. Up to now these new techniques have been exploited only in the context of model checking algorithms: it would be interesting to investigate the possibility of maintaining the same efficiency in the CTMC generation phase.

4.2   De-composing SWN Models

In this section the analysis methods proposed in [37, 38, 23] are briefly summarized: the main idea behind these methods consists of combining two orthogonal techniques for coping with the state space explosion problem that often arises when the CTMC of a complex SWN model is generated for performance analysis. The first technique is *state aggregation*, which in the context of SWN can be obtained using the SRG generation algorithm, the second technique is *decomposition* of a model into submodels and representation of the CTMC infinitesimal generator (the matrix of state transition rates) as an expression of much smaller generators, obtained from each submodel in isolation and combined through tensor algebra (Kronecker) operators (see [25, 24] for the application of this technique to GSPN models). In the latter technique, the CTMC is solved using its *decomposed* representation, so that the space required to generate and solve the CTMC is considerably smaller than that required to store the complete CTMC explicitly.

Two way of (de)composing a SWN model are considered: synchronous and asynchronous. In both cases the component submodels need to be extended in order to compute from each of them in isolation a local SRG. The method is not completely general: in fact it can be applied to a decomposed SWN model only if some constraints are satisfied: the constraints are needed to ensure that the lumped CTMC resulting from Kronecker composition of local lumped CTMCs is equivalent to the lumped CTMC corresponding to the SRG of the whole model. The subclass of SWN models that satisfy the constraints may appear as rather restricted, but it is representative of a relevant class of models from interesting application domains. In [37, 38] both semantic conditions (defined on the state space) and syntactic (sufficient) conditions (defined on the net structure and its decomposition) are defined to ensure that the method can be applied.

These composition/decomposition methods have been implemented (see Sec. 5) and partially integrated in GreatSPN: the most difficult part for the application of the method is that currently no automatic support is provided to decompose the model, build the extended subnets, and check that the syntactic conditions for the applicability of the method are satisfied. Some theoretical results about structural analysis of SWNs are available, but more work is needed to adapt them to this specific problem, and to develop an automatized (or at least semi-automatized) procedure for the application of these methods.

5   Tools for the Design and Analysis of SWN Models

This section reports the current state of the main software tools supporting SWN analysis: these tools have been developed at the University of Piemonte Orientale (Dip. di Informatica), Alessandria, and at the University of Torino (Dip. di Informatica), Torino in Italy, at the University of Paris VI (LIP6), the University of Paris Dauphine (LAMSADE) and at the University of Reims Champagne-Ardenne in France: there is an ongoing effort to integrate these tools. A certain level of integration is already partially implemented[40].

GreatSPN extended with some new modules GreatSPN [19] is a software tool supporting GSPN and SWN models: it can be used to design a model through a GUI, to build a model by composing several submodels (this feature however cannot be performed through the GUI), and to simulate or solve the models. Both symbolic and ordinary reachability graph can be constructed, and the corresponding CTMC automatically generated and solved (both steady state and transient solutions are supported). Steady state discrete event simulation (using the batch means method and providing confidence intervals for the required performance measures) can be performed either by using symbolic marking and firing, or using the ordinary marking and firing.

In order to exploit analysis modules developed by other research groups, some translators have been implemented, that transform the internal model representation of GreatSPN in the representation expected by such analysis modules. For example a translator from GreatSPN representation for SWN models to an high level Petri net model accepted by the PROD tool has been implemented [26] to allow verification of qualitative properties in PROD.

New modules for efficient analysis have been recently implemented and interfaced with GreatSPN (although not yet fully integrated in it). *TenSWN*, developed at LAMSADE, Univ. Paris Dauphine and at the Univ. Reims Champagne-Ardenne, is a tool that applies the decomposition and aggregation solution techniques introduced in Sec. 4.2 to properly decomposed SWN models. It relies on the GreatSPN SRG generator to develop the SRG of each submodel, then it computes the CTMC and its solution through very efficient techniques and using very efficient data structures, the Multivalued Decision Diagrams (MDD) to represent the whole set of SRG states in a very compact form, and MatriX Diagrams (MXD) [20] to efficiently represent the Kronecker expression for the lumped CTMC[23]. The *WNESRG* module, implemented at the Univ. of Paris VI, builds the Extended SRG of a SWN model. A module that checks the lumpability of the CTMC of the SWN model with respect to the aggregation induced by the ESRG, refines the ESRG if necessary, and generates the final lumped CTMC is being implemented [5].

CPN-AMI and SPOT CPN-AMI is a Petri Net based CASE tool for the verification of parallel systems developed at LIP6, Univ. of Paris VI. It includes a graphical interface (Macao), and integrates several solution modules developed at the Univ. of Paris VI, as well as in other research institutions, for example it integrates several solution modules of GreatSPN. The most recent tools integrated in CPN-AMI are a module for automatically discovering the asymmetries of a model expressed with extended SWN syntax, and translating it into a proper SWN model: the module uses the technique presented in [49]. It also includes a model checking tool based on Data Decision Diagrams[21], a very efficient data structure for the state space representation, which can be used to store the SWN symbolic state space. An evolution of the tool based on a new data structure defined very recently [50] is under development. Some model checking tools integrated in CPN-AMI rely on SPOT [28], a C++ model checking library including a set of building blocks for constructing model checkers. The GreatSPN SRG generation module has been integrated (through an adaptation layer) with the SPOT library to allow model checking of LTL (Linear-Time Temporal Logic) formulas on the SRG of SWN models [48].

DrawNET The tools so far discussed, mainly deal with the problem of model analysis, while in Sec. 3 the importance of a support for SWN model construction has been highlighted, in particular user friendly tools are needed for compositional model construction. In the last years, a new tool called DrawNET has been designed, some versions have been implemented[51], and a new version is now under devlopment[33]. DrawNET comprises several components, including (1) a customizable graphical interface that can be configured to design hierarchical models expressed with user defined formalisms (multi-formalism models can also be built), (2) a suite of XML based languages to describe (a) graph based formalisms, (b) models expressed with such formalisms, and (c) the results that can be obtained by analysing such models as well as the queries to be sent to solvers to obtain such results, (3) a number of libraries that can be used to handle the formalism and model data structures defined in DrawNet, and (4) a library to implement connections between DrawNet and existing as well as new solvers. The tool has been experimented with different formalisms, including various flavours of PNs: for example the application oriented formalisms used to compose SWN models in [31] and [30] and developed within the OsMoSys framework [52], have been implemented in DrawNET, and integrated with the composer module *algebra* and the solution modules of GreatSPN.

6   Conclusions

This paper is a survey on the SWN formalism and some relevant analysis techniques that have been developed and implemented since its introduction to now. Although significant results have been achieved, there is still space for new research and implementation efforts. Some possible topics for future work concern both extensions of the formalism itself and further improvements in the analysis algorithms and tools. Concerning the formalism extensions, the most urgent ones are related with compositionality: in particular the possibility of defining parametric submodels (where also the color structure can be a parameter) and then instantiating and composing them in a flexible way would ease the model building process. This feature could be very useful also when SWN models are automatically built from other modeling formalism and could be exploited to ease the modeling task for not expert users, that may reuse existing submodels from a library and compose them through connectors tailored to the specific application field.

Another weak point is the lack of sufficiently complete structural analysis algorithms and tools to support the modeler in building correct models before proceeding to the state space generation. These tools could be useful also in support of decomposition methods, and in the development of partial order methods.

From the point of view of the state space size reduction methods, on one hand it would be worthwhile to push the Extended SRG method further along the lines already proposed (e.g. the E^2SRG), on the other hand the combination of the SRG symmetry exploitation technique with orthogonal techniques based on Decision Diagram data structures seem to have more potential for reduction, hence, the methods already used for efficient model checking should be extended to quantitative evaluation.

Finally a rich set of tools has already been implemented or are under development: more integration among the tools would ease the application and diffusion of all currently available techniques and solution modules.

References

1. M. Ajmone Marsan, G. Balbo, G. Conte, S. Donatelli, and G. Franceschinis. *Modelling with Generalized Stochastic Petri Nets*. J. Wiley, 1995.
2. C. Anglano, S. Donatelli, and R. Gaeta. Parallel architectures with regular structure: a case study in modelling using SWN. In *Proc. 5th Intern. Workshop on Petri Nets and Performance Models*, Toulouse, France, October 1993. IEEE-CS Press.
3. S. Baarir, S. Haddad, and J.M. Ilié. Exploiting partial symmetries in well-formed nets for the reachability and the linear time model checking problems. In *Proc. of WODES'04, IFAC Workshop on Discrete Event Systems*, Reims, France, Sept 2004.
4. P. Ballarini, S. Donatelli, and G. Franceschinis. Parametric stochastic well-formed nets and compositional modelling. In *In Proc. 21st International Conference on Application and Theory of Petri Nets*, Aarhus, Denmark, June 2000.
5. M. Beccuti. Tesi di laurea. trattamento delle simmetrie parziali nelle reti di Petri (in italian.). Università del Piemonte Orientale, Alessandria, Italy.
6. C. Bellettini and L. Capra. A quotient graph for asymmetric distributed systems. In *Proc. Of the 12th IEEE International Symposium on Modeling Analysis, and Simulation of Computer and Telecommunication Systems (MASCOTS04)*. IEEE-CS, 2004.
7. S. Bernardi, S. Donatelli, , and A. Horváth. Implementing compositionality for stochastic Petri nets. *Journal of Software Tools for Technology Transfer*, 3(4):417–430, Sept 2001. Special section on the pratical use of high-level Petri Nets.
8. E. Best, H. Flrishhacl ANF W. Fraczak, R. Hopkins, H. Klaudel, and E. Pelz. A class of composable high level Petri nets with an application to the semantics of B(PN)^2. In *Proc. of the 16th international conference on Application and Theory of Petri nets 1995*, Torino, Italy, 1995. Springer Verlag. Volume LNCS935.

9. P. Buchholz. Exact and ordinary lumpability in finite markov chains. *Journal of Applied Probability*, 31:59–74, 1994.

10. L. Capra. *Exploiting partial symmetries in SWN models*. PhD thesis, Università di Torino, 2001.

11. L. Capra, M. De Pierro, and G. Franceschinis. An application example of symbolic calculus for SWN structural relations. In *Proceedings of the 7th International Workshop on Discrete Event Systems, 2004, Reims, France*. Elsevier-Oxford, 2005.

12. L. Capra, M. De Pierro, and G. Franceschinis. A high level language for structural relations in well-formed nets. In *Proc. of 26th Int. Conf. on Application and Theory of Petri Nets and other models of concurrency, ATPN'05*, Miami,FL,USA, 2005. Springer Verlag. LNCS.

13. L. Capra, C. Dutheillet, G. Franceschinis, and J-M. Ilié. Towards performance analysis with partially symmetrical SWN. In *Proc. of the 9th IEEE International Symposium on Modeling Analysis, and Simulation of Computer and Telecommunication Systems (MASCOTS99)*. IEEE-CS, 1999.

14. L. Capra, C. Dutheillet, G. Franceschinis, and J-M. Ilié. Exploiting partial symmetries for markov chain aggregation. *Electronic Notes in Theoretical Computer Science*, 39(3), 2000.

15. G. Chiola, C. Dutheillet, G. Franceschinis, and S. Haddad. On Well-Formed coloured nets and their symbolic reachability graph. In *Proc. 11th Intern. Conference on Application and Theory of Petri Nets*, Paris, France, June 1990. Reprinted in *High-Level Petri Nets. Theory and Application*, K. Jensen and G. Rozenberg (editors), Springer Verlag, 1991.

16. G. Chiola, C. Dutheillet, G. Franceschinis, and S. Haddad. Stochastic well-formed coloured nets for symmetric modelling applications. *IEEE TC*, 42(11):1343–1360, November 1993.

17. G. Chiola, C. Dutheillet, G. Franceschinis, and S. Haddad. A Symbolic Reachability Graph for Coloured Petri Nets. *Theoretical Computer Science B (Logic, semantics and theory of programming)*, 176(1&2):39–65, April 1997.

18. G. Chiola, G. Franceschinis, and R. Gaeta. Modelling symmetric computer architectures by SWNs. In *Proc. of the 15th Intern. Conference on Applications and Theory of Petri Nets*, number 815 in Lecture Notes in Computer Science. Springer-Verlag, 1994.

19. G. Chiola, G. Franceschinis, R. Gaeta, and M. Ribaudo. GreatSPN 1.7: Graphical Editor and Analyzer for Timed and Stochastic Petri Nets. *Performance Evaluation, special issue on Performance Modeling Tools*, 24(1&2):47–68, November 1995.

20. G. Ciardo and A. Miner. A data structure for the efficient Kronecker solution of GSPNs. In *Proc. of the 8th Int. Workshop on Petri nets and performance models (PNPM99)*, pages 22–31, Zaragoza, Spain, September 8–10 1999. IEEE Comp. Soc. Press.

21. J-M. Couvreur, E. Encrenaz, E. Paviot-Adet, and D. Poitrenaud. Data decision diagrams for Petri net analysis. In *Proc. of Int. Conf. on Application and Theory of Petri Nets, ICATPN'02*, pages 101–120. Springer Verlag, June 2002. LNCS 2360.

22. M. De Pierro. *Structural analysis of conflicts and causality in GSPN and SWN*. PhD thesis, Università di Torino - Italia., 2004.

23. C. Delamare, Y. Gardan, and P. Moreaux. Performance evaluation with asynchronously decomposable SWN: implementation and case study. In *Proc. of the 10th Int. Workshop on Petri nets and performance models (PNPM03)*, pages 20–29, Urbana-Champaign, IL, USA, September 2–5 2003. IEEE Comp. Soc. Press.

24. S. Donatelli. Kronecker algebra and Petri nets: is it worth the effort? In *Proc. 15th Int. Conf. on Application and Theory of Petri Nets*, Zaragoza, Spain, June 1994. Springer Verlag. Volume LNCS 815.

25. S. Donatelli. Superposed Generalized Stochastic Petri Nets: definition and efficient solution. In *Proceedings of the 15th International Conference on Applications and Theory of Petri Nets*, number 815 in Lecture Notes in Computer Science, Zaragoza, Spain, 1994. Springer-Verlag.

26. S. Donatelli and L. Ferro. Validation of GSPN and SWN models through the PROD tool. In *Proc. of 12th Int. Conference on Modelling Tools and Techniques for Computer and Communication System Performance Evaluation (TOOLS 2002)*, London, UK, 2002. Springer Verlag. Volume LNCS 2324.

27. S. Donatelli and G. Franceschinis. The PSR methodology: integrating hardware and software models. In *Proc. of the 17th International Conference in Application and Theory of Petri Nets, ICATPN '96*, Osaka, Japan, june 1996. Springer Verlag. LNCS, Vol 1091.

28. A. Duret-Lutz and D. Poitrenaud. SPOT: an extensible model checking library using transition-based generalized buchi automata. In *Proc. of 12th IEEE/ACM Int. Symp. on Modeling, Analysis and Simulation of Computer and Teleccommunication Sustems (MASCOTS'04)*, Volendam, Netherlands, Sept. 2004.

29. C. Dutheillet and S. Haddad. Regular stochastic Petri nets. In *Proc. 10th Intern. Conf. Application and Theory of Petri Nets*, Bonn, Germany, June 1989.

30. G. Franceschinis, M. Gribaudo, M. Iacono, S. Marrone, N. Mazzocca, and V. Vittorini. Compositional modeling of complex systems: contact center scenarios in OsMoSys. In *Proc. 25th Int. Conf. on Application and Theory of Petri Nets*, Bologna, Italy, June 2004. Springer Verlag. Volume LNCS 3099.

31. G. Franceschinis, V. Vittorini, S. Marrone, and N. Mazzocca. Swn client-server composition operators in the OsMoSys framework. In *Proceedings 10th International Workshop on Petri Net and Performance Models (PNPM2003)*, pages 52–61, Urbana-Champaign, IL, USA, 2003. IEEE CS.

32. G. Gaeta. Efficient discrete-event simulation of colored Petri nets. *IEEE Transaction on Software Engineering*, 22(9), September 1996.

33. M. Gribaudo, D. Codetta Raiteri, and G. Franceschinis. Draw-net: a customizable multi-formalism, multi-solution tool for the quantitative evaluation of system. Submitted for publication to QEST'05, 2005.

34. S. Haddad. *Une Categorie Regulier de Reseau de Petri de Haut Niveau: Definition, Proprietes et Reductions*. PhD thesis, Lab. MASI, Universite P. et M. Curie (Paris 6), Paris, France, Oct 1987. These de Doctorat, RR87/197 (in French).

35. S. Haddad, J-M. Ilié, and K. Ajami. A model checking method for partially symmetric systems. In *Proc. of FORTE/PSTV'00*, pages 121–136. Kluwer Verlag, 2000.

36. S. Haddad, J.M. Ilié, M. Taghelit, and B. Zouari. Symbolic Reachability Graph and Partial Symmetries. In *Proc. of the 16th Intern. Conference on Application and Theory of Petri Nets*, volume 935 of *LNCS*, pages 238–257, Turin, Italy, June 1995. Springer Verlag.

37. S. Haddad and P. Moreaux. Evaluation of high level Petri nets by means of aggregation and decomposition. In *Proc. of the 6th International Workshop on Petri Nets and Performance Models*, pages 11–20, Durham, NC, USA, October 3–6 1995. IEEE Computer Society Press.

38. S. Haddad and P. Moreaux. Asynchronous composition of high level Petri nets: a quantitative approach. In *Proc. of the 17th International Conference on Application and Theory of Petri Nets*, number 1091 in LNCS, pages 193–211, Osaka, Japan, June 24–28 1996. Springer–Verlag.

39. J-M. Ilié, S. Baarir, and A. Duret-Lutz. Improving reachability analysis for partially symmetric high level Petri nets. In *Proc. of 12th IEEE/ACM Int. Symp. on Modeling, Analysis and Simulation of Computer and Teleccommunication Sustems (MASCOTS'04)*, Volendam, Netherlands, Sept. 2004.

40. J.M. Ilié, S. Baarir, M. Beccuti, C. Delamare, S. Donatelli, C. Dutheillet, G. Franceschinis, R. Gaeta, and P. Moreaux. Extended SWN solvers in greatspn. In *Proc. 1st Int. Conf. on Quantitative Evaluation of Systems (QEST04)*, Enschede, The Netherlands, Sept. 2004. (Tool presentation paper.).

41. K. Jensen. *Coloured Petri Nets, Basic Concepts, Analysis Methods and Practical Use. Volume 1.* Springer Verlag, 1992.

42. K. Jensen. *Coloured Petri Nets, Basic Concepts, Analysis Methods and Practical Use. Volume 2.* Springer Verlag, 1995.

43. Isabel C. Rojas M. *Compositional construction and Analysis of Petri net Systems.* PhD thesis, University of Edinburgh, 1997.

44. William H. Sanders and John F. Meyer. Stochastic activity networks: Formal definitions and concepts. In *European Educational Forum: School on Formal Methods and Performance Analysis*, pages 315–343. Springer Verlag, 2000. LNCS 2090.

45. C. Sibertin-Blanc. Comunicative and cooperative nets. In *Proc. of the 15th Int. Conf. on Application and Theory of Petri Nets*, Zaragoza, Spain, June 1994. LNCS 815, Springer Verlag.

46. C. Sibertin-Blanc. CoOperative Objects: Principles, use and implementation. In G. Agha, F. De Cindio, and G. Rozenberg, editors, *Concurrent Object-Oriented Programming and Petri Nets.* Springer Verlag, 2001. LNCS 2001.

47. E. Teruel, G. Franceschinis, and M. De Pierro. Well-defined generalized stochastic Petri nets: A net-level method to specify priorities. *IEEE TSE*, 29(11):962–973, November 2003.

48. Y. Thierry-Mieg. *Techniques pour le Model Checking de spécifications de haut niveau.* PhD thesis, Univ. of Paris VI, 2004.

49. Y. Thierry-Mieg, C. Dutheillet, and I. Mounier. Automatic symmetry detection in Well-formed Nets. In *Proc. of 24th Int. Conf. on Application and Theory of Petri Nets, ICATPN'03*, pages 82–101. Springer Verlag, June 2003. LNCS 2679.

50. Y. Thierry-Mieg, J-M. Ilié, and D. Poitrenaud. A symbolic symbolic state space. In *Proc. of 24th IFIP WG6.1 Int Conf. on Formal Techniques for Networked and Distributed Systems (FORTE'04)*, pages 279–291, Madrid, Spain, Sept. 2004. Springer. LNCS 3235.

51. V. Vittorini, G. Franceschinis, M. Gribaudo, M. Iacono, and N. Mazzocca. Drawnet++: Model objects to support performance analysis and simulation of complex systems. In *Proc. of 12th Int. Conference on Modelling Tools and Techniques for Computer and Communication System Performance Evaluation (TOOLS 2002)*, pages 233–238, London, UK, 2002. Springer Verlag. Volume LNCS 2324.

52. V.Vittorini, M. Iacono, N. Mazzocca, and G. Franceschinis. The osmosys approach to multi-formalism modeling of systems. *Journal of Software and System Modeling*, 3(1), March 2004.

Applications of Craig Interpolation to Model Checking

Kenneth McMillan

Cadence Berkeley Labs

A Craig interpolant [1] for a mutually inconsistent pair of formulas (A,B) is a formula that is (1) implied by A, (2) inconsistent with B, and (3) expressed over the common variables of A and B. It is known that a Craig interpolant can be efficiently derived from a refutation of $A \wedge B$, for certain theories and proof systems. For example, interpolants can be derived from resolution proofs in propositional logic, and for systems of linear inequalities over the reals [6, 4]. These methods have been recently extended to combine linear inequalities with uninterpreted function symbols, and to deal with integer models [5]. One key aspect of these procedures is that the yield quantifier-free interpolants when the premises A and B are quantifier-free.

This talk will survey some recent applications of Craig interpolants in model checking. We will see that, in various contexts, interpolation can be used as a substitute for image computation, which involves quantifier elimination and is thus computationally expensive. The idea is to replace the image with a weaker approximation that is still strong enough to prove some property.

For example, interpolation can be used to construct an inductive invariant for a transition system that is strong enough to prove a given property. In effect, it gives us an abstract image operator that can be iterated to a fixed point to obtain an invariant. This invariant contains only information actually deduced by a prover in refuting counterexamples to the property of a fixed number of steps. Thus, in a certain sense, we abstract the invariant relative to a given property. This avoids the complexity of computing the strongest inductive invariant (i.e., the reachable states) as is typically done in model checking.

This approach gives us a complete procedure for model checking temporal properties of finite-state systems that allows us to exploit recent advances in SAT solvers for the proof generation phase. Experimentally, this method is found to be quite robust for verifying properties of industrial hardware designs, relative to other model checking approaches. The same approach can be applied to infinite-state systems, such as programs and parameterized protocols (although there is no completeness guarantee in this case). For example, it is possible to verify systems of timed automata in this way, or simple infinite-state protocols, such as the N-process "bakery" mutual exclusion protocol.

Alternatively, interpolants derived from proofs can be mined to obtain predicates that are useful for predicate abstraction [7]. This approach has been used in a software model checking to verify properties of C programs with in excess of 100K lines of code [2]. Finally, interpolation can be used to approximate the

G. Ciardo and P. Darondeau (Eds.): ICATPN 2005, LNCS 3536, pp. 15–16, 2005.
© Springer-Verlag Berlin Heidelberg 2005

transition relation of a system relative to a given property [3]. This is useful in predicate abstraction, where constructing the exact abstract transition relation is prohibitive.

References

1. W. Craig. Linear reasoning: A new form of the Herbrand-Gentzen theorem. *J. Symbolic Logic*, 22(3):250–268, 1957.
2. T. A. Henzinger, R. Jhala, Rupak Majumdar, and K. L. McMillan. Abstractions from proofs. In *ACM Symp. on Principles of Prog. Lang. (POPL 2004)*, 2004. to appear.
3. R. Jhala and K. L. McMillan. Interpolant-based transition relation approximation. In *CAV 2005*, 2005.
4. J. Krajíček. Interpolation theorems, lower bounds for proof systems, and independence results for bounded arithmetic. *J. Symbolic Logic*, 62(2):457–486, June 1997.
5. K. L. McMillan. An interpolating theorem prover. In *TACAS 2004*, pages 16–30, 2004.
6. P. Pudlák. Lower bounds for resolution and cutting plane proofs and monotone computations. *J. Symbolic Logic*, 62(2):981–998, June 1997.
7. Hassen Saïdi and Susanne Graf. Construction of abstract state graphs with PVS. In Orna Grumberg, editor, *Computer-Aided Verification, CAV '97*, volume 1254, pages 72–83, Haifa, Israel, 1997. Springer-Verlag.

Towards an Algebra for Security Policies
(Extended Abstract)

Jon Pincus[1] and Jeannette M. Wing[2]

[1] Microsoft Research,
One Microsoft Way, Redmond, WA 98052
jpincus@microsoft.com
[2] Computer Science Department, Carnegie Mellon University,
5000 Forbes Avenue, Pittsburgh, PA 15213
wing@cs.cmu.edu

1 Context

Clashing security policies leads to vulnerabilities. Violating security policies leads to vulnerabilities. A system today operates in the context of a multitude of security policies, often one per application, one per process, one per user. The more security policies that have to be simultaneously satisfied, the more likely the possibility of a clash or violation, and hence the more vulnerable our system is to attack. Moreover, over time a system's security policies will change. These changes occur at small-scale time steps, e.g., using setuid to temporarily grant a process additional access rights; and at large-scale time steps, e.g., when a user changes his browser's security settings. We address the challenge of determining when a system is in a consistent state in the presence of diverse, numerous, and dynamic interacting security policies.

Formal specifications of these security policies let us pinpoint two potential problems: when security policies for different components are inconsistent and when a component does not satisfy a given security policy. We present a simple algebra for combining and changing security policies, and show how our algebraic operations can be used to explain different real-life examples of security policy clashes and violations.

2 Model and Definitions

We model security policies as access rights matrices:

$$SP \subseteq P \times O \times R$$

where P is a set of principals, O is a set of typed objects, and R is a set of rights. Principals include processes, users, applications, etc. Objects include files, directories, registry keys, communication channels, etc. Rights are type-specific: for each type of object, there are certain associated operations. For example, for a file, the operations might be open, close, read, write, and execute; for a Web service, the operations might be search, recommend, and purchase. Henceforth, when we say "security policy," we mean its underlying access rights matrix.

G. Ciardo and P. Darondeau (Eds.): ICATPN 2005, LNCS 3536, pp. 17–25, 2005.

For a given security policy, sp, principal, p, and object, o, we write $sp(p^c\,o)$ to stand for the set of rights in R that p has on o. Informally, this means that for an operation in $sp(p^c\,o)$, p has the right to invoke that operation on o. Negative rights are represented implicitly, by the absence of an explicit right in the security policy.

Definition 1. *Two security policies, sp_1 and sp_2, **clash** iff $sp_1 \neq sp_2$.*

Definition 2. *Given two security policies, sp_1 and sp_2, sp_1 **respects** sp_2 iff $\forall p \forall o \,.\, sp_1(p^c\,o) \subseteq sp_2(p^c\,o)$; otherwise, sp_1 **disrespects** sp_2.*

Combining two security policies potentially introduces vulnerabilities. Whether there is a vulnerability depends on the way in which the two security policies combine. Let $\_\oplus\_ : SP \times SP \to SP$ denote a combination operation on two security polices. There is a potential vulnerability if $sp_1 \oplus sp_2$ disrespects sp_1 or $sp_1 \oplus sp_2$ disrespects sp_2. Disrespecting security policies imply they clash.

3   An Algebra for Security Policies

Since security policies are ternary relations, i.e., sets of triples, we define combinations of security policies in terms of operations on sets. In practice, in combining two security policies we might require that we satisfy both (*And*); satisfy either (*Or*); or satisfy one but not another (*Minus*). We might simply override (*Trumps*) the second by the first.

For the following combinations, we assume that the security policies are defined over the same sets of P, O, and R.

> **spec** SecurityPolicy
> $\_And\_ : SP \times SP \to SP$
> $\_Or\_ : Sp \times SP \to SP$
> $\_Minus\_ : SP \times SP \to SP$
> $\_Trumps\_ : SP \times SP \to SP$
>
> $sp_1\ And\ sp_2 = sp_1 \cap sp_2$
> $sp_1\ Or\ sp_2 = sp_1 \cup sp_2$
> $sp_1\ Minus\ sp_2 = sp_1 \setminus sp_2$
> $sp_1\ Trumps\ sp_2 = sp_1$
> **end** SecurityPolicy

Whereas *And* and *Or* are commutative, *Trumps* and *Minus* are not, as should be clear from the equations above. The following facts follow from the definitions of *respects*, *And*, and *Trumps*:

$(sp_1\ And\ sp_2)$ respects sp_1
$(sp_1\ And\ sp_2)$ respects sp_2
$(sp_1\ Trumps\ sp_2)$ respects sp_1

When we use *Or* to combine two security policies, sp_1 and sp_2, the combination might disrespect sp_1 or disrespect sp_2; similarly, when we use *Trumps*, the combination might disrespect sp_2.

We also introduce a way for one principal to gain (*Inherit*) its rights from another, so that $Inherit(p_1\,{}^{\backprime}\,p_2)$ means p_1 inherits rights from p_2; and conversely, for one principal to grant (*Delegate*) its rights to another. We include a revocation (*Revoke*) operator to remove all rights associated with a given principal.

> **spec** ChangeSP **extends** SecurityPolicy
> $Inherit : SP \times P \times P \to SP$
> $Delegate : SP \times P \times P \to SP$
> $Revoke : SP \times P \to SP$
>
> $Inherit(sp\,{}^{\backprime}\,p_1\,{}^{\backprime}\,p_2) = sp \cup \{\langle p\,{}^{\backprime}\,o\,{}^{\backprime}\,r\rangle \,|\, \langle p_1\,{}^{\backprime}\,o\,{}^{\backprime}\,r\rangle \in sp \wedge p = p_2\}$
> $Delegate(sp\,{}^{\backprime}\,p_1\,{}^{\backprime}\,p_2) = Inherit(sp\,{}^{\backprime}\,p_2\,{}^{\backprime}\,p_1)$
> $Revoke(sp\,{}^{\backprime}\,p) = sp \rhd (Prin(sp) \setminus \{p\})$
> **end** ChangeSP

where $\rhd$ is the domain restriction operator and $Prin : SP \to Set[P]$ returns the set of principals over which a given security policy is defined. *Inherit* (and similarly *Delegate*) has the following compositional property:

$$Inherit(Inherit(sp\,{}^{\backprime}\,p_2\,{}^{\backprime}\,p_3)\,{}^{\backprime}\,p_1\,{}^{\backprime}\,p_2) \supseteq Inherit(sp\,{}^{\backprime}\,p_1\,{}^{\backprime}\,p_3)$$

which informally says "If p_2 inherits p_3's rights and then p_1 inherits p_2's rights, then p_1 inherits p_3's rights."

4   Examples

We give a series of four examples. The first (Outlook and IE) illustrates an example of two different components, each of which has their own security policy; when put together, one security policy "trumps" the other, causing a potential privacy violation as well an surprising behavior to the user. The second (Run As) illustrates a use of inheriting rights that causes potential security vulnerabilities. The third (Google Desktop Search) illustrates another use of inheritance. Finally, the fourth example (Netscape and DNS) simply shows what can happen when an informally stated security policy is ambiguous, leading to an implementation with a security vulnerability. We present the first example in detail and give the gists of the problems for the other three. The first, third, and fourth examples are also examples of *composition flaws*, when two independently designed and implemented components are combined in a way that lead to unintended interactions, which could lead to security vulnerabilities.

4.1   Outlook and IE

Many applications (e.g., email clients and browsers) have security settings that users may modify. Each configuration of the settings represents a different security policy.

Let's consider an example where we would like to block the display of embedded graphics in our mail messages. Microsoft's Outlook email client includes a setting "Don't Download Pictures" which lets the user specify this behavior. Microsoft's Internet Explorer (IE) browser includes a similar setting "Show Pictures" which lets the user control the display of graphics embedded in any HTML document. Since Outlook uses IE's HTML rendering component, these two settings interact in the case of HTML email.

Actual Behavior

The following table describes the actual behavior when reading and forwarding messages. There are four possibilities for how an image is displayed:

- **display**: Retrieve and display the image.
- **red X**: Display a small icon of a red X, with a textual comment saying "Right-click here to download pictures. To help protect your privacy, Outlook prevented automatic download of this picture from the Internet."
- **small graphic**: Display a small "unknown graphic" icon, with the same textual comment.
- **sized graphic**: Display a correctly-sized box with a small "unknown graphic" icon (and no textual comment).

As will be discussed below, the behavior in italics is a vulnerability.

IE Show Pictures	Outlook Don't Download Pictures	
	False	True
False	Read: sized graphic	Read: small graphic
	Forward: sized graphic	Forward: sized graphic
True	Read: display	Read: red X
	Forward: display	*Forward: display*

The Vulnerability and an Attack

From a security perspective, there are at least two reasons to disable graphics in email. First, downloading the graphic transfers information back to the web site containing the graphic; at a minimum, this kind of information disclosure confirms that the mail was received and viewed, e.g., allowing spammers to confirm that an email address is real. Second, disabling graphics is a *Defense in Depth* strategy that mitigates the risk of unknown exploitable bugs, e.g., buffer overruns in image rendering code. Thus, the counter-intuitive display of graphics while forwarding email is a vulnerability.

An attacker can exploit this vulnerability in a social engineering attack if he can convince a user, Alice, to forward some email containing an image. For example, consider the scenario where the attacker wants to validate whether an email address `Alice@bigco.com` corresponds to a real user (perhaps in the context of a brute-force generation of all email addresses for the domain `bigco.com` to discover which are valid addresses to resell to spammers). If Alice has downloading of images disabled, the straightforward attack of simply getting Alice to read the mail will fail. However, if the attacker knows that `bigco` has an internal policy of forwarding all "phishing" email to a central alias for followup, then the attacker simply sends an obviously fake "phishing" email containing an embedded image to `Alice@bigco.com`. Alice can read the email without any danger; but when she goes to forward it, because of the vulnerability, the image is downloaded, and the attacker confirms that the address is valid.

A Fix

The vulnerability disappears if the behavior is as follows, where the only change, shown in boldface, is to display a red X on mail forwarding instead of retrieving and displaying the image.

IE Show Pictures	Outlook Don't Download Pictures	
	False	True
False	Read: sized graphic Forward: sized graphic	Read: small graphic Forward: sized graphic
True	Read: display Forward: display	Read: red X **Forward: red X**

Using our Security Policy Formalism

More formally, we can use our algebra for security policies to characterize the security policy clash. Let *viewread* be the right associated with the operation that lets us view an embedded graphic while reading a mail message; *viewforward*, while forwarding.

Security Policy for Outlook ($SP_{Outlook}$): If the "Don't Download Pictures" setting is false (i.e., downloading pictures is ok), then all users can view embedded graphics when both reading and forwarding mail messages. More precisely:

$$\text{Download Pictures} = \text{true}$$
$$\Leftrightarrow$$
$$\forall p : user \ \forall o : embedded\_graphic$$
$$\langle p^c\, o^c\, viewread \rangle \in SP_{Outlook} \wedge \langle p^c\, o^c\, viewforward \rangle \in SP_{Outlook}$$

(To avoid a double negative, we wrote above "Download Pictures = true" rather than the more accurate "Don't Download Pictures = false".)

Security Policy for IE (SP_{IE}): If the "Show Pictures" setting is true, then all users can view embedded graphics when viewing HTML. This setting applies both when reading and forwarding mail messages. More precisely:

$$\text{Show Pictures} = \text{true}$$
$$\Leftrightarrow$$
$$\forall p : user \ \forall o : embedded\_graphic$$
$$\langle p^c\, o^c\, viewread \rangle \in SP_{IE} \wedge \langle p^c\, o^c\, viewforward \rangle \in SP_{IE}$$

When Outlook is combined with IE, then the actual behavior differentiates between the reading and forwarding cases:

Read: SP_{IE} And $SP_{Outlook}$
Forward: SP_{IE} Trumps $SP_{Outlook}$

When reading email, then the actual behavior reflects the conjunction of the security policies associated with Outlook and IE; but when forwarding, then the actual behavior reflects IE's security policy, trumping Outlook's. Thus, when forwarding, the combined security policy disrespects the Outlook security policy:

(SP_{IE} Trumps $SP_{Outlook}$) disrespects $SP_{Outlook}$

The fix reflects conjunction in both cases, regardless of whether we are reading or forwarding email:

Read: SP_{IE} *And* $SP_{Outlook}$
Forward: SP_{IE} *And* $SP_{Outlook}$

By properties of *And* (Section 3), we know that the combined security policies respect both of the individual ones.

Note that the original combination of security policies led not only to a security vulnerability but also to a confusing usability problem. Users would normally expect a single policy to hold (whether it be Outlook's, IE's, or their conjunction) regardless of what operation they perform, e.g., reading or forwarding a mail message. When the two policies combine in one way for one operation and in another way for the other, the user's mental model is inconsistent, which invariably leads to usability issues. The proposed fix also removes this inconsistency.

4.2   Run as

Suppose in a file system, *FS*, we wish to let a process, p, perform operations on behalf of (i.e., "run as") a user, u; this "run as" capability allows p to gain access temporarily to a set of files owned by u. We can model this behavior in terms of inheriting rights. For example, if u can read a file, f, and if p inherits u's rights, then p can read f. The Unix setuid mechanism and Windows impersonate privileges are implementations of this functionality.

Let SP_{FS} stand for the security policy on a given file system. Now consider the following scenario. First, u_1 executes p_1. This has the effect of giving u_1's rights to p_1:

[1] $SP_E = Inherit(SP_{FS}{}^c p_1{}^c u_1)$

Now, p_1 "runs as" u_2. This has the effect of first revoking p_1's original rights and then giving u_2's rights to p_1:

[2] $SP_{RA} = Inherit(Revoke(SP_E{}^c p_1)^c p_1{}^c u_2)$

Assuming u_1 controls p_1's behavior, then for the lifetime of p_1, u_1 gets whatever rights p_1 acquires:

[3] $SP_C = Inherit(SP_{RA}{}^c u_1{}^c p_1)$

From the compositional property of *Inherits* (Section 3), [2] and [3] leads to [4] below, and we are left in a state where u_1 has inherited u_2's rights:

[4] $SP_C \supseteq Inherit(SP_{FS}{}^c u_1{}^c u_2)$

Let's see how the use of "run as" can lead to a security vulnerability. First, we introduce a new kind of access right, a, to stand for "can run as." In our security policy, we have entries such as $\langle p^c u^c a \rangle$, which says that a process, p, "can run as" user, u.

Now consider a file system with entries such as $\langle u_1{}^c p_1.exe^c x \rangle$, which says that user u_1 has execute rights on the executable object, $p_1.exe$ (the executable associated with process p_1), as well as entries such as $\langle u_1{}^c f^c read \rangle$, which says that user u_1 has read access to file f.

A process's rights may change over time, depending on what user they are running as. Thus, a process can access a file, f, if the user the process is currently running as has access to f.

More concretely, a vulnerability can arise, if SP_{FS} consists of the following entries:

$$\langle p_1 ^\mathsf{c} * ^\mathsf{c} a \rangle$$
$$\langle u_1 ^\mathsf{c} p_1.exe ^\mathsf{c} x \rangle$$
$$\langle u_2 ^\mathsf{c} f ^\mathsf{c} read \rangle$$

The first entry is the troublesome one: it gives process p_1 the ability to run as any user. Since u_1 can execute $p_1.exe$, which creates a process p_1 with the given rights, and since p_1 can run as u_2 and u_2 has access to f, then u_1 gains access to f even though that right is absent from the security policy.

4.3   Google Desktop Search

When a `google.com` request is made, Google Desktop Search (GDS) performs a search on the local file system with the same request. The local search results include 30-40 character snippets of local files that contain the query's terms. GDS integrates its local search results with the webpage returned by `google.com`. If the query has been made by a user in a standard browser window, the user sees a webpage that contains results of the search on the local host and results from `google.com`; but if the query is made by an applet, then GDS introduces a potential security vulnerability [2].

Security Policy for Google Desktop Search (SP_{GDS}): The Google Desktop Search process has read access to all files on the local host.

Or more formally,

$$\forall f \in localhost \ \langle GDS ^\mathsf{c} f ^\mathsf{c} read \rangle \in SP_{GDS}$$

Suppose an applet, app, sends a query that GDS executes. We have the following situation:

[1] $SP_{App} = Inherit(SP_{GDS} ^\mathsf{c} app ^\mathsf{c} GDS)$

Since the applet can connect to a remote host, in particular, the host, rh, from which the applet originally came, the remote host inherits the applet's rights:

[2] $SP_{RemHost} = Inherit(SP_{App} ^\mathsf{c} rh ^\mathsf{c} app)$

By compositionality of *Inherits* again ([1] and [2] leads to [3] below), we are in the situation where the remote host inherits the rights of GDS:

[3] $SP_{RemHost} \supseteq Inherit(SP_{GDS} ^\mathsf{c} rh ^\mathsf{c} GDS)$

which means the remote host has read access to files on the local host—a security vulnerability!

The fix that Google made to address this vulnerability is to disallow the applet from seeing the results of a local search using GDS. This has the effect of invalidating the state labeled [1] above.

4.4   Netscape and DNS

Our final example shows what can go wrong when the specification of a security policy is ambiguous, leading to an implementation with a security vulnerability. Here the vulnerable component is the browser, which operates in the DNS infrastructure environment [1]. Applets running in the browser sometimes need to contact the server from which it originated. We have the following security policy on applets (which also must hold for the Google Desktop Search example):

Security Policy for Applet (SP_{applet}): An applet should connect to the same server from which it originated.

More formally, in terms of an access rights matrix, we have the following, where *conn* stands for the right for an applet a to connect to a host h:

$$\forall a : applet \; \forall h : host \; \langle a^{\smallsmile} h^{\smallsmile} conn \rangle \in SP_{applet}$$
$$\Leftrightarrow SameAs(a.OriginatingHost^{\smallsmile} h)$$

The ambiguity in the informal policy is what "same server" means. In the formal statement, it boils down to how *SameAs* is interpreted. There are two sources of ambiguity. First, does *SameAs* mean "same IP address" or "same name"? An interpretation of "same IP address" for *SameAs* seems too restrictive to support some common usage scenarios, and so Netscape chose to resolve this first ambiguity by using a check based on DNS names. Unfortunately, the name-based check raises two problems: (1) the name of a server might map to multiple IP addresses (machines) and (2) the mapping of names to IP addresses can change over time. The possibility of change over time is a second source of ambiguity: Doing a DNS lookup on a name at one point in time, e.g., when the applet is downloaded, does not guarantee the same result as doing it at a later point in time, e.g., when the applet wishes to connect to a remote host.

In more detail, here is the problem. In the effort to enforce the security policy, Netscape's original check used two DNS name lookups. Let *n2a* be the many-to-many relation that maps names to IP addresses, *From* be the name of the server from which the applet originated, and *To* be the name of the server to which the applet wishes to connect. If the lookup on both names yields a nonempty intersection of IP addresses, then the assumption is that *From* and *To* are "the same server," and we allow the connection. More succinctly, Netscape implemented this check:

if *n2a(From)* $\cap$ *n2a(To)* $\neq \emptyset$
then $\exists x \in n2a(From) \; \exists y \in n2a(To)$ such that *connect(x, y)*

where $connect : Host \times Host \to Bool$ means that for hosts h_1 and h_2 $connect(h_1^{\smallsmile} h_2)=$ *true* iff there is a connection from h_1 to h_2.

There are two problems with Netscape's check. The first problem is directly related to the second ambiguity and leads to a vulnerability: in doing the lookup on *From* at the time when the applet wants to connect to *To*, the set of IP addresses to which *From* maps may be different from the time when it was first downloaded. The second problem is simply a logical flaw: choosing some x in *n2a(From)* and some y in *n2a(To)* to establish the connection does not even guarantee that the x and y are in the (nonempty) intersection of *n2a(From)* and *n2a(To)*.

By manipulating DNS lookup (for example, by running his or her own name resolver), an attacker can effectively allow an applet to connect to any host on the network without violating the policy.

Netscape fixed the vulnerability by storing the actual IP address i of the originating server (eliminating the first lookup) and changing the intersection check to a membership check, $i \in n2a(To)$. This fix means the implementation matches the intended security policy.

5 Summary and Future Work

We sketched the foundation of a simple algebra for reasoning about security policies, viewed as access rights matrices. We also sketched the use of our algebra on four real-life examples: the Outlook and IE example shows a security policy clash; the Run As and the Google Desktop Search examples show two different uses of inheriting rights as security policies change over time; and the Netscape and DNS example shows the consequence of a violation of a security policy.

We would like to develop a security policy language and logic for expressing policies closer to the way in which software designers think about security requirements and use our algebra to show when clashes can occur or when designs violate their requirements. We are also interested in building tool support for automating our reasoning and for letting us scale our approach to large examples.

Acknowledgments

We thank Oren Dobzinski for helping us think about the Google Desktop Search example and for comments on earlier drafts of this manuscript. The second author thanks Microsoft Research for partial support of this work. She is also partially sponsored by the Army Research Office under contract no. DAAD190110485 and DAAD19-02-1-0389, the National Science Foundation under grant no. CCR-0121547 and CNS-0433540, and the Software Engineering Institute through a US government funding appropriation.

The views and conclusions contained herein are those of the authors and should not be interpreted as necessarily representing the official policies or endorsements, either expressed or implied, of the sponsoring institutions, the US Government or any other entity.

References

1. D. Dean, E.W. Felten, and D.S. Wallach, "Java Security: From HotJava to Netscape and Beyond," *Proceedings of the 1996 IEEE Symposium on Security and Privacy,* Oakland, CA, May 1996.
2. S. Nielson, S.J. Fogarty, and D.S. Wallach, "Attacks on Local Searching Tools," Technical Report TR04-445, Department of Computer Science, Rice University, December 2004.

Continuization of Timed Petri Nets: From Performance Evaluation to Observation and Control

Manuel Silva and Laura Recalde*

Dep. Informática e Ingeniería de Sistemas, Centro Politécnico Superior de Ingenieros,
Universidad de Zaragoza, María de Luna 3, E-50012 Zaragoza, Spain
{silva, lrecalde}@unizar.es

Abstract. State explosion is a fundamental problem in the analysis and synthesis of discrete event systems. Continuous Petri nets can be seen as a relaxation of discrete models allowing more efficient (in some cases polynomial time) analysis and synthesis algorithms. Nevertheless computational costs can be reduced at the expense of the analyzability of some properties. Even more, some net systems do not allow any kind of continuization. The present work first considers these aspects and some of the alternative formalisms usable for continuous relaxations of discrete systems. Particular emphasis is done later on the presentation of some results concerning performance evaluation, parametric design and marking (i.e., state) observation and control. Even if a significant amount of results are available today for continuous net systems, many essential issues are still not solved. A list of some of these are given in the introduction as an invitation to work on them.

1 Introduction

It is well-known that Discrete Event Dynamic Systems (DEDS), and in particular Petri Nets (PN), suffer the *state explosion problem*, what is particularly true when the system is "highly populated" (i.e., the initial marking is large). One way to tackle that problem is to use some kind of *relaxation. Fluidification* (or *continuization*) is a classical relaxation technique that tries to deal with the state explosion by removing the integrality constraints. The idea is analogous to that allowing the transformation of an Integer Linear Programming problem (ILP, NP-hard) into a Linear Programming problem (LP, polynomial time complexity). The systematic use of linear programming in the fundamental or "state" equation for the analysis of Petri Nets was proposed in 1987 (see [1] for a revised version of the seminal work, and [2] for a more recent survey).

In Queuing Networks (QN) approximating the clients flow with a continuous fluid flow is a classical relaxation (see, for example, [3, 4, 5, 6]). For PN, a similar relaxation was introduced and developed by R. David and co-authors, starting

---

* This work was partially supported by project CICYT and FEDER DPI2003-06376.

G. Ciardo and P. Darondeau (Eds.): ICATPN 2005, LNCS 3536, pp. 26–47, 2005.

in 1987 [7] (see [8] for a very recent perspective). These models were called *continuous Petri Nets*.

Fluidified models have advantages in the sense that they allow to obtain better analytical characterizations, or computationally more efficient algorithms. However, being an approximation, there are properties that cannot be analyzed (mutex for example), and often only partial results can be obtained with respect to their validity in the original system [9].

Even if the idea of continuization of Petri nets is well inscribed in the framework of classical relaxations, even if several interesting analysis and synthesis results have been obtained in the last years [8, 9, 10, 11], the field is still very young and essential contributions are needed. Let us just start this overview —we recognize partially biased from and towards our works— saying that many essential questions do not have a satisfactory answer, in several cases because the problem has not been addressed yet. Just quoting a few:

1. Is a given net system fluidizable? Some net models are not "approximated" at all by its fluidization, just as many differential equation systems (as chaotic models) do not admit a reasonable linearization.
2. How to define the firing policies of transitions when submerging nets systems in time (i.e., how to define the *routing decisions* and *service rates* at conflicts and at stations, respectively)? Today two servers semantics (finite or constant speed, and infinite or variable speed) are mostly used in a *deterministic* setting, but many others can be defined.
3. Provided that several firing (service and routing) semantics can be defined, which one is the best —or a good one— for a given particular case?
4. Given a timing semantics: When does a steady state exist? (The apparent mismatch being that the underlying markovian model may be ergodic, while the continuous case is oscillatory.)
5. Marking reachability in untimed net systems has today a quite reasonable characterization in algebraic terms. For timed models, even for steady state markings, only necessary conditions are well known in general. Of course, for particular classes of net systems (for example, live and bounded equal conflicts), more powerful results are available. The issue is: How to improve the characterization of which steady states can be reached from a given initial marking (and, eventually, which is a "good" control policy)?
6. Even if for *off-line design* problems some interesting results are already known, *observation* and, essentially, *control* of continuous net models require still important improvements. Lose of observability or stability requires still much work.
7. Assuming that "good" off-line designs or dynamic controls are obtained for the continuous relaxation, how to come back to a "reasonable" design or control (scheduling) in the original discrete setting? For this problem, some post-optimization strategies (eventually using metaheuristics like simulated annealing or taboo search [12] can be used, but the problem is essentially unexplored).

Therefore in this work no concluding remarks will be given, leaving the presentation relatively open. The reader is recommended to have the above open questions and many others in mind while going trough the following material. Even if the warning to existing "holes" in our basic understanding (the theory) is put here in the introduction, some hopefully interesting results are available. They provide some behavioral characterizations, sometimes in polynomial time.

The structure of the paper is as follows: In Sect. 2 both autonomous and timed continuous PN are introduced. Sect. 3 briefly compares PN with other formalisms in which similar relaxations are used (queuing networks, Forrester diagrams and positive systems). Performance evaluation of continuous timed PN is addressed in Sect. 4. The results that are obtained are applied in Sect. 5 to some synthesis problems. Sect. 6 is devoted to the study of observability. Finally, the dynamic control of continuous timed models is considered in Section 7.

2   Continuous Petri Nets: On the Relaxation of DEDS Models

2.1   Autonomous Continuous Petri Nets

We assume that the reader is familiar with PN (for notation we use the standard one, see for instance [13]).

The usual PN system, $\langle \mathcal{N}, \mathbf{m_0} \rangle$, will be said to be *discrete* so as to distinguish it from a *continuous* PN. The structure $\mathcal{N} = \langle P, T, \mathbf{Pre}, \mathbf{Post} \rangle$ of continuous Petri nets is the same as in discrete PN, the difference is in the evolution rule. In continuous PN firing is not restricted to be done in integer amounts. As a consequence the marking is not forced to be integer. More precisely, a transition t is *enabled* at $\mathbf{m}$ iff for every $p \in {}^\bullet t$, $\mathbf{m}[p] > 0$, and its *enabling degree* is $\mathrm{enab}(t, \mathbf{m}) = \min_{p \in {}^\bullet t} \{\mathbf{m}[p]/\mathbf{Pre}[p, t]\}$. The firing of t in a certain amount $\alpha \leq \mathrm{enab}(t, \mathbf{m})$ leads to a new marking $\mathbf{m}' = \mathbf{m} + \alpha \cdot \mathbf{C}[P, t]$.

The set of reachable markings in continuous PN verifies some properties that do not hold in the discrete case. For example, the set of reachable markings of a continuous system is a convex set [14].

In continuous Petri nets the firing can be done in so small quantities that a net can be "almost" in a deadlock state but never reach it. In our opinion, these *limit markings* should also be considered as reachable. Otherwise, there are nets, as the one in Fig. 1(a), that will deadlock as discrete for any initial marking, but can never be *completely* blocked as continuous. In [14] the limit reachability concept was introduced. The set of these limit-reachable markings will be denoted as $\mathrm{RS}_\mathrm{C}(\mathcal{N}, \mathbf{m_0})$. The set of solutions of the relaxed fundamental equation will be denoted as $\mathrm{LRS}_\mathrm{C}(\mathcal{N}, \mathbf{m_0})$, that is, $\mathrm{LRS}_\mathrm{C}(\mathcal{N}, \mathbf{m_0}) = \{\mathbf{m} \mid \mathbf{m} = \mathbf{m_0} + \mathbf{C} \cdot \boldsymbol{\sigma} \geq \mathbf{0}, \boldsymbol{\sigma} \geq \mathbf{0}\}$. As in discrete PN, the fundamental equation relaxation may add spurious solutions to the relaxation made at the net level, that is, $\mathrm{RS}_\mathrm{C}(\mathcal{N}, \mathbf{m_0}) \subseteq \mathrm{LRS}_\mathrm{C}(\mathcal{N}, \mathbf{m_0})$. However, and contrary to what happens in the discrete case, in most practical cases these sets are equal. More precisely, if $\mathcal{N}$ is consistent (i.e., $\exists \mathbf{x} > \mathbf{0} : \mathbf{C} \cdot \mathbf{x} = \mathbf{0}$) and every transition can

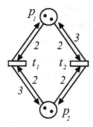
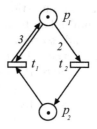

Fig. 1. (a) If limit markings are not considered, liveness of a continuous net is not sufficient for (structural) liveness of its discrete counterpart. (b) In any case, (structural) liveness of the continuous net is not necessary for (structural) liveness of its discrete counterpart

be fired, $\exists \boldsymbol{\sigma} > \mathbf{0}, \mathbf{m_0} \xrightarrow{\sigma}$ (or, equivalently, $\exists \mathbf{m} > \mathbf{0}, \mathbf{m} \in \mathrm{RS}_C(\mathcal{N}, \mathbf{m_0})$), then: $\mathrm{RS}_C(\mathcal{N}, \mathbf{m_0}) = \mathrm{LRS}_C(\mathcal{N}, \mathbf{m_0})$ [14]. Since the preconditions are very weak, this means that *in practice* the relaxation at net level is "equivalent" to the relaxation at the fundamental equation level. That is, there do not exist *spurious* solutions of the fundamental equation.

Hence, properties like deadlock-freeness can be analytically studied. Moreover, structural traps ($\Theta \subseteq P$ such that $\Theta^\bullet \subseteq {}^\bullet\Theta$) do not necessarily "trap" tokens in continuous PN. That is, the behavioural counterpart of structural traps is not true anymore: in continuous PN, traps may be emptied. Moreover, if every transition of the net system can be fired, then every T-semiflow can be fired in isolation [14]. This has an important consequence: *behavioural and structural synchronic relations [15] coincide.*

Fluidification means simplifying the model, assuming that this will allow to use computationally less expensive techniques and get more information about the system. However, one has to keep on mind that those results refer to the continuous PN, and do not always provide "useful" information about the underlying discrete model. Since continuous PN are a relaxation of discrete PN, for those properties based on universal (existential) quantifiers the continuous PN will provide sufficient (necessary) conditions. For example, if the continuous PN is bounded, so will be the discrete PN. For a marking to be reachable in the discrete model, reachability in the continuous one must be guaranteed. However, for those properties formulated interleaving universal and existential quantifiers the analysis of the continuous PN may not provide information about the behaviour of its discrete counterpart. For example, liveness (deadlock-freeness) of the continuous PN is neither necessary nor sufficient for liveness (deadlock-freeness) of the discrete PN. Nevertheless, to be fair one should take into account that maybe the only problem of the discrete net is that it does not have enough tokens. In fact, it can be proved that any (lim-)live continuous PN is structurally live as a discrete PN, although not necessarily live, i.e., the structure of the net is "correct", although the marking may be "not large enough" [14]. On the other hand, a live discrete net may be so only with a particular marking, and any increase of the marking makes it non-live (see Fig. 1(b)). That kind of nets will never go well with continuization, since continuization can be interpreted as if the marking were multiplied by a very large number (infinite populations).

Continuization leads to "easier to analyze" models compared to the discrete models. Nevertheless, the price that has been payed for the relaxation is that some properties of discrete PN cannot be observed in continuous systems, for example mutex relationship, since this property is based on the notion of disjunctive resources, which is lost in the continuous models. Also the distinction between reversibility and existence of home states is lost. This clearly extends to some monopoly and fairness situations.

2.2   Timed Continuous Petri Nets

A simple and interesting way to introduce time in discrete PN is to assume that all the transitions are timed with exponential probability distribution function (pdf). This way, a purely markovian performance model is obtained, for which, due to the memoryless property, the state of the underlying Markov chain is the very marking of the autonomous PN [16].

For the timing interpretation of continuous PN we will use a first order (or deterministic) approximation of the discrete case [17], assuming that the delays associated to the firing of transitions can be approximated by their mean values. Notice that for "congested" systems, this approximation is valid for any pdf — applying the central limit theorem. Here, for simplicity, *immediate* transitions will only be used for "free" conflicts that will be solved according to (marking and time independent) routing rates $\mathbf{R}$.

Different semantics have been defined for continuous timed transitions, the two most important being *infinite server* (or *variable speed*) and *finite server* (or *constant speed*) [18, 17]. Under finite server semantics, the flow of t_i has just an upper bound, $\boldsymbol{\lambda}[t_i]$ (the number of servers times the speed of a server). Then $\mathbf{f}(\tau)[t_i] \leq \boldsymbol{\lambda}[t_i]$ (knowing that at least one transition will be in saturation, that is, its utilization will be equal to 1). Under infinite server semantics, the flow through a timed transition t is the product of the speed, $\boldsymbol{\lambda}[t]$, and the instantaneous enabling of the transition, i.e., $\mathbf{f}[t] = \boldsymbol{\lambda}[t] \cdot \mathrm{enab}(t, \mathbf{m}) = \boldsymbol{\lambda}[t] \cdot \min_{p \in \bullet t}\{\mathbf{m}[p]/\mathbf{Pre}[p, t]\}$. In both cases piecewise linear differential systems are obtained.

For discrete PN infinite server semantics is *more general*, since it allows to implement finite server semantics by adding a place marked with as many tokens as the number of servers. However, this *does not* represent finite-servers semantics if these tokens are interpreted as fluids. In the continuous case the two evolution rules are related to different relaxations of the model. A transition is like an station in QN, thus "the meeting point" of clients and servers. Assuming that there may be many or few of each one of them, fluidization can be considered for clients, for servers or for both. Table 1 represents the four theoretically possible cases. Finite server semantics corresponds at conceptual level to a *hybrid* behaviour: fluidization is applied only to clients, while servers are kept as discrete, counted as a finite number (the firing speed is bounded by the

Table 1. The four cases for possible continuization of a transition [11]. The third one corresponds to *delays* in QN

Clients	Servers	Semantics of the transition
few (D)	few (D)	Discrete transition
many (C)	few (D)	Finite server semantics (bounds to firing speed)
few (D)	many (C)	Discrete transition (servers become *implicit places*)
many (C)	many (C)	Infinite server semantics (speed is enabling-driven)

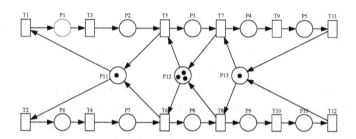

Fig. 2. A simplification of a production line in [20]

product of the speed of a server and the number of servers in the station). On the other hand, infinite server semantics really relaxes both clients and servers.

In both cases, although the fluidization is total, models are hybrid in the sense that they are piecewise linear systems, in which switching among the embedded linear systems is not externally driven as in [19], but internally through the minimum operators. If the PN belongs to the join-free subclass (i.e., transitions cannot have more than one input place), its fluidization generates a linear differential equation system.

The simple (asymmetric choice) model in Fig. 2 represents a production system with two production lines (it is part of an example studied in [20]). It has been analyzed using infinite server semantics and finite server semantics (single-server). In both cases, as the marking of the net is multiplied by a constant k, the throughput (normalized in the infinite server case) approaches to the one obtained when the net is seen as continuous (Table 2). In this case, the continuous net under infinite server semantics is a better approximation of the behaviour of the original discrete system. The finite server semantics disregards the restrictions due to the shared resources, that in this example are quite important unless the system is heavily loaded (large k). Since the two semantics are not comparable, an immediate question is: Do they provide really different performance measures? The answer is positive, being possible to have large differences in the throughput (finite server semantics being usually more optimistic). Therefore, for a particular case, which continuous semantics is better? How much error can we have? For these questions we have no definitive answer, and it is not clear if some in depth understanding can be obtained; nevertheless, an experimental analysis of benchmark examples from the literature is being considered [20]. As a

Table 2. Steady state throughput of the system in Fig. 2, assuming each operation takes 1 t.u. (For k=10 and k=50, markovian simulations are used.)

k	Reachable markings	Infinite servers		Single server
		Thr. of every t_i	Thr./k	Thr. of every t_i
1	250	0.172	0.172	0.172
2	6300	0.366	0.183	0.303
3	67375	0.564	0.188	0.399
5	2159136	0.966	0.193	0.528
10	?	1.96	0.196	0.693
50	?	9.97	0.199	0.91
...	...	...	...	...
Continuous			0.2	1

preliminary remark, it seems that in most cases infinite server semantics provides a better approximation, although it seems difficult to obtain a characterization of the cases in which this happens. Moreover, other models —particularly useful for population dynamic problems— obtained through decoloration of colored models lead to a different semantics in which the "min" operator is replaced by a *product* [9] (which naturally keeps positive the decolored model).

3   Alternative Formalisms

The approximation of a discrete event model by a continuous one is not new, and can be found in different formalisms.

Deterministic first-order approximations have been long used in QN [21, 3, 22, 4]. The fluid QN arises as a limit, in the sense of functional strong law of large numbers, of the stochastic network with the appropriate scaling. In [23] it was proven that the fluid models of certain QN could be used to analyze the (positive) recurrence of their discrete counterparts. In the last years, this has been extended to open or closed multiclass networks under different policies. Fluid models have also been applied in the synthesis of controls or scheduling (see, for example, [5, 6, 24, 25, 26]), proving that the optimal policy of the fluid networks can be somehow translated to a "good" policy for some discrete QN. In this setting, the emphasis has been usually put on the rigorous mathematical justification, even if that meant that fluid models were applied to "narrow" net classes.

Comparisons of PN and (monoclass) QN can be seen in [27, 28, 29]. Both formalisms are in essence *bipartite*: Places and transitions for PN; Queues and stations for QN. From an structural point of view, the main differences are the possible simultaneous existence in a single PN model of arc weights, attributions, choices, forks and joins, and the possible absence of local conservation rules when transitions are fired. QN are in essence timed models, that may be provided with very rich routing service and queuing disciplines, while PN can be studied also as autonomous (idea of non-determinism). Moreover, as it was be pointed out in

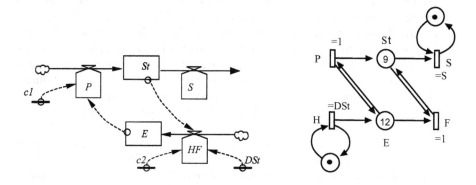

Fig. 3. A system described using FD and PN [33]. In the Forrester diagram, St and E are deposits with levels, while P, S and HF are valves regulating the flows

Subsect. 2.2, different timed interpretations of the net lead to different firing/flow policies. The firing logic of PN is of the type *consumption/production*, a kind of generalization of the classical *client/servers* in QN (Jackson, Gordon-Newell).

Forrester Diagrams (FD) appeared in the Systems Dynamics framework, and is a well-known, essentially continuous timed formalism for modelling certain classes of DES [30, 31]. FD provide a graphic representation of the system, that corresponds to an hydrodynamic interpretation. A comparison of PN and FD can be found in [32, 33]. For example, Fig. 3 represents both a FD and a PN model of an example. It is a production system that maintains a certain stock (St, with initial value 9) and a number of employees (E, with initial value 12). Products decrease due to sales, which are constant in time (S=12), and increases with production (P), which is proportional to the number of employees (here with constant 1). Employees change due to hiring or firing (HF), which is proportional to the difference between the desired stock (DSt=10) and the actual stock (St). Two remarks with respect to the PN model of the system: (1) variables have to be positive, hence variable HF has been split in two, hiring and firing; (2) to represent that sales are constant, or that firing depends on the stock, self-loops have been added with large enough arc weights to guarantee that these places always define the enabling degree (here weight 1 is enough). Although the continuous net is not structurally bounded, it is bounded with this timing. That is, complementary places ($\overline{St}, \overline{E}$) with a large enough marking could be added without changing the behaviour. Continuous PN and FD provide a graphical support to generate systems of differential equations easily and there is a clear correspondence among their main types of nodes — place/level and transition/valve (or firing-speed/flow-variable). However, there are some differences [32, 33]:

– *Marking of places vs levels.* In FD each level corresponds to a state variable. Although in PN places are essentially state variables, redundancies may exist due to token conservation laws. Particular cases are structural implicit places. Levels may be negative, but markings are in essence positive.

- *Transitions vs. valves: flows.* The evolution of flows in FD takes place according to the information that valves receive from the whole state of the system, through the information channels. The evolution of the PN takes place according to the information that each transition receives from its input places. Thus, FD separate the *material* and the *information* flows, and evolve according to global information of the system. On the other hand, PN have only a flow of material that carries the information implicitly, and evolve according to information that, in standard uses, is local to each transition (its input places).
- In FD synchronizations are not explicitly modelled: there exist no elements to represent "rendez-vous", and must be simulated by means of flow equations.
- In FD material is strictly conservative around the valves (the relationship among input and output flow is always 1:1), while in PN weighted conservation is often found.

As QN, FD are timed models. In FD the methodological analysis is basically focused to simulation, although there are some researching groups who also did go into the mathematical analysis of the system, basically sensibility, bifurcations and qualitative analysis (see for example [34]). Contrarily to QN, in FD the relationship between the solution of the continuous model and the original discrete model has not received much attention.

Under both, finite and infinite server semantics, "unforced" PN models are *positive systems* in Luenberger sense [35, 36], that is, the non-negativity constraints on the marking are redundant. A particularly interesting case of positive systems are *compartmental systems*, which are composed of a finite number of subsystems (compartments), interacting by exchanging material among the compartments and with the environment [37, 38, 39].

An immediate similarity between PN and compartmental systems is that both allow representations based on *graphs*. However, PN are bipartite graphs, while compartmental models have a single kind of nodes. In discrete PN there are two different kinds of nodes: OR nodes (attributions/choices), and AND nodes (joins/forks). Nevertheless, in continuous PN (under infinite server semantics) the forward OR node in homothetic conflicts (if between t_1 and t_2, $\mathbf{Pre}[P, t_1] = \alpha \cdot \mathbf{Pre}[P, t_2]$) is transformed into a "+" operation. Therefore, choices can be seen as *flow splitters*.

As in FD, in compartmental systems there is a strong "strict" conservation law: matter is not created, although it may "evaporate" and disappear if the system is not (output) closed. In PN such kind of constraint does not exist.

Another difference between PN and the graphs associated to compartmental systems is the arc weights. However, this is not a real generalization in the case of continuous nets *without synchronizations* (join-free nets), since for any of those nets, an equivalent one exists with arc weights one [36]. In other words, weights in continuous models without synchronizations constitute a modelling convenience (i.e., do not add theoretical expressive power).

Putting all together, strongly connected and conservative PN without synchronizations are equivalent, from the modelling point of view, to closed linear compartmental systems [36].

The above consideration of alternative "fluidified" formalisms is a source of opportunities in order to bridge the results in the analysis or synthesis of continuous PN models with that in "close" or related paradigms.

4   Performance Evaluation

Analyzing the performance of a continuous PN both in the transient and in the steady state involves integrating a set of differential equations. In theory it is possible to solve it analytically: solve the linear differential equations defined by the initial marking, and study among the different "minimums" associated to the synchronizations which one will be reached before, then repeat the process. In practice the existence of many differential equations, and many synchronizations makes this "artisanal" approach unfeasible, although a numerical version can be easily implemented in a computer, for example using Matlab. The equations that define the behaviour of the system in Fig. 4(a) are:

$$\mathbf{f}(\tau)[t_1] = \boldsymbol{\lambda}[t_1] \cdot \mathbf{m}(\tau)[p_1]$$
$$\mathbf{f}(\tau)[t_2] = \boldsymbol{\lambda}[t_2] \cdot \mathbf{m}(\tau)[p_2]$$
$$\mathbf{f}(\tau)[t_3] = \boldsymbol{\lambda}[t_3] \cdot \min(\mathbf{m}(\tau)[p_2], \mathbf{m}(\tau)[p_3])$$
$$\mathbf{f}(\tau)[t_4] = \boldsymbol{\lambda}[t_4] \cdot \min(\mathbf{m}(\tau)[p_3], \mathbf{m}(\tau)[p_4])$$

If it is only the steady state we care about, some results and techniques have been developed (see [10]). First of all, it has to be remarked that, in general, there is no guarantee about the existence of a steady state. For example, the net in Fig. 3(b) oscillates indefinitely without, even asymptotically, approaching to a steady state [33](see Fig. 5). To the best of our knowledge, the existence

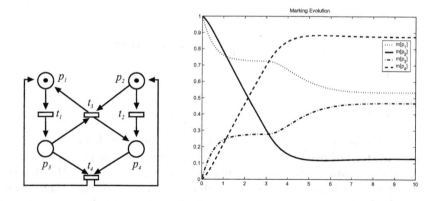

Fig. 4. Continuous PN and the evolution of the marking with $\boldsymbol{\lambda} = [0.5\,0.1\,1\,0.3]$

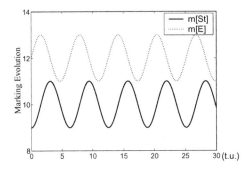

Fig. 5. The net system in Fig. 3(b) does not have a steady state

of a steady state had always be assumed, even for hybrid nets. In the future some work will be needed for the development of at least necessary or sufficient conditions. In the rest of this section some results for the computation of the steady state will be given, being assumed the existence of a steady state marking under infinite server semantics. Nevertheless, it should be pointed out that the markovian discrete counterpart of this model is ergodic. Let $\mathbf{m}_{ss}$ be the steady state marking of a bounded continuous net system: $\mathbf{m}_{ss} = \lim_{\tau \to \infty} \mathbf{m}(\tau)$. Then, for every $\tau > 0$ it must be true:

$$\dot{\mathbf{m}}(\tau) = \mathbf{C} \cdot \mathbf{f}(\tau)$$

$$\mathbf{f}[t](\tau) = \boldsymbol{\lambda}[t] \cdot \min_{p \in {}^{\bullet}t} \left\{ \frac{\mathbf{m}[p](\tau)}{\mathbf{Pre}[p,t]} \right\} \quad \forall \text{ non-immediate transition } t \tag{1}$$

$$\mathbf{R} \cdot \mathbf{f}(\tau) = \mathbf{0}$$

$$\mathbf{m}(0) = \mathbf{m_0}$$

Using $\boldsymbol{\phi}$ as an approximation of $\mathbf{f}$ in the steady state, and $\boldsymbol{\mu}$ as an approximation of the marking in the steady state, the above equations can be relaxed as follows:

$$\boldsymbol{\mu} = \mathbf{m_0} + \mathbf{C} \cdot \boldsymbol{\sigma}$$

$$\boldsymbol{\phi}[t] = \boldsymbol{\lambda}[t] \cdot \min_{p \in {}^{\bullet}t} \left\{ \frac{\boldsymbol{\mu}[p]}{\mathbf{Pre}[p,t]} \right\} \quad \forall \text{ non-immediate transition } t \tag{2}$$

$$\mathbf{R} \cdot \boldsymbol{\phi} = \mathbf{0}$$

$$\mathbf{C} \cdot \boldsymbol{\phi} = \mathbf{0}$$

$$\boldsymbol{\mu}, \boldsymbol{\sigma}, \boldsymbol{\phi} \geq \mathbf{0}\}$$

With this relaxation we have replaced the condition of being a reachable marking with that of being a solution of the fundamental equation. That is, we are loosing the information about the feasibility of the transient path. Observe that the system is non-linear ("min" operator) and a unique solution is not guaranteed. For example, for the net system in Fig. 6 with $\boldsymbol{\lambda} = [2, 1, 1]$, any marking $[10 - 5 \cdot \alpha, 4 \cdot \alpha - 3, \alpha, \alpha]$, with $1 \leq \alpha \leq 5/3$, verifies (2), and all of

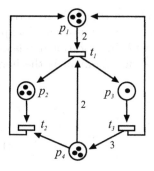

Fig. 6. A net system for which, with $\boldsymbol{\lambda} = [2, 1, 1]$, infinite solutions of (2) exist

them lead to different throughput. Maximizing the flow of a transition, an *upper bound* of the throughput is obtained:

$$\max\{\phi[t_1] \mid \boldsymbol{\mu} = \mathbf{m_0} + \mathbf{C} \cdot \boldsymbol{\sigma}$$

$$\phi[t] = \boldsymbol{\lambda}[t] \cdot \min_{p \in {}^\bullet t}\left\{\frac{\boldsymbol{\mu}[p]}{\mathbf{Pre}[p, t]}\right\} \ \forall \text{ non-immediate transition } t$$

$$\mathbf{R} \cdot \boldsymbol{\phi} = \mathbf{0} \tag{3}$$

$$\mathbf{C} \cdot \boldsymbol{\phi} = \mathbf{0},$$

$$\boldsymbol{\mu}, \boldsymbol{\sigma}, \boldsymbol{\phi} \geq \mathbf{0}\}$$

Notice that the solution of (3) is always dynamically "reachable" in the sense that with a suitable initial distribution of the tokens inside the P-semiflows, this throughput can be obtained (for instance, with the steady state distribution). Nevertheless, the programming problem in (3) is not easy to solve due to the "min" operator, that makes it *non linear*. The problem is that there is no way to know in advance which of the input places will restrict the flow, and so, a kind of branch and bound algorithm is used to solve it [10]. The idea is to solve the LPP defined by the system of (in)-equalities that appears choosing one input place per transition as the one which gives the minimum. If the marking does not correspond to a steady state (i.e., there is at least one transition such that all its input places have "too many" tokens) choose one of the synchronizations and solve the set of LPP that appear when each one of the input places are assumed to be defining the flow. That is, build a set of LPP by adding an equation that relates the marking of each input place with the flow of the transition. These subproblems become children of the root search node. The algorithm is applied recursively, generating a tree of subproblems. If an optimal steady state marking is found to a subproblem, it is a possible steady state marking, but not necessarily globally optimal. Since it is feasible, it can be used to prune the rest of the tree: if the solution of the LPP for a node is smaller than the best known feasible solution, no globally optimal solution can exist in the subspace of the feasible region represented by the node. Therefore, the node can be removed from consideration. The search proceeds until all nodes have been solved or pruned.

Some results have been developed that help to further prune the tree [10], but nevertheless, worst cases may be computationally expensive.

This suggests to go on with an additional relaxation, knowing that some accuracy may be lost. Since the minimum is the hardest point, that equation in (3) can be replaced with:

$$\phi_{ss}[t] = \lambda[t] \cdot \frac{\mu_{ss}[p]}{\mathbf{Pre}[p,t]} \quad \text{if} \quad p = {}^{\bullet}t \tag{4}$$

$$\phi_{ss}[t] \leq \lambda[t] \cdot \frac{\mu_{ss}[p]}{\mathbf{Pre}[p,t]} \quad \forall p \in {}^{\bullet}t \quad \text{otherwise} \tag{5}$$

$$\frac{\phi_{ss}[t_i]}{\lambda[t_i]} = \frac{\phi_{ss}[t_j]}{\lambda[t_j]} \quad \forall \, t_i, t_j \text{ in EQ relation} \tag{6}$$

This way we have a single linear programming problem (LPP), thus computation is of polynomial time complexity. Unfortunately, the LPP provides in general a non tight bound, i.e, the solution may be non reachable for any distribution of the tokens verifying the P-semiflow load conditions, $\mathbf{y} \cdot \mathbf{m_0}$. This may occur because it may be the case that for that solution none of the input places of a synchronization really restricts the flow of that transition. When this happens, the marking cannot define the steady state (the flow of that transition would be larger). See [10] for conditions that guarantee in some systems the reachability of the bound.

A slight relaxation of that LPP, using inequalities in all the transitions, leads to a result that had been obtained long before for discrete nets. For bounded *discrete* net systems, an upper bound of the throughput of one transition can be obtained by means of that LPP [40, 41].

$$\max\{\phi[t_i] \mid \boldsymbol{\mu} = \mathbf{m_0} + \mathbf{C} \cdot \boldsymbol{\sigma}$$

$$\phi[t] \leq \lambda[t] \cdot \frac{\mu[p]}{\mathbf{Pre}[p,t]} \quad \forall p \in {}^{\bullet}t$$

$$\mathbf{R} \cdot \boldsymbol{\phi} = \mathbf{0} \tag{7}$$

$$\mathbf{C} \cdot \boldsymbol{\phi} = \mathbf{0},$$

$$\boldsymbol{\mu}, \boldsymbol{\sigma}, \boldsymbol{\phi} \geq \mathbf{0}\}$$

For nets in which the steady state visit ratio can be deduced from the structure and the rates of the transitions (mono T-semiflow reducible nets), it can be proved that both statements are equivalent (in the sense that if one solution with inequality exists, another one with the same throughput verifies the equality) [10].

5   Parametric Design

In engineering, analysis techniques frequently guide in the definition of synthesis methods. Assuming approximate computation of performance with efficient algorithms, the problem of designing the best set of resources (best $\mathbf{m_0}$), the best routing policy (best $\mathbf{R}$), the best type of machines (machine selection problem, appearing indirectly as determining the best $\boldsymbol{\lambda}$) can be straightforwardly stated. Observe that these are not "on-line" control problems, but "off-line" design problems in which parametric optimizations are being solved. A recent presentation of some "easily" (i.e., polynomial time) solvable problems of this type can be found in [11]. One of the basic statements is as follows: Let $\mathbf{g}\cdot\mathbf{f}-\mathbf{w}\cdot\mathbf{m}-\mathbf{b}\cdot\mathbf{m_0}$, be the profit function to be optimized, where, $\mathbf{g}$ represents a gain vector associated to flows/throughput, $\mathbf{w}$ is the *cost* vector due to immobilization to maintain the production flow (e.g., due to the levels in stores), and vector $\mathbf{b}$ represent *depreciations or amortization of the initial investments* (e.g., due to the size of the stores, number of machines, ...). Assume also upper bounds in the use of the resources ($\mathbf{V}\cdot\mathbf{m_0}\leq\mathbf{k}$), i.e., its availability is limited.

This kind of optimization problem admits a particularly elegant and efficient solution if the LPP, stated in Sect. 4 lead to the exact value (otherwise *upper bounds are obtained*). As was previously mentioned, this happens, for example, for structurally live and bounded equal conflict (EQ) nets (its characterization can be computed polynomially through the rank theorem [9]). For simplicity, in the sequel of this section *let us assume that nets are structurally live and bounded EQ (thus mono-T-semiflow reducible), and conflicts among immediate transitions are solved according to routing rates, $\mathbf{R}$.* The following LPP can be written:

$$\begin{aligned}
\max\{\ &\mathbf{g}\cdot\boldsymbol{\phi}-\mathbf{w}\cdot\boldsymbol{\mu}-\mathbf{b}\cdot\boldsymbol{\mu}_0\\
\text{s.t.}\ \ &\boldsymbol{\mu}=\boldsymbol{\mu}_0+\mathbf{C}\cdot\boldsymbol{\sigma}\\
&\phi[t]\leq\boldsymbol{\lambda}[t]\cdot\tfrac{\mu[p]}{\mathbf{Pre}[p,t]}\quad\forall p\in{}^\bullet t\\
&\phi[t]=\boldsymbol{\lambda}[t]\cdot\tfrac{\mu[p]}{\mathbf{Pre}[p,t]}\quad\text{if }p={}^\bullet t\\
&\mathbf{C}\cdot\boldsymbol{\phi}=\mathbf{0}\\
&\mathbf{R}\cdot\boldsymbol{\phi}=\mathbf{0}\\
&\boldsymbol{\phi},\boldsymbol{\mu},\boldsymbol{\mu}_0\geq\mathbf{0}\\
&\mathbf{V}\cdot\boldsymbol{\mu}_0\leq\mathbf{k}\}
\end{aligned}\tag{8}$$

In other cases the problem to be solved is which are the minimum cost resources ($\mathbf{b}\cdot\mathbf{m_0}$) that guarantee a certain throughput (see Problem 5 in [11]). The routing matrix $\mathbf{R}$ may be also the parameters to be optimized, looking for best production mix, or better internal routing at the factory. A simple case for optimizing a profit function w.r.t. the routing $\mathbf{R}$ is the following example: Maximize $\mathbf{g}\cdot\boldsymbol{\phi}-\mathbf{w}\cdot\boldsymbol{\mu}-\mathbf{b}\cdot\boldsymbol{\mu}_0$, with respect to the routing. The following LPP computes an optimal flow vector, $\boldsymbol{\phi}$, being $\mathbf{R}$ free.

$$\begin{aligned}
\max\{\ &\mathbf{g}\cdot\boldsymbol{\phi}-\mathbf{w}\cdot\boldsymbol{\mu}-\mathbf{b}\cdot\boldsymbol{\mu}_0\\
\text{s.t.}\ \ &\boldsymbol{\mu}=\boldsymbol{\mu}_0+\mathbf{C}\cdot\boldsymbol{\sigma}\\
&\phi[t]\leq\boldsymbol{\lambda}[t]\cdot\tfrac{\mu[p]}{\mathbf{Pre}[p,t]}\quad\forall p\in{}^\bullet t\\
&\phi[t]=\boldsymbol{\lambda}[t]\cdot\tfrac{\mu[p]}{\mathbf{Pre}[p,t]}\quad\text{if }p={}^\bullet t\\
&\mathbf{C}\cdot\boldsymbol{\phi}=\mathbf{0}\\
&\boldsymbol{\mu}\geq\mathbf{0}\}
\end{aligned}\tag{9}$$

Once LPP (9) has been solved, the computation of the routing matrix $\mathbf{R}$, is straightforward, just proceed free-choice by free-choice. Assuming for simplicity that choices are binary: $\phi_1/\phi_2 = r_1/r_2$, and $r_1 + r_2 = 1$.

If all free conflicts are solved with immediate transitions, and $\mathbf{g} = \mathbf{1}, \mathbf{w} = \mathbf{b} = \mathbf{0}$, this LPP is analogous to the one stated in [42], assuming boundedness. Even if in this last case nets are P-timed (i.e., with delays associated to places), and conflicts are solved according to a stationary routing policy (a simplifying preselection policy, which in practice is equivalent a net without conflicts), and have different transient behaviour, their steady state is the same.

6   Observation

In order to control a dynamic system, frequently it is necessary to know its current state. Sensors can be used to get information from the plant, but often some of the variables cannot be directly measured, either because it is not physically possible, or because of its cost. If the information that can be obtained from the system allows to estimate the value of a variable, that variable is said to be observable (with that instrumentation), and the estimate constitutes the observation. The *observability problem*, i.e., the characterization of which state variables are observable and its observation, has been studied both for continuous systems (in particular linear systems) and for discrete event systems. Some results related to observability of discrete event models can be found in [43, 44].

With respect to continuous systems, observability is quite a classical problem, for which easy to understand and general results were obtained for time-invariant linear systems in the sixties of the last century [45, 35]. The contribution of inputs to the evolution of a linear system can be easily computed and subtracted from the total output. Therefore, observability of linear systems can be studied using its unforced counterpart.

A time-invariant linear system can be expressed as $\dot{\mathbf{x}}(\tau) = \mathbf{A} \cdot \mathbf{x}(\tau) + \mathbf{B} \cdot \mathbf{u}(\tau), \mathbf{y}(\tau) = \mathbf{S} \cdot \mathbf{x}(\tau)$, where $\mathbf{y}$ represents its output, that is, "what is seen" of it. A linear system is said to be *observable* iff knowing $\mathbf{y}(\tau)$, it is possible to compute its initial state $\mathbf{x}(\tau_0)$. That is, iff knowing $\mathbf{y}(\tau)$ the equation

$$\mathbf{y}(\tau) = \mathbf{S} \cdot e^{\mathbf{A} \cdot \tau} \cdot \mathbf{x}(\tau_0)$$

can be solved for every $\mathbf{x}(\tau_0)$. It can be seen that this is equivalent to matrix $\vartheta = (\mathbf{S}^T | (\mathbf{S}\mathbf{A})^T | \cdots | (\mathbf{S}\mathbf{A}^{n-1})^T)^T$ having full rank. This result is known as the *observability theorem* and the matrix is known as the *observability matrix* [45]. For linear systems, the observable subspace can be characterized algebraically. Intuitively, a system state estimation can be *theoretically* obtained from the output signal and the computation of its derivatives.

Hence observability is completely characterized for nets that can be described with a linear system, that is, join-free nets (i.e., nets that do not have synchronizations). If the net has synchronizations, there are several linear systems that may define the evolution of the system, depending on which is the place that restricts each transition. The observability theorem has been extended to general piecewise linear systems. The complete system is observable iff the pairwise intersection of different observable subspaces is trivial, that is, the joint observability matrix of each pair of linear system has full rank [46]. However, continuous PN have a characteristic that these general piecewise systems do not have: the change of one linear system to another one is triggered by the continuous state (the marking). This makes the observability of continuous PN a more simple issue [47]: if the system passes through an observable linear system, its marking at that moment can be observed. And since it is deterministic, it is possible to simulate it backwards and deduce the initial marking.

Notice also that observability of a synchronization will not be possible in general unless *all its input places* are measured (it might be possible to measure one place only if it were timed implicit). Moreover, observability cannot be extended forward (the output flow of a transition does not provide information to deduce the marking of the next place). Hence, the problem can be tackled by measuring the places in synchronizations. The net that remains removing those places and their input and output arcs—eventually composed of several unconnected subnets—is join-free. For these subnets, the observability theorem can be applied. Hence, given a set of measured places it is not difficult to prove whether the net is observable or not.

Nets without *synchronizations and attributions* ($p \in P$ is an attribution if $|{}^\bullet p| > 1$) can be observed just measuring the "final" places (places without output arcs) or measuring one (any) place if it is a weighted cycle. As a direct consequence, it can be stated that a weighted T-system is observable for any initial marking iff all synchronization places are measured, or, in the case of a cycle, one arbitrary place is measured [48]. For this kind of nets, the rates of

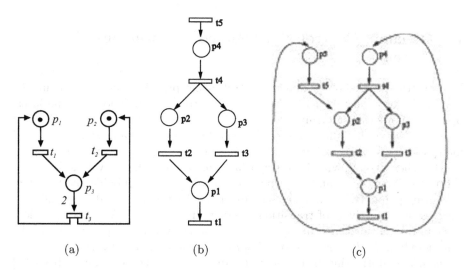

Fig. 7. Observability in nets with attributions depends on both the rates, and the structure

the transitions do not have any influence on the observability of the system. *Attributions* on the other hand, force to study the rates of the net. Observe for example the net in Fig. 7(a), and assume that p_3 is measured. The system is observable iff the rates of t_1 and t_2 are not equal, i.e. $\boldsymbol{\lambda}[t_1] \neq \boldsymbol{\lambda}[t_2]$. Intuitively, if they have the same rate, it is not possible to distinguish which part of the flow is coming from each place [47]. Moreover, this is not a local problem, but a global one. On the one hand, it is not just the rates of the input transitions of the attribution that have to be taken into account. For example, the net in Fig. 7(b) is not observable if $\boldsymbol{\lambda}[t_2] = \boldsymbol{\lambda}[t_3]$, but neither is if $\boldsymbol{\lambda}[t_4] = \frac{2 \cdot \boldsymbol{\lambda}[t_2] \cdot \boldsymbol{\lambda}[t_3]}{\boldsymbol{\lambda}[t_2] + \boldsymbol{\lambda}[t_3]}$. On the other hand, attributions are not "independent". For example, the net in Fig. 7(c) with $\boldsymbol{\lambda}[t_2] \neq \boldsymbol{\lambda}[t_3]$ and $\boldsymbol{\lambda}[t_4] = \boldsymbol{\lambda}[t_5] = \frac{2 \cdot \boldsymbol{\lambda}[t_2] \cdot \boldsymbol{\lambda}[t_3]}{\boldsymbol{\lambda}[t_2] + \boldsymbol{\lambda}[t_3]}$ is observable if p_4 is measured, but not if p_5 is measured. For any other value of $\boldsymbol{\lambda}[t_4] = \boldsymbol{\lambda}[t_5]$ it is observable measuring either p_4 or p_5. A related "design" problem is to determine minimal cost observability. That is, when a cost is assigned to measuring each place, which is the best selection of the places guaranteeing that the system is observable? To apply the previous result would mean to solve a combinatorial set of observability problems. Nevertheless this number can be greatly reduced in many cases applying the following property [48]: Let p and p' be such that there is a path from p' to p without synchronizations or attributions. Then

- p' can be deduced from the observation of p.
- If the net is not observable measuring a set of places containing place p, it cannot be observable if p is replaced by p'.

An algorithm, and its application to an example showing how the combinatory is reduced can be found in [48].

7   Dynamic Control: On "Forced" Continuous Net Systems

In order to speak about dynamic control, some previous questions should be answered. For example, *what to control*? According to the adopted time interpretation, flows through transitions should be controlled, both w.r.t. *routing* and *service*. Observe that this is not really something new; the same strategy is used for QN, were servers activity and routing of customers are controlled; analogously, when dealing with Forrester Diagrams, the opening of valves has to be controlled. Now the second question, *how to control*? The only idea is to control at routing points (what may be complex at non free-choices) and, eventually, to slow down the activity of transitions (servers in a station). As a last question, it should be decided *how to express* the control. Two main approaches can be considered: *multiplicative* (the speed of t is controlled as $\alpha \cdot \boldsymbol{\lambda}[t]$, with $\alpha \in (0, 1)$) or *additive* (subtracting $\mathbf{u}$, $\mathbf{0} \leq \mathbf{u} \leq \mathbf{f}$). In any case, the flow can go from $\mathbf{f}[t]$ to $\mathbf{0}$. That is, the control can locally slow down the activity of transitions. It is not the moment to discuss that issue in detail, let us just say that they are "in essence" equivalent. Our choice here is to use the additive formulation. Proceeding in that way, using $\mathbf{u}$ as the *slow down control vector*, the fundamental equation is now: $\dot{\mathbf{m}} = \mathbf{C} \cdot (\mathbf{f}(\mathbf{m}) - \mathbf{u})$, were $\mathbf{0} \leq \mathbf{u} \leq \mathbf{f}(\mathbf{m})$.

The above statement suggests two different remarks: (1) the system is *not positive* anymore in the classical (and restrictive) sense of [35, 37] (see [36]); (2) the slowing down action is *dynamically bounded* by the actual state (marking) of the system.

Some results are already known for controllability in the previous framework [49]. For the present purpose let us just point out that if all transitions are controllable, reachability in timed models is equivalent to reachability in the underlying untimed models [50]. In other words, if marking $\mathbf{m}$ is reachable in the untimed model $\langle \mathcal{N}, \mathbf{m_0} \rangle$, there exists a way of controlling the transitions for reaching it in the controlled timed model.

Notice that it will take infinite time to reach a marking that empties a place, unless perhaps if it were already empty in the initial marking. However, that kind of markings will always have at least one transition with 0 throughput, hence they are not very interesting as steady state markings. Once again, this problem is not something new: the loading of a capacity in a basic RC-electrical circuit cannot be "complete" in finite time. Nevertheless, engineers use the classical concept of response time (at 5%, 3%, 1%) in order to have a practical view of the duration of the transient behaviour.

Let us assume in the sequel that *all* the transitions are controllable, and let us concentrate first in the steady state control. A first remark is that given a net and a constant steady state control, $\mathbf{u}$, there may exist several markings, perhaps with different flows, that may be steady state markings. For example, for the system depicted in Fig. 6, with $\boldsymbol{\lambda} = [2, 1, 1]$, and $\mathbf{u} = \mathbf{0}$, marking $[5, 1, 1, 1]$ (with flow $[1, 1, 1]$) and marking $[5/3, 11/3, 5/3, 5/3]$, (with flow $[5/3, 5/3, 5/3]$) can be both steady states. Hence, a first interesting step is to obtain the optimum steady state, and a control action for it. A LPP similar to the one in (9) can be used for that [50]:

$$
\begin{aligned}
\max\{ &\ \mathbf{g} \cdot \boldsymbol{\phi} - \mathbf{w} \cdot \boldsymbol{\mu} - \mathbf{b} \cdot \boldsymbol{\mu_0} \\
\text{s.t.} &\ \ \boldsymbol{\mu} = \boldsymbol{\mu_0} + \mathbf{C} \cdot \boldsymbol{\sigma} \\
&\ \ \phi[t] = \boldsymbol{\lambda}[t] \cdot \frac{\mu[p]}{\mathbf{Pre}[p, t]} - \mathbf{v}[p, t] \quad \forall p \in {}^{\bullet}t \\
&\ \ \mathbf{C} \cdot \boldsymbol{\phi} = \mathbf{0} \\
&\ \ \boldsymbol{\mu}, \boldsymbol{\sigma} \geq \mathbf{0} \\
&\ \ \mathbf{v} \geq \mathbf{0} \}
\end{aligned}
\tag{10}
$$

The only difference is that now a set of slack variables $\mathbf{v}[p, t]$ have been added in the equation that relates the marking with the flow. If $|{}^{\bullet}t| = 1$, the slack variable $\mathbf{v}[{}^{\bullet}t, t]$ represents the control action of the transition. In general, it can be seen that $\mathbf{u}[t] = \min_{p \in {}^{\bullet}t} \mathbf{v}[p, t]$ is an appropriate control input.

With respect to the transient, the use of $\mathbf{u}$ as the single reference input is not enough in general to reach the optimal flow in the net. In [11, 50] a schema as the one in Fig. 8 is proposed, in which the control action depends on the

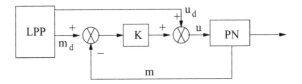

Fig. 8. Control schema

steady state control action, and the difference between the actual marking and the steady state one.

One approach that has been used in the literature to face the problem of optimal control of hybrid systems is to approximate them by discrete-time systems, and represent them as Mixed Logical Dynamical (MLD) systems [51]. Usually, in a MLD system the time step is constant. Time discretization has two important drawbacks: (1) The length of the sampling period is not easy to define. There exists a tradeoff between accuracy (short sampling period) and computational speed (long sampling period). In fact, the complexity typically grows exponentially with the number of switching variables, and these, for a given time interval, are inversely proportional to the length of the sampling period; (2) It is assumed that events can occur only at time instants that are multiple of the sampling period. In fact, it would be desirable to deal with a model that requires a minimum number of steps (samples) without losing accuracy.

In [52] it was seen that the behaviour of finite server semantics continuous PN system could be described by means of an MLD system. Moreover, since PN are event-driven systems, it could be a continuous-time event-driven MLD, instead of one that evolves with a fixed time step. Observe that this approach has two interesting advantages: (1) Event-discretization does not imply loss of accuracy: The marking evolution of a continuous PN is linear between events, and so it can be determined from the marking of the net at the event instants. (2) The number of steps is minimized: A step happens only when it is really required (an event happens).

Different kinds of optimal control problems can be solved by means of the explained event-driven approach, for example: reaching a target marking in minimum time, i.e., time optimal control, maximizing the steady state throughput, or maximizing an optimization function in which several different parameters are involved [52].

Some preliminary efforts are also being made to transform the optimal control problem into a multi-parametric quadratic program, and apply the techniques developed for this kind of systems [19].

Acknowledgments

We would like to thank the following people for their contributions to this research line: E. Teruel, J. Júlvez, E. Jiménez, C. Mahulea and D. Rodríguez.

References

1. Silva, M., Colom, J.M.: On the computation of structural synchronic invariants in P/T nets. In Rozenberg, G., ed.: Advances in Petri Nets 1988. Volume 340 of Lecture Notes in Computer Science. Springer (1988) 387–417

2. Silva, M., Teruel, E., Colom, J.M.: Linear algebraic and linear programming techniques for the analysis of net systems. In Rozenberg, G., Reisig, W., eds.: Lectures in Petri Nets. I: Basic Models. Volume 1491 of Lecture Notes in Computer Science. Springer (1998) 309–373

3. Kleinrock, L.: Queuing Systems, Volume II: Computer Applications. Volume 2. Wiley (1976)

4. Mandelbau, A., Chen, H.: Discrete flow networks: Bottleneck analysis and fluid approximations. Mathematical Operations Research **16** (1991) 408–446

5. Chen, H., Yao, D.: Fundamentals of Queueing Networks. Performance, Asymptotics and Optimization. Volume 46 of Applications of Mathematics. Stochastic Modelling and Applied Probability. Springer (2001)

6. Cassandras, C.G., Sun, G., Panayiotou, C.G., Wardi, Y.: Perturbation analysis of multiclass stochastic fluid models. In: 15th IFAC World Congress, Barcelona, Spain (2002)

7. David, R., Alla, H.: Continuous Petri nets. In: Proc. of the 8th European Workshop on Application and Theory of Petri Nets, Zaragoza, Spain (1987) 275–294

8. David, R., Alla, H.: Discrete, Continuous, and Hybrid Petri Nets. Springer-Verlag (2004)

9. Silva, M., Recalde, L.: Petri nets and integrality relaxations: A view of continuous Petri nets. IEEE Trans. on Systems, Man, and Cybernetics **32** (2002) 314–327

10. Júlvez, J., Recalde, L., Silva, M.: Steady-state performance evaluation of continuous mono-t-semiflow Petri nets. Automatica **41** (2005) 605–616

11. Silva, M., Recalde, L.: On fluidification of Petri net models: from discrete to hybrid and continuous models. Annual Reviews in Control **28** (2004) 253–266

12. Nakamura, M., Silva, M.: An iterative linear relaxation and tabu search approach to minimum initial marking problems of timed marked graphs. In: Procs. of European Control Conference, ECC99, Aachen, Germany (1999)

13. Silva, M.: Introducing Petri nets. In: Practice of Petri Nets in Manufacturing. Chapman & Hall (1993) 1–62

14. Recalde, L., Teruel, E., Silva, M.: Autonomous continuous P/T systems. In Donatelli, S., Kleijn, J., eds.: Application and Theory of Petri Nets 1999. Volume 1639 of Lecture Notes in Computer Science., Springer (1999) 107–126

15. Silva, M.: Towards a synchrony theory for P/T nets. In Voss, K., et al., eds.: Concurrency and Nets. Springer (1987) 435–460

16. Molloy, M.K.: Performance analysis using stochastic Petri nets. IEEE Trans. on Computers **31** (1982) 913–917

17. Recalde, L., Silva, M.: Petri Nets fluidification revisited: Semantics and steady state. European Journal of Automation APII-JESA **35** (2001) 435–449

18. Alla, H., David, R.: Continuous and hybrid Petri nets. Journal of Circuits, Systems, and Computers **8** (1998) 159–188

19. Bemporad, A., Giua, A., Seatzu, C.: An iterative algorithm for the optimal control of continuous-time switched linear systems. In Silva, M., Giua, A., Colom, J., eds.: WODES 2002: 6th Workshop on Discrete Event Systems, Zaragoza, Spain, IEEE Computer Society (2002) 335–340

20. Mahulea, C., Rodríguez, D., Recalde, L., Silva, M.: Choosing server semantics for continuous Petri nets: An examples driven approach. Research report, Dep. Informática e Ingeniería de Sistemas, Universidad de Zaragoza, María de Luna, 1, 50018 Zaragoza, Spain (2005) Forthcoming.
21. Newell, G.F.: Applications of Queueing Theory, Second Edition. Chapman and Hall (1982)
22. Vandergraft, J.: A fluid model of networks of queues. Management Science **29** (1983) 1198–1208
23. Dai, J.: On positive Harris recurrence of multiclass queueing networks: a unified approach via fluid limit models. Annals of Applied Probability **5** (1995) 49–77
24. Weiss, G.: Scheduling and control of manufacturing systems — a fluid approach. In: Proceedings of the 37 Allerton Conference. (1999) 577–586
25. Meyn, S.P.: Sequencing and routing in multiclass queueing networks. part I: Feedback regulation. SIAM Journal on Control and Optimization **40** (2002) 741–776
26. Meyn, S.P.: Sequencing and routing in multiclass queueing networks part II: Workload relaxations. SIAM Journal on Control and Optimization **42** (2003) 178–217
27. Vernon, M., Zahorjan, J., Lazowska, E.D.: A comparison of performance Petri nets and queueing network models. In: 3rd Int. Workshop on Modelling Techniques and Performance Evaluation, Paris, France (1987)
28. Silva, M., Campos, J.: Performance models based on Petri nets. In: Proceedings of the IMACS/IFAC Second International Symposium on Mathematical and Intelligent Models in System Simulation, Brussels, Belgium. (1993) 14–21
29. Chiola, G.: Petri nets versus queueing netwoks. [53] chapter 4 121–134
30. Forrester, J.W.: Industrial Dynamics. MIT Press, Cambridge, Mass. (1961)
31. Forrester, J.W.: Urban Dynamics. Productivity Press (1969)
32. Jiménez, E., Recalde, L., Silva, M.: Forrester diagrams and continuous Petri nets: A comparative view. In: Proc. of the 8th IEEE Int. Conf. on Emerging Technologies and Factory Automation (ETFA 2001). (2001) 85–94
33. Jiménez, E., Júlvez, J., Recalde, L., Silva, M.: Relaxed continuous views of discrete event systems: considerations in Forrester diagrams and Petri nets. In: Proc. of the Int. Conf. on Systems, Man and Cybernetics (SMC 2004), The Hague, The Netherlands (2004)
34. Mosekilde, E., Aracil, J., Allen, P.: Instabilities and chaos in nonlinear dynamic systems. Systems Dynamics Review **4** (1988) 14–55
35. Luenberger, D.G.: Introductions to Dynamic Systems: Theory, Models and Applications. John Wiley and Sons, New York (1979)
36. Silva, M., Recalde, L.: Unforced continuous Petri nets and positive systems. In Benvenuti, L., Santis, A.D., Farina, L., eds.: Positive Systems. Proceedings of the First Multidisciplinary International Symposium on Positive Systems: Theory and Applications (POSTA 2003). Volume 294 of LNCIS., Rome, Italy, Springer (2003) 55–62
37. Farina, L., Rinaldi, S.: Positive Linear Systems. Theory and Applications. Pure and Applied Mathematics. John Wiley and Sons, New York (2000)
38. Benevenuti, L., Farina, L.: Positive and compartmental systems. IEEE Transacions on Automatic Control **47** (2002) 370–373
39. Walter, G., Contreras, M.: Compartmental Modeling With Networks. Birkhauser Boston (1999)
40. Chiola, G., Anglano, C., Campos, J., Colom, J., Silva, M.: Operational analysis of timed Petri nets and application to the computation of performance bounds. In: Procs. of the 5th Int. Workshop on Petri Nets and Performance Models (PNPM93), Toulouse, France, IEEE Computer Society Press (1993) 128–137

41. Campos, J.: Performance bounds. [53] chapter 17 587–636
42. Gaujal, B., Giua., A.: Optimal routing of continuous timed Petri nets. In: 15th IFAC World Congress, Barcelona, Spain (2002)
43. Ramirez-Trevino, A., Rivera-Angel, I., Lopez-Mellado, E.: Obsevability of discrete event systems modeled by interpreted Petri nets. IEEE Trans. on Robotics and Automation **19** (2003) 557–565
44. Giua, A., Seatzu, C.: Observability of Place/Transition nets. IEEE Trans. on Automatic Control **47** (2002) 1424–1437
45. Luenberger, D.G.: Observers for multivariable systems. IEEE Trans. on Automatic Control **AC-11** (1966) 190–197
46. Vidal, R., Chiuso, A., Soatto, S., Sastry, S.: Observability of linear hybrid systems. In: Hybrid Systems: Computation and Control, 6th International Workshop, HSCC 2003 , April 3-5, 2003, Proceedings. Volume 2623 of Lecture Notes in Computer Science., Prague, Czech Republic, Springer (2003) 526–539
47. Júlvez, J., Jiménez, E., Recalde, L., Silva, M.: On observability in timed continuous Petri net systems. In G. Franceschinis, J.K., Woodside, M., eds.: Proc. of the 1st Int. Conf. on the Quantitative Evaluation of Systems (QEST 2004), Enschede, The Netherlands, IEEE Computer Society Press (2004) 60–69
48. Mahulea, C., Recalde, L., Silva, M.: Optimal observability for continuous Petri nets. In: 16th IFAC World Congress, Prague, Czech Republic (2005) To appear.
49. Jiménez, E., Júlvez, J., Recalde, L., Silva, M.: On controllability of timed continuous petri net systems: the join free case. Research report, Dep. Informática e Ingeniería de Sistemas, Universidad de Zaragoza, María de Luna, 1, 50018 Zaragoza, Spain (2005) Submitted to CDC-ECC'05.
50. Mahulea, C., Ramirez-Trevino, A., Recalde, L., Silva, M.: On the control of continuous Petri nets under infinite server semantics. Research report, Dep. Informática e Ingeniería de Sistemas, Universidad de Zaragoza, María de Luna, 1, 50018 Zaragoza, Spain (2005)
51. Bemporad, A., Morari, M.: Control of systems integrating logic, dynamics, and constraints. Automatica **35** (1999) 407–427
52. Júlvez, J., Bemporad, A., Recalde, L., Silva, M.: Event-driven optimal control of continuous Petri nets. In: 43rd IEEE Conference on Decision and Control (CDC 2004), Paradise Island, Bahamas (2004)
53. Balbo, G., Silva, M., eds.: Proc. of Human Capital and Mobility MATCH— Performance Advanced School. In Balbo, G., Silva, M., eds.: Performance Models for Discrete Event Systems with Synchronozations: Formalisms and Analysis Techniques, Jaca, Spain (1998)

Genetic Process Mining

W.M.P. van der Aalst, A.K. Alves de Medeiros, and A.J.M.M. Weijters

Department of Technology Management, Eindhoven University of Technology,
P.O. Box 513, NL-5600 MB, Eindhoven, The Netherlands
{w.m.p.v.d.aalst, a.k.medeiros, a.j.m.m.weijters}@tm.tue.nl

Abstract. The topic of process mining has attracted the attention of both researchers and tool vendors in the Business Process Management (BPM) space. The goal of process mining is to *discover* process models from event logs, i.e., events logged by some information system are used to extract information about activities and their causal relations. Several algorithms have been proposed for process mining. Many of these algorithms cannot deal with concurrency. Other typical problems are the presence of duplicate activities, hidden activities, non-free-choice constructs, etc. In addition, real-life logs contain noise (e.g., exceptions or incorrectly logged events) and are typically incomplete (i.e., the event logs contain only a fragment of all possible behaviors). To tackle these problems we propose a completely new approach based on genetic algorithms. As can be expected, a genetic approach is able to deal with noise and incompleteness. However, it is not easy to represent processes properly in a genetic setting. In this paper, we show a genetic process mining approach using the so-called *causal matrix* as a representation for individuals. We elaborate on the relation between Petri nets and this representation and show that genetic algorithms can be used to discover Petri net models from event logs.

Keywords: Process Mining, Petri Nets, Genetic Algorithms, Process Discovery, Business Process Intelligence, Business Activity Monitoring.

1   Introduction

Buzzwords such as Business Process Intelligence (BPI) and Business Activity Monitoring (BAM) illustrate the practical interest in techniques to extract knowledge from the information recorded by today's information systems. Most information systems support some form of logging. For example, Enterprise Resource Planning (ERP) systems such as SAP R/3, PeopleSoft, Oracle, JD Edwards, etc. log transactions at various levels. Any Workflow Management (WfM) system records audit trails for individual cases. The Sarbanes-Oxley act is forcing organizations to log even more information. The availability of this information triggered the need for process mining techniques that analyze event logs.

The goal of process mining is to extract information about processes from transaction logs [3]. We assume that it is possible to record events such that (i) each event refers to an *activity* (i.e., a well-defined step in the process), (ii)

G. Ciardo and P. Darondeau (Eds.): ICATPN 2005, LNCS 3536, pp. 48–69, 2005.
© Springer-Verlag Berlin Heidelberg 2005

Table 1. An event log (audit trail)

case id	activity id	originator	timestamp	case id	activity id	originator	timestamp
case 1	activity A	John	09-3-2004:15.01	case 3	activity E	Pete	10-3-2004:12.50
case 2	activity A	John	09-3-2004:15.12	case 3	activity F	Carol	11-3-2004:10.12
case 3	activity A	Sue	09-3-2004:16.03	case 4	activity D	Pete	11-3-2004:10.14
case 3	activity D	Carol	09-3-2004:16.07	case 3	activity G	Sue	11-3-2004:10.44
case 1	activity B	Mike	09-3-2004:18.25	case 3	activity H	Pete	11-3-2004:11.03
case 1	activity H	John	10-3-2004:09.23	case 4	activity F	Sue	11-3-2004:11.18
case 2	activity C	Mike	10-3-2004:10.34	case 4	activity E	Clare	11-3-2004:12.22
case 4	activity A	Sue	10-3-2004:10.35	case 4	activity G	Mike	11-3-2004:14.34
case 2	activity H	John	10-3-2004:12.34	case 4	activity H	Clare	11-3-2004:14.38

each event refers to a *case* (i.e., a process instance), (iii) each event *can* have a *performer* also referred to as *originator* (the actor executing or initiating the activity), and (iv) events *can* have a *timestamp* and are totally ordered. Table 1 shows an example of a log involving 18 events and 8 activities. In addition to the information shown in this table, some event logs contain more information on the case itself, i.e., data elements referring to properties of the case.

Event logs such as the one shown in Table 1 are used as the starting point for mining. We distinguish three different perspectives: (1) the process perspective, (2) the organizational perspective and (3) the case perspective. The *process perspective* focuses on the control-flow, i.e., the ordering of activities. The goal of mining this perspective is to find a good characterization of all possible paths, e.g., expressed in terms of a Petri net or Event-driven Process Chain (EPC). The *organizational perspective* focuses on the originator field, i.e., which performers are involved and how are they related. The goal is to either structure the organization by classifying people in terms of roles and organizational units or to show relation between individual performers (i.e., build a social network [2]). The *case perspective* focuses on properties of cases. Cases can be characterized by their path in the process or by the originators working on a case. However, cases can also be characterized by the values of the corresponding data elements. For example, if a case represents a replenishment order it is interesting to know the supplier or the number of products ordered.

The process perspective is concerned with the "How?" question, the organizational perspective is concerned with the "Who?" question, and the case perspective is concerned with the "What?" question. In this paper we will focus completely on the process perspective, i.e., the ordering of the activities. This means that here we ignore the last two columns in Table 1. (Although the timestamps determine the order of events (activities) in a case, the actual timestamps are not used during mining.) For the mining of the other perspectives we refer to [3] and http://www.processmining.org. Note that the ProM tool described in this paper is able to mine the other perspectives and can also deal with other issues such as transactions, e.g., in the ProM tool we consider different event types such as "schedule", "start", "complete", "abort", etc. However, for reasons of simplicity we abstract from this in this paper and consider activities to be atomic as shown in Table 1.

If we abstract from the other perspectives, Table 1 contains the following information: case 1 has event trace A, B, H, case 2 has event trace A, C, H, case 3 has event trace A, D, E, F, G, H, and case 4 has event trace A, D, F, E, G, H. If we analyze these four sequences we can extract the following information about the process (assuming some notion of completeness and no noise). The underlying process has 8 activities (A, B, ..., H). A is always the first activity to be executed and H is always the last one. After A is executed, activities B, C or D can be executed. In other words, after A, there is a *choice* in the process and only one of these activities can be executed next. When B or C are executed, they are followed by the execution of H (see cases 1 and 2). When D is executed, both E *and* F can be executed in any order. Since we do not consider explicit parallelism, we assume E and F to be concurrent (see cases 3 and 4). Activity G synchronizes the parallel branches that contain E and F. Activity H is executed whenever B, C or G has been executed. Based on these observations, the Petri net shown in Figure 1 is a good model for the event log containing the four cases. Note that each of the four cases can be "reproduced" by the Petri net shown in Figure 1, i.e. the Petri net contains all observed behavior. In this particular case, also the reverse holds, i.e., all possible firing sequences of the Petri net shown in Figure 1 are contained in the log. Generally, this is not the case since in practice it is unrealistic to assume that all possible behavior is always contained in the log, cf. the discussion on completeness in [4].

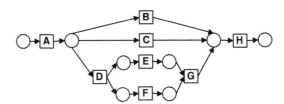

Fig. 1. Petri net discovered based on the event log in Table 1

Existing approaches for mining the process perspective [3, 4, 5, 6, 11, 13, 18] have problems dealing with issues such as duplicate activities, hidden activities, non-free-choice constructs, noise, and incompleteness. The problem with *duplicate activities* occurs when the same activity can occur at multiple places in the process. This is a problem because it is no longer clear to which activity some event refers. The problem with *hidden activities* is that essential routing decisions are not logged but impact the routing of cases. *Non-free-choice* constructs are problematic because it is not possible to separate choice from synchronization. We consider two sources of *noise*: (1) incorrectly logged events (i.e., the log does not reflect reality) or (2) exceptions (i.e., sequences of events corresponding to "abnormal behavior"). Clearly noise is difficult to handle. The problem of *incompleteness* is that for many processes it is not realistic to assume that all possible behavior is contained in the log. For processes with many alternative routes and parallelism, the number of possible event traces is typically exponential in the number of activities, e.g., a process with 10 binary choices in a sequence will have $2^{10}(= 1024)$ possible event sequences and a process with 10 activities in parallel will have even $10!(= 3628800)$ possible event sequences.

We can consider process mining as a search for the most appropriate process out of the search space of candidate process models. Mining algorithms can use different strategies to find the most appropriate model. Two extreme strategies can be distinguished (i) *local strategies* primarily based on a step by step building of the optimal process model based on local information, and (ii) *global strategies* primarily based on a one strike search for the optimal model. Most process mining approaches use a local strategy. An example of a such a local strategy is used by the α-algorithm [4] where only local information about binary relations between events is used. A genetic search is an example of a global search strategy; because the quality or fitness of a candidate model is calculated by comparing the process model with all traces in the event log the search process takes place at a global level. For a local strategy there is no guarantee that the outcome of the locally optimal steps (at the level of binary event relations) will result in a globally optimal process model. Hence, the performance of such local mining techniques can be seriously hampered when the necessary information is not locally available because one erroneous example can completely mess up the derivation of a right model. Therefore, we started to use genetic algorithms.

In this paper, we present a *genetic algorithm to discover a Petri net given a set of event traces*. Genetic algorithms are adaptive search methods that try to mimic the process of evolution [9]. These algorithms start with an initial population of individuals (in this case process models). Populations evolve by selecting the fittest individuals and generating new individuals using genetic operators such as *crossover* (combining parts of two of more individuals) and *mutation* (random modification of an individual). Our initial work on applying genetic algorithms to process mining [14] shows that a direct representation of individuals in terms of a Petri net is not a very convenient. First of all, the Petri net contains places that are not visible in the log. (Note that in Figure 1 we cannot assign meaningful names to places.) Second, the classical Petri net is not a very convenient notation for generating an initial population because it is difficult to apply simple heuristics. Third, the definition of the genetic operators (crossover and mutation) is cumbersome. Finally, the expressive power of Petri nets is in some cases too limited (combinations of AND/OR-splits/joins). Therefore, we use an internal representation named *casual matrix*. However, we use Petri nets to give semantics to this internal representation and adopt many ideas from Petri nets (e.g., playing the token game to measure fitness). Moreover, in this paper we focus on the relation between the casual matrix and Petri nets.

The remainder of this paper is organized as follows. First, we discuss some related work (Section 2) and start with some preliminaries (Section 3). Then, in Section 4, we present the causal matrix as our internal representation. In Section 5 we explore the relation between the causal matrix and Petri nets.

Section 6 introduces the genetic algorithm and in Section 7 some experimental results are given. Finally, we conclude the paper.

2   Related Work

The idea of process mining is not new [3, 4, 2, 5, 6, 11, 13, 18]. Most of the scientific papers aim at the control-flow perspective, although a few focus on other perspectives such as the organizational perspective [2]. It is also interesting to note that some commercial tools such as ARIS PPM offer some limited form of process mining as discussed in this paper. However, most tools in the BPI/BAM arena focus on key performance indicators such as flow time and frequencies.

Given the many papers on mining the process perspective it is not possible to give a complete overview. Instead we refer to [3, 4]. Historically, Cook et al. [6] and Agrawal et al. [5] started to work on the problem addressed in this paper. Herbst et al. [11] took an alternative approach which allows for dealing with duplicate activities. The authors of this paper have been involved in different variants of the so-called α-algorithm [4, 18]. Each of the approaches has its pros and its cons. Most approaches that are able to discover concurrency have problems dealing with issues such as duplicate activities, hidden activities, non-free-choice constructs, noise, and incompleteness.

There have been some papers combining Petri nets and genetic algorithms, cf. [12, 15, 16, 17]. However, these papers do not try to discover a process model based on some event log.

For readers familiar with Petri net theory, it is important to discuss the relation between this work and the work on regions [8]. The seminal work on regions investigates which transition systems can be represented by (compact) Petri nets (i.e., the so-called synthesis problem). Although there are related problems such as duplicate transitions, etc., the setting is quite different because our notion of completeness is much weaker than perfect knowledge of the underlying transition system. We assume that the log contains only a fraction of the possible behavior, as mentioned in the introduction.

The approach in this paper is the first approach using genetic algorithms for process discovery. Some more details about the experimental/genetic-side of this approach can be found in a technical report [14]. The goal of using genetic algorithms is to tackle problems such as duplicate activities, hidden activities, non-free-choice constructs, noise, and incompleteness, i.e., overcome the problems of some of the traditional approaches. However, in this paper we focus on the initial idea and the representation rather than a comparison with existing non-genetic algorithms.

3   Preliminaries

This section briefly introduces the basic *Petri net* terminology and notations, and also discusses concepts such as *WF-nets* and *soundness*.

Definition 1 (Petri net). *A Petri net is a triple (P, T, F). P is a finite set of places, T is a finite set of transitions $(P \cap T = \emptyset)$, and $F \subseteq (P \times T) \cup (T \times P)$ is a set of arcs (flow relation).*

For any relation/directed graph $G \subseteq N \times N$ we define the preset $\bullet n = \{(m_1, m_2) \in G \mid n = m_2\}$ *and postset* $n\bullet = \{(m_1, m_2) \in G \mid n = m_1\}$ *for any node* $n \in N$. We use $\overset{G}{\bullet} n$ or $n \overset{G}{\bullet}$ to explicitly indicate the context G if needed. Based on the flow relation F we use this notation as follows. $\bullet t$ denotes the set of input places for a transition t. The notations $t\bullet$, $\bullet p$ and $p\bullet$ have similar meanings, e.g., $p\bullet$ is the set of transitions sharing p as an input place. Note that we do not consider multiple arcs from one node to another.

At any time a place contains zero or more *tokens*, drawn as black dots. This state, often referred to as *marking*, is the distribution of tokens over places, i.e., $M \in P \to \mathbb{N}$. For any two states M_1 and M_2, $M_1 \leq M_2$ iff for all $p \in P$: $M_1(p) \leq M_2(p)$. We use the standard *firing rule*, i.e., a transition t is said to be *enabled* iff each input place p of t contains at least one token, an enabled transition may *fire*, and if transition t fires, then t *consumes* one token from each input place p of t and *produces* one token for each output place p of t.

Given a Petri net (P, T, F) and a state M_1, we have the standard notations for a transition t that is enabled in state M_1 and firing t in M_1 results in state M_2 (notation: $M_1 \overset{t}{\to} M_2$) and a firing sequence $\sigma = t_1 t_2 t_3 \ldots t_{n-1}$ leads from state M_1 to state M_n via a (possibly empty) set of intermediate states (notation: $M_1 \overset{\sigma}{\to} M_n$). A state M_n is called *reachable* from M_1 (notation $M_1 \overset{*}{\to} M_n$) iff there is a firing sequence σ such that $M_1 \overset{\sigma}{\to} M_n$. Note that the empty firing sequence is also allowed, i.e., $M_1 \overset{*}{\to} M_1$.

In this paper, we will focus on a particular type of Petri nets called *WorkFlow nets* (WF-nets) [1, 7, 10].

Definition 2 (WF-net). *A Petri net $PN = (P, T, F)$ is a WF-net (Workflow net) if and only if:*

(i) There is one source place $i \in P$ such that $\bullet i = \emptyset$.

(ii) There is one sink place $o \in P$ such that $o\bullet = \emptyset$.

(iii) Every node $x \in P \cup T$ is on a path from i to o.

A WF-net represents the life-cycle of a case that has some initial state represented by a token in the unique input place (i) and a desired final state represented by a token in the unique output place (o). The third requirement in Definition 2 has been added to avoid "dangling transitions and/or places". In the context of workflow models or business process models, transitions can be interpreted as *tasks* or *activities* and places can be interpreted as *conditions*. Although the term "WorkFlow net" suggests that the application is limited to workflow processes, the model has wide applicability, i.e., any process where each case has a life-cycle going from some initial state to some final state fits this basic model.

The three requirements stated in Definition 2 can be verified statically, i.e., they only relate to the structure of the Petri net. To characterize desirable dynamic properties, the notation of *soundness* has been defined [1, 7, 10].

Definition 3 (Sound). *A procedure modelled by a WF-net PN = (P, T, F) is sound if and only if:*

(i) *For every state M reachable from state i, there exists a firing sequence leading from state M to state o. Formally:* $\forall_M (i \xrightarrow{*} M) \Rightarrow (M \xrightarrow{*} o)$.[1]

(ii) *State o is the only state reachable from state i with at least one token in place o. Formally:* $\forall_M (i \xrightarrow{*} M \ \wedge \ M \geq o) \Rightarrow (M = o)$.

(iii) *There are no dead transitions in (PN, i). Formally:* $\forall_{t \in T} \exists_{M, M'} \ i \xrightarrow{*} M \xrightarrow{t} M'$.

Note that the soundness property relates to the dynamics of a WF-net. The first requirement in Definition 3 states that starting from the initial state (state i), it is always possible to reach the state with one token in place o (state o). The second requirement states that the moment a token is put in place o, all the other places should be empty. The last requirement states that there are no dead transitions (activities) in the initial state i.

4   Causal Matrix

After these preliminaries we return to the goal of this paper: genetic process mining. In order to apply a genetic algorithm we need to represent individuals. Each individual corresponds to a possible process model and its representation should be easy to handle. Our initial idea was to represent processes directly by Petri nets. Unfortunately, Petri nets turn out to be a less convenient way to represent processes in this context. The main reason is that in Petri nets there are places whose existence cannot be derived from the log, i.e., events only refer to the active components of the net (transitions). Because of this it becomes more difficult to generate an initial population, define genetic operators (crossover and mutation), and describe combinations of AND/OR-splits/joins. Note that given a log it is very easy to discover the activities and therefore the transitions that exist in the Petri net. However, enforcing certain routings by just connecting transitions through places is complex (if not impossible). Therefore, we will use a different internal representation. However, this representation and its semantics are closely linked to Petri nets as will be shown in Section 5.

Table 2 shows the internal representation of an individual used by our genetic mining approach. This so-called *causal matrix* defines the causal relations between the activities and in case of multiple input or output activities, the logic is depicted. Consider for example the row starting with A. This row shows that

---

[1] Note that there is an overloading of notation: the symbol i is used to denote both the *place i* and the *state* with only one token in place i. The same holds for o.

Table 2. A causal matrix is used for the internal representation of an individual

				INPUT					
	true	A	A	A	D	D	$E \wedge F$	$B \vee C \vee G$	
$\rightarrow$	A	B	C	D	E	F	G	H	OUTPUT
A	0	1	1	1	0	0	0	0	$B \vee C \vee D$
B	0	0	0	0	0	0	0	1	H
C	0	0	0	0	0	0	0	1	H
D	0	0	0	0	1	1	0	0	$E \wedge F$
E	0	0	0	0	0	0	1	0	G
F	0	0	0	0	0	0	1	0	G
G	0	0	0	0	0	0	0	1	H
H	0	0	0	0	0	0	0	0	true

Table 3. A more succinct encoding of the individual shown in Table 2

ACTIVITY	INPUT	OUTPUT
A	{}	{{B, C, D}}
B	{{A}}	{{H}}
C	{{A}}	{{H}}
D	{{A}}	{{E}, {F}}
E	{{D}}	{{G}}
F	{{D}}	{{G}}
G	{{E}, {F}}	{{H}}
H	{{B, C, G}}	{}

there is a not a causal relation between A and A (note the first 0 in the row), but there is a causal relation between A and B (note the first 1 in this row). The next two entries in the row show that there are also causal relations between A and C and A and D. The last element in the row shows the routing logic, i.e., $B \vee C \vee D$ indicates that A is followed by B, C, or D. The column labelled 'OUTPUT' shows the logic relating an activity to causally following activities. The first row below 'INPUT' shows the logic relating an activity to causally preceding activities. Note that the input condition of A is $true$, i.e., no input needed. Activity G has $E \wedge F$ as input condition, i.e., both E and F need to complete in order to enable G. Activity H has $B \vee C \vee G$ as input condition, i.e., B, C, or G needs to complete in order to enable H.

Table 3 shows a more convenient notation removing some of the redundancies. Note that the 0 and 1 entries in Table 2 can be trivially derived from the input and output conditions. Moreover, we assume that we can write the logical expressions in a normal form, e.g., $\{\{B, C, D\}\}$ corresponds to $B \vee C \vee D$, $\{\{E\}, \{F\}\}$ corresponds to $E \wedge F$, and $\{\{A, B\}, \{C, D\}\}$ corresponds to $(A \vee B) \wedge (C \vee D)$. In fact, the logical expression is represented by a set of sets corresponding to a conjunction of disjunctions, i.e., a kind of Conjunctive Normal Form (CNF).[2]

---

[2] Note that unlike the conjunctive normal form we do not allow for negation and also do not allow for "overlapping" disjunctions, cf. Definition 4.

Let us now formalize the notion of a *causal matrix*.

Definition 4 (Causal Matrix). *A Causal Matrix is a tuple* $CM = (A, C, I, O)$, *where*

- *A is a finite set of activities,*
- $C \subseteq A \times A$ *is the causality relation,*
- $I \in A \to \mathcal{P}(\mathcal{P}(A))$ *is the input condition function,*[3]
- $O \in A \to \mathcal{P}(\mathcal{P}(A))$ *is the output condition function,*

such that

- $C = \{(a_1, a_2) \in A \times A \mid a_1 \in \bigcup I(a_2)\}$,[4]
- $C = \{(a_1, a_2) \in A \times A \mid a_2 \in \bigcup O(a_1)\}$,
- $\forall_{a \in A} \ \forall_{s,s' \in I(a)} \ s \cap s' \neq \emptyset \ \Rightarrow \ s = s'$,
- $\forall_{a \in A} \ \forall_{s,s' \in O(a)} \ s \cap s' \neq \emptyset \ \Rightarrow \ s = s'$,
- $C \cup \{(a_o, a_i) \in A \times A \mid a_o \overset{C}{\bullet} = \emptyset \ \wedge \ \overset{C}{\bullet} a_i = \emptyset\}$ *is a strongly connected graph.*

The mapping of Table 3 onto $CM = (A, C, I, O)$ is straightforward (the latter two columns represent I and O). Note that C can be derived from both I and O. Its main purpose is to ensure consistency between I and O. For example, if a_1 has an output condition mentioning a_2, then a_2 has an input condition mentioning a_1 (and vice versa). This is enforced by the first two constraints. The third and fourth constraint indicate that some activity a may appear only once in the conjunction of disjunctions, e.g., $\{\{A, B\}, \{A, C\}\}$ is not allowed because A appears twice. The last requirement has been added to avoid that the causal matrix can be partitioned in two independent parts or that nodes are not on a path from some source activity a_i to a sink activity a_o.

5 Relating the Causal Matrix and Petri Nets

In this section we relate the causal matrix to Petri nets. We first map Petri nets (in particular WF-nets) onto the notation used by our genetic algorithms. Then we consider the mapping of the causal matrix onto Petri nets.

5.1 Mapping a Petri Net onto a Causal Matrix

The mapping from an arbitrary Petri net to its corresponding causal matrix illustrates the expressiveness of the internal format used for genetic mining. First, we give the definition of the mapping $\Pi_{PN \to CM}$.

Definition 5 ($\Pi_{PN \to CM}$). *Let* $PN = (P, T, F)$ *be a Petri net. The mapping of* PN *is a tuple* $\Pi_{PN \to CM}(PN) = (A, C, I, O)$, *where*

---

[3] $\mathcal{P}(A)$ denotes the powerset of some set A.
[4] $\bigcup I(a_2)$ is the union of the sets in set $I(a_2)$.

- $A = T$,
- $C = \{(t_1, t_2) \in T \times T \mid t_1 \bullet \cap \bullet t_2 \neq \emptyset\}$,
- $I \in T \to \mathcal{P}(\mathcal{P}(T))$ such that $\forall_{t \in T}\ I(t) = \{\bullet p \mid p \in \bullet t\}$,
- $O \in T \to \mathcal{P}(\mathcal{P}(T))$ such that $\forall_{t \in T}\ O(t) = \{p \bullet \mid p \in t \bullet\}$.

Let PN be the Petri net shown in Figure 1. It is easy to check that $\Pi_{PN \to CM}(PN)$ is indeed the causal matrix in Table 2. However, there may be Petri nets PN for which $\Pi_{PN \to CM}(PN)$ is not a causal matrix. The following lemma shows that for the class of nets we are interested in, i.e., WF-nets, the requirement that there may not be two different places in-between two activities is sufficient to prove that $\Pi_{PN \to CM}(PN)$ represents a causal matrix as defined in Definition 4.

Lemma 1. *Let $PN = (P, T, F)$ be a WF-net with no duplicate places in between two transitions, i.e., $\forall_{t_1, t_2 \in T}\ |t_1 \bullet \cap \bullet t_2| \leq 1$. $\Pi_{PN \to CM}(PN)$ represents a causal matrix as defined in Definition 4.*

Proof. Let $\Pi_{PN \to CM} = (A, C, I, O)$. Clearly, $A = T$ is a finite set, $C \subseteq A \times A$, and $I, O \in A \to \mathcal{P}(\mathcal{P}(A))$. $C = \{(a_1, a_2) \in A \times A \mid a_1 \in \bigcup I(a_2)\}$ because $a_1 \in \bigcup I(a_2)$ if and only if $a_1 \bullet \cap \bullet a_2 \neq \emptyset$. Similarly, $C = \{(a_1, a_2) \in A \times A \mid a_2 \in \bigcup O(a_1)\}$. $\forall_{a \in A}\ \forall_{s, s' \in I(a)}\ s \cap s' \neq \emptyset \Rightarrow s = s'$ because $\forall_{t_1, t_2 \in T}\ |t_1 \bullet \cap \bullet t_2| \leq 1$. Similarly, $\forall_{a \in A}\ \forall_{s, s' \in O(t)}\ s \cap s' \neq \emptyset \Rightarrow s = s'$. Finally, it is easy to verify that $C \cup \{(a_o, a_i) \in A \times A \mid a_o \bullet = \emptyset \wedge \bullet a_i = \emptyset\}$ is a strongly connected graph.   □

The requirement $\forall_{t_1, t_2 \in T}\ |t_1 \bullet \cap \bullet t_2| \leq 1$ is a direct result of the fact that in the conjunction of disjunctions in I and O, there may not be any overlaps. This restriction has been added to reduce the search space of the genetic mining algorithm, i.e., the reason is more of a pragmatic nature. However, for the success of the genetic mining algorithm such reductions are of the utmost importance.

5.2 A Naive Way of Mapping a Causal Matrix onto a Petri Net

The mapping from a causal matrix onto a Petri net is more involved because we need to "discover places" and, as we will see, the causal matrix is slightly more expressive than classical Petri nets.[5] Let us first look at a naive mapping.

Definition 6 ($\Pi_{CM \to PN}^{N}$)**.** *Let $CM = (A, C, I, O)$ be a causal matrix. The naive Petri net mapping is a tuple $\Pi_{CM \to PN}^{N}(CM) = (P, T, F)$, where*

- $P = \{i, o\} \cup \{i_{t,s} \mid t \in A \wedge s \in I(t)\} \cup \{o_{t,s} \mid t \in A \wedge s \in O(t)\}$,
- $T = A \cup \{m_{t_1, t_2} \mid (t_1, t_2) \in C\}$,
- $F = \{(i, t) \mid t \in A \wedge \overset{C}{\bullet} t = \emptyset\} \cup \{(t, o) \mid t \in A \wedge t \overset{C}{\bullet} = \emptyset\} \cup \{(i_{t,s}, t) \mid t \in A \wedge s \in I(t)\} \cup \{(t, o_{t,s}) \mid t \in A \wedge s \in O(t)\} \cup \{(o_{t_1,s}, m_{t_1, t_2}) \mid (t_1, t_2) \in C \wedge s \in O(t_1) \wedge t_2 \in s\} \cup \{(m_{t_1, t_2}, i_{t_2, s}) \mid (t_1, t_2) \in C \wedge s \in I(t_2) \wedge t_1 \in s\}$.

---

[5] Expressiveness should not be interpreted in a formal sense but in the sense of convenience when manipulating process instances, e.g., crossover operations.

The mapping $\Pi^N_{CM \to PN}$ maps activities onto transitions and adds input places and output places to these transitions based on functions I and O. These places are local to one activity. To connect these local places, one transition m_{t_1,t_2} is added for every $(t_1, t_2) \in C$. Figure 2 shows a causal matrix and its naive mapping $\Pi^N_{CM \to PN}$ (we have partially omitted place/transition names).

$ACTIVITY$	$INPUT$	$OUTPUT$
A	{}	{{C, D}}
B	{}	{{D}}
C	{{A}}	{}
D	{{A, B}}	{}

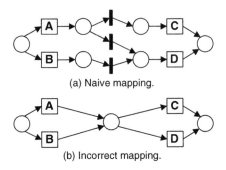

(a) Naive mapping.

(b) Incorrect mapping.

Fig. 2. A causal matrix (left) and two potential mappings onto Petri nets (right)

Figure 2 shows two WF-nets illustrating the need for "silent transitions" of the form m_{t_1,t_2}. The dynamics of the WF-net shown in Figure 2(a) is consistent with the causal matrix. If we try to remove the silent transitions, the best candidate seems to be the WF-net shown in Figure 2(b). Although this is a sound WF-net capturing incorporating the behavior of the WF-net shown in Figure 2(a), the mapping is *not* consistent with the causal matrix. Note that Figure 2(b) allows for a firing sequence where B is followed by C. This does not make sense because $C \notin \bigcup O(B)$ and $B \notin \bigcup I(C)$. Therefore, we use the mapping given in Definition 6 to give Petri-net semantics to causal matrices.

It is easy to see that a causal matrix defines a WF-net. However, note that the WF-net does not need to be sound.

Lemma 2. *Let* $CM = (A, C, I, O)$ *be a causal matrix.* $\Pi^N_{CM \to PN}(CM)$ *is a WF-net.*

Proof. It is easy to verify the three properties mentioned in Definition 2. Note that the "short-circuited" C is strongly connected and that each m_{t_1,t_2} transition makes a similar connection in the resulting Petri net. □

Figure 3 shows that despite the fact that $\Pi^N_{CM \to PN}(CM)$ is a WF-net, the introduction of silent transitions may introduce a problem. Figure 3(b) shows the WF-net based on Definition 6, i.e., the naive mapping. Clearly, Figure 3(b) is not sound because there are two potential deadlocks, i.e., one of the input places of E is marked and one of the input places of F is marked but none of them is enabled. The reason for this is that the choices introduced by the silent transitions are not "coordinated" properly. If we simply remove the silent transitions, we obtain the WF-net shown in Figure 3(a). This network is consistent with the causal matrix.

ACTIVITY	INPUT	OUTPUT
A	$\{\}$	$\{\{B\},\{C,D\}\}$
B	$\{\{A\}\}$	$\{\{E,F\}\}$
C	$\{\{A\}\}$	$\{\{E\}\}$
D	$\{\{A\}\}$	$\{\{F\}$
E	$\{\{B\},\{C\}\}$	$\{\{G\}\}$
F	$\{\{B\},\{D\}\}$	$\{\{G\}\}$
G	$\{\{E\},\{F\}\}$	$\{\}$

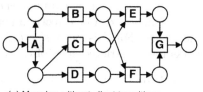

(a) Mapping without silent transitions.

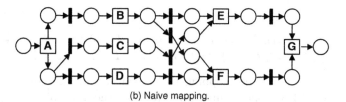

(b) Naive mapping.

Fig. 3. Another causal matrix (left) and two potential mappings onto Petri nets (right)

This can easily be checked because applying the mapping $\Pi_{PN \to CM}$ defined in Definition 5 to this WF-net yields the original causal matrix shown in Figure 3.

Figures 2 and 3 show a dilemma. Figure 2 demonstrates that silent transitions are needed while Figure 3 proves that silent transitions can be harmful. There are two ways to address this problem taking the mapping of Definition 6 as a starting point.

First of all, we can use *relaxed soundness* [7] rather than soundness [1]. This implies that we only consider so-called sound firing sequences and thus avoid the two potential deadlocks in Figure 3(b). See [7] for transforming a relaxed sound WF-net into a sound one.

Second, we can change the firing rule such that silent transitions can only fire if they actually enable a non-silent transition. The enabling rule for non-silent transitions is changed as follows: *a non-silent transition is enabled if each of its input places is marked or it is possible to mark all input places by just firing silent transitions*, i.e., silent transitions only fire when it is possible to enable a non-silent transition. Note that non-silent and silent transitions alternate and therefore it is easy to implement this semantics in a straightforward and localized manner.

In this paper we use the second approach, i.e., a slightly changed enabling/-firing rule is used to specify the semantics of a causal matrix in terms of a WF-net. This semantics allows us also to define a notion of fitness required for the genetic algorithms. Using the Petri-net representation we can play the "token game" to see how each event trace in the log fits the individual represented by a causal matrix.

5.3   A More Sophisticated Mapping

Although not essential for the genetic algorithms, we elaborate a bit on the dilemma illustrated by figures 2 and 3. The dilemma shows that the causal net representation is slightly more expressive than ordinary Petri nets. (Note the earlier comment on expressiveness!) Therefore, it is interesting to see which causal matrices can be directly mapped onto a WF-net without additional silent transitions. For this purpose we first define a mapping $\Pi_{CM \to PN}^{R}$ which only works for a restricted class of causal matrices.

Definition 7 ($\Pi_{CM \to PN}^{R}$). *Let $CM = (A, C, I, O)$ be a causal matrix. The restricted Petri net mapping of is a tuple $\Pi_{CM \to PN}^{R}(CM)$, where*

- $X = \{(T_i, T_o) \in \mathcal{P}(A) \times \mathcal{P}(A) \mid \forall_{t \in T_i} \, T_o \in O(t) \, \wedge \, \forall_{t \in T_o} \, T_i \in I(t)\}$,
- $P = X \cup \{i, o\}$,
- $T = A$,
- $F = \{(i, t) \mid t \in T \wedge \overset{C}{\bullet} t = \emptyset\} \cup \{(t, o) \mid t \in T \wedge t \overset{C}{\bullet} = \emptyset\} \cup \{((T_i, T_o), t) \in X \times T \mid t \in T_o\} \cup \{(t, (T_i, T_o)) \in T \times X \mid t \in T_i\}$.

If we apply this mapping to the causal matrix shown in Figure 3, we obtain the WF-net shown in Figure 3(a), i.e., the desirable net without the superfluous silent transitions. However, in some cases the $\Pi_{CM \to PN}^{R}$ does not yield a WF-net because some connections are missing. For example, if we apply $\Pi_{CM \to PN}^{R}$ to the causal matrix shown in Figure 2, then we obtain a result where there are no connections between A, B, C, and D. This makes sense because there does not exist a corresponding WF-net. This triggers the question whether it is possible to characterize the class of causal matrices for which $\Pi_{CM \to PN}^{R}$ yields the correct WF-net.

Definition 8 (Simple). *Let $CM = (A, C, I, O)$ be a causal matrix. CM is simple if and only if* $\forall_{t_A, t_B \in T} \, \forall_{T_A \in O(t_A)} \, \forall_{T_B \in O(t_B)} \, \forall_{t_C \in (T_A \cap T_B)} \, \forall_{T_C \in I(t_C)} \{t_A, t_B\} \subseteq T_C \Rightarrow T_A = T_B$ *and* $\forall_{t_A, t_B \in T} \, \forall_{T_A \in I(t_A)} \, \forall_{T_B \in I(t_B)} \, \forall_{t_C \in (T_A \cap T_B)} \, \forall_{T_C \in O(t_C)} \{t_A, t_B\} \subseteq T_C \Rightarrow T_A = T_B$.

Clearly the causal matrix shown in Figure 3 is simple while the one in Figure 2 is not. The following lemma shows that $\Pi_{CM \to PN}^{R}$ provides indeed the correct mapping if the causal matrix is simple.

Lemma 3. *Let $CM = (A, C, I, O)$ be a causal matrix. If CM is simple, then each of the following properties holds:*

(i) $\forall_{(t_1, t_2) \in C} \, \exists_{T_1, T_2 \in \mathcal{P}(A)} \, t_1 \in T_1 \wedge t_2 \in T_2 \wedge (\forall_{t \in T_1} \, T_2 \in O(t)) \wedge (\forall_{t \in T_2} \, T_1 \in I(t))$,

(ii) $\Pi_{CM \to PN}^{R}(CM)$ *is a WF-net, and*

(iii) $\Pi_{PN \to CM}(\Pi_{CM \to PN}^{R}(CM)) = CM$.

Proof. We only provide a sketch of the full proof (a more detailed proof is beyond the scope of this paper). The first property can be derived by using the following observation: $(t_1, t_2) \in C$ iff $\exists_{T_2 \in O(t_1)} t_2 \in T_2$ iff $\exists_{T_1 \in O(t_2)} t_1 \in T_1$. Hence there is exactly one T_1 and T_2 such that $t_1 \in T_1$, $t_2 \in T_2$, $T_2 \in O(t_1)$, and $T_1 \in O(t_2)$. For $t \in T_1$ we need to prove that $T_2 \in O(t)$. This follows from the definition of simple by taking $t_A = t_1$ and $t_B = t$. The other cases are similar. The second property follows from the first one because if $(t_1, t_2) \in C$ then a connecting place between t_1 and t_2 is introduced by the set X. The rest of the proof is similar to the proof of Lemma 2. The third property can be shown using similar arguments. Note that no information is lost during the mapping onto the WF-net $\Pi^R_{CM \to PN}(CM)$ and that $\Pi_{PN \to CM}$ retranslates the sets T_i and T_o in the places of X to functions I and O.   □

In this section, we discussed the relation between the representation used by our genetic algorithm and Petri nets. We used this relation to give semantics to our representation. It was shown that this representation is slightly more expressive than Petri nets because any WF-net can be mapped into causal matrix while the reverse is only possible after introducing silent transitions and modifying the firing rule or using relaxed soundness. We also characterized the class of causal matrices that can be mapped directly. In the next sections, we will demonstrate the suitability of the representation for genetic process mining.

6   Genetic Algorithm

In this section we explain how our genetic algorithm (GA) works. Figure 4 describes its main steps. In the following subsections (6.1 – 6.3) we roughly explain the most important building blocks of our genetic approach: (i) the initialization process, (ii) the fitness measurement, and (iii) the genetic operators. For a more detailed explanation about the algorithm we refer to [14].

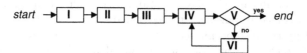

Fig. 4. Main steps of our genetic algorithm. (I) Read the event log. (II) Calculate dependency relations among activities. (III) Build the initial population. (IV) Calculate individuals' fitness. (V) Stop and return the fittest individuals? (VI) Create next population by using the genetic operators

6.1   Initial Population

The initial population is randomly built by the genetic algorithm. As explained in Section 4, individuals are causal matrices. When building the initial population, we roughly follow Definition 4. Given a log, all individuals in any population of the genetic algorithm have the same set of activities (or tasks) A. This set contains the tasks that appear in the log. However, the causality relation C and the condition functions I and O may be different for every individual in the population. Additionally, to guide the GA algorithm during the building of the initial population, the initialization of the causality relation C is supported by the dependency measure heuristics [18]. The motivation behind this heuristic is simple. If, in the event log, the pattern $t_1 t_2$ appears frequently and $t_2 t_1$ only as an exception, then there is a high probability that t_1 and t_2 are in the causality relation (i.e., $(t_1, t_2) \in C$). The conditions functions I and O are randomly

Table 4. Two randomly created individuals for the log in Table 1

<table>
<tr><th colspan="3">Individual1</th><th colspan="3">Individual2</th></tr>
<tr><th>ACTIVITY</th><th>INPUT</th><th>OUTPUT</th><th>ACTIVITY</th><th>INPUT</th><th>OUTPUT</th></tr>
<tr><td>A</td><td>{}</td><td>{{B, C, D}}</td><td>A</td><td>{}</td><td>{{B, C, D}}</td></tr>
<tr><td>B</td><td>{{A}}</td><td>{{H}}</td><td>B</td><td>{{A}}</td><td>{{H}}</td></tr>
<tr><td>C</td><td>{{A}}</td><td>{{H}}</td><td>C</td><td>{{A}}</td><td>{{H}}</td></tr>
<tr><td>D</td><td>{{A}}</td><td>{{E}}</td><td>D</td><td>{{A}}</td><td>{{E, F}}</td></tr>
<tr><td>E</td><td>{{D}}</td><td>{{G}}</td><td>E</td><td>{{D}}</td><td>{{G}}</td></tr>
<tr><td>F</td><td>{}</td><td>{{G}}</td><td>F</td><td>{{D}}</td><td>{{G}}</td></tr>
<tr><td>G</td><td>{{E}, {F}}</td><td>{{H}}</td><td>G</td><td>{{E}, {F}}</td><td>{{H}}</td></tr>
<tr><td>H</td><td>{{C, B, G}}</td><td>{}</td><td>H</td><td>{{C}, {B}, {G}}</td><td>{}</td></tr>
</table>

built. As a result, the initial population can have any individual in the search space defined by a set of activities A. The higher the amount of tasks that a log contains, the bigger this search space. Given the event log in Table 1, Table 4 shows two individuals that could be created during the initialization.

6.2   Fitness Calculation

If an individual in the genetic population correctly describes the registered behavior in the event log, the fitness of that individual will be high. In our approach the fitness is strongly related to the number of correctly parsed traces from the event log. Note that in case of noisy situation, we cannot aim at mining a process model that can correctly parse *all* traces, because the traces with noise cannot also be parsed by the desired model.

The parsing technique we use for the causal matrix is very similar to the firing rule for Petri nets as discussed in Section 5.2. We use the naive semantics with silent transitions that only fire when needed and simply play the "token game". When the activity to be parsed is not enabled, the parsing process does not stop. The problem is registered and the parsing proceeds as if the activity was enabled (conceptually, this is equivalent to adding the necessary missing tokens in the Petri net to enable the activity and, then, firing it). We adopt this parsing semantics because it is more robust to noisy logs and it gives more information about the fitness of the complete process models (i.e not biased to only the first part of the process model). In a noise-free situation, the fitness of a model can be 1 (or 100%) (i.e. all traces can be parsed). In practical situations, the fitness value ranges from 0 to 1. The exact fitness of an individual to a given log is given by the formula:

$$fitness = 0.40 \times \frac{allParsedActivities}{numberOfActivitiesAtLog} + 0.60 \times \frac{allProperlyCompletedLogTraces}{numberOfTracesAtLog}$$

where: *numberOfActivitiesAtLog* is the number of activities in the log. For instance, the log shown in Table 1 has 18 activities. *numberOfTracesAtLog* is the number of log traces, e.g., in Table 1 there are 4. *allParsedActivities* is the sum of parsed activities (i.e. activities that could fire without the artificial addition of tokens) for all log traces. *allProperlyCompletedLogTraces* is the number of log traces that were properly parsed (i.e. after the parsing the end place is the only one to be marked).

6.3   Genetic Operations

We use elitism, crossover and mutation to build the population elements of the next genetic generation. Elitism means that a percentage of the fittest individuals in the current generation is copied into the next generation. Crossover and mutation are the basic genetic operators. Crossover creates new individuals (offsprings) based on the fittest individuals (parents) in the current population. So, crossover recombines the fittest material in the current population in the hope that the recombination of useful material in one of the parents will generate an even fitter population element. The mutation operation will change some minor details of a population element. The hope is that the mutation operator will insert new useful material in the population. The genetic algorithm (GA) stops when: (i) it finds an individual whose fitness is 1; *or* (ii) it computes n generations, where n is the maximum number of generation that is allowed; *or* (iii) the fittest individual has not changed for $n/2$ generations in a row. If none of these conditions hold, the GA creates a new population as follows:

Input: *current population, elitism rate, crossover rate and mutation rate*
Output: *new population*

1. *Copy "elitism rate × population size" of the best individuals in the current population to the next population.*
2. *While there are individuals to be created do:*

   (a) *Use tournament selection to select parent1.*

   (b) *Use tournament selection to select parent2.*

   (c) *Select a random number r between 0 (inclusive) and 1 (exclusive).*

   (d) *If r less than the crossover rate:*
       then do crossover with parent1 and parent2. This operation generates two offsprings: offspring1 and offspring2.
       else offspring1 equals parent1 and offspring2 equals parent2.

   (e) *Mutate offspring1 and offspring2. (This step is only needed if the mutation rate is non-zero.)*

   (f) *Copy offspring1 and offspring2 to the new population.* [6]

3. *Return the new population.*

---

[6] Note: If the population size is n and the new population has already $n-1$ individuals, then only *offspring1* is copied into this new population.

Tournament Selection. To select a parent the tournament selection algorithm randomly selects 5 individuals and returns the fittest individual among the five ones.

Crossover. The most important and complex genetic operation in our genetic approach is the crossover operation. Starting point of the crossover operation are the two parents (i.e. parent1 and parent2). The result of applying the crossover operation are two offsprings (offspring1 and offspring2). First, the crossover algorithm randomly selects an activity t to be the *crossover point*. This means that the INPUT and OUTPUT of t in parent1 will be recombined with the INPUT and OUTPUT of t in parent2. Second, parent1 is copied to offspring1 and parent2 to offspring2. Third, the algorithm randomly selects a *swap point* for the INPUT(t) sets in both offsprings and another *swap point* for the OUTPUT(t) sets. The swap point determines which subsets of the INPUT/OUTPUT of t in the offspring are going to be *swapped* (or exchanged). A random swap point is chosen for every INPUT/OUTPUT and offsprings. The respective INPUT and OUTPUT sets of the crossover point at the two offsprings are then recombined by interchanging the subsets from the swap point until the end of the set. The recombined INPUT/OUTPUT sets are then checked to make sure that they are proper partitions. Finally, the two offsprings undergo a repair operation called *update related elements*.

Update Related Elements. When the parents (and consequently the offsprings) have different causal matrices, the crossover operation may generate inconsistencies. Note that the boolean expression may contain activities whose respective cell in the causal matrix is zero. Similarly, an activity may not appear in the boolean expression after the crossover and the causal matrix still has a non-zero entry for it. So, after the INPUT/OUTPUT sets have being recombined, we need to check the consistency of the recombined sets with respect to the other activities boolean expressions and the causal matrix. When they are inconsistent, we need to update the causal matrix and the related boolean expressions of the other activities. As an example, assume *Parent1* equals *Individual1* and *Parent2* equals *Individual2* in Table 4. These two parents undergo crossover and mutation to generate two offsprings. Let activity D be the randomly selected crossover point. Since INPUT1(D) equals INPUT2(D), the crossover has no real effect for D's INPUT. Let us look at the D's OUTPUT sets. Both D's OUTPUT sets have a single subset, so the only possible swap point to select equals 0, i.e., before the first and only element. After swapping the subsets *Offpring1* (*Parent1* after crossover) has INPUT1(D)= $\{\{A\}\}$ and OUTPUT1(D)= $\{\{E, F\}\}$. Note that OUTPUT1(D) now contains F. So, the *update related elements* algorithm makes INPUT1(F)= $\{\{D\}\}$. *Offspring2* is updated in a similar way. The two offsprings are shown in Table 5.

Mutation. The mutation works on the INPUT and OUTPUT boolean expressions of an activity. For every activity t in an individual, a new random number

Table 5. Example of two offsprings that can be produced after a crossover between the two individuals in Table 4

	Offspring1			*Offspring2*	
ACTIVITY	INPUT	OUTPUT	ACTIVITY	INPUT	OUTPUT
A	$\{\}$	$\{\{B,C,D\}\}$	A	$\{\}$	$\{\{B,C,D\}\}$
B	$\{\{A\}\}$	$\{\{H\}\}$	B	$\{\{A\}\}$	$\{\{H\}\}$
C	$\{\{A\}\}$	$\{\{H\}\}$	C	$\{\{A\}\}$	$\{\{H\}\}$
D	$\{\{A\}\}$	$\{\{E,F\}\}$	D	$\{\{A\}\}$	$\{\{E\}\}$
E	$\{\{D\}\}$	$\{\{G\}\}$	E	$\{\{D\}\}$	$\{\{G\}\}$
F	$\{\{D\}\}$	$\{\{G\}\}$	F	$\{\}$	$\{\{G\}\}$
G	$\{\{E\},\{F\}\}$	$\{\{H\}\}$	G	$\{\{E\},\{F\}\}$	$\{\{H\}\}$
H	$\{\{C,B,G\}\}$	$\{\}$	H	$\{\{C\},\{B\},\{G\}\}$	$\{\}$

r is selected. Whenever r is less than the "mutation rate" [7], the subsets in INPUT(t) are randomly merged or split. The same happens to OUTPUT(t). As an example, consider *Offspring1* in Table 5. Assume that the random number r was less than the mutation rate for activity D. After applying the mutation, OUTPUT(D) changes from $\{\{E,F\}\}$ to $\{\{E\},\{F\}\}$. Note that this mutation does not change an individual's causal relations, only its AND-OR/join-split may change.

7   Some Experimental Results

To test our approach we applied the algorithm to many examples, mostly artificially generated and some based on real-life logs [14]. In this paper we only consider two WF-nets; one with 8 and one with 12 activities. Both models contain concurrency and loops. The WF-net with 8 activities corresponds to the first example in this paper, i.e., the Petri net shown in Figure 1. The other WF-net represents a completely different process model. For each model we generated 10 different event logs with 1000 traces. Without noise the rediscovering of the underlying WF-nets was no problem for the genetic algorithm. To test the behavior of the genetic algorithm for event logs with noise, we used 6 different noise types: *missing head, missing body, missing tail, missing activity, exchanged activities* and *mixed noise*. If we assume a event trace $\sigma = t_1...t_{n-1}t_n$, these noise types behave as follows. *Missing head, body* and *tail* randomly remove subtraces of activities in the head, body and tail of σ, respectively. The head goes from t_1 to $t_{n/3}$. The body goes from $t_{(n/3)+1}$ to $t_{(2n/3)}$. The tail goes from $t_{(2n/3)+1}$ to t_n. *Missing activity* randomly removes *one* activity from σ. *Exchanged activities* exchange two activities in σ. *Mixed noise* is a fair mix of the other 5 noise types. Real life logs will typically contain mixed noise. However, the separation between

---
[7] The mutation rate determines the probability that an individual's task undergoes mutation.

the noise types allow us to better assess how the different noise types affect the genetic algorithm.

For every noisy type, we generated logs with 5%, 10% and 20% of noise. So, every process model in our experiments has $6 \times 3 = 18$ noisy logs. For each event-log the genetic algorithms runs 10 experiments with a different random initialization. The populations had 500 individuals and were iterated for at most 100 generations. The crossover rate was 1.0 and the mutation rate was 0.01. The elitism rate was 0.01. Details about the experiments and results can be found in [14]. Here we summarize the main findings.

Overall the higher the noise percentage, the lower the probability the algorithm will come up with the original WF-net. In this particular setting the algorithm can always handle the *missing tail* noise type. This is related to the high impact of *proper completion* in our fitness measure. Also the *exchanged activities* noise type does not harm the performance of the algorithm. This is related to the heuristic that is used during the initialization of the population. *Missing head* impacts more the algorithm because our fitness does not (yet!) punish individuals with missing tokens. *Missing body* and *missing activity* noise types are the most difficult to handle. This is also related to the heuristics during the building of the initial population because the removal of activities generates $t_1 t_2$ "fake" subtraces that will not be counter-balanced by subtraces $t_2 t_1$. Consequently, the probability that the algorithm will causally relate t_1 and t_2 is increased. Tables 6 and 7 contain the results obtained for the noisy logs of the two process models with 8 and 12 activities.

Table 6. Results of applying the genetic algorithm for noisy logs of the process model with 8 activities, i.e., Figure 1. The ratio relates the number of times the algorithm found the correct model by the number of times the algorithm ran

			noise type			
noise percentage	Missing head	Missing tail	Missing body	Missing activity	Exchanged activities	Mixed noise
5%	10/10	10/10	0/10	1/10	9/10	1/10
10%	10/10	10/10	1/10	1/10	5/10	3/10
20%	0/10	10/10	0/10	0/10	0/10	2/10

Tables 6 and 7 show that our approach works well for relatively simple examples. Moreover in contrast to most of the existing approaches it is able to deal with noise. To improve our approach we are now refining the fitness calculation. For instance, the fitness should consider the number of tokens that remained in the individual after the parsing is finished as well as the number of tokens that needed to be added during the parsing.

The generic mining algorithm presented in this paper is supported by a plugin in the ProM framework (cf. http://www.processmining.org). Figure 5 shows screenshot of the plugin. This screenshot presents the result for the process model with 8 activities in terms of Petri nets and EPCs.

Table 7. Results of applying the genetic algorithm for noisy logs of the process model with 12 activities

noise percentage	noise type					
	Missing head	Missing tail	Missing body	Missing activity	Exchanged activities	Mixed noise
5%	10/10	10/10	0/10	0/10	10/10	2/10
10%	1/10	10/10	0/10	0/10	9/10	3/10
20%	1/10	10/10	0/10	0/10	8/10	2/10

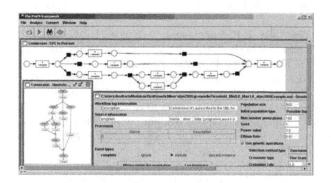

Fig. 5. A screenshot of the GeneticMiner plugin in the ProM framework analyzing the event log in Table 1 and generating the correct WF-net, i.e., the one shown in Figure 1

8   Conclusion

In this paper we presented our first experiences with a more global mining technique (e.g. a genetic algorithm). For convenience we did not use Petri nets for the internal representation of the individuals in a genetic population. Instead we used causal matrices which are slightly more expressive. We elaborated on the relationships between our representation and Petri nets. The bottom line is that any WF-net can be mapped onto our notation and also some constructs that require duplicate activities (i.e., two transitions with the same label) or silent transitions (i.e., steps not visible in the log) can be discovered. This way the approach overcomes some of the problems with earlier algorithms. For example, it is possible to mine non-free choice constructs. The main added value of using a genetic algorithm is the ability to deal with noise and incompleteness. At this point in time we are fine-tuning our genetic algorithms based on many artificial examples (i.e., process mining based on simulation logs) and real-life examples. Experimental results on event logs with noise point out that we are on the right track on our quest to develop a genetic algorithm that mines process models. Our next steps will focus on further improvements of the fitness measurement so that it gives a better indication of the optimal fit between a process model and an event-log.

References

1. W.M.P. van der Aalst. Business Process Management Demystified: A Tutorial on Models, Systems and Standards for Workflow Management. In J. Desel, W. Reisig, and G. Rozenberg, editors, *Lectures on Concurrency and Petri Nets*, volume 3098 of *Lecture Notes in Computer Science*, pages 1–65. Springer-Verlag, Berlin, 2004.
2. W.M.P. van der Aalst and M. Song. Mining Social Networks: Uncovering Interaction Patterns in Business Processes. In J. Desel, B. Pernici, and M. Weske, editors, *International Conference on Business Process Management (BPM 2004)*, volume 3080 of *Lecture Notes in Computer Science*, pages 244–260. Springer-Verlag, Berlin, 2004.
3. W.M.P. van der Aalst, B.F. van Dongen, J. Herbst, L. Maruster, G. Schimm, and A.J.M.M. Weijters. Workflow Mining: A Survey of Issues and Approaches. *Data and Knowledge Engineering*, 47(2):237–267, 2003.
4. W.M.P. van der Aalst, A.J.M.M. Weijters, and L. Maruster. Workflow Mining: Discovering Process Models from Event Logs. *IEEE Transactions on Knowledge and Data Engineering*, 16(9):1128–1142, 2004.
5. R. Agrawal, D. Gunopulos, and F. Leymann. Mining Process Models from Workflow Logs. In *Sixth International Conference on Extending Database Technology*, pages 469–483, 1998.
6. J.E. Cook and A.L. Wolf. Discovering Models of Software Processes from Event-Based Data. *ACM Transactions on Software Engineering and Methodology*, 7(3):215–249, 1998.
7. J. Dehnert and W.M.P. van der Aalst. Bridging the Gap Between Business Models and Workflow Specifications. *International Journal of Cooperative Information Systems*, 13(3):289–332, 2004.
8. A. Ehrenfeucht and G. Rozenberg. Partial (Set) 2-Structures — Part 1 and Part 2. *Acta Informatica*, 27(4):315–368, 1989.
9. A.E. Eiben and J.E. Smith. *Introduction to Evolutionary Computing*. Natural Computing. Springer-Verlag, Berlin, 2003.
10. K. van Hee, N. Sidorova, and M. Voorhoeve. Soundness and Separability of Workflow Nets in the Stepwise Refinement Approach. In W.M.P. van der Aalst and E. Best, editors, *Application and Theory of Petri Nets 2003*, volume 2679 of *Lecture Notes in Computer Science*, pages 335–354. Springer-Verlag, Berlin, 2003.
11. J. Herbst. A Machine Learning Approach to Workflow Management. In *Proceedings 11th European Conference on Machine Learning*, volume 1810 of *Lecture Notes in Computer Science*, pages 183–194. Springer-Verlag, Berlin, 2000.
12. H. Mauch. Evolving Petri Nets with a Genetic Algorithm. In E. Cantu-Paz and J.A. Foster et al., editors, *Genetic and Evolutionary Computation (GECCO 2003)*, volume 2724 of *Lecture Notes in Computer Science*, pages 1810–1811. Springer-Verlag, Berlin, 2003.
13. A.K.A. de Medeiros, W.M.P. van der Aalst, and A.J.M.M. Weijters. Workflow Mining: Current Status and Future Directions. In R. Meersman, Z. Tari, and D.C. Schmidt, editors, *On The Move to Meaningful Internet Systems 2003: CoopIS, DOA, and ODBASE*, volume 2888 of *Lecture Notes in Computer Science*, pages 389–406. Springer-Verlag, Berlin, 2003.
14. A.K.A. de Medeiros, A.J.M.M. Weijters, and W.M.P. van der Aalst. Using Genetic Algorithms to Mine Process Models: Representation, Operators and Results. BETA Working Paper Series, WP 124, Eindhoven University of Technology, Eindhoven, 2004.

15. J.H. Moore and L.W. Hahn. Petri Net Modeling of High-Order Genetic Systems Using Grammatical Evolution. *BioSystems*, 72(1-2):177–86, 2003.

16. J.H. Moore and L.W. Hahn. An Improved Grammatical Evolution Strategy for Hierarchical Petri Net Modeling of Complex Genetic Systems. In G.R. Raidl et al., editor, *Applications of Evolutionary Computing, EvoWorkshops 2004*, volume 3005 of *Lecture Notes in Computer Science*, pages 63–72. Springer-Verlag, Berlin, 2004.

17. J.P. Reddy, S. Kumanan, and O.V.K. Chetty. Application of Petri Nets and a Genetic Algorithm to Multi-Mode Multi-Resource Constrained Project Scheduling. *International Journal of Advanced Manufacturing Technology*, 17(4):305–314, 2001.

18. A.J.M.M. Weijters and W.M.P. van der Aalst. Rediscovering Workflow Models from Event-Based Data using Little Thumb. *Integrated Computer-Aided Engineering*, 10(2):151–162, 2003.

The (True) Concurrent Markov Property and Some Applications to Markov Nets

Samy Abbes*

Institute for Systems Research,
A.V. Williams Building, University of Maryland,
College Park, MD 20742, USA

Abstract. We study *probabilistic safe Petri nets*, a probabilistic extension of safe Petri nets interpreted under the true-concurrent semantics. In particular, the likelihood of processes is defined on partial orders, not on firing sequences.

We focus on *memoryless* probabilistic nets: we give a definition for such systems, that we call Markov nets, and we study their properties. We show that several tools from Markov chains theory can be adapted to this true-concurrent framework. In particular, we introduce *stopping operators* that generalize stopping times, in a more convenient fashion than other extensions previously proposed. A Strong Markov Property holds in the concurrency framework. We show that the Concurrent Strong Markov property is the key ingredient for studying the dynamics of Markov nets. In particular we introduce some elements of a recurrence theory for nets, through the study of *renewal* operators. Due to the concurrency properties of Petri nets, Markov nets have global and local renewal operators, whereas both coincide for sequential systems.

1   Introduction

In the context of a continuously growing interest of the scientific community for *distributed systems*, Petri nets in general, and their true-concurrent dynamics in particular, become a paradigm used in various application areas. Good examples are found in systems theory, where more and more Petri nets models are proposed for the management of complex concurrent systems such as telecommunication networks and services [1, 8].

In turn, studies motivated by various applications can bring back new conceptions and results about Petri nets. This is in particular the case of probabilistic Petri nets that have interested both computer scientists [13, 12] and scientists from systems theory [4]. I intentionally choose the term of *probabilistic* Petri net to emphasize the difference with *stochastic* Petri nets. The latter refers to processes where a real time parameter t describes the random evolution of a Petri net—in this model, concurrency is interpreted through an interleaving semantics. The purpose of probabilistic Petri nets is different. The dynamics of a

---

* Work supported by IRISA (France) and ISR (USA).

G. Ciardo and P. Darondeau (Eds.): ICATPN 2005, LNCS 3536, pp. 70–89, 2005.

probabilistic Petri net is directly defined through *random discrete partial orders*, in order to fit the true-concurrent semantics of the underlying Petri net model.

The following advantages have been recognized to the true-concurrency semantics. First, for large scale concurrent systems, the true-concurrency semantics, by identifying different interleavings of a same process, saves a lot of computational complexity: true-concurrency tackles the "state explosion" problem. Second, events of a distributed system such as a telecommunication network with real asynchronous components may obey only to local clocks, without reference to any global clock [5]. This corresponds to events partially ordered, and thus to a true-concurrent semantics, not to an interleaving semantics. Statistical treatment of systems, such as state estimation or learning of parameters, need to have at hand a probabilistic model, including results on the asymptotic dynamics of systems. We contribute in this paper to the set up of this theory, within the model of safe Petri nets.

True-concurrent processes of a Petri net, i.e. Mazurkiewicz traces of firing sequences, can be regarded as configurations of the unfolding of the net [9, 7]. Configurations are partially ordered by the relation of set-inclusion (traces are prefix from one another). Setting up a true-concurrent random dynamics for a safe Petri net is done by considering a probability measure $\mathbb{P}$ on the space Ω of *maximal configurations* of the unfolding of the net [13, 4, 12]. Referring to the net as to a concrete device, the meaning of the so obtained probability space $(\Omega, \mathbb{P})$ is as follows: Let v be a finite trace of the net. The $\mathbb{P}$-probability for v to occur in an execution of the net is $\mathbb{P}(A)$, where A is the subset of Ω defined by: $A = \{\omega \in \Omega : \omega \supseteq v\}$. This extends the framework of sequential discrete events random processes studied from both the mathematical (e.g., [11]) and the Computer Science (e.g., [10]) viewpoints.

The topic of this paper is the following: Can we go further in the generalization, and extend to concurrent systems both the definition and some properties of *Markovian models* such as finite Markov chains in discrete time (DTMC)? We demonstrate that the answer is "yes": we introduce a new definition for *Markov nets*, and we obtain qualitative results on their dynamics by studying their recurrence properties. This leads to elegant results with simple formulation, yet with some strong hidden mathematical background.

Our definition of Markov nets relies on the intuitive concept of memoryless systems. Here, the state of the system is the marking of the net. Hence a Markov net will be a probabilistic net such that the probabilistic future of a finite process v, ending to a marking M, only depends on M, and not on the entire process v. Starting from this definition, we follow the classical theory of Markov chains, adapting to concurrent systems several concepts and tools from this theory. The effective construction of Markov nets is known for a large class of safe Petri nets, including free-choice and confusion-free nets [2, 3]. But it is rather a technical construction, more complicated than the classical construction of Markov chains based on a transition matrix; therefore, in this paper, we will assume that the existence of Markov nets is an established fact, in order to focus on their properties.

A first basic result for Markov chains is the so-called *Strong Markov Property*, a formula which says in a condensed form that the system is indeed memory-less. This formula relies on the notion of *stopping time*. We adapt the notion of stopping time to true-concurrent systems, introducing *stopping operator* for nets—this part was already challenging, since our framework does not posses a global clock. Then we formulate and prove the Strong Markov property for Markov nets. The remaining of the paper is devoted to illustrate the use of the Strong Markov Property for concurrent systems. We present some elements of a *recurrence theory* in the framework of concurrent systems: We prove that the initial marking of a net has probability 0 or 1 to return infinitely often in an execution, an extension of the same well-known result for Markov chains with state instead of marking. Stopping operators, the Markov property for concurrent systems and its application to the recurrence properties of nets are the main contributions of this paper, beside an abstract definition of Markov nets.

Hence, the new techniques that we introduce allow to extend existing results from sequential to concurrent systems. But other developments are allowed where the concurrency properties of the Petri nets model play a more significant role. Due to lack of place, we only shortly introduce these properties that take into account the local characteristics of concurrent systems. This constitutes elements of a *local recurrence* theory, to be distinguished from the above recurrence, that appears *a posteriori* as a *global recurrence*. For sequential systems, global and local recurrences coincide, but not anymore for concurrent systems.

The paper is organized as follows. In §2 we recall the background from Probability and from finite Markov chains theory. We set up a symmetric framework for concurrent systems in §3, introducing Markov nets and ending with the statement of the Concurrent Strong Markov property. Then §4 is devoted to the application of this new Markov property to elements of a recurrence theory for Markov nets. Finally, §5 discusses some perspectives.

2   Background on Probability and Expectation

Notations for Usual Sets. We denote by $\mathbb{N}$ and $\mathbb{R}$ respectively, the sets of non-negative integers and of real numbers. We consider the following extensions of $\mathbb{N}$ and $\mathbb{R}$:

$$\overline{\mathbb{N}} = \mathbb{N} \cup \{+\infty\}, \qquad \overline{\mathbb{R}} = \mathbb{R} \cup \{-\infty, +\infty\}. \tag{1}$$

σ-Algebra and Probability Spaces. Let Ω be a set, a family $\mathcal{F}$ of subsets of Ω is said to be a **σ-algebra** of Ω if $\mathcal{F}$ is closed under countable intersection, if $A \in \mathcal{F} \Rightarrow \Omega \setminus A \in \mathcal{F}$, and if $\emptyset \in \mathcal{F}$. The pair $(\Omega, \mathcal{F})$ is called a **measurable space**, and the elements of $\mathcal{F}$ constitute the **measurable sets** of Ω. The set $\overline{\mathbb{R}}$ defined by Eq. (1) is equipped with its Borel σ-algebra, generated by the Euclidean topology on $\mathbb{R}$. If $(\Omega, \mathcal{F})$ and $(\Omega', \mathcal{F}')$ are two measurable spaces, a mapping $f : \Omega \to \Omega'$ is **$\mathcal{F}$-measurable** (or simply, measurable) if $f^{-1}(A) \in \mathcal{F}$ for all $A \in \mathcal{F}'$.

Let $(\Omega, \mathcal{F})$ be a measurable space. A **probability measure** on $(\Omega, \mathcal{F})$ is a function $\mathbb{P} : \mathcal{F} \to \mathbb{R}$ with $\mathbb{P}(\Omega) = 1$ and $\mathbb{P}(A) \geq 0$ for all $A \in \mathcal{F}$, and such that for every countable family $(A_n)_{n \in \mathbb{N}}$ of measurable sets, if $i \neq j \Rightarrow A_i \cap A_j = \emptyset$, then: $\mathbb{P}\left(\bigcup_{n \in \mathbb{N}} A_n\right) = \sum_{n \in \mathbb{N}} \mathbb{P}(A_n)$. The triple $(\Omega, \mathcal{F}, \mathbb{P})$ is called a **probability space**.

We follow some traditional language conventions that are convenient when dealing with probability spaces. Measurable functions are called **random variables**. If X is a real-valued random variable, its integral under measure $\mathbb{P}$ is called its expectation, and is denoted $\mathbb{E}(X)$. We also write $\{X = 0\}$ to denote the set of elements $\omega \in \Omega$ such that $X(\omega) = 0$, and $\mathbb{P}(X = 0)$ stands for "the probability that $X = 0$", i.e.:

$$\mathbb{P}(X = 0) = \mathbb{P}(\{X = 0\}) = \mathbb{P}(\{\omega \in \Omega : X(\omega) = 0\}).$$

With a slight and classical abuse of terminology, we identify a random variable X and the class of random variables X' that differ from X only on a set of probability zero, i.e. the class of X' such that $\mathbb{P}(X \neq X') = 0$.

Finally, for A a subset of Ω, we use the notation $\mathbf{1}_A$ to denote the **characteristic** function of A, defined by:

$$\mathbf{1}_A(\omega) = \begin{cases} 1, & \text{if } \omega \in A \\ 0, & \text{if } \omega \notin A \end{cases}.$$

Then $\mathbf{1}_A$ is measurable if and only if A is measurable, in which case $\mathbb{P}(A) = \mathbb{E}(\mathbf{1}_A)$.

Conditional Expectation. We first recall the definition of conditional expectation w.r.t. a measurable subset. Let $(\Omega, \mathcal{F}, \mathbb{P})$ be a probability space, let A be a measurable subset of Ω, and assume that $\mathbb{P}(A) > 0$. Then the following formula defines a probability $\mathbb{P}_A$ on $(\Omega, \mathcal{F})$, called **probability conditional on** A:

$$\forall B \in \mathcal{F}, \quad \mathbb{P}_A(B) = \frac{\mathbb{P}(B \cap A)}{\mathbb{P}(A)}.$$

The probability $\mathbb{P}_A(\cdot)$ is usually denoted by $\mathbb{P}(\cdot \mid A)$.

We now recall the definition of conditional expectation w.r.t. σ-algebras (see e.g. [6]). Let $(\Omega, \mathcal{F}, \mathbb{P})$ be a probability space, let X be a nonnegative real random variable, and let $\mathcal{F}' \subseteq \mathcal{F}$ be a sub-σ-algebra of $\mathcal{F}$. A classical result states that there is a unique $\mathcal{F}'$-measurable random variable X' characterized by:

$$\forall A \in \mathcal{F}', \quad \mathbb{E}(\mathbf{1}_A X) = \mathbb{E}(\mathbf{1}_A X').$$

X' is called the **expectation of X conditional on** $\mathcal{F}'$, and is denoted by $X' = \mathbb{E}(X \mid \mathcal{F}')$. Intuitively, X' is the best $\mathcal{F}'$-measurable approximation of X.

In the sequel, we will use the two following properties of conditional expectation:

1. For every nonnegative random variable X and sub-σ-algebra $\mathcal{G}$ of $\mathcal{F}$, we have:

$$\mathbb{E}(X) = \mathbb{E}(\mathbb{E}(X \mid \mathcal{G})). \tag{2}$$

2. For every sub-σ-algebra $\mathcal{G}$ of $\mathcal{F}$ and nonnegative random variables X, Y, if Y is $\mathcal{G}$-measurable, then:

$$\mathbb{E}(XY \mid \mathcal{G}) = Y\mathbb{E}(X \mid \mathcal{G}). \tag{3}$$

Sequential Probabilistic Processes. Consider a finite set S, thought of as a state space. We define a **process** over S as a finite or infinite sequence of elements of S. If $v = (s_0, \ldots, s_n)$ is a finite process, we say that s_n is the **end state** of v, and we denote it $s(v) = s_n$. We denote by Ω the set of infinite processes over S, i.e. Ω is the infinite product set $\Omega = S^{\mathbb{N}}$. For each integer $n \geq 0$, we denote by X_n the nth projection $\Omega \to S$, so that we have:

$$\forall \omega \in \Omega, \quad \omega = (X_0(\omega), X_1(\omega), \ldots).$$

For each integer $n \geq 0$, consider the finite σ-algebra $\mathcal{F}_n$ of Ω spanned by the subsets of the form:

$$\{X_0 = s_0, \ldots, X_n = s_n\},$$

with $(s_0, \ldots, s_n)$ ranging over S^{n+1}. The **product σ-algebra** $\mathcal{F}$ on Ω is defined as the smallest σ-algebra that contains all $\mathcal{F}_n$, for $n \geq 0$.

We define a **probabilistic process** over S as a pair $(S, \mathbb{P})$, where $\mathbb{P}$ is a probability on $(\Omega, \mathcal{F})$. If there is an element $s_0 \in S$ such that $X_0 = s_0$, we say that s_0 is the **initial state** of the probabilistic process $(S, \mathbb{P})$. Let $v = (s_0, \ldots, s_n)$ be a finite process, and consider the measurable subset of Ω defined by:

$$\mathcal{S}(v) = \{X_0 = s_0, \ldots, X_n = s_n\}. \tag{4}$$

We define the **likelihood** of v by: $p(v) = \mathbb{P}(\mathcal{S}(v))$. Intuitively, $p(v)$ is the probability of v to occur in an execution of the system. Be aware however that the likelihood function does not define a probability on the set of finite processes, since it does not sum to 1.

Sequential Probabilistic Future and Markov Chains. Let $(S, \mathbb{P})$ be a probabilistic process, and let v be a finite process over S with $p(v) > 0$, with v given by $v = (s_0, \ldots, s_n)$. Recall the definition (4) of $\mathcal{S}(v)$, and consider the measurable mapping $\rho_v : \mathcal{S}(v) \to \Omega$ defined by:

$$\rho_v(s_0, \ldots, s_n, X_{n+1}, X_{n+2}, \ldots) = (s_n, X_{n+1}, X_{n+2}, \ldots).$$

The mapping ρ_v let us define a probability $\mathbb{P}^v$ on $(\Omega, \mathcal{F})$ as follows:

$$\forall A \in \mathcal{F}, \quad \mathbb{P}^v(A) = \mathbb{P}(\rho_v^{-1}(A) \mid \mathcal{S}(v)), \tag{5}$$

where $\mathbb{P}(\cdot \mid \mathcal{S}(v))$ is the probability conditional on $\mathcal{S}(v)$. We call the new probabilistic process $(S, \mathbb{P}^v)$ the **probabilistic future** of process v. We denote by the symbol $\mathbb{E}^v$ the expectation on Ω under probability $\mathbb{P}^v$. By construction, $s_n = s(v)$ is the initial state of the probabilistic future of v.

We say that $(S, \mathbb{P})$ is a **Markov chain** if, for every pair (v, v') of finite processes over S with $p(v), p(v') > 0$, we have:

$$s(v) = s(v') \Rightarrow \mathbb{P}^v = \mathbb{P}^{v'} . \tag{6}$$

Equation (6) formulates the intuition that, for Markov chains, the probabilistic future of a process only depends on the current state of the system, i.e. state $s(v)$, and not on the entire history the process v. As a consequence, it makes sense to denote by $\mathbb{P}^s$ and $\mathbb{E}^s$ the probability and the expectation starting from state s, and defined by:

$$\mathbb{P}^s = \mathbb{P}^v, \quad \mathbb{E}^s = \mathbb{E}^v,$$

for any finite process v with positive likelihood and with s as end state[1].

Sequential Shift Operators. Stopping Times and the Markov Property. Consider again the measurable space $(\Omega, \mathcal{F})$ constructed as above from finite set S, and define the pointwise transformation $\theta : \Omega \to \Omega$ as follows:

$$\forall \omega \in \Omega, \quad \theta(\omega) = (X_1(\omega), X_2(\omega), \ldots) . \tag{7}$$

Transformation θ is called the **shift operator**. The iterates of θ are traditionally denoted by θ_n, for $n \geq 0$, i.e. $\theta_0 = \mathrm{Id}$ and $\theta_n = \theta_{n-1} \circ \theta$ for $n \geq 1$. Furthermore, assume that $T : \Omega \to \mathbb{N}$ is an integer random variable. We denote by θ_T the pointwise transformation $\Omega \to \Omega$ that "iterates T times θ", which is defined by:

$$\forall \omega \in \Omega, \quad \theta_T(\omega) = \theta_{T(\omega)}(\omega) . \tag{8}$$

We shall authorize T to take an infinite value, so that T is defined $\Omega \to \overline{\mathbb{N}}$, with $\overline{\mathbb{N}} = \mathbb{N} \cup \{\infty\}$. If $T(\omega) = \infty$, then $\theta_T(\omega)$ is not defined. A random variable $T : \Omega \to \overline{\mathbb{N}}$ is called a **stopping time** if for every $n \geq 0$ (see e.g. [6, 11]):

$$\{T = n\} \text{ is a } \mathcal{F}_n\text{-measurable subset of } \Omega. \tag{9}$$

We will see below the particular role of stopping times in the analysis of Markov chains. To make the notion intuitive, we mention a typical example of stopping time, the **hitting time** T_x of a given state x: for $\omega \in \Omega$, $T_x(\omega)$ is the smallest integer n such that $X_n(\omega) = x$, if such an integer exists, otherwise $T_x(\omega) = \infty$.

A stopping time T defines a sub-σ-algebra of $\mathcal{F}$, denoted by $\mathcal{F}_T$, as follows:

$$\forall A \in \mathcal{F}, \quad A \in \mathcal{F}_T \Leftrightarrow \forall n \geq 0, \ A \cap \{T = n\} \in \mathcal{F}_n . \tag{10}$$

We can now state the so-called "strong Markov property", a basic tool in the analysis of Markov chains (see for instance [11]): Let $(S, \mathbb{P})$ be a Markov chain over a finite set S. For every stopping time T, and for every nonnegative random variable $h : \Omega \to \overline{\mathbb{R}}$, the following identity holds:

$$\mathbb{E}\left(h \circ \theta_T \mid \mathcal{F}_T\right) = \mathbb{E}^{X_T}(h) , \tag{11}$$

---

[1] With our definition, the transition matrix P of the chain can be retrieved as follows: the sth row of matrix P is the probability vector $\mathbb{P}^s(X_1 = s')$, for s' ranging over S.

where, by convention, both members identically vanish on $\{T = \infty\}$. The second member must be understood as the composition of the two functions $\omega \in \Omega \mapsto X_{T(\omega)}(\omega)$ and $s \in S \mapsto \mathbb{E}^s(h)$.

Instead of showing the consequences that are usually obtained from the Markov property, we will instead show how the previous notions generalize in a concurrent framework. After having established a Concurrent Markov property, we will directly derive in the concurrent framework some of its classical consequences as it is usually done in the sequential framework.

3   Probabilistic Safe Petri Nets

We now analyze the case of concurrent systems, within the model of safe Petri nets. We will try to set up a probabilistic framework symmetric to the one that we have introduced above in §2.

Safe Petri Nets and Unfoldings. True-Concurrent Dynamics. We assume basic knowledge of the reader on Petri nets, true-concurrent dynamics and unfoldings of safe Petri nets, such as set up in [9]. We consider a safe and finite Petri net $\mathcal{N} = (P, T, F, M_0)$, where P and T respectively denote the sets of places and transitions, F stands for the flow relation, and M_0 is the initial marking of the net. We denote by $(\mathcal{U}, \lambda)$ the **unfolding** of $\mathcal{N}$, where $\mathcal{U}$ is the universal occurrence net associated with $\mathcal{N}$, and $\lambda : \mathcal{U} \to \mathcal{N}$ is the canonical **labeling** mapping, with the slight abuse of notations that identifies a net and its set of nodes. According to the usual terminology, places of the unfolding are called conditions.

The causality relation in $\mathcal{U}$ is denoted by $\preceq$. The set of $\preceq$-minimal nodes of $\mathcal{U}$ is called the **initial cut** of $\mathcal{U}$, and we denote it by c_0. We recall that c_0 is in bijection with M_0 through λ. We say that a node x of a subset $A \subseteq \mathcal{U}$ is **terminal** in A if x is a maximal node of A, the maximality being defined w.r.t. the causality relation $\preceq$. The set of terminal nodes of A is denoted by $\gamma(A)$.

We denote by $\mathcal{M}$ the set of **reachable markings** of $\mathcal{N}$, and for $M \in \mathcal{M}$, we note with M as an exponent all objects related to net (P, T, F, M): $\mathcal{N}^M$ stands precisely for this net, $(\mathcal{U}^M, \lambda^M)$ for the unfolding of $\mathcal{N}^M$, etc.

We will analyze the dynamics of net $\mathcal{N}$ through the dynamics of its unfolding $\mathcal{U}$. Define a **configuration** of $\mathcal{U}$ as a conflict-free prefix of $\mathcal{U}$, containing the initial cut of $\mathcal{U}$ and with conditions only as terminal nodes. Configurations are partially ordered by set inclusion. For those readers used to deal with Mazurkiewicz traces instead, let us recall that both conceptions are equivalent, as stated by [9–Prop. 6], in the sense that every the posets of finite traces and of finite configurations are isomorphic. The **end marking** of a finite configuration v is the marking reached by any finite sequence linearizing v. We denote this marking by $m(v)$, and it is well known that $m(v)$ is given by:

$$m(v) = \lambda\big(\gamma(v)\big),$$

where $\gamma(v)$ denotes the set of terminal nodes (actually, conditions) of v.

Probabilistic Petri Nets. To introduce a measurable space that will support a probability measure and model a probabilistic dynamics of a safe Petri net, it is not suitable to consider the set of infinite configurations of the unfolding. Indeed, one branch of a configuration may be infinite, whereas other branches remain finite, and this introduces non natural choices. A more convenient sample space is found by considering the set of **maximal configurations** of the unfolding, the maximality being defined w.r.t. the set inclusion.

Hence, considering a safe Petri net $\mathcal{N}$ with unfolding $(\mathcal{U}, \lambda)$, we denote by Ω the set of maximal configurations of $\mathcal{U}$. Elements of Ω are generically denoted by ω; from time to time, we call an element ω an **execution** of the net. The properties of occurrence nets, combined with an application of Zorn's lemma, show that every configuration is subset of a maximal configuration, and in particular Ω is non empty. For v a configuration of $\mathcal{U}$, we define the **shadow** of v as the following subset of Ω:

$$\mathcal{S}(v) = \{\omega \in \Omega : \omega \supseteq v\}.$$

The terminology of *shadow* is justified by thinking of $\mathcal{U}$ as lightened from its initial cut, and of Ω as to the *boundary at infinity* of $\mathcal{U}$.

We say that a shadow $\mathcal{S}(v)$ is a **finitary shadow** if v is a finite configuration— be aware that $\mathcal{S}(v)$ is not a finite set however. The σ-algebra on Ω to be considered is the σ-algebra spanned by the finitary shadows $\mathcal{S}(v)$. We denote this σ-algebra by $\mathcal{F}$, so that $\mathcal{F}$ is the smallest σ-algebra of Ω that makes measurable all the finitary shadows—and then every shadow is $\mathcal{F}$-measurable. We define a **probabilistic net** as a pair $(\mathcal{N}, \mathbb{P})$, where $\mathbb{P}$ is a probability measure on the measurable space $(\Omega, \mathcal{F})$. This definition includes the case of probabilistic sequential processes defined in §2, provided that they have an initial state (this later technical restriction could be easily removed).

We immediately derive the notion of **likelihood** of a configuration v: the likelihood $p(v)$, associated with probability $\mathbb{P}$, is the probability of configuration v to occur in an execution of the system, and is defined by:

$$p(v) = \mathbb{P}\big(\mathcal{S}(v)\big).$$

Probabilistic Future. Markov nets. From this definition of a probabilistic concurrent process, defining the probabilistic future is straightforward. Notice however the slight difference with the sequential case, where all futures are defined on the same measurable space. This could be done as well, but the following definition is more suitable.

Let v be a finite configuration over a safe petri net $\mathcal{N}$, and assume that v has positive likelihood. Then the shadow $\mathcal{S}(v)$ is naturally equipped with the conditional probability $\mathbb{P}(\cdot \mid \mathcal{S}(v))$. It is straightforward to show (Cf. for instance [2–Ch. 3]) that the shadow $\mathcal{S}(v)$ is isomorphic, as a measurable space, to the space of maximal configurations of the unfolding $\mathcal{U}^{m(v)}$ of net $\mathcal{N}^{m(v)}$. Denoting by $\Omega^{m(v)}$ the set of maximal configurations of $\mathcal{U}^{m(v)}$, the isomorphism $\phi_v : \mathcal{S}(v) \to \Omega^{m(v)}$ is given by:

$$\forall \omega \in \mathcal{S}(v), \quad \phi_v(\omega) = (\omega \setminus v) \cup \gamma(v). \tag{12}$$

Combined with the conditional probability $\mathbb{P}(\cdot \mid \mathcal{S}(v))$, ϕ_v is the key to define a probability $\mathbb{P}^v$ on $\Omega^{m(v)}$, by setting (remark the analogy with the function ρ_v defined in (5) for sequential systems):

$$\forall A \in \mathcal{F}^{m(v)}, \quad \mathbb{P}^v(A) = \mathbb{P}\big(\phi_v^{-1}(A) \mid \mathcal{S}(v)\big),$$

where $\mathcal{F}^{m(v)}$ is the σ-algebra on $\Omega^{m(v)}$. We define then the **probabilistic future** of finite configuration v as the probabilistic net $(\mathcal{N}^{m(v)}, \mathbb{P}^v)$. Following our reformulation (6) of Markov chains, we introduce the following definition of Markov nets.

Definition 1 (Markov net). *Let $(\mathcal{N}, \mathbb{P})$ be a probabilistic net. We say that $(\mathcal{N}, \mathbb{P})$ is a* **Markov net** *if, for every pair (v, v') of finite configurations with positive likelihoods, the following holds:*

$$m(v) = m(v') \Rightarrow \mathbb{P}^v = \mathbb{P}^{v'}. \tag{13}$$

As for Markov chains, and from Eq. (13), it makes sense for a Markov net $(\mathcal{N}, \mathbb{P})$ and for a reachable marking m, to use the notations:

$$\mathbb{P}^m, \quad \mathbb{E}^m, \tag{14}$$

to respectively denote the probability $\mathbb{P}^v$ and the expectation $\mathbb{E}^v$ for any finite configuration v with positive likelihood, and such that $m = m(v)$, provided that such a v exists.

Table 1 summarizes and compares the definitions introduced so far, and emphasizes the symmetry between sequential and concurrent systems.

Table 1. *Comparison of sequential and concurrent probabilistic processes*

	Sequential systems	Concurrent systems
Finite machine	Finite state space S	Finite safe Petri net $\mathcal{N}$
State	Element of S	Marking of $\mathcal{N}$
Finite process	Finite sequence of states	Finite configuration
State reached by a finite process v	State $s(v)$	Marking $m(v)$
Space Ω	$\Omega = \{\text{Infinite sequences}\}$	$\Omega = \{\text{Maximal configurations of } \mathcal{U}\}$
Probabilistic system	$(S, \mathbb{P})$	$(\mathcal{N}, \mathbb{P})$
Probabilistic future of finite process v	$(S, \mathbb{P}^v)$	$(\mathcal{N}^{m(v)}, \mathbb{P}^v)$
Markovian system	$s(v) = s(v') \Rightarrow \mathbb{P}^v = \mathbb{P}^{v'}$	$m(v) = m(v') \Rightarrow \mathbb{P}^v = \mathbb{P}^{v'}$

Example 1. Although we do not provide in this paper a general construction for Markov nets (see [2, 3] for this topic), let us indicate an example. Consider the

Petri net depicted in Fig. 1 (top), some pages forward. Let v be the configuration depicted at bottom, that we write $v = (acbdbe)$ (different interleavings make this writing non unique). The *choices* involved in this configurations are the following: 1) the first choice between a and d, which gives a, 2) the first choice between b and b', which gives b, 3) the second choice between a and d, which gives d, and 4) the second choice between b and b', which gives b. Remark that, due to the true-concurrency semantics, we do not have to answer a question like: "What about the speed of the token coming from C to D? Does it influence the choice between a and d?" This simply has no meaning in the true-concurrency semantics. Hence the configuration v is the successive arrival of 4 choices. Although the net presents concurrent events, there is never concurrent choices—found in more sophisticated examples. We fix two probabilistic parameters q_1 and q_2, with q_1 the probability of firing a versus d on the one hand, and q_2 the probability of firing b versus b' on the other hand. Then we set the likelihood $p(v)$ by:

$$p(v) = q_1 \times q_2 \times (1 - q_1) \times q_2 .$$

This construction of the likelihood function p could have been done for any finite configuration. By a measure-theoretic extension argument, we conclude that there is a unique probability $\mathbb{P}$ on Ω with the likelihood p. Since each time we encounter a choice, we always use the *same* probabilistic parameter, it is intuitively clear (and can be shown rigorously) that the probabilistic net so constructed is indeed Markovian.

Stopping Operators. Still following and adapting the theory of sequential probabilistic processes, we wish to establish a Strong Markov property. For this, we need to formulate an adequate definition of stopping times for the concurrency framework. By an adequate definition, we mean a definition that:

1. reduces to usual stopping times if the concurrent system is actually a sequential system,
2. is general enough to deal in particular with "hitting times" in the concurrent framework,
3. is not too much general, so that the Markov property still holds.

It should be noticed that requirement 1 is not enough. For example, stopping times of [4] satisfy this requirement, but they are not compliant with the second requirement. We have thus proposed in [2] an other extension of stopping times to concurrent systems, that is reproduced below.

Intuitively, for a sequential probabilistic process, a stopping time T (recall that T is a integer random variable satisfying Eq. (9)) is set up in order to evaluate state X_T, or, equivalently, the finite process $(X_0, \ldots, X_T)$. Hence the "abstract" order $\{0, \ldots, T\}$ is lifted into the "concrete" order $\{X_0, \ldots, X_T\}$. The sequential framework takes benefit from the fact that all the "concrete" orders $\{X_0, \ldots, X_n\}$ corresponding to stopped executions of the system, can be abstractly seen as embedded in a same total order, the canonical chain of integers. This must be revised for concurrent systems, since different executions, supported by different partial orders, cannot be superimposed anymore. This suggests to forget about the abstract universal order, and to only retain the concrete orders. Whence the following definition.

Definition 2 (stopping operator). *Let $(\mathcal{N}, \mathbb{P})$ be a probabilistic safe Petri net, and denote by $\mathcal{W}$ the poset of configurations of the unfolding of $\mathcal{N}$. We say that a random variable $V : \Omega \to \mathcal{W}$ is a* **stopping operator** *if V satisfies the two following properties:*

1. *$\forall \omega \in \Omega : \quad V(\omega) \subseteq \omega \quad (V(\omega)$ is a prefix of $\omega)$,*
2. *$\forall \omega, \omega' \in \Omega : \quad \omega' \supseteq V(\omega) \Rightarrow V(\omega') = V(\omega)$.*

We associate with V the σ-algebra $\mathcal{F}_V$, defined by:

$$\forall A \in \mathcal{F}, \quad A \in \mathcal{F}_V \iff \forall \omega, \omega' \in \Omega, \quad \omega \in A, \omega' \supseteq V(\omega) \Rightarrow \omega' \in A. \quad (15)$$

Point 1 of Def. 2 derives from the above discussion. The signification of Point 2 will be clear when discussing below the case of renewal operators, as a generalization of hitting times introduced in §2 (cf. §3, Example 2). Our stopping operators include in particular stopping times from [4]. If the safe Petri net considered actually simulates a sequential system, it is readily checked (Cf. [2–Ch. 5, Prop. II-4.7]) that there is a one-to-one association between stopping times in the classical sense, and stopping operators of Def. 5. The association is defined as follows—note the coherence with the above discussion:

$$\text{for } T \text{ a stopping time, set:} \quad V_T = (X_0, \ldots, X_T),$$

$$\text{for } V \text{ a stopping operator, set } T_V \text{ such that:} \quad V = (X_0, \ldots, X_{T_V}).$$

Moreover, the associated σ-algebras from Eq. (15) for V_T and from Eq. (10) for T_V coincide.

Shift Operators and the Concurrent Markov Property. In order to set up a strong Markov property for concurrent systems, we need to adapt the notion of shift operators. If T is a stopping time defined for a sequential system, the shift operator θ_T is defined by "iterating T times θ", where $\theta : \Omega \to \Omega$ is the canonical shift operator, defined by Eq. (7). In the absence of a canonical shift operator for concurrent systems, we can still define shift operators adapted to stopping operators, as we detail next.

The following definition is based on a simple observation. Consider a safe Petri net $\mathcal{N}$ with associated object $\mathcal{U}, \Omega$, etc. Let v be a finite configuration of $\mathcal{U}$. Recall that $\Omega^{m(v)}$ denotes the space of maximal configurations of $\mathcal{U}^{m(v)}$, and that we have at our disposal the isomorphism of measurable spaces $\phi_v : \mathcal{S}(v) \to \Omega^{m(v)}$, defined by Eq. (12). In particular, if V is a stopping operator, it follows from Point 1 in Def. 2 that we have $\omega \in \mathcal{S}(V(\omega))$, and thus $\phi_{V(\omega)}(\omega)$ is well defined if $V(\omega)$ is finite.

Definition 3 (shift operator). *For V a stopping operator, the* **shift operator** *θ_V associated with V is the mapping defined by:*

$$\forall \omega \in \Omega, \quad \theta_V(\omega) = \phi_{V(\omega)}(\omega),$$

if $V(\omega)$ is finite, $\theta_V(\omega)$ is undefined otherwise.

In particular, remark that we always have, if $V(\omega)$ is finite:

$$\theta_V(\omega) \in \Omega^{m(V(\omega))} . \tag{16}$$

It seems that we now have all ingredients to formulate the Strong Markov property: stopping operators, their associated σ-algebras and shift operators. A last item is still needed, however. In the usual Markov property (11), real-valued random variable h is to be seen as a test function. Remark that, in Eq. (11), because of the action of the shift θ_T, $h : \Omega \to \mathbb{R}$ also acts on the probabilistic futures of configurations. For concurrent systems, the unfolding formalism makes it more convenient to consider that futures starting from different makings have different sample spaces Ω's. We are thus prompted to introduce the following definition of *test functions* for concurrent systems.

Definition 4 (test functions). *Let $\mathcal{N}$ be a safe Petri net, and let $\mathcal{M}$ denote the set of reachable markings of $\mathcal{N}$. We define a **test function** as a finite collection $h = (h_m)_{m \in \mathcal{M}}$, where $h_m : \Omega^m \to \mathbb{R}$ is a real-valued measurable function for each $m \in \mathcal{M}$.*

We say that test function $h = (h_m)_{m \in \mathcal{M}}$ is nonnegative if every h_m is nonnegative, for m ranging over $\mathcal{M}$.

The Strong Markov property for concurrent systems takes then the following form. Recall the notion $\mathbb{E}^m$ from Eq. (14).

Theorem 1 (Concurrent Markov Property). *Let $(\mathcal{N}, \mathbb{P})$ be a Markov net. The following identity holds for every stopping operator V and for every nonnegative test function $h = (h_m)_{m \in \mathcal{M}}$:*

$$\mathbb{E}(h \circ \theta_V \mid \mathcal{F}_V) = \mathbb{E}^{m \circ V}(h_{m \circ V}) , \tag{17}$$

where, by convention, both members vanish on $\{\omega \in \Omega : V(\omega) \text{ is not finite}\}$. The right member of Eq. (17) must be understood as the composition of the mappings $\omega \mapsto m(V(\omega))$ and $m \in \mathcal{M} \mapsto \mathbb{E}^m(h_m)$, whereas the notation $h \circ \theta_V$ stands for the real-valued random variable defined on $\{V \text{ is finite}\}$ by:

$$h \circ \theta_V(\omega) = h_{m(V(\omega))}(\theta_V(\omega)) ,$$

which is well defined according to Eq. (16).

The proof of Th. 1 is found in [2–Ch. 5]. The remaining of the paper is devoted to illustrate how the Concurrent Markov property can be applied to derive results on the dynamics of Markov nets.

4   Global and Local Renewals for Markov Nets

This section is devoted to the application of the Concurrent Markov property to the renewal properties of Markov nets. As we shall see, we can derive that the initial marking of a Markov net has probability either 0 or 1 to return infinitely often—a precise definition of the return of the initial marking is given below—, which is a generalization of a well known result for Markov chains. With this result, we demonstrate that the formalism introduced above successfully overcomes the absence of a global clock in probabilistic Petri nets. But we do not make use of the *concurrency* properties of the models. Nevertheless, it is already a first interesting result, showing that we are not helpless in the framework of probabilistic concurrent systems. Finer results, that make a specific use of the concurrency properties of Petri nets, are discussed at the end of the section.

Global Renewal Operator. We define a stopping operator, called renewal operator, that gives in some sense the *first return* of the initial marking. We first recall an easy and well-known result, that makes an essential use of the safeness of the net. Recall that, for v a finite configuration of $\mathcal{U}$, $\gamma(v)$ denotes the set of terminal conditions of v.

Lemma 1. *Let $\mathcal{U}$ be the unfolding of a safe Petri net $\mathcal{N}$.*

1. *Let v, v' be two finite and compatible configurations of $\mathcal{U}$. The following formula holds, where $\mathrm{Min}_{\preceq}(A)$ denotes the set of $\preceq$-minimal nodes of a subset $A \subseteq \mathcal{U}$:*
$$\gamma(v \cap v') = \mathrm{Min}_{\preceq}\big(\gamma(v) \cup \gamma(v')\big).$$

2. *Let M be a marking of $\mathcal{N}$, and let u be a configuration of $\mathcal{U}$. Denote by $\mathcal{W}_0$ the set of finite configurations of $\mathcal{U}$, and set:*
$$C(u) = \{v \in \mathcal{W}_0 \ : \ v \subseteq u, \quad m(v) = M\}.$$

   Then $C(u)$ is a lattice.

Recall that c_0 denotes the initial cut of $\mathcal{U}$, and that M_0 denotes the initial marking of $\mathcal{N}$. Keep the notation $\mathcal{W}_0$ from Lemma 1 to denote the set of finite configurations of $\mathcal{U}$, and set:
$$\forall \omega \in \Omega, \quad D(\omega) = \{v \in \mathcal{W}_0 \ : \ v \subseteq \omega, \ m(v) = M_0, \ \gamma(v) \cap c_0 = \emptyset\}.$$

It follows from Lemma 1 that $D(\omega)$ is stable under finite intersections. Thus, if non empty, $D(\omega)$ admits a unique minimal element, that belongs to $D(\omega)$, whence the following definition.

Definition 5 (Global renewal operator). *Let $\mathcal{W}$ denote the set of configurations of unfolding $\mathcal{U}$. We define the mapping $R : \Omega \to \mathcal{W}$ as follows:*
$$R(\omega) = \begin{cases} \min\big(D(\omega)\big), & \text{if } D(\omega) \neq \emptyset, \\ \omega, & \text{otherwise.} \end{cases}$$

*R is called the **global renewal** operator of $\mathcal{N}$, or **renewal** operator for short.*

Intuitively, for each ω, $R(\omega)$ is the smallest sub-configuration of ω that returns back to the initial marking, *making all the tokens move in $\mathcal{N}$*, if such configuration exists. It must be compared in the sequential framework with the hitting times introduced in §2.

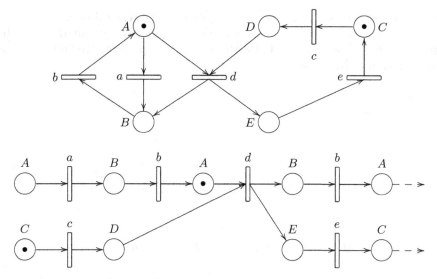

Fig. 1. Top, a safe Petri net $\mathcal{N}$. Bottom, a partial execution of net $\mathcal{N}$ that illustrates the renewal operator

Example 2. Consider the net $\mathcal{N}$ depicted at top of Fig. 1, and any maximal configuration ω that contains the configuration depicted at bottom of Fig. 1. Let p be the prefix of this configuration with events labeled by (ab), and ending with the conditions indicated by the tokens in Fig. 1. Then p does *not* constitute the renewal $R(\omega)$ since the token in C has not moved, although the marking reached by p is indeed the initial marking. The renewal $R(\omega)$ is given instead by the complete configuration depicted, and containing the events labeled by $(abcdbe)$. It is intuitively clear that the renewal $R(\omega)$ is the same for any ω containing this configuration: this is precisely the meaning of Point 2 in the definition of stopping operators (Def. 2). Whence the following lemma.

Lemma 2. *The renewal operator R is a stopping operator.*

Proof. We first check that, by construction, $D(\omega) \subseteq \omega$ for all $\omega \in \Omega$. It remains thus to check point 2 of Def. 2, i.e.:

$$\forall \omega, \omega' \in \Omega : \quad \omega' \supseteq R(\omega) \Rightarrow R(\omega') = R(\omega). \tag{18}$$

Let $\omega, \omega' \in \Omega$, and assume that $\omega' \supseteq R(\omega)$. According to Def. 5, we have to distinguish two cases:

First case: $D(\omega) = \emptyset$. Then $R(\omega) = \omega$, and thus $\omega' \supseteq \omega$. Since ω is maximal, it implies $\omega' = \omega$, and thus $R(\omega) = R(\omega')$.

Second case: $D(\omega) \neq \emptyset$. Set $v = R(\omega)$. Then, since $\omega' \supseteq v$, we have $D(\omega') \neq \emptyset$ and $v \in D(\omega')$. By minimality, it implies that $R(\omega') \subseteq v$. Symmetrically, we find that $R(\omega) \subseteq R(\omega')$ and thus finally: $R(\omega) = R(\omega')$.

Eq. (18) is satisfied in both cases: the proof of Lemma 2 is complete. $\qquad\square$

The Successive Renewal Operators. Having defined the renewal operator, we are brought to iterate the renewal process. This is achieved without difficulty by using the shift operator associated with the renewal operator. We first make the following simple observation.

Lemma 3. *Denote by θ_R the shift operator associated with the renewal operator R. Then we have:*

$$\forall \omega \in \Omega, \quad R(\omega) \notin \Omega \Rightarrow \theta_R(\omega) \in \Omega.$$

Proof. Let $\omega \in \Omega$, and assume that $R(\omega) \notin \Omega$. From Def. 5, it follows that $D(\omega) \neq \emptyset$ and thus $R(\omega)$ is finite. According to Eq. (16), it implies:

$$\theta_R(\omega) \in \Omega^{m(R(\omega))},$$

but $m\big(R(\omega)\big) = M_0$ by construction, and $\Omega^{M_0} = \Omega$, hence finally: $\theta_R(\omega) \in \Omega$, what was to be shown. $\qquad\square$

Consider then the following inductive construction. Start from an element $\omega \in \Omega$ such that $R(\omega) \notin \Omega$. Then $\theta_R(\omega)$ represents the *tail* of ω, after having subtracted the beginning $R(\omega)$. Since $\theta_R(\omega) \in \Omega$, according to Lemma 3 above, we can apply again the renewal operator to $\theta_R(\omega)$, to obtain the element $R \circ \theta_R(\omega)$. Since $R(\omega)$ *ends* with marking M_0, whereas $R \circ \theta_R(\omega)$ *begins* with marking M_0, we can form their catenation in the unfolding $\mathcal{U}$, that we denote by:

$$R(\omega) \oplus R \circ \theta_R(\omega),$$

and that corresponds indeed to the catenation of any pair of linearization sequences of configurations $R(\omega)$ and $R \circ \theta_R(\omega)$.

Continuing this inductive construction, we are brought to state the following generic formula, illustrated by Fig. 2.

$$S_1 = R, \quad S_{n+1} = S_n \oplus R \circ \theta_{S_n}.$$

A more precise definition is as follows.

Definition 6. *Denote by c_0 the initial cut of $\mathcal{U}$, and denote by $\mathcal{W}$ the set of configurations of $\mathcal{U}$. We define the **successive renewal operators** as the sequence of mappings $S_n : \Omega \to \mathcal{W}$, given by:*

$$S_0 = c_0, \quad S_{n+1}(\omega) = \begin{cases} \omega, & \text{if } S_n(\omega) \in \Omega, \\ S_n(\omega) \oplus R \circ \theta_{S_n}(\omega), & \text{if } S_n(\omega) \notin \Omega. \end{cases}$$

Remark that we have $S_1 = R$. Generalizing Lemma 2, the following result holds.

Lemma 4. *For each integer $n \geq 0$, S_n is a stopping operator.*

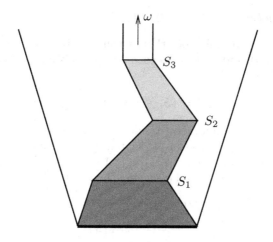

Fig. 2. *The successive renewal operators applied to an element ω*

Proof. We proceed by induction on integer n. The case $n = 0$ is trivial. Assume that S_n is a stopping operator for some integer $n \geq 0$. By construction, $S_{n+1}(\omega) \subseteq \omega$, so it remains to check Point 2 of Def. 2 applied to S_{n+1}. Let $\omega, \omega' \in \Omega$, and assume that $\omega' \supseteq S_{n+1}(\omega)$. Without loss of generality, we can assume that $S_n(\omega) \notin \Omega$, otherwise $S_{n+1}(\omega') = \omega = S_{n+1}(\omega)$ and we are done. Then $\omega' \supseteq S_n(\omega)$, and since S_n is a stopping operator according to the induction hypothesis, it implies:

$$S_n(\omega') = S_n(\omega). \tag{19}$$

It follows that $\theta_{S_n}(\omega') \supseteq R \circ \theta_{S_n}(\omega)$. But since R is a stopping operator according to Lemma 2, and since $\theta_{S_n}(\omega')$ and $\theta_{S_n}(\omega)$ are two elements of Ω by Lemma 3, we obtain that $R \circ \theta_{S_n}(\omega) = R \circ \theta_{S_n}(\omega')$. Together with Eq. (19) and Def. 6, it implies that $S_{n+1}(\omega) = S_{n+1}(\omega')$, which completes the proof. □

Recurrent Nets. Now that we have defined the successive renewal operators, that are the successive first returns to the initial marking—in the sense of Def. 5—the natural question that arises is: will the successive renewal operators actually define *non trivial* renewals? Indeed, operator S_n falls into the trivial value $S_n(\omega) = \omega$ as soon as there is no n^{th} return to the initial marking in execution ω. This suggests the following definition.

Definition 7. *We say that execution $\omega \in \Omega$ has no trivial renewal if:*

$$\forall n \geq 1, \quad S_n(\omega) \notin \Omega.$$

Theorem 2. *Let $(\mathcal{N}, \mathbb{P})$ be a Markov net. Then there are only two possibilities:*

1. *Elements $\omega \in \Omega$ have probability 1 to have no trivial renewal.*
2. *Elements $\omega \in \Omega$ have probability 0 to have no trivial renewal.*

Case 1 holds if and only if $\mathbb{P}(R \notin \Omega) = 1$.

Proof. We follow the formal proof that is usually given for sequential systems (see for instance [11]), and that works with our formalism for concurrent systems. From Def. 6, we have:

$$\{S_{n+1} \notin \Omega\} = \{S_n \notin \Omega\} \cap \{R \circ \theta_{S_n} \notin \Omega\}. \tag{20}$$

Consider for $j \geq 0$ the nonnegative random variables $h_j = \mathbf{1}_{\{S_j \notin \Omega\}}$. Let $n \geq 0$ and let $\mathcal{G} = \mathcal{F}_{S_n}$ be the σ-algebra associated with stopping operator S_n as in Def. 2. Eq. (20) can be written as: $h_{n+1} = h_n \mathbf{1}_{\{R \circ \theta_{S_n} \notin \Omega\}}$. Applying successively properties (2) and (3) to h_{n+1} and σ-algebra $\mathcal{G}$, we get, since h_n is $\mathcal{G}$-measurable:

$$\mathbb{P}(S_{n+1} \notin \Omega) = \mathbb{E}(h_{n+1})$$
$$= \mathbb{E}\big(\mathbb{E}(h_{n+1} \,|\, \mathcal{G})\big)$$
$$= \mathbb{E}\big(h_n \mathbb{E}(\mathbf{1}_{\{R \circ \theta_{S_n} \notin \Omega\}} \,|\, \mathcal{G})\big). \tag{21}$$

By the Concurrent Markov property (Th. 1) applied to stopping operator S_n and to any test function (Def. 4) that extends the nonnegative function $\mathbf{1}_{\{R \circ \theta_{S_n} \notin \Omega\}}$, we have:

$$\mathbb{E}(\mathbf{1}_{\{R \circ \theta_{S_n} \notin \Omega\}} \,|\, \mathcal{G}) = \mathbb{E}^{m(S_n)}(\mathbf{1}_{\{R \notin \Omega\}}) = \mathbb{P}(R \notin \Omega),$$

the later equality since $m(S_n) = M_0$ on $\{S_n \notin \Omega\}$. Setting $a = \mathbb{P}(R \notin \Omega)$, and using Eq. (21), we get:

$$\mathbb{P}(S_{n+1} \notin \Omega) = a\mathbb{E}(h_n) = a\mathbb{P}(S_n \notin \Omega). \tag{22}$$

We distinguish thus two cases. If $a = 1$, then $\mathbb{P}(S_n \notin \Omega) = 1$ for all $n \geq 1$, which implies:

$$\mathbb{P}\big(\bigcap_{n \geq 1}\{S_n \notin \Omega\}\big) = 1,$$

this is case 1 in Th. 2. Otherwise we have $a < 1$, and then, from Eq. (22):

$$\sum_{n \geq 1} \mathbb{P}(S_n \notin \Omega) < \infty.$$

By the Borel-Cantelli Lemma (Cf. for instance [6]), it implies that $\{S_n \notin \Omega\}$ has probability 0 to occur for infinitely many integers n, which is case 2 of Th. 2. □

Comment on Th. 2. In case 1 of Th. 2, we will say that $(\mathcal{N}, \mathbb{P})$ is **recurrent**. Then recurrent nets reduce to recurrent Markov chains in case of a Markov net that reduces to a sequential system, and thus simulates a Markov chain. It follows that Th. 2 extends a well-known result on Markov chains, where "marking" must be replaced by "state".

A practical recurrence criterion is given by the following result. It is convenient to say that a marking m is **$\mathbb{P}$-reachable** if there is a finite configuration v such that $p(v) > 0$ and $m = m(v)$. $\gamma(v)$ denotes as usual the set of terminal conditions of v.

Proposition 1. *A Markov net* $(\mathcal{N}, \mathbb{P})$ *is recurrent if and only if:*

1. *there is a configuration v such that $p(v) > 0$ and $c_0 \cap \gamma(v) = \emptyset$, and*
2. *for every $\mathbb{P}$-reachable marking m, M_0 is $\mathbb{P}^m$-reachable.*

Example 3. Consider again the net depicted in Fig. 1. We have seen in §3, Example 1, that this net can be made a Markov net, by using two probabilistic parameters q_1 and q_2. These parameters correspond respectively to the probabilities of "local choices" between a and d on the one hand, and between b and b' on the other hand. Assume that both parameters are non-degenerated, i.e. $p, q \notin \{0, 1\}$. Then, using Prop. 1, one sees that net $\mathcal{N}$ is recurrent.

The probabilistic framework is well adapted to state the recurrence properties of $\mathcal{N}$. Indeed, although $\mathcal{N}$ is recurrent, there exists executions $\omega \in \Omega$ with trivial renewal (i.e., that return only finitely many times to the initial marking), for instance $\omega = (c\,ab\,ab\,ab\ldots)$. But, as stated by Th. 2, these executions are "rare": all together, they have probability zero.

Example 4. Consider the net depicted in Fig. 3. An analysis of configurations similar to the one explained about the previous example can also be done. As for the previous example, we derive from this analysis the construction of a Markov net from some local probabilistic parameters. One of these parameters is the probability of firing b versus a, say q_0. As soon as $q_0 > 0$, the net is non-recurrent. If moreover $q_0 < 1$, the random number of renewals, say N, has the geometric law of parameter $(1 - q_0)$, so that $\mathbb{P}(N = n) = (1 - q_0)q_0^n$ for $n \geq 0$. Remark that the law of N only depends on the *local* probabilistic parameter that concerns the transitions a and b.

Local Renewal for Nets. So far we have shown that our formalism allows to free the probabilistic framework from any global clock, but still keeping qualitative results on the probabilistic behavior of systems. One can argue that the results presented above do not take benefit from the concurrency *properties* of the model, they only deal with the *problems* brought by concurrency! Yes... It is

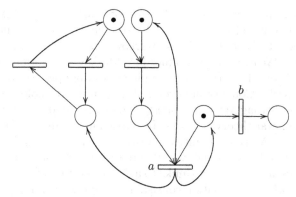

Fig. 3. *A non recurrent net*

however possible to obtain positive results due to concurrency, in particular with the notion of *local renewal*. This new topic is a refinement of the global renewal presented above. The techniques for dealing with both renewals are the same: the basic tool is still the Concurrent Strong Markov property. We introduce local renewal on an example.

Consider again the recurrent net $\mathcal{N}$ depicted in Fig. 1. Extending the discussion of §3, Example 1, the dynamics of $\mathcal{N}$ can be seen as a partially ordered succession of *local choices*. For instance, the component $E \to e \to C$ can be seen as a trivial choice, but still as a choice. As an other example, the choices made between b and b' are non trivial local choices. The *local renewal* is concerned by the arrivals of local choices. Consider for instance the local choices made between b and b'. As a consequence of the safeness of $\mathcal{N}$, the different arrivals of these choices, inside a same execution ω, are *totally ordered*. The sub-configurations of ω that lead to the successive local choices $\{b, b'\}$ constitute the *successive local renewals* associated to the choice $\{b, b'\}$. As for the successive global renewals, it is shown that local renewals are stopping operators.

Local renewal has the following properties: the finiteness of the global renewal guarantees the finiteness of the local renewal—hence iterate local renewals are well defined for recurrent nets. If the net simulates a sequential system, global and local renewals coincide. Finally, if we extend the construction of Markov nets detailed in §3, Example 1, obtained from the decomposition of configurations through local choices, the local choices are precisely determined by the successive local renewals: the local renewals are the random configurations that lead to the local choices. Moreover, the local decisions performed by the net, and associated to the successive occurrences of the same local choice, constitute a sequence of i.i.d. random variables. This quite intuitive result does not hold for general Markov nets. It is the basis for instance of a local performance evaluation, as well as a basic tool for a statistical estimation of parameters.

5   Conclusion and Perspectives

This paper has introduced a new definition of Markov nets. Markov nets are a special case of probabilistic nets, which are true-concurrent random systems based on the model of safe Petri nets. Markov nets are defined as memory-less probabilistic nets. We have also introduced for Markov nets notions adapted from Markov chains theory. In particular, stopping operators play the same role than stopping times, and a Concurrent Strong Markov property holds. The efficiency of the Concurrent Strong Markov property is demonstrated by establishing some elements of a renewal theory for nets. In particular, recurrent nets extend the notion of recurrent Markov chains. Interesting enough, nets have tow kinds of renewal: a global and a local one, whereas both coincide for Markov chains.

For further research, one first thinks to a decomposition of nets into recurrent components. Another continuation consists in studying the relationship between stochastic Petri nets and probabilistic Petri nets. In particular, can a probabilistic Petri net be seen as the "uniformization" of a stochastic Petri net, generalizing the relationship between discrete and continuous time Markov chains? Finally, we currently work on a compositional theory for Markov nets. Indeed, it is well known that synchronization of sequential systems leads to concurrent systems—for example, synchronizing transition systems bring safe Petri nets. A probabilistic interpretation of the synchronization could furnish elements for a validation *a posteriori* of our results.

Acknowledgments. This work would not have been possible without the ideas and the help of A. Benveniste, S. Haar and E. Fabre. I wish to thank them here.

References

1. *Proc. of the 16th Conference on Mathematical Theory of Networks and Systems (MTNS 2004)*, Leuven, Belgium, 2004. ISBN 90-5682-517-8.
2. S. Abbes. *Probabilistic model of distributed and concurrent systems. Limit theorems and Applications.* PhD thesis, Université de Rennes 1, IRISA. Available from `ftp://ftp.irisa.fr/techreports/theses/2004/abbes.pdf`.
3. S. Abbes and A. Benveniste. Branching cells as local states for event structures and nets: probabilistic applications. In V. Sassone, editor, *Proc. of FOSSACS 05*, volume 3441 of *LNCS*, pages 95–109, 2005. Extended version available as Research Report INRIA RR-5347, `http://www.inria.fr/rrrt/rr-5347.html`.
4. A. Benveniste, S. Haar, and E. Fabre. Markov nets: probabilistic models for distributed and concurrent systems. *IEEE Trans. on Aut. Control*, 48(11):1936–1950, Nov. 2003.
5. A. Benvensite, E. Fabre, S. Haar, and C. Jard. Distributed monitoring of concurrent and asynchronous systems. In *Proc. of CONCUR 03*, volume 2761 of *LNCS*, pages 1–26, 2003.
6. L. Breiman. *Probability*. SIAM, 1992.
7. J. Engelfriet. Branching processes of Petri nets. *Acta Informatica*, 28:575–591, 1991.
8. C. Girault and R. Valk. *Petri nets for systems engeneering*. Springer, 2003.
9. M. Nielsen, G. Plotkin, and G. Winskel. Petri nets, event structures and domains, part 1. *T.C.S.*, 13:86–108, 1980.
10. M. O. Rabin. Probabilistic automata. *Inf. and Control*, 6(3):230–245, 1963.
11. D. Revuz. *Markov chains*. Number 11 in N.-H. Math. Library. North-Holland, 1975.
12. D. Varacca, H. Völzer, and G. Winskel. Probabilistic event structures and domains. In *Proc. of CONCUR 04*, number 3170 in LNCS, pages 481–496, 2004.
13. H. Völzer. Randomized non-sequential processes. In *Proc. of CONCUR 01*, volume 2154 of *LNCS*, pages 184–201, 2001.

On the Equivalence Between Liveness and Deadlock-Freeness in Petri Nets

Kamel Barkaoui[1], Jean-Michel Couvreur[2], and Kais Klai[2]

[1] CEDRIC-CNAM - Paris, France
barkaoui@cnam.fr
[2] LaBRI - Universit de Bordeaux, France
{couvreur, kais.klai}@labri.fr

Abstract. This paper deals with the structure theory of Petri nets. We define the class of P/T systems namely K-systems for which the equivalence between controlled-siphon property (cs property), deadlock freeness, and liveness holds. Using the new structural notions of ordered transitions and root places, we revisit the non liveness characterization of P/T systems satisfying the cs property and we define by syntactical manner new and more expressive subclasses of K-systems where the interplay between conflict and synchronization is relaxed.

Keywords: structure theory, liveness, deadlock-freeness, cs-property.

1 Introduction

Place / Transition (P/T) systems are a mathematical tool well suited for the modelling and analyzing systems exhibiting behaviours such as concurrency, conflict and causal dependency among events. The use of structural methods for the analysis of such systems presents two major advantages with respect to other approaches: the state explosion problem inherent to concurrent systems is avoided, and the investigation of the relationship between the behaviour and the structure (the graph theoretic and linear algebraic objects and properties associated with the net and initial marking) usually leads to a deep understanding of the system. Here we deal with liveness of a marking, i.e. , the fact that every transition can be enabled again and again. It is well known that this behavioural property is as important as formally hard to treat. Although some structural techniques can be applied to general nets, the most satisfactory results are obtained when the interplay between conflicts and synchronization is limited. An important theoretical result is the controlled siphon property[3]. Indeed this property is a condition which is necessary for liveness and sufficient for deadlock-freeness. The aim of this work is to define and recognize structurally a class of P/T systems, as large as possible, for which the equivalence between liveness and deadlock freeness holds. In order to reach such a goal, a deeper understanding of the causes of the non equivalence between liveness and deadlock-freeness is required.

This paper is organized as follows. In section 2, we recall the basic concepts and notations of P/T systems. In section 3, we first define a class of P/T systems, namely

G. Ciardo and P. Darondeau (Eds.): ICATPN 2005, LNCS 3536, pp. 90–107, 2005.
© Springer-Verlag Berlin Heidelberg 2005

K-systems, for which the equivalence between controlled-siphon property (cs property), deadlock freeness, and liveness holds. In section 4, we revisit the structural conditions for the non liveness under the cs property hypothesis. In section 5, we define by a syntactical manner several new subclasses of K-systems where the interplay between conflict and synchronization is relaxed. Such subclasses are characterized using the new structural notions of ordered transitions and root places. In section 6, we define two other subclasses of K-systems based on T-invariants. We conclude in section 5 with a summary of our results and a discussion of an open question.

2   Basic Definitions and Notations

This section contains the basic definitions and notations of Petri nets' theory [11] which will be needed in the rest of the paper.

2.1   Place/Transition Nets

Definition 1. *A P/T net is a weighted bipartite digraph $N = \langle P, T, F, V \rangle$ where:*

- *$P \neq \emptyset$ is a finite set of node places;*
- *$T \neq \emptyset$ is a finite set of node transitions;*
- *$F \subseteq (P \times T) \cup (T \times P)$ is the flow relation;*
- *$V : F \to \mathbb{N}^+$ is the weight function (valuation).*

Definition 2. *Let $N = \langle P, T, F, V \rangle$ be a P/T net.*
The preset of a node $x \in (P \cup T)$ is defined as $^\bullet x = \{y \in (P \cup T) s.t. (y, x) \in F\}$,
The postset of a node $x \in (P \cup T)$ is defined as $x^\bullet = \{y \in (P \cup T) s.t. (x, y) \in F\}$,
The preset (resp. postset) of a set of nodes is the union of the preset (resp. postset) of its elements.
The sub-net induced by a sub-set of places $P' \subseteq P$ is the net $N' = \langle P', T', F', V' \rangle$ defined as follows:

- *$T' = {}^\bullet P' \cup P'^\bullet$,*
- *$F' = F \cap ((P' \times T') \cup (T' \times P'))$,*
- *V is the restriction of V on F'.*

The sub-net induced by a sub-set of transitions $T' \subseteq T$ is defined analogously.

Definition 3. *Let $N = \langle P, T, F, V \rangle$ be a P/T net.*

- *A shared place p $(\mid p^\bullet \mid \geq 2)$ is said to be homogenous iff: $\forall t, t' \in p^\bullet$, $V(p, t) = V(p, t')$.*
- *A place $p \in P$ is said to be non-blocking iff: $p^\bullet \neq \emptyset \Rightarrow Min_{t \in {}^\bullet p}\{V(t, p)\} \geq Min_{t \in p^\bullet}\{V(p, t)\}$.*
- *If all shared places of P are homogenous, then the valuation V is said to be homogenous.*

The valuation V of a P/T net N can be extended to the application W from $(P \times T) \cup (T \times P) \to \mathbb{N}$ defined by:
$\forall u \in (P \times T) \cup (T \times P)$, $W(u) = V(u)$ if $u \in F$ and $W(u) = 0$ otherwise.

Definition 4. *The matrix C indexed by $P \times T$ and defined by $C(p, t) = W(t, p) - W(p, t)$ is called the* incidence matrix *of the net.*
An integer vector $f \neq 0$ indexed by P ($f \in \mathbf{Z}^P$) is a P-invariant *iff $f^t.C = 0^t$.*
An integer vector $g \neq 0$ indexed by T ($g \in \mathbf{Z}^T$) is a T-invariant *iff $C.g = 0$.*
$\|f\| = \{p \in P / f(p) \neq 0\}$ *(resp. $\|g\| = \{t \in t / g(t) \neq 0\}$) is called the* support of *f (resp. of g).*
We denote by $\|f\|^+ = \{p \in P / f(p) > 0\}$ and by $\|f\|^- = \{p \in P / f(p) < 0\}$.
N is said to be conservative *iff there exists a P-invariant f such that $\|f\| = \|f\|^+ = P$.*

2.2 Place/Transition Systems

Definition 5. *A* marking *M of a P/T net $N = \langle P, T, F, V \rangle$ is a mapping $M : P \to \mathbb{N}$ where $M(p)$ denotes the number of tokens contained in place p. The pair $\langle N, M_0 \rangle$ is called a P/T* system *with M_0 as initial marking.*
A transition $t \in T$ is said to be enabled *under M, in symbols $M \xrightarrow{t}$, iff $\forall p \in {}^\bullet t$: $M(p) \geq V(p, t)$. If $M \xrightarrow{t}$, the transition t may* occur, *resulting in a new marking M', in symbols $M \xrightarrow{t} M'$, with: $M'(p) = M(p) - W(p, t) + W(t, p)$, $\forall p \in P$.*
The set of all reachable markings, in symbols $R(M_0)$, is the smallest set such that $M_0 \in R(M_0)$ and $\forall M \in R(M_0)$, $t \in T$, $M \xrightarrow{t} M' \Rightarrow M' \in R(M_0)$.
If $M_0 \xrightarrow{t_1} M_1 \xrightarrow{t_2} \ldots M_{n-1} \xrightarrow{t_n}$, then $\sigma = t_1 t_2 \ldots t_n$ is called an occurrence se-quence.

In the following, we recall the definition of some basic behavioural properties.

Definition 6. *Let $\langle N, M_0 \rangle$ be a P/T system.*
A transition $t \in T$ is said to be dead *for a marking $M \in R(M_0)$ iff $\nexists M^* \in R(M)$ s.t. $M^* \xrightarrow{t}$.*
A marking $M \in R(M_0)$ is said to be a dead marking *iff $\forall t \in T$, t is dead for M.*
$\langle N, M_0 \rangle$ is weakly live *(or deadlock-free) for M_0 iff $\forall M \in R(M_0)$, $\exists t \in T$ such that $M \xrightarrow{t}$ ($\langle N, M_0 \rangle$ has no dead marking).*
A transition $t \in T$ is said to be live *for M_0 iff $\forall M \in R(M_0)$, $\exists M' \in R(M)$ such that $M' \xrightarrow{t}$ (t is not live iff $\exists M' \in R(M_0)$ for which t is dead).*
$\langle N, M_0 \rangle$ is live *for M_0 iff $\forall t \in T$, t is live for M_0.*
A place $p \in P$ is said to be marked *for $M \in R(M_0)$ iff $M(p) \geq Min_{t \in p^\bullet} \{V(p, t)\}$.*
A place $p \in P$ is said to be bounded *for M_0 iff $\exists k \in \mathbb{N}$ s.t. $\forall M \in R(M_0)$, $M(p) \leq k$. $\langle N, M_0 \rangle$ is* bounded *iff $\forall p \in P$, p is bounded for M_0.*
If N is conservative then $\langle N, M_0 \rangle$ is bounded for any initial marking M_0.

2.3 Controlled Siphon Property

A key concept of structure theory is the siphon.

Definition 7. *Let $\langle N, M_0 \rangle$ be a P/T system.*
A nonempty set $S \subseteq P$ is called a siphon *iff ${}^\bullet S \subseteq S^\bullet$. Let S be a siphon, S is called* minimal *iff it contains no other siphon as a proper subset.*

In the following, we assume that all P/T nets have *homogeneous valuation*, and $V(p)$ denotes $V(p, t)$ for a any $t \in p^\bullet$.

Definition 8. *A siphon S of a P/T system $N = \langle P, T, F, V \rangle$ is said to be controlled iff:*
S is marked at any reachable marking i.e. $\forall M \in R(M_0)$, $\exists p \in S$ s.t. p is marked.

Definition 9. *A P/T system $\langle N, M_0 \rangle$ is said to be satisfying the controlled-siphon property (cs-property) iff each minimal siphon of $\langle N, M_0 \rangle$ is controlled.*

In order to check the cs-property, two main structural conditions (*sufficient but not necessary*) permitting to determine whether a given siphon is controlled are developed in [3, 9]. These conditions are recalled below.

Proposition 1. *Let $\langle N, M_0 \rangle$ be a P/T system and S a siphon of $\langle N, M_0 \rangle$. If one of the two following conditions holds, then S is controlled:*

1 $\exists R \subseteq S$ such that $R^\bullet \subseteq {}^\bullet R$, R is marked at M_0 and places of R are non-blocking (siphon S is said to be containing a trap R).
2 $\exists$ a P-invariant $f \in \mathbf{Z}^P$ such that $S \subseteq \|f\|$ and $\forall p \in (\|f\|^- \cap S)$, $V(p) = 1$, $\|f\|^+ \subseteq S$ and $\sum_{p \in P}[f(p).M_0(p)] > \sum_{p \in S}[f(p).(V(p) - 1)]$.

A siphon controlled by the first (resp. second) mechanism is said to be trap-controlled (resp. invariant controlled).

Now, we recall two well-known basic relations between liveness and the cs-property [3]. The first states that the cs-property is a sufficient deadlock-freeness condition, the second states that the cs-property is a necessary liveness condition.

Proposition 2. *Let $\langle N, M_0 \rangle$ be a P/T system. The following property holds:*
$\langle N, M_0 \rangle$ satisfies the cs-property $\Rightarrow \langle N, M_0 \rangle$ is weakly live (deadlock-free).

Proposition 3. *Let $\langle N, M_0 \rangle$ be a P/T system. The following property holds:*
$\langle N, M_0 \rangle$ is live $\Rightarrow \langle N, M_0 \rangle$ satisfies the cs-property.

Hence, for P/T systems where the cs-property is a sufficient liveness condition, there is an equivalence between liveness and deadlock freeness. In the following section, we define such systems and propose basic notions helping for their recognition.

3   K-Systems

In this section, we first introduce a new class of P/T systems, namely *K-systems*, for which the equivalence between liveness and deadlock freeness holds. Before, let us establish some new concepts and properties related to the causality relationship among dead transitions.

Definition 10. *Let $\langle N, M_0 \rangle$ be a P/T system. A reachable marking $M^* \in R(M_0)$ is said to be stable iff $\forall t \in T$, t is either live ore dead for M^*. Hence, T is is partitioned into two subsets $T_D(M^*)$ and $T_L(M^*)$, and for which all transitions of $T_L(M^*)$ are live and all transitions of $T_D(M^*)$ are dead.*

Proposition 4. *Let $\langle N, M_0 \rangle$ be a weakly live but not live P/T system. There exists a reachable stable marking M^* for which $T_D \neq \emptyset$ and $T_L \neq \emptyset$.*

Proof. trivial, otherwise the net is live $(T = T_L)$ or not weakly live $(T = T_D)$.

Remark: This partition is not necessarily unique but there exists at least one. It is important to note that T_D is maximal in the sense that all transitions that do not belong to T_D, will never become dead.

Definition 11. *Let $N = \langle P, T, F, V \rangle$ be a P/T net, $r \in P$, $t \in r^\bullet$. r is said to a be a root place for t iff $r^\bullet \subseteq p^\bullet$, $\forall p \in {}^\bullet t$.*

An important feature of root places is highlighted in the following proposition.

Proposition 5. *Let $N = \langle P, T, F, V \rangle$ be a P/T net, $r \in P$, $t \in r^\bullet$. If r is a root place for t then $\forall t' \in r^\bullet$, ${}^\bullet t \subseteq {}^\bullet t'$.*

Proof. Let t be a transition having r as a root place and let t' be a transition in $r^\bullet$. Now, let p be a place in ${}^\bullet t$ and let as show that $p \in {}^\bullet t'$:
Since r is a root place for t and $p \in {}^\bullet t$ then we have $r^\bullet \subseteq p^\bullet$ and hence $t' \in r^\bullet$ implies that $t' \in p^\bullet$, equivalently $p \in {}^\bullet t'$.

Given a transition t, $Root(t)_N$ denotes the set of its root places in N. When the net is clear from the context, this set is simply denoted by $Root(t)$.

Definition 12. *Let t be a transition of T. If $Root(t) \neq \emptyset$, t is said to be an ordered transition iff $\forall p, q \in {}^\bullet t, p^\bullet \subseteq q^\bullet$ or $q^\bullet \subseteq p^\bullet$.*

Remark: An ordered transition has necessarily a root but one transition admitting a root is not necessarily ordered. P/T Systems where all transitions are ordered are called ordered systems. Consider the Figure 1, one can check that $Root(t_1) = \{a\}$, $Root(t_2) = \{b\}$, $Root(t_3) = \{e\}$ and $Root(t_4) = \{d\}$. Transitions t_1, t_3, t_4 are ordered but not t_2.

Proposition 6. *Let $\langle N, M_0 \rangle$ be a not live P/T system. Let r be a root of a transition t: $t \in T_D \Rightarrow r^\bullet \cap T_L = \emptyset$ (i.e. $r^\bullet \subseteq T_D$).*

Proof. As ${}^\bullet t \subset {}^\bullet t'$ for every t' of $r^\bullet$: t, dead for M, can never be enabled, a fortiori t' can not be enabled.

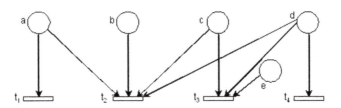

Fig. 1. Illustration: a not ordered transition

Also, we can state the following : if all input transitions of a place are dead, then all its output transitions are dead.

Proposition 7. *Let $\langle N, M_0 \rangle$ be a not live P/T system.*
Let p be a place of P: ${}^{\bullet}p \cap T_L = \emptyset \Rightarrow p^{\bullet} \cap T_L = \emptyset$.

Proof. Suppose that the proposition is not true. In this case, there exists a place p with all input transitions in T_D (${}^{\bullet}p \cap T_L = \emptyset$) and at least one output transition t_v in T_L ($p^{\bullet} \cap T_D \neq \emptyset$). Since t_v is live, after a finite number of firings, place p becomes non marked because all its input transitions are dead. So t_v becomes dead. This contradicts that $t \in T_L$ (and maximality of T_D).

Proposition 8. *Let $\langle N, M_0 \rangle$ be a not live P/T system.*
Let p be a bounded place of P: $p^{\bullet} \cap T_L = \emptyset \Rightarrow {}^{\bullet}p \cap T_L = \emptyset$.

Definition 13. *Let $\langle N, M_0 \rangle$ be a P/T system. $\langle N, M_0 \rangle$ is a K-system iff for all stable markings M^*, $T_D(M^*) = T$ or $T_L = T$. The above property is called the K-property.*

Remark:
According to the previous definition, one can say that the K-systems contain all the live systems and a subclass of not deadlock-free systems. One can then deduce the following theorem.

Theorem 1. *Let $\langle N, M_0 \rangle$ be a P/T system. $\langle N, M_0 \rangle$ is a K-system. Then the three following assertions are equivalent:*

- *(1) $\langle N, M_0 \rangle$ is deadlock free,*
- *(2) $\langle N, M_0 \rangle$ satisfies the cs-property,*
- *(3) $\langle N, M_0 \rangle$ is live.*

Proof. $\Rightarrow$ Note first that we immediately have $(3) \Rightarrow (2) \Rightarrow (1)$ using proposition 2 and 3. The proof is then reduced to show that deadlock freeness is a sufficient liveness condition for K-systems. Assume that the K-system $\langle N, M_0 \rangle$ is not live then by definition it is not deadlock free (since $T_D(M^*) = T$ for each stable marking M^*).
$\Leftarrow$ The converse consists to prove the following implication:
$((1) \Rightarrow (3)) \Rightarrow \langle N, M_0 \rangle$ is a K-system
Assume that $\langle N, M_0 \rangle$ is not a K-system. Then, by definition, there exists a stable marking $m*$ for which $T_D \neq \emptyset$ and $T_L \neq \emptyset$. Hence, $\langle N, M_0 \rangle$ is deadlock free but not live, which contradicts $((1) \Rightarrow (3))$.

The Definition13 of K-systems is a behavioural one. In the following part of this paper, we deal with the problem of recognizing, in a structural manner, the membership of a given P/T system in the class of K-systems.

4   Structural Non-liveness Characterization

In this section, we highlight some intrinsical properties of systems satisfying the cs-property but not live. Our idea is to characterize a "topological construct" making possible the simultaneous existence of dead and live transitions for such systems.

Lemma 1. *Let $\langle N, M_0 \rangle$ be a P/T system satisfying the cs-property but not live. Let M^* be a reachable stable marking. There exists $t^* \in T_D$ such that:*
$\forall p \in {}^\bullet t^*$ *such that* ${}^\bullet p \cap T_L = \emptyset$, $M(p) = M^*(p) \geq V(p, t^*) \ \forall M \in R(M^*)$.

Proof. Suppose that $\forall t \in T_D$, there exists $p_t \in {}^\bullet t$ with ${}^\bullet p \cap T_L = \emptyset$ and $M^*(p_t) < V(p_t, t^*)$. Let $S = \{p_t, t \in T_D\}$. By construction, ${}^\bullet S \subseteq T_D$ and $T_D \subseteq S^\bullet$ (for all $p_t \in S$, ${}^\bullet p_t \cap T_L = \emptyset$). So S is a siphon. Since $\forall p_t \in S$, $M^*(p_t) < V(p_t, t)$, S is non marked for M^* ($M^* \in R(M^*)$) and hence the cs-property hypothesis is denied. Using now the proposition 7, (if a place p has no live input transition then all output transitions of p are dead), one can deduce that the marking of such places does not change for all reachable markings from M^*.

Theorem 2. *Let $\langle N, M_0 \rangle$ be a P/T system satisfying the cs-property but not live. Let M^* be a reachable stable marking.*
There exists a non ordered transition $t^* \in T_D$ and $\forall M \in R(M^*)$, $\exists p \in {}^\bullet t^*$ s.t. $M(p) < V(p, t^*)$.

Proof. Let t^* be a transition satisfying the previous lemma 1. Let us denote by $L_P(t^*)$ the subset of shared places included in ${}^\bullet t^*$ and defined as follows:
$L_P(t^*) = \{p \in {}^\bullet(t^*)$ s.t. ${}^\bullet p \cap T_L \neq \emptyset$ and $p^\bullet \cap T_L \neq \emptyset\}$.
We first prove that $L_P(t^*) \neq \emptyset$ ($L_p(t^*) \subseteq {}^\bullet t^*$). Suppose that $L_P = \emptyset$: any input place of t^* having a live input transition (there exists at least one otherwise t^* will be enabled at M^* using proposition 7). As the other input places of t^* are such that their pre-conditions on t^* are satisfied at M^* and remain satisfied (proposition 7), we can reach a marking M from M^* such that $t*$ would be enabled at M. This contradicts that t^* is dead for M^*. Moreover, t^* is not ordered otherwise $L_P(t^*) = \{p_1, \ldots, p_m\}$ ($|L_P| = m$) can be linearly ordered. Without loss of generality we may assume that $p_1{}^\bullet \subseteq \cdots \subseteq p_m{}^\bullet$. Then there exists a marking M' reachable from M^* for which a transition $t \in p_1{}^\bullet \cap T_L$ and t^* are enabled (homogenous valuation). This contradicts that t^* is dead for M^*. Since $L_P(t^*)$ ($\subset {}^\bullet t^*$) has no root place we deduce that t^* is not ordered.
Finally, $\forall M \in R(M^*)$, $L_P(t^*)$ contains a non marked place otherwise t^* would be not dead for M^*.

From the previous theorem (theorem 2) one can derive easily the following result.

Theorem 3. *Let $\langle N, M_0 \rangle$ be an ordered P/T system. The two following statements are equivalent:*

- *(1) $\langle N, M_0 \rangle$ satisfies the cs-property,*
- *(2) $\langle N, M_0 \rangle$ is live.*

This last result permits us to highlight the structural and behavioural unity between subclasses of ordered P/T systems i.e. (not necessarily bounded) asymmetric choice systems [3] (AC), Join Free (JF) systems, Equal Conflict (EC) systems[13], and Extended Free Choice (EFC) nets. Let us recall that, for theses subclasses, except AC nets, the cs-property is reduced to the well-known Commoner's property [2], [1], [5], [6] and the liveness monotonicity [3]) holds.

In the following, we show how to exploit this material in order to recognize structurally other subclasses of K-systems, with non ordered transition, for which the equivalence between deadlock-freeness and liveness hold. Such structural extensions are based on the two following concepts: the notion of root places as a relaxation of the strong property of ordered transitions and the covering of non ordered transitions by invariants.

5   Dead-Closed Systems

From our better understanding of requirements which are at the heart of non equivalence between deadlock-freeness and liveness, we shall define new subclasses of K-systems for which membership problem is always reduced to examining the net without requiring any exploration of the behaviour.

Let t be a transition of a P/T system, we denote by $D(t)$ the set of transitions defined as follows: $D(t) = \{t' \in T \text{ s.t. } t \in T_D \Rightarrow t' \in T_D\}$

This set is called the *dead closure* of the transition t. In fact, $D(t)$ contains all transitions that are dead once t is assumed to be dead.

In the following, we show how one can compute structurally a subset $D_{Sub}(t)$ of $D(t)$ for any transition t.

Given a transition t_0, we set $D_{Sub}(t_0) = \{t_0\}$ and enlarge it using the three following structural rules related to propositions 6, 7, and 8 respectively:

R_1.   Let p be a root place of t, $t \in D_{Sub}(t_0) \Rightarrow p^\bullet \subseteq D_{Sub}(t_0)$
R_2.   Let p be a place of P, $^\bullet p \subseteq D_{Sub}(t_0) \Rightarrow p^\bullet \subseteq D_{Sub}(t_0)$
R_3.   Let p be a bounded place of P, $p^\bullet \subseteq D_{Sub}(t_0) \Rightarrow {}^\bullet p \subseteq D_{Sub}(t_0)$.

Formally, $D_{Sub}(t_0)$ is defined as the smallest subset of T containing t_0 and fulfilling rules R_i ($i = 1 \ldots 3$). When the computed subsets $D_{Sub}(t)$ are all equal to T, we deduce that the system is a K-system.

Definition 14. *Let $\langle N, M_0 \rangle$ be a P/T system. $\langle N, M_0 \rangle$ is said to be a dead-closed system if for every transition t of N: $D_{Sub}(t) = T$.*

The algorithm5 [4] computes the subset $D_{Sub}(t)$ for a given transition t. Its complexity is similar to classical graph traversal algorithms. An overall worst-case complexity bound is $\emptyset(|P| \times |T|)$.

Theorem 4. *Let $\langle N, M_0 \rangle$ be a dead-closed system. Then $\langle N, M_0 \rangle$ is a K-system.*

Proof. The proof is obvious since the computed set $D_{Sub}(t)$ for every transition t is a subset of $D(t)$.

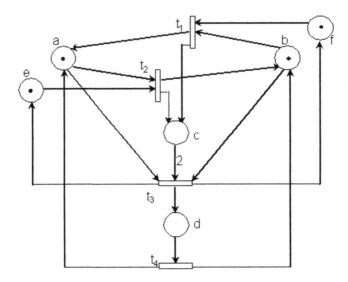

Fig. 2. An example of K system

Using theorem 1 one can deduce the following result.

Corollary 1. *Let $\langle N, M_0 \rangle$ be a dead-closed system. The three following state-ments are equivalent:*

- *(1) $\langle N, M_0 \rangle$ is deadlock free,*
- *(2) $\langle N, M_0 \rangle$ satisfies the cs-property,*
- *(3) $\langle N, M_0 \rangle$ is live.*

Consider the net of Figure2, note first that it is a conservative net. One can check, by applying the algorithm *computing $D(t)$*, that it is a dead-closed system. It contains the four following minimal siphons: $S_1 = \{a, b, d\}$, $S_2 = \{e, c, f\}$, $S_3 = \{e, b, d\}$ and $S_4 = \{a, f, d\}$. For any initial marking (e.g. $M_0 = a+b+e+f$) satisfying the four following conditions: $a+b+d > 0$, $e+c+f > 0$, $e+b+d-f > 0$ and $a + f + d - e > 0$, this net satisfies the cs-property and hence is live.

5.1   Root Systems

Here, we define a subclass of dead-closed systems called *Root Systems* exploiting in particular the causality relationships among output transitions of root places. Before, we define a class of P/T nets where each transition admits a root place, such nets are called *Root nets*.

Definition 15. *Let $N = \langle P, T, F, V \rangle$ be a P/T net. N is a root net iff $\forall t \in T$, $\exists$ a place $r \in P$ which is a root for t.*

Every transition t of N has (at least) a root place, but it is not necessarily or-dered. Thus, ordered nets are strictly included in root nets.

Algorithm 5.1 Computing $D_{Sub}(t)$

1: **Input:** a transition t; // t is assumed to be dead
2: **Output:** $D_{Sub}(t)$, a set of transitions; // $D(t)$
3: **Variable** $D_t marked$: a set of transitions//
4: **Begin**
5: $D_{Sub}(t) \leftarrow \{t\}$;
6: $D_t marked \leftarrow \emptyset$
7: **for** $(D_t marked \leftarrow \emptyset; (D_{Sub}(t) \setminus D_t marked) \neq \emptyset; D_t marked \leftarrow D_t marked \cup \{t\})$
   do
8:    get t from $D_{Sub}(t) \setminus D_t marked$;
9:    **if** r is root place **then**
10:       $D_{Sub}(t) \leftarrow D_{Sub}(t) \cup r^\bullet$; //application of R_1
11:       **for each** $(p \in t^\bullet)$ **do**
12:          **if** $(^\bullet p \subseteq D_{Sub}(t))$ **then**
13:             $D_{Sub}(t) \leftarrow D_{Sub}(t) \cup p^\bullet$; //application of R_2
14:          **end if**
15:       **end for**
16:       **for for each** $(p \in {}^\bullet t)$ **such that** $(p$ is bounded$)$ **do**
17:          **if** $p^\bullet \subseteq D_{Sub}(t)$ **then**
18:             $D_{Sub}(t) \leftarrow D_{Sub}(t) \cup {}^\bullet p$;// application of R_3
19:          **end if**
20:       **end for**
21:    **end if**
22: **end for**
23: **End**

The class of *Root nets* is extremely large, we have to add some structural constraints in order to recognize structurally their membership in the class of dead-closed systems.

Given a root net N, we first define a particular subnet called *Root component* based on the set of the root places of N. The *Root component* is slightly different from the subnet induced by the root places: It contains all root places and adjacent transitions. But, a (root) place p admits an output transition t in the *root subnet* if and only if p is a root place for t.

Definition 16. *Let $N = \langle P, T, F, V \rangle$ be a root net and $Root_N$ be the set of its root places.*
The Root component *of N is the net $N'^* = \langle Root_N, T^*, F^*, V^* \rangle$ defined as follows:*

- $T^* = Root_N{}^\bullet = T$,
- $F^* \subseteq (F \cap ((Root_N \times T^*) \cup (T^* \times Root_N)))$, *s.t.* $(p, t) \in F^*$ *iff* $(p, t) \in F$ *and p is a root place for t, and $(t, p) \in F^*$ iff $(t, p) \in F$*
- V' *is the restriction of V on F^*.*

Definition 17. *Let $\langle N, M_0 \rangle$ be P/T system.*
$\langle N, M_0 \rangle$ is called a Root System iff N is a root net and its root component N^* is conservative and strongly connected.

Theorem 5. *Let $\langle N, M_0 \rangle$ be a Root-system.*
$\langle N, M_0 \rangle$ *is a dead-closed system.*

Proof. Note first that the subnet N^* contains all transitions of N (N is weakly ordered). Let us show that $D(t) = T$ for all transition $t \in T$ in N^*. Let t and t' be two transitions and suppose that t is dead. Since N^* is strongly connected, there exists a path $\mathcal{P}_{t' \to t} = t' r_1 t_1 \ldots t_n r_n t$ leading from t' to t s.t. all the places r_i ($i \in \{1 \ldots\}$) are root places. Let us reason by recurrence on the length $\mid \mathcal{P}_{t' \to t} \mid$ of $\mathcal{P}_{t' \to t}$.

 - $\mid \mathcal{P}_{t' \to t} \mid = 1$: Obvious
 - Suppose that the proposition is true for each path $\mathcal{P}_{t' \to t}$ with $\mid \mathcal{P}_{t' \to t} \mid = n$.
 - Let $\mathcal{P}_{t' \to t}$ be a $n + 1$-length path leading from t' to t.
   Using proposition 6 (or rule R_1), one can deduce that all output transitions of r_n are dead. Now, since r is a bounded place, we use proposition 8 (or R_3) to deduce that all its input transition are dead and fortiori the transition t_n (the last transition before t in the path) is dead. Now the path $\mathcal{P}_{t' \to t_n}$ satisfies the recurrence hypothesis. Consequently, one can deduce that t' is dead as soon as t_n is dead.

The following corollary is a direct consequence of theorem 5, theorem 4 and theorem 1 respectively.

Corollary 2. *Let $\langle N, M_0 \rangle$ be a Root-system. The three following assertions are equivalent:*

 - *(1) $\langle N, M_0 \rangle$ is deadlock free,*
 - *(2) $\langle N, M_0 \rangle$ satisfies the cs-property,*
 - *(3) $\langle N, M_0 \rangle$ is live.*

Example: The K-system (dead-closed) of figure 2 is not a Root system. Indeed, its root component N^* (Figure 3) is not strongly connected.

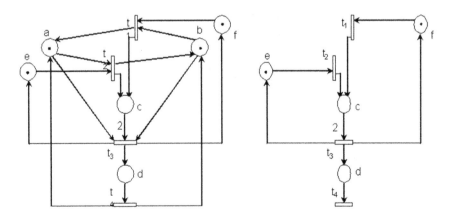

Fig. 3. An example of non Root but K system

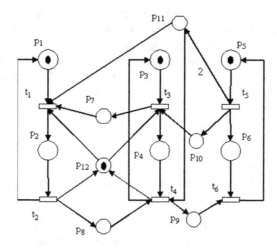

Fig. 4. An example of Root system

However, the non ordered system $\langle N, M_0 \rangle$ of Figure4 (t_1 is not ordered and p_{11}, p_{12} are not root places) is a Root system. In fact, the corresponding root component is conservative and strongly connected.

Let us analyze structurally the corresponding net N. One can check that N admits the eight following minimal siphons: $S_1 = \{p_5, p_6\}$, $S_2 = \{p_3, p_4\}$, $S_3 = \{p_1, p_2\}$, $S_4 = \{p_5, p_{10}, p_4, p_9\}$, $S_5 = \{p_5, p_{10}, p_7, p_2, p_8, p_9\}$, $S_6 = \{p_{12}, p_2, p_4\}$, $S_7 = \{p_3, p_7, p_2, p_8\}$ and $S_8 = \{p_5, p_{11}, p_9\}$. These siphons are invariant-controlled for any initial marking satisfying the following conditions: $p_5 + p_6 > 0$, $p_3 + p_4 > 0$, $p_1 + p_2 > 0$, $p_5 + p_{10} + p_4 + p_9 > 0$, $p_5 + p_{10} + p_7 + p_2 + p_8 + p_9 > 0$, $p_{12} + p_2 + p_4 > 0$, $p_3 + p_7 + p_2 + p_8 > 0$ and $p_{11} + 2.p_9 + 2.p_5 - p_3 - p_7 > 0\}$. Such conditions hold for the chosen initial marking $M_0 = p_1 + p_3 + p_5 + p_{12}$. Consequently, $\langle N, M_0 \rangle$ satisfies the cs-property i.e. live (according to theorem 5).

Obviously, the structure of N^* is a sufficient but not a necessary condition to ensure the K-property (and its membership in the class of K-systems). However, by adding structure to the subnet induced by the Root places (considered as modules) one can provide methods for synthesis of live K-systems.

In the following section, we first prove that the class of dead-closed systems is closed by a particular synchronization through asynchronous buffers. Then, this result will be used to extend the subclass of dead-closed systems structurally analyzable.

5.2   SDCS: Synchronized Dead-Closed Systems

In this section we prove that the class of dead-closed systems admits an interesting feature: it is closed by a particular synchronization through asynchronous buffers. The obtained class is a modular subclass of P/T nets called *Synchronized dead-closed Systems* (SDCS). By *modular* we emphasize that their definition is oriented to a bottom-up modelling methodology or structured view: individual agents, or modules, in the system are identified and modelled independently by means of live (i.e. cs-property) dead-closed systems (for example root systems), and the global model is obtained by synchronizing these modules through a set of places, the *buffers*. Such building process was already be used to define the class of Deterministically Synchronized Sequential Processes (DSSP) (see [10], [7] [12] for successive generalization) where elementary modules are simply live and safe state machines and where the interplay between conflict and synchronization is limited compared to dead-closed systems.

Definition 18. *A P/T system $\langle N, M_0 \rangle$, with $N = \langle P, T, F, V \rangle$, is a Synchronized dead-closed System (or simply an SDCS) if and only if P is the disjoint union $P_1, \ldots, P_n$ and B, T is the disjoint union $T_1, \ldots, T_n$, and the following holds:*

- *(1) For every $i \in \{1, \ldots, n\}$, let $N_i = \langle P_i, T_i, F_{\lfloor ((P_i \times T_i) \cup (T_i \times P_i))},$ $V_{\lfloor ((P_i \times T_i) \cup (T_i \times P_i))} \rangle$. Then $\langle N_i, m_{0 \lfloor P_i} \rangle$ is a live dead-closed system.*
- *(2) For every $i, j \in \{1, \ldots, n\}$, if $i \neq j$ then $V_{\lfloor ((P_i \times T_i) \cup (T_i \times P_i))} = \boldsymbol{0}$.*
- *(3) For each module N_i, $i \in \{1, \ldots, n\}$:*
  - *(a) $\exists$ (a buffer) $b \in B$ s.t. $b^\bullet \subseteq T_i$ (a private output buffer),*
  - *(b) $\forall b \in B$, b preserves the sets of root places of N_i (i.e., $\forall t \in T_i$, $Root(t)_{N_i} \subseteq Root(t)_N$).*
- *(4) Let $B' \subseteq B$ denotes the set of the output private buffers of N, then there exists a subset $B'' \subseteq B'$ such that the subnet induced by the dead-closed systems (N_i, $i \in \{1, \ldots, n\}$) and the buffers of B'' is conservative and strongly connected.*

Actually, we synchronize dead-closed system in such a way that we preserve the K-property (i.e. the equivalence between deadlock-freeness and liveness). Contrary to the DSSP modules, competition between those of an SDCS system is allowed, as long as the sets of root places of modules are preserved by composition(3.*b*) (but not necessarily the set of equal conflicts). After composition, a buffer can be a root place in the composed net but it cannot take the place of another one. Moreover, no restriction is imposed on the connection nature of the buffers. This allows modules to compete for resources. A second feature of SDCS class which enlarge the description power of DSSP is the fact that a given buffer does not have to be a output (destination) private as long as it exists such a buffer for each module (3.*a*).

Hence, one can easily prove that the class of SDCS represents a strict generalization of conservative and strongly connected DSSP systems. Moreover, when we compose dead-closed systems, or even root systems, the obtained system remains dead-closed.

Figure 5 illustrates an example of SDCS system. This system is composed of two modules, N_1 and N_2 (enclosed by the dashed lines) communicating through three buffers b_1, b_2 and b_3. Each module is represented by a Root system (N_1 is not a state machine). Also, each buffer is not restrained to respect internal modules conflict as long as it preserves their root places. For instance, the buffer b_1 doesn't respect the conflict between transitions t_1 and t_3 of N_1 ($V(b_1, t_1) = 1$

but $V(b_1, t_3) = 0$) but it preserves the root place p_1 of t_1. This system is not a Root-system since its root component N^*, induced here by N_1, N_2 and the buffers b_2 and b_3, (the buffer b_1 is not a root place), is strongly connected but not conservative (the buffer b_3 is not structurally bounded). However, this system is an SDCS since, with notations of definition 18 (4), the subset $B'' = \{b_1, b_2\}$ allows the condition (4) of to be satisfied.

The following theorem states that the class of SDCS is a subclass of dead-closed Systems. This means that when we synchronize several dead-closed systems as described in definition18 we obtain a dead-closed system.

Theorem 6. *Let $\langle N, M_0 \rangle$ be an SDCS system. Then $\langle N, M_0 \rangle$ is a dead-closed system.*

Proof. Let t and t' be two transitions of N and suppose that t is dead. Let N_n and N_1 be the modules containing t and t' respectively. Since the subnet induced by modules and output private buffers is strongly connected, there exists an (elementary) path $\mathcal{P}_{N_1 \to N_n} = N_1 b_1 \ldots b_{n-1} N_n$ leading from N_1 to N_n and each b_i ($i \in \{1, \ldots, n-1\}$) is a buffer having N_{i+1} as output private. Let us reason by induction on the number of modules N_i ($i \in \{1, \ldots, n\}$) involved in the path, Let us note $| \mathcal{P}_{N_i} |$ such a number.

- $| \mathcal{P}_{N_i} | = 0$: i.e. t and t' belong to the same module N_1. Since N_1 is a dead-closed system, one can use Theorem 4 (N_1 is also a K-system) to deduce that t' is dead.

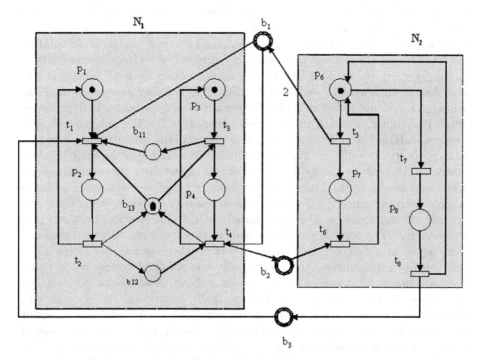

Fig. 5. An example of SKS system

- Suppose that the proposition is true for each path $\mathcal{P}_{N_1 \to N_n}$ involving less than n modules.
- Let $\mathcal{P}_{N_1 \to N_n}$ be a path leading from N_1 to N_n and passing through n modules.

  Consider b_{n-1}, the output private buffer of the module N_n. Using Theorem 5 one can deduce that all transitions in N_n are dead. Now, since b_{n-1} is a bounded place, we use proposition 8 to deduce that all its input transitions are dead and fortiori those of the module N_{n-1} (the module that appears before N_n in the path) are dead. The subpath leading from N_1 to N_{n-1} involves $n-1$ modules and hence satisfies the recurrence hypothesis. Consequently, one can deduce that t' is dead as soon as an input transition of b_{n-1} (belonging to N_{n-1}) is dead.

Corollary 3. *Let $\langle N, M_0 \rangle$ be an synchronized dead-closed system. The three following assertions are equivalent:*

- *(1) $\langle N, M_0 \rangle$ is deadlock free,*
- *(2) $\langle N, M_0 \rangle$ satisfies the cs-property,*
- *(3) $\langle N, M_0 \rangle$ is live.*

The previous corollary is a direct consequence of theorem 6, theorem4 and theorem1 respectively.

The main practical advantage of the definition of the SDCS class systems is that the equivalence between deadlock freeness and liveness can be preserved when we properly synchronize several dead-closed systems. A larger subclass based on the root nets structure can be obtained by applying the basic building process of the SDCS in a recursive way, i.e. modules can be Root systems, SDCS (or simply synchronized root systems) or more complex systems defined in this way. We are then able to revisit and extend the building process of the class of modular systems called multi-level deterministically synchronized processes (DS)*SP systems proposed in [8] which generalizes DSSP. Such a result will permit to enlarge the subclass of K-systems, structurally recognizable, for which the cs-property is a sufficient liveness condition. One can follow the same building process of (DS)*SP by taking live root systems as elementary modules (instead of safe and live state machines). We synchronize these root systems leading to an (root-based system) SDCS which is not necessarily a Root system. Then, one can take several (root-based system based) SDCS and synchronize them in a the same way. The resulting net, that is dead-closed system, can be considered as an agent in a further interconnection with other agents, etc. Doing so, a multi-level synchronization structure is built: the obtained system is composed of several agents that are coupled through buffers; these agents may also be a set of synchronized agents, etc. This naturally corresponds to systems with different levels of coupling: low level agents are tightly coupled to form an agent in a higher level, which is coupled with other agents, and so on. The class of systems thus obtained is covered by dead-closed systems, but largely generalizes strongly connected and conservative (DS)*SP (for which the deadlock freeness is a sufficient liveness condition [8]). From this view, the system of Figure5 can be viewed as "multilevel" SDCS where the module N_1 is composed of two submodules (Root systems) communicating through three buffers b_{11}, b_{12} and b_{13} (which is not output private but it preserves the set of the root places).

6 Other Subclasses Based on T-Invariants

Finally, we define two other subclasses of live K-systems, exploiting the fact that in every infinite occurrence sequence there must be a repetition of markings under boundedness hypothesis. We denote by T_{no} the subset of non ordered transitions. Nets of the first class are bounded and satisfy the following structural condition : the support of each T-invariant contains all non-ordered transitions. This class includes the one T-invariants nets from which (ordinary) bounded nets covered by T-invariants can be approximated as proved in [9].

Theorem 7. *Let $\langle N, M_0 \rangle$ be a P/T system such that:*

(i) N is conservative
(ii) $\forall$ T-invariant j: $T_{no} \subseteq \|j\|$.

$\langle N, M_0 \rangle$ is live if and only if $\langle N, M_0 \rangle$ satisfies the controlled-siphon property.

Proof. Assume that $\langle N, M_0 \rangle$ satisfies the cs-property but is not live. According to theorem 2, $T_D \neq \emptyset$ and $T_L \neq \emptyset$. Consider the subnet induced by T_L. This subnet is live and bounded for M^*. There exists necessarily an occurrence sequence for which count-vector is a T-invariant j and $T_{no} \not\subset \|j\|$. This contradicts condition *(ii)*.

Now, we define a last subclass of non-ordered systems (systems having a non ordered transition) where the previous structural condition *(ii)* is refined as follows: for any non-ordered transition t, we can not get a T-invariant on the subnet induced by $T \setminus D(t)$.

Theorem 8. *Let $\langle N, M_0 \rangle$ be a non-ordered system satisfying the two following conditions:*

(i) N is conservative
(ii) $\forall$ T-invariant j and $\forall t \in T_{no}$: $(\|j\| \cap D(t)) \neq \emptyset$

$\langle N, M_0 \rangle$ is live if and only if $\langle N, M_0 \rangle$ satisfies the controlled-siphon property.

Proof. Let $\langle N, M_0 \rangle$ be satisfying the cs-property but not live. Consider the subnet induced by T_L. This subnet is live and bounded. Hence, there exists a T-invariant j corresponding to an occurrence sequence in the subnet and do not cover neither the (not ordered) transition t^* nor any transition in $D(t^*)$ $((\|j\| \cap D(t^*)) = \emptyset)$. This contradicts condition *(ii)*.

Remark: Note that the non ordered system (t_3 is not ordered) of Figure 2 can also be recognized structurally as a K-system since conditions *(i)* and *(ii)* of Theorem 8 $(D(t_3) = T)$ are satisfied.

7   Conclusion

The aim of this paper was to deepen into the structure theory on P/T systems,namely K-systems, for which the equivalence between controlled-siphon property, deadlock freeness, and liveness holds. Using the new structural concepts of ordered transitions and root places, we present a refined characterization of the non-liveness condition under cs property hypothesis. Such result permits us to revisit from a new perspective some well known results and to structurally characterize new and more expressive subclasses of K-systems. This work poses a challenging question: What are the structural mechanisms ensuring a siphon to be controlled other than based on trap or p-invariant concept? The interest of a positive answer is a broader decision power of controlled siphon property in particular for systems where the purely algebraic methods such rank theorem [5] are important.

Acknowledgements. The authors wish to thank anonymous referees for their useful comments.

References

1. K. Barkaoui, J-M. Couvreur, and C. Dutheillet. On liveness in extended non self-controlling nets. In *Proceeding of the 16th International Conference on Application and Theory of Petri Nets, Turin, June 1995.*, pages 25–44, 1995.

2. K. Barkaoui and M. Minoux. A polynomial-time graph algorithm to decide liveness of some basic classes of bounded petri nets. In Jensen, K., editor, *Lecture Notes in Computer Science; 13th International Conference on Application and Theory of Petri Nets 1992, Sheffield, UK*, volume 616, pages 62–75. Springer-Verlag, June 1992.

3. K. Barkaoui and J-F-P. Peyre. On liveness and controlled siphons in petri nets . In *Proceeding of the 16th International Conference on Application and Theory of Petri Nets, Osaka, June 1996.*, 1996.

4. K. Barkaoui and B. Zouari. On liveness and deadlock freeness in petri nets . In *Proceeding of the 6th International Symposium in Programming Systems, Algiers, May 2003.*, 2003.

5. J. Desel. A proof of the rank theorem for extended free choice nets. In Jensen, K., editor, *Lecture Notes in Computer Science; 13th International Conference on Application and Theory of Petri Nets 1992, Sheffield, UK*, volume 616, pages 134–153. Springer-Verlag, June 1992.

6. J. Esparza and M. Silva. A polynomial-time algorithm to decide liveness of bounded free choice nets. *Theor. Comput. Sci.*, 102(1):185–205, 1992.

7. E. Teruel L. Recalde and M. Silva. Modeling and analysis of sequential processes that cooperate through buffers. *IEEE Transactions on Robotics and Automation*, 11:267–277, 1985.

8. E. Teruel L. Recalde and M. Silva. Structure theory of multi-level deterministically synchronized sequential processes. *Theor. Comput. Sci.*, 254(1-2):1–33, 2001.

9. K. Lautenbach and H. Ridder. Liveness in bounded petri nets which are covered by T-invariants. In Valette, R., editor, *Lecture Notes in Computer Science; Application and Theory of Petri Nets 1994, Proceedings 15th International Conference, Zaragoza, Spain*, volume 815, pages 358–375. Springer-Verlag, 1994.
10. W. Reisig. Deterministic buffer synchronization of sequential processes. *Acta Informatica 18*, pages 117–134, 1982. NewsletterInfo: 13.
11. W. Reisig. *Petri nets: an introduction*. Springer-Verlag New York, Inc., 1985.
12. Y. Souissi. Deterministic systems of sequential processes: A class of structured petri nets. In *Proceedings of the 12th International Conference on Application and Theory of Petri Nets, 1991, Gjern, Denmark*, pages 62–81, June 1991. NewsletterInfo: 39.
13. Enrique Teruel and Manuel Silva. Structure theory of equal conflict systems. *Theor. Comput. Sci.*, 153(1&2):271–300, 1996.

Extremal Throughputs in Free-Choice Nets

Anne Bouillard[1], Bruno Gaujal[1], and Jean Mairesse[2]

[1] LIP (UMR CNRS, ENS Lyon, INRIA, Université Claude Bernard Lyon 1),
École Normale Supérieure de Lyon,
46, allée d'Italie - 69364 Lyon Cedex 07 - France
{abouilla, bgaujal}@ens-lyon.fr
[2] LIAFA, CNRS - Université Paris 7,
Case 7014 - 2, place Jussieu - 75251 Paris Cedex 5 - France
mairesse@liafa.jussieu.fr

Abstract. We give a method to compute the throughput in a timed live and bounded free-choice Petri net under a total allocation (*i.e.* a 0-1 routing). We also characterize and compute the conflict-solving policies that achieve the smallest throughput in the special case of a 1-bounded net. They do not correspond to total allocations, but still have a small period.

1 Introduction

Petri nets are logical objects, originally and above all. However, the interest of Petri nets for modelling purposes has induced the need for timed and stochastic extensions of the model. Performance evaluation then becomes a central issue, and the throughput is arguably the main performance indicator.

Consider now a live and bounded free-choice Petri net (LBFC). Such Petri nets realize a good compromise between modelling power and mathematical tractability, see [6] for several striking examples of the latter. Assume that the Petri net is timed with a timing specified by a constant real-valued firing time for each transition. To remove the undeterminism of the behavior of the Petri net, a policy for the resolution of all the conflicts needs to be decided. Once it is chosen, all the enabled transitions start to fire as soon as possible, and the time that elapses between the beginning and the completion of the firing of a transition is equal to the firing time. Therefore, the timed evolution of the Petri net is completely determined.

Our goal is to study the global *activity* or *throughput* or *firing rate of the transitions* of the Petri net in a sense to be made precise later on. Of course, the activity depends on the chosen policy for resolving conflicts.

In a free-choice Petri net, one may view a conflict-solving policy as a set of local functions associated with conflict places, and assigning tokens to output transitions. The simplest class of policies consists of the so-called *0-1 policies*: for a conflict place p, allocate all the tokens to a fixed transition. Zero-one policies are called *total allocations* in [6]. The next simplest class of policies is, arguably, the *periodic*

G. Ciardo and P. Darondeau (Eds.): ICATPN 2005, LNCS 3536, pp. 108–127, 2005.
© Springer-Verlag Berlin Heidelberg 2005

ones: for a conflict place p, allocate the tokens to the output transitions according to some fixed periodic pattern. Obviously, 0-1 policies are also periodic policies.

In this paper, we address the following natural questions:

- A. Given a periodic policy, is the activity explicitly computable?
- B. Consider the set of all possible, arbitrarily complex, policies for resolving conflicts. Is the infimum, resp. supremum, of the activity over this set attained by a 0-1 or a periodic policy? Can we explicitly determine the policies realizing the infimum, resp. supremum?

For both questions, we are also concerned by the algorithmic complexity of the computations.

Consider first Question A. It is known that the activity is explicitly computable when the timings are rational-valued [3]. The solution relies on the construction of a very large finite graph G in which a state incorporates three different types of information: the current marking; the remaining time before completion for the currently firing transitions; and the current position of the cursor within the periods for the periodic policy. The timed behavior is ultimately periodic and the period corresponds to an elementary circuit in the graph G. The activity is computed along this circuit.

The method has two major drawbacks. First, it is not efficient from an algorithmic point of view. Indeed, the graph G is in general much larger than the reachability (marking) graph whose size may already be exponential in the one of the Petri net. Second, it does not provide much insight on the structure of the timed behavior.

Here, we show that both restrictions can be overcome in the special case of a 0-1 policy: the live part of the Petri net becomes a disjoint union of event graphs. Consequently, the activity can be computed in polynomial (cubic) time in the size of the Petri net, using classical results on the throughput of timed event graphs [1, 4, 5]. Furthermore, the previous restriction on having rational timings is not necessary anymore for 0-1 policies.

Consider now Question B and assume that the timings are rational-valued for simplicity. Using a simplified version of the above graph G, in which the periodic policy is not coded anymore, one can easily prove that the supremum and the infimum of the activity are obtained for periodic policies. The drawbacks are the same as before: the time complexity, and the lack of structural insight. Concerning the latter, the method does not allow to answer the question: is the supremum or infimum attained by 0-1 policies?

Presumably against the intuition, we exhibit an example of a live and 1-bounded Petri net with firing times all equal to one, and for which the infimum is attained only by non-0-1 periodic policies. More generally, we show that for 1-bounded Petri nets with general $(0, \infty)$-valued timings, the infimum is attained by a periodic policy which may not be 0-1 but which can be characterized at the net level and which has a very small period (*i.e.* bounded by the total number of tokens). The same result fails to hold for a k-bounded Petri net, $k \geq 2$ where the general structure of infimum policies is not understood. An example is given of a 2-bounded net with timings all equal to one, for which the infimum is attained by periodic policies which are not 0-1 nor have small periods. To be complete, let us mention that the general structure of supremum policies is not well understood, even for 1-bounded nets. It is easy to build examples of 1-bounded LBFC with rational timings for which the supremum is attained only by non-0-1 Sturmian-like periodic policies, as well as examples with irrational timings for which the supremum is attained only by Sturmian-like non-periodic policies, see [9] and Example 6.2.

In order to obtain the above results, we use three different types of building blocks:

- The theory of timed event graphs;
- A structural result stating that the live part of a LBFC with a total allocation is a disjoint union of T-components; and that, given a T-component, there exists a total allocation making this T-component the only live part of the LBFC;
- The notion of Token-Transition invariants. It is a refinement of the classical notion of T-invariants with a dynamical flavor to it, since it "follows" the evolution of a token.

The first point is very classical [1, 4, 5], while the other two may be original and of some interest by their own.

The paper is organized as follows. The known results on Question A appear in Section 3. In Section 4, we study the 0-1 policies in detail. Section 5 introduces the TT-invariants. Section 6 is devoted to Question B. In particular, we characterize the policies which provide the infimum throughput for a 1-bounded net in Subsection 6.3.

Due to lack of space, all the proofs are not included. The missing proofs can be found in the corresponding research report [2].

2   Notations and Preliminaries

Denote by $\mathbb{N}$ the nonnegative integers, and by $\mathbb{N}^*$ the positive integers. Given a set T and a subset S, denote by $\chi_S : T \to \{0, 1\}$ the characteristic function of S in T defined by: $\chi_S(u) = 1$ if $u \in S$ and $\chi_S(u) = 0$ if $u \in T \backslash S$.

A *net* is a bipartite directed graph $(\mathcal{P}, \mathcal{T}, \mathcal{F})$ with $\mathcal{P} \cup \mathcal{T}$ as the set of nodes $(\mathcal{P} \cap \mathcal{T} = \emptyset)$ and $\mathcal{F} \subseteq (\mathcal{P} \times \mathcal{T}) \cup (\mathcal{T} \times \mathcal{P})$ as the set of arcs. A *Petri net* is a quadruple $(\mathcal{P}, \mathcal{T}, \mathcal{F}, M)$, where $(\mathcal{P}, \mathcal{T}, \mathcal{F})$ is a net and M is a map from $\mathcal{P}$ to $\mathbb{N}$. We sometimes write the Petri net $\mathcal{N}$ as $(\mathcal{N}, M)$ to emphasize the special role of M. The elements of $\mathcal{P}$ are called *places* and are represented by circles and those of $\mathcal{T}$ are called *transitions* and represented by rectangles. The function M is called the *(initial) marking* of the net and is represented by tokens in places. Let $x \in \mathcal{P} \cup \mathcal{T}$ be a node. We denote by ${}^\bullet x$ the set of its predecessors and by $x^\bullet$ the set of its successors. We also set ${}^\bullet X = \cup_{x \in X} {}^\bullet x$ and $X^\bullet = \cup_{x \in X} x^\bullet$. A transition is *conflicting* if one of its input place has at least two successors. Otherwise, the transition is *non-conflicting*.

The marking evolves according to the *firing rule*. A transition t is *enabled* if: $\forall p \in {}^{\bullet}t$, $M(p) \geq 1$. An enabled transition can *fire*, and then the marking becomes M' with $M'(p) = M(p) - \chi_{{}^{\bullet}t}(p) + \chi_{t^{\bullet}}(p)$.

If the marking M' is obtained from M by firing the transition t, we write $M \xrightarrow{t} M'$. If M' is obtained by successively firing $\sigma = t_1 t_2 \cdots t_n \in T^*$, we write $M \xrightarrow{\sigma} M'$. The sequence σ is called ad *(admissible) firing sequence*. Finally, if M' can be reached from M by firing some sequence, we write $M \rightarrow M'$. The set of the *reachable markings* of M is $\mathcal{R}(\mathcal{G}, M) = \mathcal{R}(M) = \{M' \mid M \rightarrow M'\}$.

A Petri net is *live* if for every transition t and every reachable marking M_1 there exists a marking M_2, reachable from M_1, that enables t. A Petri net is *deadlock-free* if there exists no reachable marking in which no transition is enabled. A Petri net is *k-bounded*, $k \in \mathbb{N}$, if for every reachable marking, the number of tokens in a place is less or equal to k. A Petri net is bounded if it is k-bounded for some k. A net $\mathcal{N}$ is *structurally live* if there exists a marking M such that the Petri net $(\mathcal{N}, M)$ is live. A net $\mathcal{N}$ is *well-formed* if there is a marking M that makes the Petri net $(\mathcal{N}, M)$ live and bounded.

An *event graph* is a (Petri) net where: $\forall p \in \mathcal{P}$, $|{}^{\bullet}p| = |p^{\bullet}| = 1$. A *state machine* is a (Petri) net where: $\forall t \in T$, $|{}^{\bullet}t| = |t^{\bullet}| = 1$. A *free-choice (Petri) net* is a (Petri) net where: $\forall (p, t) \in \mathcal{P} \times T$, $(p, t) \in \mathcal{F} \Rightarrow (p^{\bullet} = \{t\})$ or $({}^{\bullet}t = \{p\})$. We use the acronym *LBFC* for a live and bounded free-choice Petri net. A *choice-free (Petri) net* is a (Petri) net where: $\forall p \in \mathcal{P}$, $|p^{\bullet}| = 1$.

The *incidence matrix* of a Petri net is $N \in \mathbb{Z}^{\mathcal{P} \times T}$ with $N_{p,t} = \chi_{t^{\bullet}}(p) - \chi_{{}^{\bullet}t}(p)$. Let $\sigma \in T^*$ be a firing sequence. The *commutative image* (or *Parikh vector*) of σ is $\overrightarrow{\sigma} = (|\sigma|_t)_{t \in T}$, the vector of the number of occurrences of each transition t in σ. If $M \xrightarrow{\sigma} M'$, then the equation $M' = M + N\overrightarrow{\sigma}$ is satisfied.

Invariants of Petri nets. A column vector $J \in \mathbb{N}^T \setminus \{(0, \ldots, 0)^T\}$ (resp. $I \in \mathbb{N}^{\mathcal{P}} \setminus \{(0, \ldots, 0)^T\}$) is a *T-invariant* (resp. *S-invariant*) if $NJ = 0$ (resp. $I^T N = 0$). A T-invariant (resp. S-invariant) is *minimal* if it is minimal for the component-wise ordering among all the T-invariants (resp. S-invariants). A subnet $\mathcal{N}'$ of the net $\mathcal{N}$ with the set of nodes X is a *T-component* (resp. *S-component*) if for every transition t of X, ${}^{\bullet}t \cup t^{\bullet} \subseteq X$ (resp. for every place p of X, ${}^{\bullet}p \cup p^{\bullet} \subseteq X$) and $\mathcal{N}'$ is a strongly connected event graph (resp. state machine). If $(\mathcal{P}_1, T_1, \mathcal{F}_1)$ is a T-component (resp. S-component) of the net $\mathcal{N}$, then χ_{T_1} (resp. $\chi_{\mathcal{P}_1}$) is a minimal T-invariant (resp. S-invariant) of $\mathcal{N}$. For a well-formed free-choice net, the converse is true: if J is a minimal T-invariant (resp. S-invariant), set $T_1 = \{t \in T \mid J_t \neq 0\}$ (resp. $\mathcal{P}_1 = \{p \in \mathcal{P} \mid J_p \neq 0\}$) and $\mathcal{P}_1 = {}^{\bullet}T_1 \cup T_1^{\bullet}$ (resp. $T_1 = {}^{\bullet}\mathcal{P}_1 \cup \mathcal{P}_1^{\bullet}$), then $(\mathcal{P}_1, T_1, \mathcal{F}_1)$ is a T-component (resp. S-component). See for instance [6–Prop. 5.7, Prop. 5.14, Th. 5.17].

A set of T-components (resp. S-components) forms a *T-cover (S-cover)* if every node belongs to one of these components. Well-formed free-choice nets are covered by T-components and also by S-components ([6–Theorems 6.6 and 5.18]).

We will also need the following result.

Theorem 2.1. *[6–Theorem 5.9] Let p be a place of a live and bounded free-choice Petri net $(\mathcal{N}, M)$. The bound of p is $\min\{\sum_{s\in\mathcal{P}_1} M(s) \mid p\in\mathcal{P}_1, (\mathcal{P}_1, \mathcal{T}_1, \mathcal{F}_1)$ is a S-component of $\mathcal{N}\}$.*

Clusters. The *cluster* $[x]$ of $x \in \mathcal{P}\cup\mathcal{T}$ is the smallest subset of $\mathcal{P}\cup\mathcal{T}$ such that: (i) $x \in [x]$; (ii) $p \in \mathcal{P}, p \in [x] \Rightarrow p^\bullet \in [x]$; (iii) $t \in \mathcal{T}, t \in [x] \Rightarrow {}^\bullet t \in [x]$. The set of all the clusters of a net defines a partition of the nodes of the net. For free-choice nets, each cluster contains only one place or only one transition.

Blocking marking. Let $(\mathcal{N}, M)$ be a Petri net and t a non-conflicting transition of $\mathcal{N}$. A *blocking marking* of the transition t is a reachable marking such that the only enabled transition is t. If t is a conflicting transition, a *blocking marking* of t is a reachable marking such that the only enabled transitions belong to the cluster $[t]$.

In [8–Theorem 3.1], it is shown that in a connected LBFC, for any transition b, there exists a unique blocking marking M_b. Moreover, M_b is reachable from any other reachable marking without firing b.

Timed and routed nets. A timed Petri net is a Petri net in which timings have been added on places and transitions. With no loss of generality, we only consider timings on the transitions, and not on the places. We also consider non-null timings. This is assumed for convenience. The results of the paper could be generalized with null timings, under the assumption that it is not possible to have an infinite number of firings occurring in 0 time. Set $\mathbb{R}_+^* = (0, +\infty)$. A timed Petri net is denoted by $(\mathcal{N}, M, \tau)$ with $(\mathcal{N}, M)$ a Petri net and $\tau \in (\mathbb{R}_+^*)^{\mathcal{T}}$ the vector of the timings. The timed semantics is the following one. Consider a transition t with timing τ_t which gets enabled at instant d. If the transition t is fired, the firing occurs as follows:

- At time d, the firing begins. A token is frozen in each input place of t and cannot enable another transition.
- At time $d+\tau_t$, the firing ends. The frozen tokens are removed from the input places of t and one token is added in each output place of t.

Observe that it is possible for a transition to have several ongoing firings at a given instant. The resulting evolution is called *as soon as possible (asap)*, since a firing transition begins to fire as soon as it is enabled.

Any conflict-solving policy may be viewed as a set of local *routing functions* at each conflicting place. The global routing function is a vector $u = (u_p)_{p\in\mathcal{P}}$ where u_p is a function from $\mathbb{N}^*$ to $p^\bullet$. The k-th token arriving in place p (we consider the tokens in place p in the initial marking as the first arriving tokens) can only enable the transition $u_p(k)$. So the notion of enabled transition is modified by the routing function. A transition can be fired if all its input places contain a token which is routed to that transition. We denote by $(\mathcal{N}, M, u)$ a routed Petri net with routing u and by $(\mathcal{N}, M, \tau, u)$ a timed and routed Petri net.

A marking is *reachable* for a routed Petri net $(\mathcal{N}, M, u)$ if it is reachable for $(\mathcal{N}, M)$ via a firing sequence compatible with u. The notions of boundedness and liveness of $(\mathcal{N}, M, u)$ are defined accordingly.

If the Petri net $(\mathcal{N}, M)$ is bounded, so is the routed Petri net $(\mathcal{N}, M, u)$. If the Petri net $(\mathcal{N}, M)$ is live, $(\mathcal{N}, M, u)$ is not necessarily live, nor deadlock-free. However, if $(\mathcal{N}, M)$ is a LBFC, the routed net $(\mathcal{N}, M, u)$ cannot have any deadlock, because choices and synchronizations are separated. Hence, if a routed free-choice net is not live, we can always define its non-empty live part.

A routing $u = (u_p)_{p \in \mathcal{P}}$ is *periodic* if u_p is a periodic function for every p. A routing u is *0-1* if: $\forall p \in \mathcal{P}$, u_p is a constant function. A 0-1 routing is called a *total allocation* in [6].

3   Throughput in Routed Free-Choice Petri Nets

With no loss of generality, all Petri nets considered are assumed to be connected.

Consider a timed Petri net and let $\sigma = \sigma(1)\sigma(2) \cdots \in \mathcal{T}^{\mathbb{N}}$ be an infinite firing sequence. Set $\sigma_n = \sigma(1) \cdots \sigma(n)$. Consider the timed evolution starting at instant 0 and associated with σ. The *activity* $A(\sigma)$ of σ is the asymptotic average number of firings per unit of time:

$$A(\sigma) = \liminf_{n \to \infty} \frac{n}{d(n)}$$

where $d(n)$ is the first instant of completion of all the firings from σ_n.

To make this definition more general and more flexible, it is possible to "weight" the activity of each transition.

A *weight* $\alpha_t \in \mathbb{R}_+$ is associated to each transition t, we set $\alpha = (\alpha_t)_{t \in \mathcal{T}}$, and we assume that $\alpha \neq (0, \ldots, 0)$. The *throughput* $D(\sigma)$ of σ (for the weight α) is defined by:

$$D(\sigma) = \liminf_{n \to \infty} \frac{\sum_{t \in \mathcal{T}} \alpha_t (\overrightarrow{\sigma_n})_t}{d(n)} .$$

If all the weights are equal to one, the throughput is equal to the activity. On the other hand, if $\alpha_t = 1, \alpha_{t'} = 0, t' \neq t$, then the throughput measures the firing rate of transition t.

The above notion of throughput allows one to modify a Petri net without changing its throughput, for example, by replacing a transition of timing n and weight α by n transitions of timing 1 and weight α/n.

3.1   Periodic Routings

Consider a timed and routed LBFC $(\mathcal{N}, M, \tau, u)$ with a *periodic* routing and *integer* firing times (rational firing times can be treated in a similar way).

The state of the Petri net at time t is a triple (M_t, R_t, U_t) where M_t is the marking at time t, R_t is the remaining firing time of all the current firings at time t and U_t is the current routing decision in all the routing places. Observe that the number of states is finite and bounded by $(k+1)^{|\mathcal{P}|} \times F^{k|\mathcal{T}|} \times L^{|\mathcal{P}|}$, where k is a bound on the number of tokens per place, F is a bound on the firing times of all the transitions and L is a bound on the period of the routing at each place.

Since the behavior of the net is deterministic, the net jumps from one state to its unique successor at each time-step.

The state space being finite, there exists a state which is visited twice for the first time, and the whole behavior becomes periodic from that point on. This shows that the throughput exists and can be computed along the periodic behavior of the net. However, this computation may have a very high complexity (in time and in space) because the state space is potentially huge.

A construction similar to the above one is proposed in [3].

3.2   A Particular Case: Event Graphs

In a live and bounded event graph $(\mathcal{G}, M)$, there is no routing place, hence no routing. In that case, it is useless to sweep the whole state space to compute the throughput. It is well-known that the firing rate is the same for all transitions and the throughput is given by:

$$D = \frac{\sum_{t \in \mathcal{T}} \alpha_t}{\rho(\mathcal{G}, M)}, \quad \text{where} \quad \rho(\mathcal{G}, M) = \max_{c \text{ circuit of } \mathcal{G}} \frac{\sum_{t \in c} \tau_t}{\sum_{p \in c} M(p)}. \tag{1}$$

The throughput can be computed in cubic time using Karp's algorithm, see for instance [1]. The constant $\rho(\mathcal{G}, M)$ is usually called the *cycle time* of $(\mathcal{G}, M)$ (see [1, 4, 5] for details).

4   Zero-One Policies

In this section, we consider 0-1 routing policies instead of arbitrary periodic routings. We show that all the combinatorial difficulties of periodic routings can be overcome for 0-1 routings.

4.1   Total Allocations and 0-1 Policies

An *allocation* is a function u from a set of clusters C to $\mathcal{T}$ such that: $\forall c \in C$, $u(c) \in c$. A transition is *allocated* if it belongs to the image of u. An allocation is *total* if it is defined on all clusters. An allocation *points to* C if for every place p not belonging to C, there exists a path π from p to a place of C such that every transition along the path π is allocated.

A firing sequence σ agrees with an allocation $u : C \to \mathcal{T}$ if it does not contain any transition t such that $[t] \in C$ and $t \neq u([t])$.

Lemma 4.1. *[6–Lemma 6.5] Let C be a set of clusters of a strongly connected free-choice Petri net $\mathcal{N}$ and let $\bar{C}$ be the complementary set of C in the clusters of $\mathcal{N}$. Then there exists an allocation u defined on $\bar{C}$ that points to C, and if M*

is a bounded marking and $M \xrightarrow{\sigma}$ is an infinite sequence that agrees with u, then some transition of C is fired an infinite number of times in σ.

In this paper, we see total allocations as 0-1 routing policies in the places: each place routes all its tokens to its unique allocated output transition.

When u is a 0-1 routing, $(\mathcal{N}, u)$ is a free-choice Petri net where all the transitions which are not allocated can be removed. Therefore, exactly one output transition remains for each place. We obtain a choice-free Petri net. We first study general choice-free nets before giving a characterization of the choice-free nets obtained as the live part of a free-choice Petri net with a 0-1 routing.

4.2   Choice-Free Nets

A *siphon* is a set of places R such that $^{\bullet}R \subseteq R^{\bullet}$. A *trap* is a set of places R such that $R^{\bullet} \subseteq {}^{\bullet}R$.

Lemma 4.2. *Consider a strongly connected and live choice-free Petri net. It is an event graph if and only if it is bounded.*

A connected live and bounded Petri net is strongly connected. So it follows from the above lemma, that a connected and live choice-free Petri net is either unbounded, or bounded in which case it is a strongly connected event graph.

4.3   Live Part of a LBFC with a 0-1 Routing

Consider a LBFC $(\mathcal{N}, M)$ with a 0-1 routing. Let $\widetilde{\mathcal{N}}$ be the net obtained by removing all the transitions (together with their input and output arcs) which are not chosen by the 0-1 routing. This means that the net $\widetilde{\mathcal{N}}$ may not be strongly connected anymore (some places may have no inputs). Figure 1 shows the construction of $\widetilde{\mathcal{N}}$ on an example.

The Petri net $(\widetilde{\mathcal{N}}, M)$ is choice-free and bounded, on the other hand it may not be live. We are interested in characterizing the live part of $(\widetilde{\mathcal{N}}, M)$. This live part may depend on M. More precisely, let M' be a reachable marking of $(\mathcal{N}, M)$, let $(\mathcal{N}_1, M_1)$ be the live part of $(\widetilde{\mathcal{N}}, M)$ and let $(\mathcal{N}_2, M_2)$ be the live part of $(\widetilde{\mathcal{N}}, M')$. We may have $\mathcal{N}_1 \neq \mathcal{N}_2$, as well as $\mathcal{N}_1 = \mathcal{N}_2, \mathcal{R}(\mathcal{N}_1, M_1) \neq \mathcal{R}(\mathcal{N}_2, M_2)$.

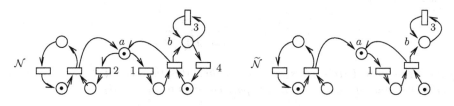

Fig. 1. The net $\widetilde{\mathcal{N}}$ is constructed by removing all non-allocated transitions. The 0-1 routing sends all tokens to transition 1 in routing place a and all tokens to transition 3 in routing place b. Transitions 2 and 4 are removed to construct $\widetilde{\mathcal{N}}$

The net $\widetilde{\mathcal{N}}$ can be decomposed into non-trivial maximal strongly connected components (mscc). There are two kinds of such components: the final components and the non-final components.

Lemma 4.3. *The final mscc are T-components of $\mathcal{N}$. The non-final mscc are event graphs, but not T-components of $\mathcal{N}$. The final mscc may or may not be live in $\widetilde{\mathcal{N}}$, the non-final mscc are not live in $\widetilde{\mathcal{N}}$.*

Lemma 4.4. *Let $(\mathcal{G}, M)$ be a live and 1-bounded event graph. Then*
- *if one token is removed, the event graph is not live anymore,*
- *if one token is added, the event graph is not 1-bounded anymore.*

As a consequence of Lemma 4.3 and Lemma 4.4, we have the following theorem.

Theorem 4.1. *Let $(\mathcal{N}, M_0)$ be a LBFC with a 0-1 routing. Let $\widetilde{\mathcal{N}}$ be the net obtained from $\mathcal{N}$ by removing the arcs and transitions which are not selected by the routing. The live part of $(\widetilde{\mathcal{N}}, M_0)$ is a non-empty disjoint union of strongly connected event graphs $(\mathcal{G}_1, M_1), \ldots, (\mathcal{G}_k, M_k)$, where each $\mathcal{G}_i$ is a T-component of $\mathcal{N}$. If $(\mathcal{N}, M_0)$ is 1-bounded, or if $k = 1$, then the sets of reachable markings $\mathcal{R}(\mathcal{G}_1, M_1), \ldots, \mathcal{R}(\mathcal{G}_k, M_k)$, do not depend on the 0-1 routing such that the live part of $(\mathcal{N}, M_0)$ consists of $\mathcal{G}_1, \ldots, \mathcal{G}_k$.*

If $\mathcal{G}$ is a T-component of $\mathcal{N}$, then there exists a 0-1 routing such that the live part of $(\widetilde{\mathcal{N}}, M_0)$ is precisely $(\mathcal{G}, M)$ for some marking M.

4.4   Throughput of 1-Bounded Free-Choice Nets with 0-1 Routings

We now have all the ingredients to prove the main result.

Theorem 4.2. *Consider a LBFC $(\mathcal{N}, M_0)$ with a 0-1 routing. Assume that the live part is $(\mathcal{G}, M)$, where $\mathcal{G}$ is a single T-component corresponding to the T-invariant $J_{\mathcal{G}}$. Then the throughput does not depend on the 0-1 routing such that $\mathcal{G}$ is the live part, and it is equal to*

$$D_{\mathcal{G}} = \frac{\alpha^T J_{\mathcal{G}}}{\rho(\mathcal{G}, M)}.$$

Assume that the live part is $(\mathcal{G}_1, M_1), \cdots, (\mathcal{G}_k, M_k)$, where $\mathcal{G}_i$ is a T-component. Then the throughput is

$$D = \sum_{i=1}^{k} D_{\mathcal{G}_i} = \sum_{i=1}^{k} \frac{\alpha^T J_{\mathcal{G}_i}}{\rho(\mathcal{G}_i, M_i)}.$$

If the free-choice Petri net is 1-bounded, then this throughput does not depend on the 0-1 routing such that the live part consists of $\mathcal{G}_1, \ldots, \mathcal{G}_k$.

Proof. The proof easily follows from (1) and Theorem 4.1. Observe that the cycle time $\rho(\mathcal{G}, M)$ of an event graph depends on M only through the token count of circuits, see (1).

5   Token-Transition-Invariants

Let $(\mathcal{N}, M)$ be a Petri net. Let $\sigma = \sigma_1 \cdots \sigma_k$ be a firing sequence. We say that $\sigma' = \sigma'_1 \cdots \sigma'_h$, $h \leq k$ is a *subsequence* of σ if there exists an increasing function $f : \{1, \ldots, h\} \to \{1, \ldots, k\}$ such that $\sigma'_i = \sigma_{f(i)}$ for all $i \in \{1, \ldots, h\}$.

Definition 5.1 (compatible firing sequence). *Let* $\pi = p_1 t_1 p_2 t_2 \cdots p_\ell t_\ell$, $p_i \in \mathcal{P}$, $t_i \in \mathcal{T}$ *be a path of* $\mathcal{N}$ *and* M *be a marking that marks* p_1. *Let* σ *be an admissible firing sequence of* $(\mathcal{N}, M)$. *The sequence* σ *is* compatible *with* π *if the first subsequence of* σ *in* $[t_1][t_2] \cdots [t_\ell]$ *is* $t_1 t_2 \cdots t_\ell$.

In the above definition, *first* has the following meaning. Order all subsequences of σ according to the point-wise ordering of the increasing functions used to defined them $(f \leq g$ is $f(i) \leq g(i)$ for all i in the domain of $f)$. *First* means smallest according to this ordering.

In other words, an admissible firing sequence σ is compatible with a path in the Petri net if all the transitions along that path are fired and in that order when σ is fired. This means that the token which was initially in place p_1 successively enters places $p_2, \ldots, p_\ell$ when σ is fired.

Definition 5.2 (Token-Transition-Invariant). *Let* $c = p_1 t_1 p_2 t_2 \cdots p_\ell t_\ell$ *be a circuit of* $\mathcal{N}$ *and* M *be a marking that marks* p_1. *A vector* $J \in \mathbb{N}^T$ *is a Token-Transition-invariant (or TT-invariant) generated by* c *and the marking* M *if it is a T-invariant and if it is the commutative image of an admissible firing sequence compatible with* c.

A TT-invariant J *generated by* c *and the marking* M *is* minimal *if for every other TT-invariant* J' *generated by* c *and the marking* M, J' *is not smaller than* J.

A TT-invariant generated by c *is* minimal *if for every other TT-invariant* J' *generated by* c *and some marking,* J' *is not smaller than* J.

In words, a TT-invariant is a T-invariant such that one token has moved along a circuit and is back to its original place when the corresponding sequence is fired (hence the name).

In spite of what the definition suggests, TT-invariants generated by c do not depend on the initial marking: if the commutative image of $\sigma_1 \cdot \sigma_2$ is a TT-invariant for c, so is the commutative image of $\sigma_2 \cdot \sigma_1$, since it is a firing sequence from the marking M' such that $M \xrightarrow{\sigma_1} M'$.

However, unlike general T-invariants, TT-invariants depend on the set of reachable markings. We will see in the following that they actually mainly depend on the maximal possible number of tokens in circuit c.

The following Lemma characterizes minimal TT-invariants in event graphs, where things are easy.

Lemma 5.1. *Let* c *be an elementary circuit of a live and 1-bounded event graph containing* n *tokens. The minimal TT-invariant generated by* c *is* $(n, \ldots, n) \in \mathbb{N}^T$.

Proof. Recall that the T-invariants of an event graph are of the form $(x, \ldots, x)$, $x \in \mathbb{N} \setminus \{0\}$, see [6–Prop. 3.16]. Now, let J be a minimal TT-invariant associated with $c = p_1 t_1 \cdots$, and let σ be a corresponding compatible firing sequence. Since the Petri net is 1-bounded, tokens along the circuit c cannot overtake each other. Hence, when the token initially in p_1 is back to p_1, after the firing of σ, we know that transition t_1 must have fired n times.

Example 5.1. Figure 2 shows the evolution of an event graph containing 2 tokens in circuit c. We look at the minimal TT-invariant generated by c. Using Lemma 5.1, the minimal TT-invariant is $(2, 2, 2, 2, 2, 2, 2, 2, 2)$. The white token (as well as the black one) is back to its original place (Figure 2(c)).

The minimal T-invariant is $(1, 1, 1, 1, 1, 1, 1, 1, 1)$. Note that, after a single firing of every transition (Figure 2(b)), the marking is unchanged , but the white token has switched its position with the black one. After firing every transition again (Figure 2(c)), the white token is back in the right place.

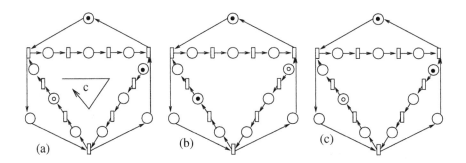

Fig. 2. TT-invariant in an event graph

We now characterize the minimal TT-invariants generated by a circuit of a live and 1-bounded free-choice Petri net. This is a more difficult case.

Let us first give an algorithm to build every minimal TT-invariant (Lemma 5.2 and Proposition 5.1). The following results (Lemma 5.3 and Proposition 5.2) show that a minimal TT-invariant generated by a circuit c is the sum of at most n minimal T-invariants, where n is the maximal number of tokens that c may contain.

Lemma 5.2. *Let c be a circuit of $\mathcal{N}$ a live and 1-bounded free-choice net, and b be a transition of c. Recall that M_b is the unique blocking marking associated with b. For every minimal TT-invariant J, there exists a firing sequence σ compatible with c such that $J = \vec{\sigma}$ that can be fired from M_b.*

Proposition 5.1. *Let c be a circuit of $\mathcal{N}$, a live and 1-bounded free-choice net. Every minimal TT-invariant generated by c is found by applying Algorithm 1.*

Algorithm 1 : Construction of minimal TT-invariants

Input: $\mathcal{N}$, a live and 1-bounded free-choice net, $c = t_1 \cdots t_k, t_{k+1} = t_1$ circuit of $\mathcal{N}$.

Output: A minimal TT-invariant generated by c.

$\sigma \leftarrow \varepsilon$;

for all $i \in \{1, \cdots, k\}$ **do**

   $\sigma_i \leftarrow$ a minimal firing sequence from M_{t_i} to $M_{t_{i+1}}$ in $\mathcal{N}$;

$\sigma \leftarrow \sigma_1 \ldots \sigma_k$;

Return $\vec{\sigma}$.

In the algorithm, a firing sequence $\sigma : M \xrightarrow{\sigma} M'$ is a *minimal* firing sequence if it does not contain any subsequence $\sigma' : M \xrightarrow{\sigma'} M'$. Such a minimal firing sequence has no reason to be unique. Hence the algorithm may yield several different outputs for a given input.

We now show that a minimal TT-invariant generated by an elementary circuit is the sum of n minimal T-invariants, where n is the maximal number of tokens in c (it is given by the number of tokens in c under the blocking marking of any transition of c). Lemma 5.3 is a lemma used to prove Proposition 5.2.

Lemma 5.3. *Let c be an elementary circuit of $\mathcal{N}$, a live and 1-bounded free-choice net and n be the maximal number of tokens in c. There exists a minimal TT-invariant generated by c that is n times the same minimal T-invariant.*

Proposition 5.2. *Let c be an elementary circuit of $\mathcal{N}$, a live and 1-bounded free-choice net and n be the maximal number of tokens in c. Every minimal TT-invariant generated by c is the sum of n minimal T-invariants.*

Example 5.2. Figure 3 illustrates Proposition 5.2: consider the circuit $c = 1, 2, 3, 5, 6, 7, 9, 10, 11$. The blocking marking of transition 1 is $\{d, g, k\}$. To reach the blocking marking of transition 2, the two possible firing sequences are $1, 9, 10, 11$ or $1, 9, 12$. To reach the blocking marking of transition 3, 2 is fired, and to reach

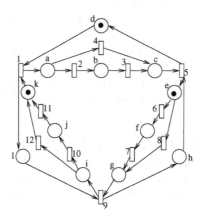

Fig. 3. Example of a free-choice net

the blocking marking of transition 5, 3 is fired. By symmetry of the net, there are also two possible minimal firing sequences from M_5 to M_6, that are $5, 1, 2, 3$ or $5, 1, 4$ and from M_9 to M_{10}, $9, 5, 6, 7$ or $9, 5, 8$. Every other minimal firing sequence from a blocking marking to the next on circuit c is made of only one transition. Then, there are 8 minimal TT-invariants (for three subsequences, there are two possibilities). For example, $1, 9, 10, 11, 2, 3, 5, 1, 4, 6, 7, 9, 5, 8, 10, 11$ is a firing sequence compatible with c whose commutative image is a minimal TT-invariant, which can be written as $I + J$, where I and J are minimal T-invariants, of respective support $1, 2, 3, 5, 6, 7, 9, 10, 11$ and $1, 4, 5, 8, 9, 10, 11$.

Lemma 5.4. *Let C be a circuit in a live and 1-bounded free-choice net and J be a minimal TT-invariant of C. The circuit is composed of the elementary circuits $c_1, \ldots, c_k$. Then there exists k minimal TT-invariants, $J_1, \ldots, J_k$ respectively generated by $c_1, \ldots, c_k$ such that $J = \sum_{i=1}^{k} J_i$.*

6   Extremal Throughputs

We now consider all possible, arbitrarily complex, routing policies and try to address the following questions: Can one compute the routing policy which yields the best or worst throughput? Are the best and worst policies periodic? Are the best and worst policies 0-1?

6.1   Dominance of Periodic Policies

In this section we show that the best and worst throughputs in LBFC with rational firing times are achieved by periodic policies, and we provide an algorithm to construct them. The construction is very close to the one given in Section 3.1 and is basically the one in [3].

Consider a timed LBFC with non-null rational timings $(\tau_t)_{t \in \mathcal{T}}$ and with a vector of weights $\alpha = (\alpha_t)_{t \in \mathcal{T}}$, see Section 3. Let $x \in \mathbb{Q}$ be such that $\tau_t = k_t x, k_t \in \mathbb{N} \setminus \{0\}$, for all $t \in \mathcal{T}$. Build a new LBFC by replacing each transition by a path: a transition with timing τ_t and weight α_t is replaced by k_t transitions of timing 1 and weight $\alpha_t / (k_t x)$. Consider an infinite firing sequence σ for the original Petri net and the corresponding firing sequence σ' in the new Petri net. Then the throughputs, defined as in Section 3, associated with σ and σ' coincide. Therefore we can, without loss of generality, consider LBFC with timings all equal to 1.

Asap Marking Graph of a Free-Choice Net. Consider a timed LBFC $(\mathcal{N}, M, \tau)$ and assume that all the timings are equal to 1: $\forall t \in \mathcal{T}, \tau_t = 1$. Denote by N the incidence matrix. In a given marking, a transition may be enabled several times. So the transitions that can be fired simultaneously form a multi-set.

The *asap marking graph* is defined as follows:

$Q \leftarrow \{M\}$; Arc $\leftarrow \emptyset$; $\tilde{Q} \leftarrow \{M\}$;
while $\tilde{Q} \neq \emptyset$ **do**
  Pick $M' \in \tilde{Q}$;
  for all maximal multi-set U of transitions that can be fired simultaneously
  from M' **do**
    $M'' \leftarrow M' + N.U$;
    if $M'' \notin Q$ **then**
      $Q \leftarrow Q \cup \{M''\}$; $\tilde{Q} \leftarrow \tilde{Q} \cup \{M''\}$;
      Arc $\leftarrow$ Arc $\cup \{M' \rightarrow M''$, with label and weight $[U \mid \sum_{t \in U} \alpha_t]\}$;
  $\tilde{Q} \leftarrow \tilde{Q} \setminus \{M'\}$;

The above construction stops because the Petri net is bounded. So the *asap* marking graph is finite.

All the *as soon as possible* (*asap*) evolutions of the Petri net can be read on this graph, hence its name. In this graph, the *weight* of a path is the sum of the weights of the arcs. The *average weight* of a path is its weight divided by its length (number of arcs).

Theorem 6.1. *Let* $(\mathcal{N}, M, \tau)$ *be a timed LBFC with non-null rational timings. The minimal and maximal throughputs are obtained for periodic routings.*

Proof. Let $(\tilde{\mathcal{N}}, \widetilde{M})$ be the LBFC obtained from $(\mathcal{N}, M)$ after duplicating the transitions such that every transition in the new Petri net has timing 1. Consider the *asap* marking graph of $(\tilde{\mathcal{N}}, \widetilde{M})$. Let c be a circuit of the *asap* graph of maximum average weight. The maximal throughput can be reached by following this circuit.

Also, the minimal throughput can be reached by following the circuit of minimal average weight.

This shows that the corresponding routings are periodic. Indeed the routing can be deduced from the labels along the circuit of the *asap* marking graph. In particular, the period is smaller than the length of the circuit.

Example 6.1. Consider the Petri net of Figure 3. Every transition has timing 1 and weight 1. The *asap* marking graph is represented in Figure 4.

The minimal average weight of a circuit is $15/9$, given by the circuit $\{(dek), (afl), (cgl), (chi), (dej), (dgk), (agl), (bhi), (chk)\}$. This gives the minimal throughput and the routing to reach it. The maximal average weight is attained for the circuit $\{(dek), (agl), (chi)\}$. And the maximal throughput is 2.

When the graph is built, computing the throughput can be made in cubic time in the number of nodes of the *asap* marking graph. But this graph can have an exponential size in the size of the original Petri net. One reason is that the number of markings of the net can be exponential in the size of the net. The other reason is that transitions are duplicated to build the Petri net with timings 1 starting from a Petri net with rational timings. Moreover, this method gives no information about the structure of the corresponding extremal policies.

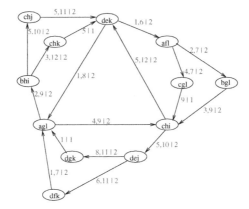

Fig. 4. Asap marking graph corresponding to the Petri net of Figure 3

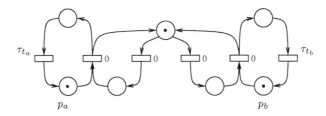

Fig. 5. One-bounded Petri net for which the optimal routing is not periodic

6.2   Non-rational Timings

For general non-rational timings, the maximal throughput is not always periodic, even for one-bounded nets, as shown in the following example.

Example 6.2. Look at Figure 5. The value of the firing times are given next to the transitions on the Figure. We have $\tau_{t_a}/\tau_{t_b} \notin \mathbb{Q}$. This model has been studied in [9]. The optimal routing is Sturmian aperiodic. The best routing consists in choosing the left (t_a) or right transition (t_b) depending on whether one token appears first in p_a or in p_b. If t_a has fired n_a times and t_b n_b times, it suffices to compare $n_a\tau_a$ and $n_b\tau_b$. The non-periodicity comes from the irrationality of the ratio of the timings.

The same Petri net can be considered with non-null rational timings approximating the ones in Figure 6.2. In this case, the maximal throughput is achieved for a Sturmian-like periodic routing policy. In particular, it is not possible to give an absolute bound for the period.

The above behaviors contrast sharply with the results to be proved in Section 6.3 on the minimal throughput for 1-bounded nets.

For general non-rational timings, we do not know if the minimal throughput is always attained by periodic policies. In the next section, we show however that it is the case for the subclass of 1-bounded Petri nets.

6.3   Minimal Throughput in 1-Bounded Free-Choice Nets

Consider a timed live and 1-bounded free-choice net $(\mathcal{N}, M, \tau)$. In this part, we show that the minimal throughput is obtained for a periodic routing even for general non-rational timings. Furthermore, we give a precise insight on the structure of the periodic routing reaching the minimal throughput. Roughly it corresponds to a critical TT-invariant.

From Theorem 4.2, we can easily deduce the following lemma:

Lemma 6.1. *The worst 0-1 routing can be chosen among those that make only one T-component live.*

Critical circuit. Suppose again that the timings are 1. The worst routing can be read on the *asap* marking graph by considering an elementary circuit, $c = (M_0, \cdots, M_{T-1})$, of minimal average weight. The length of c is T, the period of the evolution.

Let t_0 be a live transition of the net appearing in the label between states M_0 and M_1 of the circuit c. We build a path in the Petri net with final extremity t_0 in the following recursive way.

If the path $t_i, \cdots, t_1, t_0$ is built, we choose the transition t_{i+1} in the label U_i of the arc between $M_{T-i-1[T]}$ and $M_{T-i[T]}$ and such that $t_{i+1} \in {}^{\bullet\bullet}t_i$.

We stop the construction when we find $m \in \mathbb{N}^*$ and a transition t_j such that $t_j = t_{j-mT}$. Consider the circuit of the Petri net corresponding to the sequence of transitions $t_j, t_{j-1}, \ldots, t_{j-mT+1}$, and denote it by $\mathcal{C}_c$. The length of this circuit is mT by construction.

Let U_i be the set of transitions whose firing leads from $M_{T-i-1[T]}$ to $M_{T-i[T]}$. (Here U_i is a set and not a multi-set, because the Petri net is 1-bounded.)

Let K be the commutative image of $\sigma = U_{j-1} \cdots U_{j-mT}$. By construction, K is a T-invariant. Also by construction, $(t_{j-1} \cdots t_{j-mT})$ is the first sub-word of σ belonging to $[t_{j-1}], \ldots, [t_{j-mT}]$. So, K is a TT-invariant generated by $\mathcal{C}_c$.

Since K is a T-invariant associated with a firing sequence following a minimal weight circuit in the *asap* marking graph, we deduce that the worst throughput $\underline{D}$ of $(\mathcal{N}, M, \tau)$ satisfies:

$$\underline{D} = \frac{\alpha^T K}{mT} .$$

Since K is a TT-invariant for $\mathcal{C}_c$, there exists J_c a minimal TT-invariant for $\mathcal{C}_c$ such that $K \geq J_c$. The circuit $\mathcal{C}_c$ may not be elementary. In full generality, it is composed of, say, k elementary circuits $c_1, \cdots, c_k$ of length $\ell_1, \cdots, \ell_k$, with $\sum_{i=1}^{k} \ell_i = mT$. By Lemma 5.4, we have $J_c = \sum_{i=1}^{k} J_i$ where J_i is a minimal TT-invariant of c_i. Then,

$$\underline{D} = \frac{\alpha^T K}{mT} \geq \frac{\alpha^T J_c}{mT} = \frac{\alpha^T (\sum_{i=1}^{k} J_i)}{\sum_{i=1}^{k} \ell_i} = \frac{\sum_{i=1}^{k} \alpha^T J_i}{\sum_{i=1}^{k} \ell_i} \geq \min_{i=1}^{k} \frac{\alpha^T J_i}{\ell_i} .$$

Now, by definition, $\underline{D} \leq \min_{i=1}^{k}(\alpha^T J_i)/\ell_i$. Therefore, we can find an elementary circuit c of the Petri net, of length ℓ and of associated minimal TT-invariant J, such that $\underline{D} = \alpha^T J/\ell$. Such a circuit is called a *critical circuit* of the Petri net.

We are now ready to state the main result of this section. Set $\mathbb{R}_+^* = (0, +\infty)$.

Theorem 6.2. *Consider a timed live and 1-bounded free-choice net with general timings in $\mathbb{R}_+^*$. Let α be a weight on the transitions. The minimal throughput is obtained for a periodic routing. For each place, the period of the routing function is bounded by the maximal number of tokens in the net.*

Example 6.3. Consider again the example of Figure 3. Form the corresponding asap marking graph shown on Figure 4, $(1, 2, 3, 5, 6, 7, 9, 10, 11)$ is the critical circuit. Since every transition has weight 1, the firing sequences of minimal weight from the blocking marking of a transition of c to the blocking marking of the next transition of c are (starting from the blocking marking of transition 1 which is $\{d, k, g\}$): $(1, 9, 12)$; (2); (3); $(5, 1, 4)$; (6); (7); $(9, 5, 8)$; (10); (11). The minimal TT-invariant achieving the minimal throughput is composed of two minimal T-invariant of support $1, 2, 3, 5, 6, 7, 9, 10, 11$ and $1, 4, 5, 8, 9, 12$. The minimal throughput is then $15/9$ as anounced before. As for the worst routing policy, it can be obtained directly from the minimal TT-invariant: $u_a = (2, 4)^\infty, u_e = (8, 6)^\infty$ and $u_i = (12, 10)^\infty$.

6.4   Algorithm to Compute a Routing that Minimizes the Throughput

Consider a timed live and 1-bounded free-choice net $(\mathcal{N}, M, \tau)$. Let N be the incidence matrix. Let Clusters be the set of clusters. Define the matrix $K \in \{0, 1\}^{\text{Clusters} \times T}$ such that $K_{a,t} = 1$ if t belongs to the cluster a and $K_{a,t} = 0$ otherwise.

Let S be a subset of T. Let **Lightest-T-invariant**(S) be the algorithm that computes a minimal T-invariant of minimal weight that contains the transitions of S. It is the solution of the following linear programming problem:

Algorithm 2 : Lightest-T-invariant

Input: $S \subseteq T$
Minimize $\alpha^T \cdot I$
With constraints $N \cdot I = 0$; $K \cdot I \leq (1, \ldots, 1)^T$; $I \geq \chi_S$.

When the algorithm is called when S is a set of transitions belonging to the same T-component, the condition $K \cdot I \leq (1, \ldots, 1)^T$ ensures that the output is a set of disjoint minimal T-invariants, with only one containing S.

This algorithm runs in polynomial time since it is a linear program over $\mathbb{Q}$. Due to the form of the constraints, the solution will always be $\{0, 1\}$-valued.

Let us define the functions **Blocking-Marking**(t), **Cycle-time**(A), and **Timing**(c).

- **Blocking-Marking**(t) computes the blocking marking of transition t. This marking can be computed in time $O(|\mathcal{T}|^2)$ when the Petri net is an event graph, and in time $O(|\mathcal{T}|^3)$ for the general free-choice case, see [8].
- **Cycle-time**(A) computes the cycle time of a (max,+) matrix A. Here, it is used for matrices of dimension at most $|\mathcal{P}|$. Then, the time complexity is $O(|\mathcal{P}|^3)$ using Karp's algorithm, see for instance [1].
- **Timing**(c) computes the sum of the firing times of the transitions along the circuit. The time complexity is linear.

In the algorithm below, the (max,+) representation of the behavior of live and 1-bounded free-choice nets is used. For every transition b, A_b is the (max,+) matrix representing the time behavior of the firing of b (see [7] for more details). The symbol $\otimes$ denotes the multiplication of matrices in the (max,+) algebra. This operation can be done in cubic time in the dimension of the matrices. Here, it is used for matrices of dimension at most $|\mathcal{P}|$.

Using the previous results, we get the following theorem.

Theorem 6.3. *Algorithm 3 finds the minimal throughput of a timed live and 1-bounded free-choice Petri net.*

Algorithm 3 : Worst-routing

Input: $(\mathcal{N}, M, \tau)$ a timed $(\tau \in (\mathbb{R}_+^*)^{\mathcal{T}})$ 1-bounded LBFC; and $\alpha = (\alpha_t)_{t \in \mathcal{T}} \neq (0, \ldots, 0)$, a weight vector.
for all $b \in \mathcal{T}$ **do**
   $M_b \leftarrow$ **Blocking-Marking**(b);
for all $b, b' \in \mathcal{T}$ such that $\exists p,\ b \to p \to b'$ **do**
   $J \leftarrow$**Lightest-T-invariant**(b, b');
   $\sigma_{bb'} \leftarrow b, t_1, \cdots, t_m$ minimal firing sequence from M_b to $M_{b'}$ with transitions of J;
   $A_{bb'} \leftarrow A_b \otimes A_{t_1} \otimes \cdots \otimes A_{t_m}$;
   $\alpha_{bb'} \leftarrow \alpha_b + \alpha_{t_1} + \cdots + \alpha_{t_m}$;
Throughput $\leftarrow +\infty$;
Tmin $\leftarrow \emptyset$;
for all elementary circuit $c = t_1 \cdots t_k$ of $\mathcal{N}$ **do**
   $A \leftarrow A_{t_1 t_2} \otimes A_{t_2 t_3} \otimes \cdots \otimes A_{t_k t_1}$;
   $\alpha \leftarrow \alpha_{t_1 t_2} + \cdots + \alpha_{t_k t_1}$;
   if cycle-time$(A) = $ **Timing**(c) **then**
      if $\alpha/$**cycle-time**$(A) <$Throughput **then**
         $J_{\min} \leftarrow \overrightarrow{\sigma_{t_1 t_2} \cdots \ldots \sigma_{t_k t_1}}$;
         Throughput$\leftarrow \alpha/$**Timing**(c);

At each iteration in the first loop, the time complexity is $O(|\mathcal{T}|^3)$, and there are $|\mathcal{T}|$ iterations. Consider now the second loop. There are at most $|\mathcal{T}|^2$ iterations in the loop. At each iteration, we need to find a minimal sequence from M_b to $M_{b'}$ in a T-component. This can be done in time $O(|\mathcal{T}|^2)$. The length of a minimal sequence is of order $O(|\mathcal{T}|^2)$, see [8]. Hence the matrix $A_{bb'}$ can be computed in time $O(|\mathcal{P}|^3|\mathcal{T}|^2)$. At each iteration of the last loop, the time complexity is $O(|\mathcal{P}|^3|\mathcal{T}|)$ and there are as many iterations as elementary circuits in the net. Therefore, the total time complexity is $O(\mathcal{C}|\mathcal{P}|^3|\mathcal{T}|)$, where $\mathcal{C}$ is the number of elementary circuits. Since the number of elementary circuits can be exponential in the number of places $(O(2^{|\mathcal{T}|}))$, the time complexity is exponential in the worst case. As for the space complexity, it remains polynomial in the size of the Petri net.

For comparison, consider the method of computation given (for rational timings) in Section 6.1 and which uses the *asap* marking graph. The size of the *asap* marking graph is exponential in the size of the original Petri net, more precisely its size is $O(2^{|\mathcal{P}|})$. So the complexity in time is $O((2^{|\mathcal{P}|})^3) = O(8^{|\mathcal{P}|})$ and the space complexity is at least $O(2^{|\mathcal{P}|})$. Observe that these complexities are evaluated without taking into account the necessity of transforming the Petri net with rational timings into an equivalent one with timings equal to 1, see Section 6.1. This transformation makes both time and space complexity of the classical method even worse.

Finally, remark that our construction also gives some insight on the period of the worse policy. Since the critical TT-invariant associated with the critical circuit (with n tokens) is the sum of n T-invariants, this means that the period of the worse routing policy in each routing place is periodic with a period $n \leq |\mathcal{P}|$. This is several order of magnitude smaller than the period that can be deduced from the classical algorithm which is exponential $O(2^{|\mathcal{P}|})$.

6.5   Bounded Nets

If one considers a live and k-bounded free-choice net with $k \geq 2$, then the previous constructions for 1-bounded nets do not work anymore. If the timings are rational, the worst throughput is reached for periodic routings (Theorem 6.1), but the period is not bounded by the number of tokens in a circuit, as shown by the following example.

Example 6.4. Figure 6(a) represents a 2-bounded free-choice net where the period of the worst routing is greater that the number of tokens in any circuit. Furthermore the critical circuit for this routing is not elementary.

All the timings are equal to 1, as well as the weights. The routing in p_1 which gives the minimal throughput is, $(t1t4t4t4)^\infty$, and the periodic evolution is given in Figure 6(b).

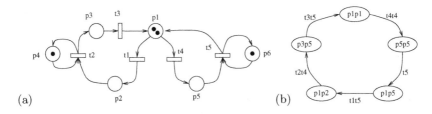

Fig. 6. Free-choice net and its worst evolution

The throughput of this evolution is 9/5, whereas if only the left (or right) event graph is live the throughput is 2. The period of the routing function of place p_1 is greater than the number of tokens in the circuits containing this place (2 tokens), as opposed to what happens in the case of 1-bounded nets. In this example, by changing the timings and/or increasing the number of transitions, but keeping the same number of tokens, it is possible to increase the period of the worst routing function arbitrarily. Therefore, bounding the period of the worst routing seems difficult, even for 2-bounded nets.

References

1. F. Baccelli, G. Cohen, G.J. Olsder, and J.P. Quadrat. *Synchronization and Linearity*. John Wiley & Sons, New York, 1992.
2. A. Bouillard, B. Gaujal and J. Mairesse. *Extremal throughput in free-choice nets*. Research Report RR LIP 2005-14, ENS Lyon, France, 2005.
3. J. Carlier and P. Chretienne. Timed Petri net schedules. In *Advances in Petri Nets*, number 340 in LNCS, pages 62–84. Springer Verlag, 1988.
4. P. Chretienne. *Les Réseaux de Petri Temporisés*. PhD thesis, Université Paris VI, Paris, 1983.
5. G. Cohen, D. Dubois, J.P. Quadrat, and M. Viot. A linear system-theoretic view of discrete-event processes and its use for performance evaluation in manufacturing. *IEEE Trans. Automatic Control*, 30:210–220, 1985.
6. J. Desel and J. Esparza. *Free Choice Petri Nets*, volume 40 of *Cambridge Tracts in Theoretical Comp. Sc.* Cambridge Univ. Press, 1995.
7. S. Gaubert and J. Mairesse. Modeling and analysis of timed Petri nets using heaps of pieces. *IEEE Trans. Aut. Cont.*, 44(4):683–698, 1999.
8. B. Gaujal, S. Haar and J. Mairesse. Blocking a transition in a free-choice net and what it tells about its throughput. *Journal of Computer and System Sciences*, 66(3):515-548, 2003.
9. J. Mairesse and L. Vuillon. Asymptotic behavior in a heap model with two pieces. *Theoret. Comput. Sci.*, 270:525–560, 2002.

A Framework to Decompose GSPN Models*

Leonardo Brenner, Paulo Fernandes**, Afonso Sales, and Thais Webber

Pontifícia Universidade Católica do Rio Grande do Sul,
Av. Ipiranga, 6681 – 90619-900 – Porto Alegre, Brazil
{lbrenner, paulof, asales, twebber}@inf.pucrs.br

Abstract. This paper presents a framework to decompose a single
GSPN model into a set of small interacting models. This decomposition
technique can be applied to any GSPN model with a finite set of tangible markings and a generalized tensor algebra (Kronecker) representation
can be produced automatically. The numerical impact of all the possible decompositions obtained by our technique is discussed. To do so we
draw the comparison of the results for some practical examples. Finally,
we present all the computational gains achieved by our technique, as well
as the future extensions of this concept for other structured formalisms.

1 Introduction

It is common knowledge in the research community the advantages in using
the GSPN (*Generalized Stochastic Petri Nets*) formalism [2] to model complex
systems, *i.e.*, systems with both parallel and synchronous behavior. For a quite
long time, the main limitation to use the GSPN formalism was the absence of
an efficient numerical support to handle useful, and consequently large, models.
Ciardo and Trivedi's work [14] brought a first approach that could be employed
to decompose a single model into components. However, their approach does
not mention any specific storage or numerically suitable solution technique. The
need of a theoretical tool to represent such structured models naturally leads
to *Tensor Algebra* representations [4, 8, 3, 15]. The term *Tensor Algebra* is being
employed in this paper, but many authors still prefer the classic denomination
Kronecker Algebra chosen in honor of Leopold Kronecker.

The first complete approaches in this direction were the works of Donatelli
in the SGSPN (*Superposed GSPN*) formalism [16, 17]. By complete, we understand that it was proposed a complete framework to: construct a SGSPN model
by assembling synchronized components; generate a Markovian descriptor, *i.e.*,
a tensor algebra formula, as the infinitesimal generator of the equivalent Markov
chain; and consequently an efficient way to solve it (functional elements). However, the SGSPN formalism could only be used to model a rather small class
of GSPN models which comply to the restrictive rules of generation defined by

---

* This work was partially funded by CNPq/Brazil.
** Corresponding author. The order of authors is merely alphabetical.

Donatelli, *i.e.*, a SGSPN model is composed of a set of standard GSPN models which interact only through a set of synchronized transitions.

The SGSPN application scope restriction, and the consequent disadvantages in terms of numerical performance, suggests the use of other formalisms that could be closer to the tensor representation, such as SAN (*Stochastic Automata Networks*) [26]. At this time, the solution through the shuffle algorithm used in SAN [18, 7] presents an efficient solution with reasonable memory needs. Evidently, the use of other structured storage and solution techniques instead of the tensor algebra also presents good alternatives to the limited scope of the SGSPN formalism. This is the case of the quite impressive techniques based on MDD (*Multi-valued Decision Diagrams*) [22] and MxD (*Matrix Diagrams*) [21] proposed by Ciardo and Miner. Furthermore, the MDD and MxD techniques are very efficient to solve very sparse models, *i.e.*, models with a huge product state space and a comparatively small number of reachable states. In fact, we believe that the techniques based on tensor algebra are still worthy, at least considering the new improvements to handle tensor structures [5, 10].

This paper presents a study about the decomposition of a very general class of GSPN models, exploiting the description power of the GSPN formalism. It also proposes a memory efficient tensor algebra format to describe the components and their interactions. As the first step, we formally define the class of GSPN models in which our technique can be applied. The proposed decomposition technique and the consequent tensor format representation are generalizations of the SGSPN formalism [17], but we extend the application scope following the ideas firstly advanced in [14] and employed later in [20]. The new contribution of our work is justified by the numerical impact of the decomposition choices on the storage demands.

We are specifically interested to handle models with a really large reachable state space. Buchholz, Ciardo, Donatelli, Kemper, and Miner [9, 23, 11] already present very efficient methods to deal with absolutely huge models (*e.g.*, 9.18×10^{626} states in 1000 dining philosophers example [22]), but with considerably fewer reachable states. Our decomposition technique intends to split a GSPN model into subnets providing a structured representation. Regardless the number of subnets, we will always have the same reachable state space. With many (small) subnets, we have a very structured (and therefore memory efficient) representation, but also a product state space much larger than the reachable state space. With few (large) subnets, we have a less structured representation, but a product state space equal or a little bit larger than the reachable state space. In fact, we intend to compare possible decompositions in order to show the trade off between many small and few large subnets.

In addition, we point out the underestimated benefits of the use of guards in the GSPN formalism, which can be clearly demonstrated by the tensor format proposed in our work. We do not pay much attention in this paper to the computational cost to solve the tensor representations. The recent evolutions in pure tensor solutions [5, 10], the promising ideas of parallel implementations, and the MDD and MxD techniques [12] suggest many changes in the computational cost in a near future. We focus our interest in the memory savings due to our decomposition techniques and the corresponding tensor format.

The next section briefly presents the theoretical tool used to the tensor representation: Classical (CTA) and Generalized Tensor Algebra (GTA). Section 3 describes the GSPN formalism, the application scope of our technique and the scope of the technique proposed for SGSPN. Section 4 presents our decomposition technique and the corresponding tensor format. Section 5 draws some considerations about the possible choices of decomposition. Section 6 presents some modeling examples in order to discuss numerical issues about the decomposition technique. Finally, the conclusion summarizes our contribution and suggests the still vast future work to be done.

2   Tensor Algebra

In this section, the concepts of Classical Tensor Algebra [3, 15] and Generalized Tensor Algebra [25, 18] are briefly presented.

2.1   CTA - Classical Tensor Algebra

The tensor product of two matrices: A of dimensions $(\rho_1 \times \gamma_1)$ and B of dimensions $(\rho_2 \times \gamma_2)$ is a tensor with dimensions $(\rho_1\rho_2 \times \gamma_1\gamma_2)$ which may be considered as consisting of $\rho_1\gamma_1$ blocks each having dimensions $(\rho_2\gamma_2)$, $i.e.$, the dimensions of B. To specify a particular element, it suffices to specify the block in which the element occurs and the position within that block of the element under consideration. Thus, as mentioned previously, element c_{36} $(a_{11}b_{02})$ is in the $(1, 1)$ block and at position $(0, 2)$ of that block. The tensor $C = A \otimes B$ is defined by assigning to the element of C that is in the (k, l) position of block (i, j), the value $a_{ij}b_{kl}$, $i.e.$, $c_{[ik][jl]} = a_{ij}b_{kl}$. The *tensor sum* of two *square* matrices A and B is defined in terms of tensor products as:

$$A \oplus B = A \otimes I_{n_B} + I_{n_A} \otimes B$$

where n_A is the order of A; n_B is the order of B; I_{n_i} is the identity matrix of order n_i; and "$+$" represents the usual operation of matrix addition. Since both sides of this operation (matrix addition) must have identical dimensions, it follows that tensor addition is defined for square matrices only. The value assigned to the element $c_{[ik][jl]}$ of the tensor $C = A \oplus B$ is $c_{[ik][jl]} = a_{ij}\delta_{kl} + b_{kl}\delta_{ij}$, where δ_{ij} is the element of ith row and j^{th} column of an identity matrix defined as $\delta_{ij} = 1$ for $i = j$ and $\delta_{ij} = 0$ for $i \neq j$.

2.2   GTA - Generalized Tensor Algebra

Generalized Tensor Algebra is an extension of Classical Tensor Algebra. The main distinction of GTA with respect to CTA is the addition of the concept of *functional elements*. However, a matrix can be composed of constant elements (belonging to $\mathbb{R}$) or functional elements. A functional element is a function evaluated in $\mathbb{R}$ according to a set of parameters composed of the rows of one or more matrices. Generalized tensor product is denoted by $\underset{g}{\otimes}$. The value assigned to the element $c_{[ik][jl]}$ of the tensor $C = A(\mathcal{B}) \underset{g}{\otimes} B(\mathcal{A})$ is $c_{[ik][jl]} = a_{ij}(b_k)b_{kl}(a_i)$. Generalized tensor sum is also analogous to the ordinary tensor sum, and is denoted by $\underset{g}{\oplus}$. The elements of the tensor $C = A(\mathcal{B}) \underset{g}{\oplus} B(\mathcal{A})$ are $c_{[ik][jl]} = a_{ij}(b_k)\delta_{kl} + b_{kl}(a_i)\delta_{ij}$.

3   Generalized Stochastic Petri Nets

The GSPN formalism [2] is a performance analysis tool on the graphical system representation typical of Petri Nets [27, 24]. The GSPN formalism is derived from the SPN formalism and contains two types of transitions: *timed* and *immediate*. An exponentially distributed random firing time is associated with each timed transition, whereas immediate transitions, by definition, fire in zero time. Immediate transitions always have precedence to fire over timed transitions. The GSPN models with immediate transitions can always be represented by a model with timed transitions.

In the graphical representation of a GSPN model, places are drawn as circles, timed transitions as rectangles and immediate transitions as bars. Places may contains *tokens*, which are drawn as black dots. A place is an input to a transition if an arc exists from the place to the transition. A place is an output from a transition if an arc exists from the transition to the place. A transition is enabled when all of its input places contain at least one token. Enabled transitions can fire, thus removing one token from each input place and placing one token in each output place. Additionally, a condition can be associated to enable the firing of the transitions. Such conditions are called *guards* and, with the availability of tokens in the input places, they are the only restrictions to enable the firing of a given transition. A formal description is presented as follows.

Let

$\mathcal{C}$ set of conditions associated to transitions of $\mathcal{T}$.

Definition 1. *A $\mathcal{GSPN}$ is defined by tuple $(\mathcal{P}, \mathcal{T}, \pi, I, O, W, G, M_0)$, where:*

1.1. $\mathcal{P}$ non-empty set of places;

1.2. $\mathcal{T}$ non-empty set of transitions;

1.3. $\pi \colon \mathcal{T} \to \{0, 1\}$ priority function of the transitions;

1.4. I and $O \colon \mathcal{T} \to \mathcal{P}$ input and output functions of the transitions;

1.5. $W \colon \mathcal{T} \to \mathbb{R}^+$ function that assigns a rate to each transition;

1.6. $G \colon \mathcal{T} \to \mathcal{C}$ function, called guard, that associates a necessary, but not sufficient, condition $c \in \mathcal{C}$ to the firing of each transition $t \in \mathcal{T}$;

1.7. $M_0 \colon \mathcal{P} \to \mathbb{N}$ initial marking in each place.

Definition 2. *$c \in C$ is a condition that may be associated to a transition $t \in T$, which depends on tokens of one or more $p \in P$. This condition is a function with domain on tokens of all places p and counter-domain on $\mathbb{R}$.*

A condition c defines the firing rate of a transition according to the number of tokens in a specific set of places. Although the counter-domain of c is $\mathbb{R}$, only a discrete set of values can be obtained, since the possible combination of markings of places (*i.e.*, the domain of c) is a discrete set.

Definition 3. *Set of timed transitions T_T of a $\mathcal{GSPN}$ is defined as $T_T = \{t \in T \mid \pi(t) = 0\}$.*

Definition 4. *Set of immediate transitions T_I of a $\mathcal{GSPN}$ is defined as $T_I = \{t \in T \mid \pi(t) = 1\}$.*

Definition 5. *Set of transitions T of a $\mathcal{GSPN}$ is defined as $T = T_T \cup T_I$ and $T_T \cap T_I = \emptyset$.*

Numerical Solution Restriction. Although the framework proposed in this paper could be applied to a larger class of GSPN models, we assume a single restriction in order to facilitate the stationary or transient numerical solution: the set of tangible markings of the models must be finite.

4   Framework

We present in this section a framework to decompose GSPN models. Our decomposition technique is shown in Fig. 1.

The basic idea is to decompose a GSPN model into N components $\mathcal{GSPN}^{(i)}$ ($i \in [1..N]$), where each component $\mathcal{GSPN}^{(i)}$ is viewed as a subsystem of the GSPN model. A component $\mathcal{GSPN}^{(i)}$ may not be a GSPN model. It is then necessary to know the possible tangible markings. This is done by the construction of $T\mathcal{RG}^{(i)}$ considering the possible firing of all transitions limited by: availability of tokens; guards; and maximum number of tokens in each place.

A component $\mathcal{GSPN}^{(i)}$ has an independent behavior (*local transitions*) and occasional interdependencies (*synchronized transitions* and/or transitions with guards). It is important to notice that there is a strong equivalence between the decomposition of all $T\mathcal{RG}^{(i)}$ and the $T\mathcal{RG}$ of the whole GSPN (which is the underlying Markov Chain). Nevertheless the computational cost to obtain the $T\mathcal{RG}$ from the composition of all $T\mathcal{RG}^{(i)}$ is usually too high. In fact, the memory needs can be prohibitive as will be seem in the Section 6.

In the next section, we define a component $\mathcal{GSPN}^{(i)}$ and its properties. In Section 4.2, we formally present the tensor format (*Markovian Descriptor*) used to obtain the infinitesimal generator Q of a decomposed GSPN model.

4.1   Decomposition

There is no restriction to decompose a GSPN model with a finite set of tangible marking into N components $\mathcal{GSPN}^{(i)}$ ($i \in [1..N]$). Our technique is less restrictive than the definition of the SGSPN formalism.

Definition 6.   *Each component $\mathcal{GSPN}^{(i)}$ is defined as a GSPN, i.e., it is defined by tuple ($\mathcal{P}^{(i)}$, $\mathcal{T}^{(i)}$, $\pi^{(i)}$, $I^{(i)}$, $O^{(i)}$, $W^{(i)}$, $G^{(i)}$, $M_0^{(i)}$), where:*

6.1. $\mathcal{P}^{(i)}$ non-empty set of places, such that $p^{(i)} \in \mathcal{P}^{(i)} \rightarrow p^{(i)} \in \mathcal{P}$ and $\overset{N}{\underset{i=1}{\cup}} \mathcal{P}^{(i)} = \mathcal{P}$ and $\not\exists \mathcal{P}^{(i)} = \mathcal{P}$;

6.2. $\mathcal{T}^{(i)}$ non-empty set of transitions, such that $t^{(i)} \in \mathcal{T}^{(i)} \rightarrow t^{(i)} \in \mathcal{T}$ and $\exists p \in I^{(i)}(t)$ or $\exists p \in O^{(i)}(t)$ such that $p \in \mathcal{P}^{(i)}$;

6.3. $\pi^{(i)}: \mathcal{T}^{(i)} \rightarrow \{0,1\}$ priority function of the transitions;

6.4. $I^{(i)}$ and $O^{(i)}: \mathcal{T}^{(i)} \rightarrow \mathcal{P}^{(i)*}$ input and output functions of the transitions in which $\mathcal{P}^{(i)*}$ denotes a possibly empty set of places;

6.5. $W^{(i)}: \mathcal{T}^{(i)} \rightarrow \mathbb{R}^+$ function that assigns a rate to each transition;

6.6. $G^{(i)}: \mathcal{T}^{(i)} \rightarrow \mathcal{C}$ function guard that associates a necessary, but not sufficient, condition $c^{(i)} \in \mathcal{C}$ to the firing of each transition $t^{(i)} \in \mathcal{T}^{(i)}$;

6.7. $M_0^{(i)}: \mathcal{P}^{(i)} \rightarrow \mathbb{N}$ initial marking in each place $p^{(i)} \in \mathcal{P}^{(i)}$.

It is important to notice that the set of places $\mathcal{P}^{(i)}$ is a subset of $\mathcal{P}$, as well as the set of transitions $\mathcal{T}^{(i)}$ is a subset of $\mathcal{T}$. Obviously, the subset of places $\mathcal{P}^{(i)}$ of a component $\mathcal{GSPN}^{(i)}$ cannot be the whole set of places $\mathcal{P}$, otherwise there is no decomposition. The same restriction does not apply to $\mathcal{T}^{(i)}$, since it can be identical to $\mathcal{T}$.

There is no restriction to places superposition. A place $p \in \mathcal{P}$ can be in as many subsets of places $\mathcal{P}^{(i)}$ as wanted. Obviously, the same applies to transitions. The sole restriction regards the immediate transitions that cannot be used to synchronized two or more partitions. However such restriction is a minor discomfort since all GSPN model can be described by an equivalent SPN (*i.e.*, without immediate transitions) model. Elements of tuple ($\mathcal{P}^{(i)}$, $\mathcal{T}^{(i)}$, $\pi^{(i)}$, $I^{(i)}$, $O^{(i)}$, $W^{(i)}$, $G^{(i)}$, $M_0^{(i)}$) are conservatives, *i.e.*, an element in component $\mathcal{GSPN}^{(i)}$ has the same value of the corresponding element in that original GSPN, *e.g.*, if $t^{(i)}$ correspond to t, then $W^{(i)}(t^{(i)})$ has the same value of $W(t)$.

4.2   Tensor Format

We now formally present the tensor format (*Markovian Descriptor*) used to obtain the infinitesimal generator Q of a decomposed GSPN model. As shown in Fig. 1, the decomposition technique uses the concepts of *Tangible Reachability Graph* and *Stochastic State Machine*.

So we firstly remind the classical definitions of *P-invariants*, *Reachability Set* (RS), *Tangible Reachability Set* (TRS), *Tangible Reachability Graph* (TRG) and *Stochastic State Machine* (SSM).

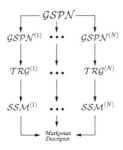

Fig. 1. Decomposition technique

Let

C 　　　incidence matrix of a GSPN (dimensions: $|\mathcal{P}| \times |\mathcal{T}|$);

c_{jk} 　　element from row j and column k of an incidence matrix.

Definition 7. *Elements of an incidence matrix C are defined by:*

7.1. $\forall p_j \in \mathcal{P}, \forall t_k \in \mathcal{T}$

$$c_{jk} = \begin{cases} +1 & \text{if } p_j \in O(t_k) \\ -1 & \text{if } p_j \in I(t_k) \\ 0 & \text{if } p_j \notin O(t_k) \text{ and } p_j \notin I(t_k) \end{cases}$$

Definition 8. *P-invariants of a GSPN are defined by vector solutions σ composed of non-negative integer: 0 and 1, given by equation $\sigma C = 0$ [27], where value 1 in i^{th} position of σ means that i^{th} place of GSPN belongs to the P-invariant.*

Let

$\mathcal{PI}$ 　　minimal set of P-invariants, where $\mathcal{PI} = \{\mathcal{PI}_1, \mathcal{PI}_2, ...\}$.

The scalar product between a P-invariant and any marking M produces a *constant*. If in a GSPN all places are covered by P-invariant, the maximum number of tokens in any place in any reachable marking is finite, and the net is said to be *bounded* [1]. Therefore, a GSPN must have all places covered by P-invariants (all places are bounded) in order to have a finite set of tangible markings. We assume a minimal set of P-invariants as a set with the smaller number of P-invariants that covers all places of the whole net.

Let

$M_i(p)$ 　number of tokens in place p in marking M_i;

$B(\mathcal{PI}_i)$ 　number of tokens in any P-invariant $\mathcal{PI}_i$ (*bound*);

$max(p)$ 　maximum number of tokens in place p defined as the minimum $B(\mathcal{PI}_i)$ for all $\mathcal{PI}_i$, where $p \in \mathcal{PI}_i$;

$M_k[t > M_l$ 　change from marking M_k to M_l due to the firing of t.

Definition 9. *Reachability Set* $\mathcal{RS}^{(i)}(M_0^{(i)})$ *of component* $\mathcal{GSPN}^{(i)}$ *is defined as the smallest set of markings, such that:*

9.1. $M_0^{(i)} \in \mathcal{RS}^{(i)}(M_0^{(i)})$;

9.2. $M_l^{(i)} \in \mathcal{RS}^{(i)}(M_0^{(i)})$, *if and only if* $\forall p^{(i)}, M_l^{(i)}(p^{(i)}) \leq max(p^{(i)})$; *and* $\exists M_k^{(i)} \in \mathcal{RS}^{(i)}(M_0^{(i)})$ *and* $\exists t \in T^{(i)}$ *such that* $M_k^{(i)}[t > M_l^{(i)}$.

Definition 10. *Tangible Reachability Set* $\mathcal{TRS}^{(i)}(M_0^{(i)})$ *of component* $\mathcal{GSPN}^{(i)}$ *is composed of all tangible markings of* $\mathcal{RS}^{(i)}(M_0^{(i)})$.

Definition 11. *Tangible Reachability Graph* $\mathcal{TRG}^{(i)}(M_0^{(i)})$ *of component* $\mathcal{GSPN}^{(i)}$ *given an initial marking* $M_0^{(i)}$ *is a labelled directed multigraph whose set of nodes* $\mathcal{TM}^{(i)}$ *is composed of markings of Tangible Reachability Set* $\mathcal{TRS}^{(i)}$ $(M_0^{(i)})$ *and whose set of arcs* $\mathcal{TARC}^{(i)}$ *is defined as follows:*

11.1. $\mathcal{TARC}^{(i)} \subseteq \mathcal{TRS}^{(i)}(M_0^{(i)}) \times \mathcal{TRS}^{(i)}(M_0^{(i)}) \times T_T^{(i)} \times T_I^{(i)*}$;

11.2. $a^{(i)} = [M_k^{(i)}, M_l^{(i)}, t_0, \sigma] \in \mathcal{TARC}^{(i)}$, *if and only if* $M_k^{(i)}[t_0 > M_1^{(i)}, \sigma = t_1, \ldots, t_n, (n \geq 0)$; *and*

11.3. $\exists M_2^{(i)}, \ldots, M_n^{(i)}$ *such that* $M_1^{(i)}[t_1 > M_2^{(i)}[t_2 > \ldots M_n^{(i)}[t_n > M_l^{(i)}$.

Definition 12. *A Stochastic State Machine (SSM) is defined by tuple* $(\mathcal{P}, \mathcal{T}, F, \Lambda)$, *where:*

12.1. $\mathcal{P}$ *set of non-empty places;*

12.2. $\mathcal{T}$ *set of non-empty transitions;*

12.3. $F \subseteq ((\mathcal{P} \times \mathcal{T}) \cup (\mathcal{T} \times \mathcal{P}))$ *with* $dom(F) \cup codom(F) = \mathcal{P} \cup \mathcal{T}$ *is the flow relation. It has to satisfy the following restriction*[1]: $\forall t \in \mathcal{T}: |^\circ t| = |t^\circ| = 1$;

12.4. $\Lambda : \mathcal{T} \to \mathbb{R}^+$, *where* $\Lambda(t)$ *is the rate of the exponential probability distribution associated to transition* t.

A decomposed GSPN model has N components $\mathcal{GSPN}^{(i)}$, where $i \in [1..N]$. Each component $\mathcal{GSPN}^{(i)}$ has a tangible reachability graph $\mathcal{TRG}^{(i)}$ (Definition 12). Each tangible reachability graph $\mathcal{TRG}^{(i)}$ has an equivalent stochastic state machine $\mathcal{SSM}^{(i)}$ such that:

1. Each node $M_j^{(i)} \in \mathcal{TM}^{(i)}$ corresponds to $p^{(i)} \in \mathcal{P}^{(i)}$ of $\mathcal{SSM}^{(i)}$;

2. Each arc $a^{(i)} \in \mathcal{TARC}^{(i)}$ corresponds to $[p^{(i)}, t^{(i)}] \in F^{(i)}$ and $[t^{(i)}, q^{(i)}] \in F^{(i)}$, if and only if exist $a^{(i)} = [M_k^{(i)}, M_l^{(i)}, t, \sigma]$ such that $M_k^{(i)}$ corresponds to place $p^{(i)} \in \mathcal{P}^{(i)}$, $M_l^{(i)}$ corresponds to place $q^{(i)} \in \mathcal{P}^{(i)}$, $t \in T_T^{(i)}$, and $\sigma \in T_I^{(i)*}$.

---

[1] $|^\circ t|$ and $|t^\circ|$ indicate the number of input and output places of t.

The transition rate of $t^{(i)} \in \mathcal{T}^{(i)}$ (obtained from $[M_k^{(i)}, M_l^{(i)}, t, \sigma]$) can be computed as $\Lambda(t).\Lambda(\sigma)^2$, where $\Lambda(t)$ is the transition rate of t. Any transition $t^{(i)} \in \mathcal{T}^{(i)}$, whose guard has dependency on markings of other components $\mathcal{GSPN}^{(j)}$ ($i, j \in [1..N]$ and $i \neq j$), has a function f multiplied by its rate. Function f is evaluated as *true* for all markings whose its guard is satisfied, and *false* otherwise. So we can now classify a transition as *local* or *synchronized*.

Let

$\mathcal{T}_l^{(i)}$ set of local transitions of component $\mathcal{SSM}^{(i)}$;

$\mathcal{T}_s^{(i)}$ set of synchronized transitions of component $\mathcal{SSM}^{(i)}$.

Definition 13. *Set of synchronized transitions $\mathcal{T}_s$ of a decomposed GSPN model is defined as $\mathcal{T}_s = \mathcal{T}_s^{(1)} \cup \mathcal{T}_s^{(2)} \cup \ldots \cup \mathcal{T}_s^{(N)}$.*

Markovian Descriptor is an algebraic formula that allows to store, in a compact form, the infinitesimal generator of an equivalent Markov chain. This mathematical formula describes the infinitesimal generator through the transition tensors of each component. Each component $\mathcal{SSM}^{(i)}$ has associated:

- 1 tensor $Q_l^{(i)}$, which has all transition rates for local transitions in $\mathcal{T}_l^{(i)}$;
- $2|\mathcal{T}_s|$ tensors $Q_{t+}^{(i)}$ and $Q_{t-}^{(i)}$, which have all transition rates for synchronized transitions in $\mathcal{T}_s^{(i)}$.

Let

$\mathcal{Q}_k^{(i)}(p^{(i)}, q^{(i)})$ tensor element $Q_k^{(i)}$ from row $p^{(i)}$ and column $q^{(i)}$, where $i \in [1..N]$ and $k \in \{l, t^+, t^-\}$;

$I_{|\mathcal{P}^{(i)}|}$ identity tensor of order $|\mathcal{P}^{(i)}|$, where $i \in [1..N]$;

$\tau_t(p^{(i)}, q^{(i)})$ occurrence rate of transition $t \in \mathcal{T}^{(i)}$, where $[p^{(i)}, t] \in F^{(i)}$ and $[t, q^{(i)}] \in F^{(i)}$;

$succ_t(p^{(i)})$ successor place $q^{(i)}$ such that $[p^{(i)}, t] \in F^{(i)}$ and $[t, q^{(i)}] \in F^{(i)}$.

Definition 14. *Tensor elements $Q_l^{(i)}$, which represent all local transitions $t \in \mathcal{T}_l^{(i)}$ of component $\mathcal{SSM}^{(i)}$, are defined by:*

14.1. $\forall p^{(i)}, q^{(i)} \in \mathcal{P}^{(i)}$ *such that* $q^{(i)} \in succ_t(p^{(i)})$ *and* $p^{(i)} \neq q^{(i)}$
$$Q_l^{(i)}(p^{(i)}, q^{(i)}) = \sum_{t \in \mathcal{T}_l^{(i)}} \tau_t(p^{(i)}, q^{(i)});$$

14.2. $\forall p^{(i)} \in \mathcal{P}^{(i)}$ *such that* $q^{(i)} \in succ_t(p^{(i)})$
$$Q_l^{(i)}(p^{(i)}, p^{(i)}) = - \sum_{t \in \mathcal{T}_l^{(i)}} \tau_t(p^{(i)}, q^{(i)});$$

---

[2] See [1] for the computation of $\Lambda(\sigma)$ (sequence of immediate transitions).

14.3. $\forall p^{(i)}, q^{(i)} \in \mathcal{P}^{(i)}$ such that $q^{(i)} \notin succ_t(p^{(i)})$ and $p^{(i)} \neq q^{(i)}$
$\quad Q_l^{(i)}(p^{(i)}, q^{(i)}) = 0.$

Let

$\quad \eta^{(t)}$ $\qquad$ set of indices i ($i \in [1..N]$) such that component $\mathcal{SSM}^{(i)}$ has at least one transition $t \in \mathcal{T}^{(i)}$;

$\quad \iota^{(t)}$ $\qquad$ index of the component $\mathcal{SSM}$ which has the transition rate of synchronized transition $t \in \mathcal{T}_s$, where $\iota^{(t)} \in [1..N]$.

Actually a transition t can be viewed as local transition if $|\eta^{(t)}| = 1$ or as synchronized transition if $|\eta^{(t)}| > 1$.

Definition 15. *Tensor elements $Q_{t+}^{(i)}$, which represent the occurrence of synchronized transition $t \in \mathcal{T}_s^{(i)}$, are defined by:*

15.1. $\forall i \notin \eta^{(t)}$
$\quad Q_{t+}^{(i)} = I_{|\mathcal{P}^{(i)}|};$
15.2. $\forall p^{(\iota^{(t)})}, q^{(\iota^{(t)})} \in \mathcal{P}^{(\iota^{(t)})}$ such that $q^{(\iota^{(t)})} \in succ_t(p^{(\iota^{(t)})})$
$\quad Q_{t+}^{(\iota^{(t)})}(p^{(\iota^{(t)})}, q^{(\iota^{(t)})}) = \tau_t(p^{(\iota^{(t)})}, q^{(\iota^{(t)})});$
15.3. $\forall i \in \eta^{(t)}$ such that $i \neq \iota^{(t)}$, $\forall p^{(i)}, q^{(i)} \in \mathcal{P}^{(i)}$ such that $q^{(i)} \in succ_t(p^{(i)})$
$\quad Q_{t+}^{(i)}(p^{(i)}, q^{(i)}) = 1;$
15.4. $\forall i \in \eta^{(t)}$, $\forall p^{(i)}, q^{(i)} \in \mathcal{P}^{(i)}$ such that $q^{(i)} \notin succ_t(p^{(i)})$
$\quad Q_{t+}^{(i)}(p^{(i)}, q^{(i)}) = 0.$

Definition 16. *Tensor elements $Q_{t-}^{(i)}$, which represent the adjustment of synchronized transition $t \in \mathcal{T}_s^{(i)}$, are defined by:*

16.1. $\forall i \notin \eta^{(t)}$
$\quad Q_{t-}^{(i)} = I_{|\mathcal{P}^{(i)}|};$
16.2. $\forall p^{(\iota^{(t)})} \in \mathcal{P}^{(\iota^{(t)})}$

$$Q_{t-}^{(\iota^{(t)})}(p^{(\iota^{(t)})}, p^{(\iota^{(t)})}) = \begin{cases} 0 & \text{if } \nexists q^{(\iota^{(t)})} \in succ_t(p^{(\iota^{(t)})}) \\ -\tau_t(p^{(\iota^{(t)})}, q^{(\iota^{(t)})}) & \text{if } \exists q^{(\iota^{(t)})} \in succ_t(p^{(\iota^{(t)})}) \end{cases}$$

16.3. $\forall i \in \eta^{(t)}$, $i \neq \iota^{(t)}$ and $\forall p^{(i)} \in \mathcal{P}^{(i)}$

$$Q_{t-}^{(i)}(p^{(i)}, p^{(i)}) = \begin{cases} 0 & \text{if } \nexists q^{(i)} \in succ_t(p^{(i)}) \\ 1 & \text{if } \exists q^{(i)} \in succ_t(p^{(i)}) \end{cases}$$

16.4. $\forall i \in \eta^{(t)}$, $\forall p^{(i)}, q^{(i)} \in \mathcal{P}^{(i)}$ and $p^{(i)} \neq q^{(i)}$
$\quad Q_{t-}^{(i)}(p^{(i)}, q^{(i)}) = 0.$

Definition 17. *Infinitesimal generator Q corresponding to the Markov chain associated to a decomposed GSPN model is represented by tensor formula called Markovian Descriptor:*

$$Q = \bigoplus_{i=1}^{N}{}_g Q_l^{(i)} + \sum_{t \in \mathcal{T}_s} \left(\bigotimes_{i=1}^{N}{}_g Q_{t+}^{(i)} + \bigotimes_{i=1}^{N}{}_g Q_{t-}^{(i)} \right) \tag{1}$$

Once a tensor sum is equivalent to a sum of particular product tensors, the *Markovian Descriptor* may be represented as:

$$Q = \sum_{j=1}^{(N+2|\mathcal{T}_s|)} \bigotimes_{i=1}^{N}{}_g\, Q_j^{(i)}, \tag{2}$$

$$\text{where } Q_j^{(i)} = \begin{cases} I_{|\mathcal{P}^{(i)}|} & \text{if } j \leq N \text{ and } j \neq i \\ Q_l^{(i)} & \text{if } j \leq N \text{ and } j = i \\ Q_{t_{(j-N)}^+}^{(i)} & \text{if } N < j \leq (N+|\mathcal{T}_s|) \\ Q_{t_{(j-(N+|\mathcal{T}_s|))}^-}^{(i)} & \text{if } j > (N+|\mathcal{T}_s|) \end{cases}$$

5   Choosing a Decomposition

In this section, we present several approaches to decompose a GSPN model. We show the necessary steps to obtain all components $\mathcal{SSM}$. Afterwards, we comment about the side effect of guards and its consequences.

Fig. 2 presents an example of a GSPN model. Based on this model, we show our decomposition technique applied in three different approaches. For all the possible decomposition approaches, the demonstration in Section 4.2 can be used to obtain the equivalent tensor algebra representation automatically.

5.1   Decomposing by Places

Firstly, we analyse a quite naive approach, which is based on decomposing a GSPN model by *places*. Each place has a maximum number of tokens K, and so we can view each place as a $\mathcal{SSM}$ with $K+1$ states. A decomposed GSPN model by *places* of Fig. 2 is presented in Fig. 3.

Each component $\mathcal{SSM}^{(i)}$ represents the possible states of place p_i of a GSPN model. Note that places p_2 and p_5 in Fig. 2 are 2-bounded, *i.e.*, there are no

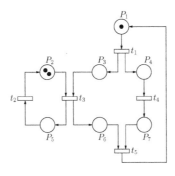

Fig. 2. An example of a GSPN model

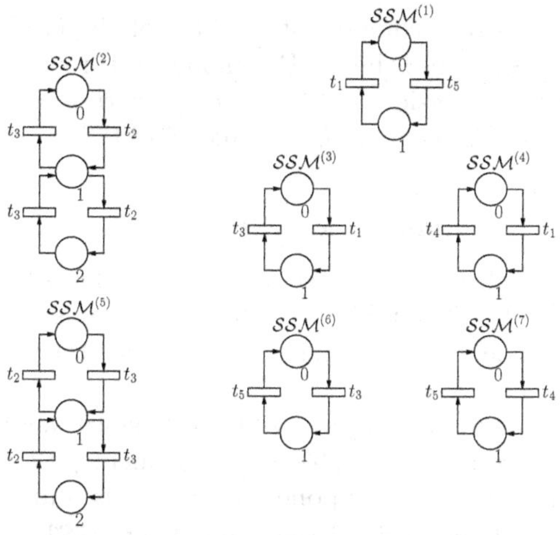

Fig. 3. Decomposed GSPN model by *places*

more than 2 tokens in each place in any marking $M \in \mathcal{RS}$. Hence $\mathcal{SSM}^{(2)}$ and $\mathcal{SSM}^{(5)}$ have three places which represent the states (0, 1 and 2) of places p_2 and p_5 respectively. Analogously, places p_1, p_3, p_4, p_6, and p_7 are 1-bounded. So $\mathcal{SSM}^{(1)}$, $\mathcal{SSM}^{(3)}$, $\mathcal{SSM}^{(4)}$, $\mathcal{SSM}^{(6)}$, and $\mathcal{SSM}^{(7)}$ have two states (0 and 1).

5.2   Decomposing by P-Invariants

Other decomposition of GSPN models is based on *P-invariants*. A P-invariant is composed of a set of places with constant token count. Fig. 4 presents a decomposed model of Fig. 2 using the *P-invariants* concept.

There are three minimal solutions of σ given by equation $\sigma C = 0$ (see Definition 8). So we can define three P-invariants to GSPN model of Fig. 2. $\mathcal{PI}_1$ has two places (p_2 and p_5), $\mathcal{PI}_2$ has three places (p_1, p_3 and p_6), and $\mathcal{PI}_3$ also has three places (p_1, p_4 and p_7).

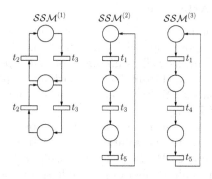

Fig. 4. Decomposed GSPN model by *P-invariants*

Each $\mathcal{PI}_i$ corresponds to a component $\mathcal{SSM}^{(i)}$ of the GSPN model. So, in this example, we can decompose the GSPN model in three components $\mathcal{SSM}$. It is important to note that, besides the transition superposition, there is a place superposition between components $\mathcal{SSM}^{(2)}$ and $\mathcal{SSM}^{(3)}$.

5.3   Decomposing as Superposed GSPN

Another possible approach to decompose a GSPN model is through transition superposition proposed by Donatelli [17]. Donatelli proposed the SGSPN formalism in which components (subsystems) interact each other through transition superposition.

Example of Fig. 2 can be decomposed by SGSPN, since there is a transition t_3 which synchronizes two components $\mathcal{GSPN}$. Component $\mathcal{GSPN}^{(1)}$ is composed of two places (p_2 and p_5), whereas component $\mathcal{GSPN}^{(2)}$ is composed of five places (p_1, p_3, p_4, p_6, and p_7). Once defined the components $\mathcal{GSPN}^{(i)}$, it is possible to obtain the equivalent components $\mathcal{SSM}^{(i)}$. Fig. 5 presents the equivalent components $\mathcal{SSM}$ of the GSPN model (Fig. 2).

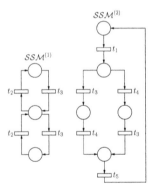

Fig. 5. Decomposed GSPN model by *SGSPN*

5.4   Decomposing Arbitrarily

It is also possible to decompose a GSPN model according to an arbitrarily chosen *semantic*. We can decompose the GSPN model of Fig. 2 as follows: markings of places p_1, p_2, p_3, and p_4, as well as markings of places p_5, p_6, and p_7. Hence component $\mathcal{SSM}^{(1)}$ is obtained from distinct markings of places p_1, p_2, p_3, and p_4 of $\mathcal{TRG}$, and component $\mathcal{SSM}^{(2)}$ is also obtained from distinct markings of places p_5, p_6, and p_7 of $\mathcal{TRG}$.

Note that the decomposition choice may privilege some features of the tensor format which is important to the solution method. In some cases, it may be important to decompose a GSPN model considering: a large or small number of components $\mathcal{SSM}$; a small number of reachable states; or even the difference between reachable and unreachable states.

5.5   Side Effect of Guards

The concept of *guards* in the GSPN formalism allows transition firing dependency according to the number of tokens in each place. Guards in GSPN are quite similar to functional elements in the SAN formalism [26, 18]. A natural decomposition among components $\mathcal{GSPN}$ of a GSPN model can be viewed through the use of guards. Therefore, guards in the GSPN models can allow to produce disconnected GSPN models, which have synchronization through the guards on theirs transitions.

As shown in Section 4.2, tensor format (*Markovian Descriptor*) of a decomposed GSPN model uses generalized tensor sum and products. GTA operators in the Markovian descriptor are used to represent the functional rates of transitions, but as long as there are no guards defined to transitions, they can be classical tensor products. Hence, guards on transitions are evaluated in *Markovian Descriptor* by GTA operators.

Another consequence of the use of guards is the possibility to define GSPN models with "disconnected parts", *i.e.*, models where there is not only a single net, but two or more nets with no arcs connecting them. In this case, there must be guards referring to places of other components in order to establish an interdependency (not a synchronization) among parts. The last example (Section 6.3) shows a net with disconnected parts and the use of guards to establish the interdependency.

6   Modeling Examples

We now present three modeling examples in order to present the approaches discussed in the previous section. The first one presents a *Structured* model, the second one describes a *Simultaneous Synchronized Tasks* model, whereas the last one shows a *Resource Sharing* model.

6.1   Structured Model

Fig. 6 presents an example of a *Structured* model composed of four submodels. The submodel i is composed of four places $(p_{a_i}, p_{b_i}, p_{c_i}, p_{d_i})$ and two local transitions. There are also four synchronized transitions responsible for the synchronization of the submodels. It is important to observe that guards were chosen to define the possible firing sequence of transitions. This model was introduced by Miner [20].

In this model, the decomposition by *places* is rather catastrophic, since there is a correlation among marking of places. It results in a quite large product state space (65, 536 states) for a rather small reachable state space (only 486 states). According to the SGSPN and P-invariants approaches, we have exactly the same decomposition and a more reasonable product state space (1, 296 states). As a general conclusion one may discard the decomposition by places approach, but this is not really a fair conclusion, since this model is quite particular. Models

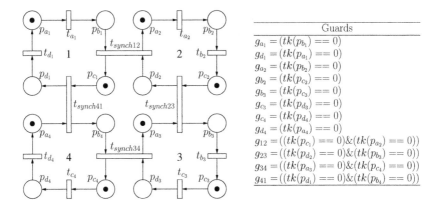

Fig. 6. Example of a structured model

with places with a larger bound (nets with more tokens) may be more interesting, as the next example will demonstrate.

6.2   Simultaneous Synchronized Tasks

Fig. 7 describes a *Simultaneous Synchronized Tasks* (SST) model in which five tasks are modeled. Such tasks have synchronization behavior among them, and those synchronization behaviors occur in different levels of the task execution.

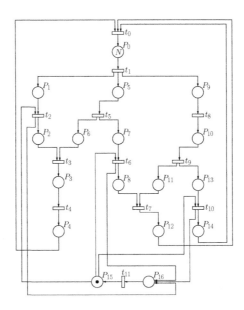

Fig. 7. Simultaneous Synchronized Tasks

Table 1. Indices of decomposed SST models

N	approach	$\#SSM$	SSM sizes	pss	rss	mem
1	Places	17	$(2 \times \cdots \times 2 \times 2 \times 2)$	1.31×10^5	98	1 KB
	P-Inv	6	$(5 \times 5 \times 5 \times 5 \times 5 \times 2)$	6.25×10^3	98	1 KB
	SGSPN	2	(49×2)	9.80×10^1	98	1 KB
3	Places	17	$(4 \times \cdots \times 4 \times 2 \times 2)$	4.29×10^9	12,100	2 KB
	P-Inv	6	$(35 \times 35 \times 35 \times 35 \times 35 \times 2)$	1.05×10^8	12,100	7 KB
	SGSPN	2	$(6,050 \times 2)$	1.21×10^4	12,100	236 KB
9	Places	17	$(10 \times \cdots \times 10 \times 2 \times 2)$	4.00×10^{15}	22,391,512	6 KB
	P-Inv	6	$(715 \times \cdots \times 715 \times 2)$	3.74×10^{14}	22,391,512	201 KB
	SGSPN	2	$(11,195,756 \times 2)$	2.24×10^7	22,391,512	672 MB
10	Places	17	$(11 \times \cdots \times 11 \times 2 \times 2)$	1.67×10^{16}	51,887,550	6 KB
	P-Inv	6	$(1,001 \times \cdots \times 1,001 \times 2)$	2.01×10^{15}	51,887,550	288 KB
	SGSPN	2	$(25,943,775 \times 2)$	5.19×10^7	51,887,550	-
20	Places	17	$(21 \times \cdots \times 21 \times 2 \times 2)$	2.72×10^{20}	18,994,747,662	12 KB
	P-Inv	6	$(10,626 \times \cdots \times 10,626 \times 2)$	2.71×10^{20}	18,994,747,662	3 MB
	SGSPN	2	$(9,497,373,831 \times 2)$	1.90×10^{10}	18,994,747,662	
21	Places	17	$(22 \times \cdots \times 22 \times 2 \times 2)$	5.48×10^{20}	29,368,986,350	13 KB
	P-Inv	6	$(12,650 \times \cdots \times 12,650 \times 2)$	6.48×10^{20}	29,368,986,350	4 MB
	SGSPN	2	$(14,684,493,175 \times 2)$	2.94×10^{10}	29,368,986,350	-
27	Places	17	$(28 \times \cdots \times 28 \times 2 \times 2)$	2.04×10^{22}	286,448,238,746	16 KB
	P-Inv	6	$(31,465 \times \cdots \times 31,465 \times 2)$	6.17×10^{22}	286,448,238,746	10 MB
	SGSPN	2	$(143,224,119,373 \times 2)$	2.86×10^{11}	286,448,238,746	-

Table 1 presents some indices to compare the decomposition alternatives. In this example, we use the following approaches: decomposing by Places (Section 5.1), decomposing by P-invariants (Section 5.2) and decomposing by Superposed GSPN (Section 5.3). N represents the number of tokens in place P_0. The number of SSM components, decomposed by all approaches, is indicated by $\#SSM$. SSM sizes represents the number of states in each component SSM. Product State Space, Reachable State Space, and memory needs to store the Markovian Descriptor[3] of the model are denoted by pss, rss, and mem respectively.

The first important phenomenon to observe in Table 1 is the increasing gains of the P-invariant and Places approaches achieved to models with large N values. For small N values, there is much waste in the product state space that is not significant compared to the memory savings, specially for the Places approach. Regarding the comparison between SGSPN and P-invariant approaches, the model with $N = 9$ is a turning point, since the memory savings are already quite significant. In fact, larger models could not even be generated using the SGSPN decomposition. The relationship between product and reachable state space for

---

[3] We do not take into account the memory needs to store neither the probability vector to compute solution, nor any special structure to represent the reachable state space.

decomposition based on P-invariants is considerably large, but we believe that an optimized solution for models with sparse reachable state space could take great advantage from this decomposition approach.

The second very impressive results taken from Table 1 is the very consistent gains of the *Places* approach. Even though the product state space waste is considerably large for smaller models (roughly one or two orders of magnitude), the difference between the product state space for the *P-invariant* and *Places* approaches becomes insignificant for the $N = 20$ model. Taking the model to its limits ($N = 21$ to 27) we observe an inversion, since the *Places* approach has a smaller product state space than the *P-invariant* approach.

6.3   Resource Sharing

Fig. 8 (a) shows a traditional example of a *Resource Sharing* (RS) model, which has N process sharing R resources. Each process i is composed of two places: S_i (*sleeping*) and U_i (*using*). Tokens in place RS represent the number of available resources, whereas they represent the number of using resources in place RU. Fig. 8 (b) is an equivalent model in which guards impose a restriction to the firing of each transition ta_i.

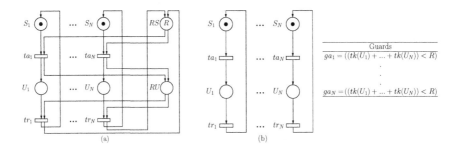

Fig. 8. (a) Resource Sharing without guards - (b) Resource Sharing with guards

Table 2 shows some indices to compare the use of guards in a GSPN model producing, in this case, an equivalent disconnected model. The indices for the model of Fig. 8 (a) are shown in the *without guards* rows, whereas indices for the model of Fig. 8 (b) are presented in the *with guards* rows. R indicates the number of tokens in place RS of Fig. 8 (a), as well as the number of available resources in the model of Fig. 8 (b). The computational cost (number of multiplications) to evaluate the product of a probability vector by the *Markovian Descriptor*[4], the memory needs and the CPU time to perform one single power iteration are denoted by *c.c.*, *mem* and *time* respectively. The numerical results for those

---

[4] The vector-descriptor multiplication is the basic operation for most of the iterative solutions, *e.g.*, Power method, Uniformization [28].

Table 2. Indices of decomposed RS models ($N = 16$)

R	model	#SSM	SSM sizes	pss	rss	c.c.	mem	time
1	without guards	17	$(2 \times \cdots \times 2 \times 2)$	131,072	17	1.34×10^8	6 KB	0.33 s
	with guards	16	$(2 \times \cdots \times 2)$	65,536	17	2.10×10^6	1 KB	0.45 s
3	without guards	17	$(2 \times \cdots \times 2 \times 4)$	262,144	697	2.73×10^8	8 KB	0.71 s
	with guards	16	$(2 \times \cdots \times 2)$	65,536	697	2.10×10^6	1 KB	0.49 s
9	without guards	17	$(2 \times \cdots \times 2 \times 10)$	655,360	50,643	6.88×10^8	14 KB	1.81 s
	with guards	16	$(2 \times \cdots \times 2)$	65,536	50,643	2.10×10^6	1 KB	0.53 s
15	without guards	17	$(2 \times \cdots \times 2 \times 16)$	1,048,576	65,535	1.10×10^9	20 KB	3.07 s
	with guards	16	$(2 \times \cdots \times 2)$	65,536	65,535	2.10×10^6	1 KB	0.53 s

examples were obtained on a 2.8 GHz Pentium IV Xeon under Linux operating system with 2 GBytes of memory.

The results in Table 2 show the decomposition based on P-invariants, since decomposition based on SGSPN could only be applied to the *without guards* model. Observe that SGSPN approach would result in exactly the same decomposition as P-invariants. The main conclusion observing this table is the absolute gains represented by the use of guards. It results in a model which has the same product state space independently from the number of resources, as well as the *pss* sizes are always smaller than those in the *without guards* models. It also has a smaller memory need and a more efficient solution (smaller computational cost).

It is a common mistake in some segments of the research community to assume that a model which requires functional evaluations (GTA operators) has a bigger CPU time to perform vector-descriptor multiplication than equivalent models described only with CTA operators. In fact, such use of guards gives to this GSPN model an efficiency as good as the one achieved by an equivalent SAN model [7]. Obviously, it happens due to the *Markovian Descriptor* representation using GTA.

7   Conclusion

The main contribution of this paper is to follow up the pioneer works of Ciardo and Trivedi [14], Donatelli [17], and Miner [20]. Our starting point is the assumption that for really large (and therefore structured) models the main difficulty is the storage of the infinitesimal generator. The solution techniques are rapidly evolving and many improvements, probably based on efficient parallel solutions, will soon enough be available. Using this assumption, we do believe that the tensor format based on Generalized Tensor Algebra has an important role to play.

For the moment, it benefits the Stochastic Automata Networks and it can also be applied to Generalized Stochastic Petri Nets. A natural future theoretical work is to expand those gains to other formalisms, such as PEPANETs [19]. A more immediate future work would be the integration of this decomposition analysis to the solvers for SAN and GSPN (PEPS [6] and SMART [13] software tools respectively). It is easy to precompute the possible decomposition with their memory and computational costs. Therefore, the integration of such precalculation may automatically suggest the best decomposition approach according to the amount of memory available.

Finally, we would like to conclude stating that the use of tensor based storage can still give very interesting results and allows the solution (which cannot be done without the storage) of larger and larger models.

References

1. M. Ajmone-Marsan, G. Balbo, G. Chiola, G. Conte, S. Donatelli, and G. Franceschinis. An Introduction to Generalized Stochastic Petri Nets. *Microelectronics and Reliability*, 31(4):699–725, 1991.
2. M. Ajmone-Marsan, G. Conte, and G. Balbo. A Class of Generalized Stochastic Petri Nets for the Performance Evaluation of Multiprocessor Systems. *ACM Transactions on Computer Systems*, 2(2):93–122, 1984.
3. V. Amoia, G. De Micheli, and M. Santomauro. Computer-Oriented Formulation of Transition-Rate Matrices via Kronecker Algebra. *IEEE Transactions on Reliability*, R-30(2):123–132, 1981.
4. R. Bellman. *Introduction to Matrix Analysis*. McGraw-Hill, New York, 1960.
5. A. Benoit, L. Brenner, P. Fernandes, and B. Plateau. Aggregation of Stochastic Automata Networks with replicas. *Linear Algebra and its Applications*, 386:111–136, July 2004.
6. A. Benoit, L. Brenner, P. Fernandes, B. Plateau, and W. J. Stewart. The PEPS Software Tool. In *Computer Performance Evaluation / TOOLS 2003*, volume 2794 of *LNCS*, pages 98–115, Urbana, IL, USA, 2003. Springer-Verlag Heidelberg.
7. L. Brenner, P. Fernandes, and A. Sales. The Need for and the Advantages of Generalized Tensor Algebra for Kronecker Structured Representations. *International Journal of Simulation: Systems, Science & Technology*, 6(3-4):52–60, February 2005.
8. J. W. Brewer. Kronecker Products and Matrix Calculus in System Theory. *IEEE Transactions on Circuits and Systems*, CAS-25(9):772–780, 1978.
9. P. Buchholz, G. Ciardo, S. Donatelli, and P. Kemper. Complexity of memory-efficient Kronecker operations with applications to the solution of Markov models. *INFORMS Journal on Computing*, 13(3):203–222, 2000.
10. P. Buchholz and T. Dayar. Block SOR for Kronecker structured representations. *Linear Algebra and its Applications*, 386:83–109, July 2004.
11. P. Buchholz and P. Kemper. Hierarchical reachability graph generation for Petri nets. *Formal Methods in Systems Design*, 21(3):281–315, 2002.
12. G. Ciardo, M. Forno, P. L. E. Grieco, and A. S. Miner. Comparing implicit representations of large CTMCs. In 4th *International Conference on the Numerical Solution of Markov Chains*, pages 323–327, Urbana, IL, USA, September 2003.
13. G. Ciardo, R. L. Jones, A. S. Miner, and R. Siminiceanu. SMART: Stochastic Model Analyzer for Reliability and Timing. In *Tools of Aachen 2001 International Multiconference on Measurement, Modelling and Evaluation of Computer-Communication Systems*, pages 29–34, Aachen, Germany, September 2001.

14. G. Ciardo and K. S. Trivedi. A Decomposition Approach for Stochastic Petri Nets Models. In *Proceedings of the 4th International Workshop Petri Nets and Performance Models*, pages 74–83, Melbourne, Australia, December 1991. IEEE Computer Society.

15. M. Davio. Kronecker Products and Shuffle Algebra. *IEEE Transactions on Computers*, C-30(2):116–125, 1981.

16. S. Donatelli. Superposed stochastic automata: a class of stochastic Petri nets with parallel solution and distributed state space. *Performance Evaluation*, 18:21–36, 1993.

17. S. Donatelli. Superposed generalized stochastic Petri nets: definition and efficient solution. In R. Valette, editor, *Proceedings of the 15th International Conference on Applications and Theory of Petri Nets*, pages 258–277. Springer-Verlag Heidelberg, 1994.

18. P. Fernandes, B. Plateau, and W. J. Stewart. Efficient descriptor - Vector multiplication in Stochastic Automata Networks. *Journal of the ACM*, 45(3):381–414, 1998.

19. J. Hillston and L. Kloul. An Efficient Kronecker Representation for PEPA models. In L. de Alfaro and S. Gilmore, editors, *Proceedings of the first joint PAPM-PROBMIV Workshop)*, pages 120–135, Aachen, Germany, September 2001. Springer-Verlag Heidelberg.

20. A. S. Miner. *Data Structures for the Analysis of Large Structured Markov Models*. PhD thesis, The College of William and Mary, Williamsburg, VA, 2000.

21. A. S. Miner. Efficient solution of GSPNs using Canonical Matrix Diagrams. In 9th *International Workshop on Petri Nets and Performance Models (PNPM'01)*, pages 101–110, Aachen, Germany, September 2001. IEEE Computer Society Press.

22. A. S. Miner and G. Ciardo. Efficient Reachability Set Generation and Storage Using Decision Diagrams. In *Proceedings of the 20th International Conference on Applications and Theory of Petri Nets*, volume 1639 of *LNCS*, pages 6–25, Williamsburg, VA, USA, June 1999. Springer-Verlag Heidelberg.

23. A. S. Miner, G. Ciardo, and S. Donatelli. Using the exact state space of a Markov model to compute approximate stationary measures. In *Proceedings of the 2000 ACM SIGMETRICS Conference on Measurements and Modeling of Computer Systems*, pages 207–216, Santa Clara, California, USA, June 2000. ACM Press.

24. T. Murata. Petri nets: Properties, analysis and applications. *Proceedings of the IEEE*, 77(4):541–580, April 1989.

25. B. Plateau. On the stochastic structure of parallelism and synchronization models for distributed algorithms. In *Proceedings of the 1985 ACM SIGMETRICS conference on Measurements and Modeling of Computer Systems*, pages 147–154, Austin, Texas, USA, 1985. ACM Press.

26. B. Plateau and K. Atif. Stochastic Automata Networks for modelling parallel systems. *IEEE Transactions on Software Engineering*, 17(10):1093–1108, 1991.

27. W. Reisig. *Petri nets: an introduction*. Springer-Verlag Heidelberg, 1985.

28. W. J. Stewart. *Introduction to the numerical solution of Markov chains*. Princeton University Press, 1994.

Modeling Dynamic Architectures
Using Nets-Within-Nets

Lawrence Cabac, Michael Duvigneau, Daniel Moldt, and Heiko Rölke

University of Hamburg, Department of Computer Science,
Vogt-Kölln-Str. 30, D-22527 Hamburg
{cabac, duvigneau, moldt, roelke}@informatik.uni-hamburg.de

Abstract. Current modeling techniques are not well equipped to design dynamic software architectures. In this work we present the basic concepts for a dynamic architecture modeling using nets-within-nets. Nets-within-nets represent a powerful formalism that allows active elements, i.e. nets, to be nested in arbitrary and dynamically changeable hierarchies. Applying the concepts from nets-within-nets, therefore, allows us to model complex dynamic system architectures in a simple way, which enables us to design the system at different levels of abstractions using refinements of net models.

Additionally to the conceptual modeling of such architecture, we provide a practical example where the concept has been successfully applied in the development of the latest release of RENEW (Version 2 of the multiformalism Petri net IDE[1]). The overall monolithic architecture has been exchanged with a system that is divided into a plug-in management system and plug-ins that provide functionality for the users. By combining plug-ins the system can be adapted to the users' needs. Through the introduction of the Petri net concepts, the new architecture is now – at runtime – dynamically extensible by registering plug-ins with the management system. The introduced architecture is applicable for any kind of architecture but most suitable for applications with dynamic structure.

Keywords: High-level Petri nets, Nets-within-nets, reference nets, RE-NEW, plug-ins, components, dynamic software architecture, modeling.

1   Introduction

Today's software systems are getting more and more complex. The amount of functionality and the number of features that are put into a system increases steadily. Moreover, features that are only loosely related are included into a system making the software more attractive for some user and at the same time too complex and bloated with features for other users. The need to switch the design to configurable or adaptable / customizable systems has been recognized for a long time.

---

[1] Integrated development environment.

G. Ciardo and P. Darondeau (Eds.): ICATPN 2005, LNCS 3536, pp. 148–167, 2005.

Many systems already provide the possibility to extend the functionality with plug-ins. Some of the software systems are component-based, leaving the user in charge of the degree of versatility of the utilized system. However, the flexibility of these systems is usually limited. Most systems only allow static configuration, some also allow to extend the functionality at runtime in a very limited way.

RENEW serves as an example for a complex system. It is an IDE for several Petri net (and Petri net related) formalisms. Examples of these are reference nets [5], P/T-nets, Workflow nets [4], Feature Structure nets [17], Timed Petri nets and Multi-agent nets [11]. RENEW has grown significantly in the last couple of years.

Many users did not need the many specialized formalisms and features in their everyday work with the tool. However, still more new extensions were on the verge to be developed, threatening that the system would grow further. While some users / developers were using the new features, others had to deal with the resulting overhead. The need for a highly customizable architecture became obvious.

The idea was born to redesign RENEW in an extensively flexible way that would solve the problems and lead to an architecture that is configurable, customizable and extensible. Moreover, the developers envisioned a system that is even more flexible: A system that is customizable at runtime, i.e. a system whose architecture is dynamically configurable.

The notion of flexibility in the re-design has to be concretized, therefore a model for a dynamic architecture is needed. The modeling process for an architecture design helps significantly to understand the architecture and the dependencies between the involved units. Modeling helps to share and discuss ideas with other software architects and to convey these ideas to developers. Furthermore, it helps to establish the concept, to find and eliminate conceptual problems and to visualize the system. However, to model a system architecture design, an established and expressive modeling technique has to be applied.

There are various architecture modeling techniques like UML [9] or architecture description languages (ADL, see [8,15]) that could be used to design our architecture. Nevertheless, we use the reference net formalism [5] which is based on Valk's nets-within-nets [16] paradigm because it combines many features that are otherwise not available in one single language: With Petri nets we have formal semantics, operational semantics and a plain graphical notation. The reference net formalism adds coverage of dynamic changes in behavior and structure as well as object-based properties like encapsulation, polymorphism and instantiation. So we can use a lot of features that would otherwise not be available within one single technique. However, the reference nets are currently not well equipped to describe static aspects of a system, like interfaces or a type hierarchy. This drawback is of minor relevance since we are interested in a dynamic architecture model in this paper.

We designed our concept model for a dynamic architecture with reference nets. The concept was successfully applied in the re-design of RENEW resulting in the current version (2.0). We achieved to transform the old—in some way extensible, but mainly monolithic—design into a highly configurable dynamic architecture.

According to our approach of implementing through model refinement we introduce the abstract concept model and successively refine it until we obtain a functional model. For the reason of efficiency, the functional model is re-implemented in *Java*. Since our models are designed and executed in RENEW it is possible to utilize both, the model implementation and the *Java* implementation.

Note that RENEW is used as modeling tool for our recursive concept model for a dynamic architecture and also as target system for the application and realization of the concept model. This self-reflective feature is one of the key aspects of our approach.

In the following section we give an introduction to reference nets. The focus lies on some features of reference nets that we use extensively in the design of our concept model. In Section 3 we present our concept model for a dynamic architecture. In Section 4 we present the realization of the concept model in RENEW and discuss some pragmatic design decisions. Section 4.4 relates our concept and modeling technique with other approaches for configurable systems.

2   Nets Within Nets

Note for the experienced reader: If you already know reference nets and are well-acquainted with their concepts we recommend that you skip this section and continue with Section 3.

Nets-within-nets are expressive high-level Petri nets that allow nets to be nested within nets in dynamical structures. In contrast to ordinary nets, where tokens are passive elements, tokens in nets-within-nets are active elements, i.e. Petri nets. In general we distinguish between two different kinds of token semantics: value semantics and reference semantics. In value semantics tokens can be seen as direct representations of nets. This allows for nested nets that are structured in a hierarchical order because nets can only be located at one location. In reference semantics arbitrary structures of net-nesting can be achieved because tokens represent references to nets. These structures can be hierarchical, acyclic or even cyclic.

In the following sections we will discuss reference nets because of three reasons. First, they are supported in the tool RENEW, second, they show the basic principles of nets-within-nets, and third, they allow for acyclic nesting structures. If the number of references for nets are reduced to one reference for a net, value semantics and reference semantics are equivalent.

2.1   Reference Nets

Reference nets [5] are object-oriented high-level Petri nets, in which tokens can be nets again. For these nets-within-nets [16], referential semantics is assumed. Tokens in one net can be references to other nets. In a simple setting of a single nesting of nets, the outer net is called system net while a token in the system net refers to an object net. Nevertheless, object nets themselves can again contain tokens that represent nets, and thus a system of nested nets can be obtained. The benefit of this feature is that the modeled system is modular and extensible. Furthermore, transitions in nets can activate and trigger the firing of transitions in other nets, just like method calls of objects, by using synchronous channels [2, 5].

RENEW (The **Reference Net Workshop** [6, 7]) combines the nets-within-nets paradigm of reference nets with the implementing power of *Java*. Here tokens can also be *Java*-objects and nets can be regarded as objects.

In comparison to the net elements of P/T-nets, reference nets offer several additional elements that increase the modeling power as well as the convenience of modeling. These additional elements include some arc types, virtual places and a declaration. Several inscription types have been added to the net elements providing functionality for the different net elements. Places can be typed and transitions can be augmented with expressions, actions, guards, synchronous channels and creation inscriptions. In the following paragraph we will focus on aspects of net instances and synchronous channels. Detailed information on nets-within-nets and reference nets can be found in [5] and [16]. It should only be mentioned briefly that we use reserve arcs as a convenient notation. In addition, we also use flexible arcs (see [10]) in our models. These are expressive arcs that can drop all elements of a collection onto a place and withdraw all (pre-known) elements of a collection from a place.

Net Instances and Synchronous Channels reference nets are object-oriented nets. Similar to objects in object-oriented programming languages, where objects are instantiations of classes, net instances are instantiations of net templates. Net templates define the structure and behavior of nets just like classes define the structure and methods of objects. While the net instance has a marking that determines its status, the net template determines only the behavior and initial marking that is common to all net instances of one type.

The paradigm of nets-within-nets introduced by Valk [16], allows tokens to be nets again. In reference nets, tokens can be anonymous, basic data types, *Java* objects or net references. Any net instance can create new net instances similar to an object creating new objects. The new net instance is marked with the initial marking according to the specification of the net template.

The notation of the creation inscription with the usage of the keyword new, to create a new instance, is displayed in Figure 1. In this example the system net has an initial marking of three integer tokens. Thus the transition can fire three times creating three new net instances.

The three new net instances are bound to the variable x and put into the output place. This is displayed in Figure 2, in which the net templates and the net instances for both nets are displayed. There is one instance of the system net and three instances of the object net.[2]

---

[2] In RENEW net instances can be identified by the names of the windows and the window background colors. Net templates have a white background color and net instances have an integer number attached to their window title bars that identifies the distinct instances. These identifying numbers are also attached to the tokens.

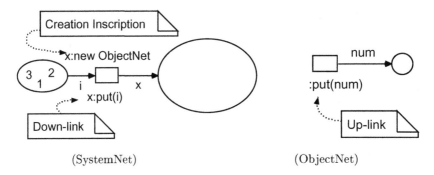

Fig. 1. Example system net and object net

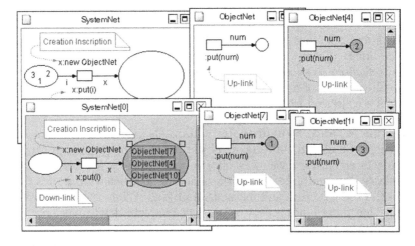

Fig. 2. A screen shot of a system net and an object net, the templates and several net instances

The different net instances are each created during one firing of the transition of the system net, which bears the creation inscription. The tokens referring to the net instances are put into the output place. In the net instance SystemNet[0] in Figure 2 these three tokens are displayed in the output place. Navigation among the net instances is done in a hypertext fashion by clicking on the reference in a net instance in order to open the referred net instance window.

For the communication between net instances, synchronous channels are used. A synchronous channel consists of two (or more) inscribed transitions. There are two types of transition inscriptions for the two ends of the synchronous channel: downlinks and uplinks. The synchronous channel forms a symmetric channel, therefore bidirectional communication is possible. Two transitions that form a synchronous channel can only fire simultaneously and only if both transitions are activated. Downlink and uplink belong to a single net or to different nets.

In both cases any object, also another net instance, can be transferred from either transition to the other. If two different net instances are involved, it is thus possible to synchronize these two nets and to transfer objects in either direction through the synchronous channel. For this the system nets must hold the references of the object nets as tokens (as in the output place in Figure 2).

The simple example of Figures 1 and 2 does not only show the creation of net instances, but also the application of synchronous channels. A synchronous channel put(.) connects the two transitions of system net and object net. The system net holds the reference to the object net instance x that is created during the firing of the transitions. The downlink x:put(i) calls the uplink in the object net :put(num). The integers are taken from the input place of the system net and bound to the variable i used as an argument in the channel inscription. Both transitions fire simultaneously and the two variables i and num are unified. Thus num is bound to the same integer as i, which finally is put into the output place of the object net. So the different numbers in the output places can distinguish the different net instances of the object net, which is the reason why we have chosen numbers as tokens in this example.

By using two (or more) parameters in a channel, information can be transfered in both directions synchronously. For instance a simple database lookup can be implemented by using this feature. One of the parameters serves as key and the other one as value. This will be used in Section 3.

2.2 RENEW

With RENEW it is possible to draw and simulate Petri nets and reference nets. The simulation engine can execute a net by creating an instance of the net. Any simulated net can instantiate other nets. Hence it is possible to produce many instances of different nets. The relationship between net template, also simply called net, and net instance can be compared to the relationship of class and object (see Section 2.1).

Editor. Figure 3 shows the graphical user interface (GUI) of RENEW, a simple Petri net in the back and a net instance.

The user interface consists of the menu bar, two palettes and a status line. The menu bar offers menus for general operations, attribute manipulations, layout adjustment and Petri net-specific operations. It also provides the possibility to control the simulation. Of the two palettes the first one consists of usual drawing tools while the second one holds the Petri net drawing tools for the creation of transitions, places, virtual places, arcs, test arcs, reserve arcs, inscriptions, names and declarations. In addition to these tools, the editor reacts in a context-sensitive manner to facilitate the drawing of nets. One example is the dropping of arcs on the background that creates a new place if the arc starts at a transition and vice versa. Another example is the right click on inscribable elements that produces an inscription for this element with a context sensitive default value.

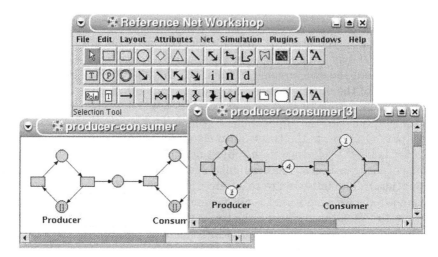

Fig. 3. RENEW GUI, Petri net and net instance (producer-consumer example)

Simulator. Net templates hold the initial marking while net instances hold the current marking. In Figure 3 the producer-consumer example has been started. In the net template (background) one of two black tokens ('[]') of the initial marking can be seen in the place labeled **Producer**. While the net instance by default only shows the number of tokens in a place it is also possible to show the contents of the places by clicking on the numbers (compare with Figure 2).

In the following section we model our concept of an dynamic architecture with reference nets. All modeling is done with RENEW, which is also the target for the realization of the concept model.

3   Concept Model

A dynamic architecture is characterized by extensibility and adaptability. In this work we conceive extensibility as a recursive feature. A system is extended by components, which again are extended by plug-ins, which are (specialized) components. Reference nets as nets-within-nets allow nets to be nested within other nets. They are, therefore, capable of modeling an extensible architecture in which a management component can act as a container for components – in our model other nets. These components again are used as containers for other components. We develop the model successively from a simple one-level view to a full-fledged plug-in-based system.

This chapter introduces the concept model for extensible systems that allows components to extend a system dynamically during runtime. The realization of this concept in Renew 2.0 is described in Section 4.

3.1   Extensibility

To construct extensible systems it is useful to get a notion of what is meant by extensibility. This is modeled here with reference nets starting on an abstract level that is then further concretized throughout this section.

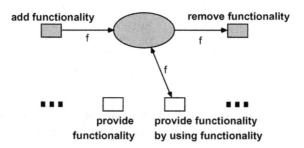

Fig. 4. Model for extensible systems

Figure 4 illustrates an extensible system in a most general way. The upper grey colored elements of the net define the extension management part of the system. The net shows the system as reference net in which the central place acts as the container for extensions. Functionality is added to the system by a synchronous channel at the transition labeled add functionality[3] and then put on the central container place. Functionality is removed by the transition labeled remove functionality.

The white transitions in the lower part are representatives for the available domain-specific functionality of the system. Some of the functionality may incorporate the functionality provided by extensions that lie in the central container place. All elements f that are extending the system are nets again, according to the nets-within-nets paradigm. An exemplary marking of the net is shown in Figure 5.

In this figure, exemplary channel inscriptions are visible. The small net tokens provide functionality by the channel uplinks :func1() or :func2(). The system provides functionality to the user through the channel uplinks :funcA() and :funcB(). However, to provide funcB, the system uses extended functionality. It specifies the extension interface, here represented by the channel downlink f:func1(). There are two available extensions providing the specified channel by an uplink, so the system may choose any of these functionality providers. Adding and removing functionality to and from the system is also accessible to the user through channel uplinks :add(f) and :remove(f).

This model leads to the concept of components. Components are units of extensibility. The net tokens that represent extensions of functionality in Figure 5

---

[3] The channel inscriptions are omitted in this figure because the focus lies on the concepts.

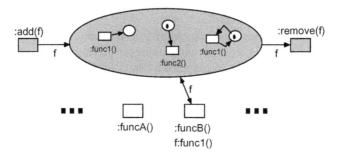

Fig. 5. Extensible system with net tokens

perfectly fit that notion. So we can define components by this net model. A textual definition of components is given by Schumacher in [13].[4]

Definition 1. *Component*
   A component is a unit of distribution that comprises executable code accompanied by appropriate documentation and provides domain-specific functionality.

3.2   Recursive Extensibility

In the current model we are able to say that the system is extensible on one level. This notion of *one-level extensibility* [14] expresses the fact that new components can be introduced to the system but these components can not be extended themselves. The concept of extension management of the system can also be applied to the components. Since the extensibility model of the previous section is built with components, we can call the extension management component management. The possibility to extend the components leads us to a notion for a recursively extensible system.

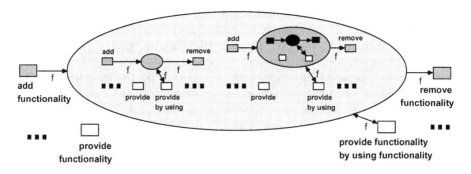

Fig. 6. Model of a recursively extensible system

---

[4] This definition is closely related to Sametinger [12].

Figure 6 shows the modified system where components may be contained within components. It can be observed that the components implement the same management interface as the system net model. A component only differs from the model of the system or any other component in the domain-specific part, which is not shown explicitly in Figure 6.

We can now regard the components that recursively extend other components as plug-ins. From Schumacher's viewpoint a system is composed of components and plug-ins. Plug-ins are special components that change the behavior of the system by changing the behavior of components. The full textual definition given in [13–p. 34] is:

Definition 2. *Plug-in Plug-ins are components that change the behavior of one or more other components in the system. This is done by using the provided interface of the components.*

With the introduction of the plug-in concept we can regard the component management as plug-in management.

However, in our net model, there is no difference between the component nets and the system net. So we cannot distinguish components from plug-ins. The system as well as all components can be extended by plug-ins. In the further refinement of the net model it is possible to observe a difference between system and components, and thus the distinction between plug-ins and components becomes significant.

Up to now, we have a hierarchical structure of the system. The extension relation is strongly tree-structured. The use of reference semantics as described in Section 2.1 enables us to relax this condition. It is possible to add one plug-in to multiple different components, so the extension relation forms an acyclic graph. Even a cyclic relation is possible, although that may lead to endless recursion.

In order to concretize the model further, we will now describe how the adding of plug-ins is introduced into the model. Figure 7 shows the simple mechanism of adding plug-ins to plug-ins. A plug-in p2 is added to another plug-in p by using the derived functionality of the component represented by the main net. Here we have a chain of channel synchronizations: When the uplink :subadd(p2) is called by some other net instance's downlink, then the downlink p:add(p2) at

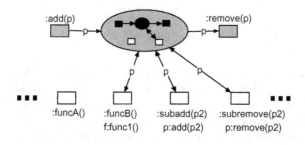

Fig. 7. A component allows the extension of its plug-ins

the same transition synchronizes with the uplink :add(p) of a third net instance. The management part of that third net instance looks the same as in Figure 7, it may also be another instance of the same net. The unloading of the plug-in is done in a similar way.

This mechanism enables the system to be recursively pluggable through plugging plug-ins into components at an arbitrary number of levels, as long as every level provides such a call-through functionality. The mechanism also shows that management of one plug-in can be seen as functionality of another plug-in. The distinction between management and functionality that we made in the first models can be dropped now.

A drawback of the current model is that the plug-in p2 cannot control its adding to or removal from selected components. It is passed as a passive object through the channels during the process. This problem will be discussed in the following section.

3.3   Communication Between Components

One of the advantages of the component-orientation is the re-usability. This means that the functionality that is offered by the plug-in is utilized by all components that need this functionality. Therefore, a component has to be able to address another component / plug-in. For this we introduce the notion of services that are offered by components to other components.

Services have to be published and made accessible for other components. Each component provides an interface by which the descriptions of the offered services are accessible. A global service directory is needed so that components can look up service provider components. We refine the abstract model of Figure 7 and introduce a net that has the management of plug-ins and their services as its only functionality. This net is called the plug-in management system (PMS).

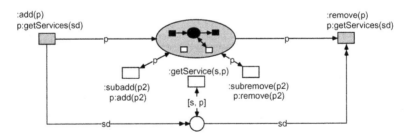

Fig. 8. A PMS that supports plug-in services

Figure 8 shows the model of a PMS with the service lookup infrastructure (SLI). When a new plug-in p is added to the system, the PMS asks for the description of the services that are offered (p:getServices(sd)) by this plug-in. This is in this net indicated by the *flexible arcs* (two arrow tips) which are able to drop all elements of a collection sd onto a place or to withdraw all elements of a collection simultaneously from a place, respectively.

Each service description is stored as a key-value tuple (`[s,p]`). Other components can get information about available services and their suppliers by using the channel uplink `:getService(s,p)`.[5]

To be able to use the lookup functionality of the PMS, all components need a reference to the PMS. Therefore we define that the PMS net should have exactly one instance. This instance can be regarded as the root of the graph of the extensibility relation. So the PMS is our explicit top-level net instance of the whole system.

Provided each plug-in has a reference to the PMS, it could also call the `subadd` and `subremove` channels of the PMS to control its own registration with other components. A more elegant approach is to use the SLI: If each extensible component declares its extension management interface as a public accessible service, potential plug-ins can query the PMS for that service and register themselves directly. So we can omit the `subadd` and `subremove` channels from the PMS (and from all other components, too).

Our current model is lacking a mechanism that passes the PMS reference to each component. The model also does not exactly determine the moment of extension registration and configuration. Therefore we enforce a life cycle for all plug-ins within the PMS.

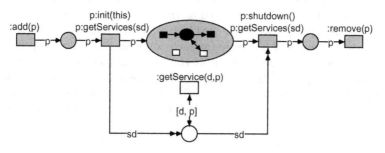

Fig. 9. A PMS with life cycle management

In Figure 9 the functionality of managing the life cycle of components is added to the PMS by two extra transitions with channel downlinks `p:init(pms)` and `p:shutdown()`). At the `init` transition, the added component gets informed about its addition to the system, receives the PMS reference and gets the chance to connect to other components that provide required services. Note that the retrieval of service descriptions has also moved from the transition where the component is added to the transition where the component is initialized. This ensures that the services of a component cannot be used by other components before the component has been properly initialized.

The introduction of the PMS and its service influences our model of a component. The updated model is shown in Figure 10. The gray management part

---

[5] Modeling database lookups is explained in Section 2.1.

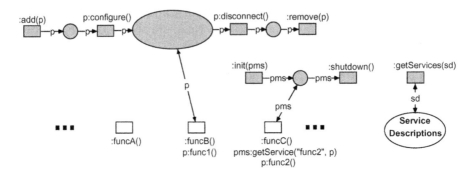

Fig. 10. Refined model of a component

has grown because of three things: First, right below the central extension place, the life cycle support has been added, the new place holds the PMS reference. Second, at the right, the channel uplink `:getServices(sd)` has been added. Here the PMS can extract information about the component. And last, left and right of the central extension place, the life cycle management of the PMS is repeated for plug-ins.

The life cycle management for plug-ins is introduced for the same reasons as the system-wide life cycle is introduced at the PMS level. However, the channel names are intentionally changed to `configure` and `disconnect` to indicate that there is exactly one global life cycle for each component. If the component also is a plug-in, it must additionally support the life cycle specific to the component where it can be plugged in. If one component can be plugged into multiple other components, it has to support each life cycle, respectively. So we now have refined the plug-in concept, we speak of plug-ins with respect to certain components.

4   RENEW Plug-in Architecture

We use RENEW itself for a case study where the plug-in concept is applied. The RENEW tool has grown enormously since its first release in 1999, and many application-specific extensions have been created in the meantime. These extensions, like a workflow engine, an agent platform or an editor for UML interaction diagrams, are themselves already grown to applications with their own extensions. Up to RENEW 1.6, all extensions were compiled into one large application. Some sets of functionality could be selected by specifying a mode at startup, but mode switching at runtime was not possible. However, this was not flexible enough: a user would normally not need all extensions at the same time, but possibly in arbitrary combinations. Altogether, RENEW is very well suited as a case study for a dynamic, recursive plug-in system.

The plug-in system along with the decomposed application has been released as RENEW 2.0 and presented from the user's point of view last year in [7]. In this section we want to show how the concepts developed in the previous section are applied to the RENEW plug-in system. First we sketch how the functionalities of the application have been decomposed into several components. We will show where the plug-in concepts can be found in some exemplary components and where the dynamics come in. Last we will mention some concessions we had to make to keep the application usable.

4.1 Functional Decomposition

From the user's point of view, RENEW comprises two main components: the simulation engine and the editor. Already in the first release it has been stated that RENEW supports multiple formalisms, since new formalisms could easily be added by implementing the appropriate compiler. Clearly it is desirable to separate each formalism into its own plug-in. A formalism management component can then provide a registry for all loaded formalisms as well as the basic functionality needed by many formalisms.

Figure 11 shows some plug-ins of the current decomposition.[6] At the bottom, there are some unnamed class libraries that are used by many or all plug-ins. Some of these libraries are integrated into the application as a plug-in of their own, but they do not provide any extension interfaces. At the right there is the main plug-in of RENEW, the simulation engine. This plug-in also includes the input and output interfaces of the simulation engine, e.g. non-graphical net representation classes, token game feedback, remote control or database backup. With this plug-in and a formalism, it is possible to execute a Petri net system (without graphical feedback).

The graphical editor comprises two plug-ins: JHotDraw and Gui. This is due to historical reasons, we decided to re-surface the JHotDraw framework that had served as the basis for the net editor. The Gui plug-in enhances the JHotDraw application by Petri net specific figures and control commands for the token game. The NetComponents plug-in serves as an example for a new plug-in that has been added after the application decomposition. It has been presented in [1] and extends the editor by a tool-bar with commonly used patterns of net elements.

The management of formalisms has been divided into two plug-ins. Formalism manages the registry of known (i.e. loaded) formalisms and provides an API to select a formalism. The FormalismGui plug-in establishes the connection to the editor by presenting the available choices to the user. It also tailors the editor's menu and tool bars to the currently selected formalism. The two white components at the top of Figure 11 represent an arbitrary formalism. The standard formalism of RENEW—the reference net formalism—is integrated in the Formalism plug-in for the time being.

The user benefits from the introduction of the dynamic plug-in system for example in a server scenario: An application that has been implemented using

---

[6] It has to be noted that the decomposition of an existing application with approximately 900 classes in 30 packages into several components is not unique and therefore some functionalities might be reassigned between components in future releases. The refactoring of RENEW is still work-in-progress.

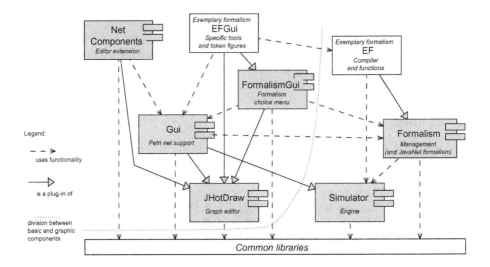

Fig. 11. Plug-ins and their dependencies as of RENEW2.0

reference nets and *Java* should normally run without the graphical net editor or animated token game. In this case, a user can start a reduced RENEW system where only the non-graphical components are installed. The Simulator and Formalism plug-ins (generally speaking, all components and common libraries right or below the dotted line in Figure 11) are sufficient to run the application. The user may also want to install a Prompt plug-in[7] so he can control the plug-in system and the simulation engine from the command line interface.

Suppose that, after the application has run undisturbed for some time, something gets stuck. The user then has the possibility to load the graphical editor and related components into the running system to debug the situation. The animated token game, although started long after the simulation setup, will show the current state of the system. So the user can search for the bug and hopefully fix it. Afterward, he can restart the simulation engine. When the application runs again, the editor and related plug-ins can be unloaded from the system and free their resources.

4.2 Applied Concepts

The RENEW plug-in system acts like the platform shown in Figure 8. There is no distinction between components and plug-ins because any component may also act as a plug-in.

The two PMS transitions that add and remove components become slightly refined in the RENEW PMS: Plug-ins can enter the system in two ways. At startup, a plug-in finder looks in a specific location for pre-installed plug-ins, and during runtime plug-ins can be loaded dynamically by supplying an URL to

---

[7] The Prompt plug-in is not included in Figure 11, but is available for download from the RENEW homepage [6].

the plug-in loader. The removal of components is realized by an `unload` command provided by the PMS.

The RENEW PMS is a flat-topped PMS as proposed in Figure 9. All plug-ins are accompanied by a description of their provided services. All components follow the life cycle shown in Figure 10. When a plug-in is loaded, the PMS calls its `init()` method. To unload a plug-in, there is a two-step process: First the PMS queries the plug-in whether it `canClose()`, afterward it may call `cleanup()` and remove it from the system.[8]

Optionally, the PMS may enforce dependencies between plug-ins. If a plug-in is also accompanied by a description of required services, the PMS will not include it in the system unless the required services are available, that is provided by some other plug-ins. Likewise, the unloading of a plug-in is prohibited as long as another plug-in requires a service provided by the plug-in to remove. A command to recursively unload all dependent plug-ins is provided. Of course, this dependency enforcement only works for static service requirements—but this is exactly what a *Java* programmer needs to ensure the availability of required class definitions.

All plug-ins in Figure 11 that are marked with two boxes at their right side, provide extension interfaces that follow the idea from Figure 10. There exist three refinements of the general idea: Either an extending plug-in provides additional implementations of existing interfaces that are seamlessly integrated into a framework, or it extends the basic plug-in by observing and reacting on events, or it just registers its own service in a database where it can be queried by other plug-ins.

In the case of the JHotDraw plug-in, other plug-ins can extend the graph editor by registering additional drawing tools, file types and menu commands. The main editor window integrates the registered functionality seamlessly. Due to the elementary functionality of the JHotDraw plug-in, all other plug-ins that provide some editing functionality are extending this plug-in (as can be seen in Figure 11).

The Simulator plug-in can be extended mainly by the notification of observer objects that can react on simulation-related events like transition firings, marking changes, or initialization and termination of the engine.

The extension interfaces of the Formalism and FormalismGui plug-ins have already been sketched above. From the technical viewpoint, these plug-ins provide all three types of extension: There is a registry of known formalisms, a notification about the choice of the default formalism, and an formalism-dependent integration of additional tools and menus into the editor.

Recursive extensibility (as introduced in Section 3.2) is represented in Figure 11 by a chain of extension arrows. An example is the EFGui plug-in which extends the FormalismGui plug-in which in turn extends the JHotDraw plug-in. Since the EFGui plug-in additionally directly extends the JHotDraw plug-in, we also have an example for the extension of multiple components by one plug-in.

---

[8] The two-step process (ask first, remove afterward) emulates the functionality of the net in Figure 9: There the plug-in also has the possibility to block its removal by not activating its `shutdown` channel uplink.

4.3   Pragmatism

The side condition that the plug-in system should not reduce the application's execution speed necessitated some pragmatic solutions. The concessions we made are restricted to the *Java* implementation of the plug-in system, the precise and concurrent Petri net semantics of the model in Section 3 have not been weakened.

The most important pragmatic decision is to not follow our usual paradigm of implementation by specification because the two uses of RENEW – as runtime engine *and* as case study – do not mix well. It would be possible to use the nets presented in Section 3 as code base for the plug-in system. By augmenting the nets with *Java* inscriptions that call the existing functionality of the application components, we would get an executable plug-in system rather easily. However, then we would have to set up the Petri net execution environment before the plug-in system in order to execute the net implementation. This would introduce a circular precedence because the simulation engine is a part of the application on top of the plug-in system. Therefore, we decided to use the insights gained from the concept model to re-implement the plug-in system in pure *Java*.

A consequence is the different behaviour of the RENEW and Java runtime environment with respect to dynamic linking. While in the reference net formalism the communicating plug-ins are linked at runtime for each communication individually, the Java virtual machine links the classes once when they are loaded. For full dynamic behaviour we would have to implement a dynamic linking layer of our own, but we decided to skip that part. The needed indirection would slow down inter-plug-in communication of repetitive jobs.

However, the `shutdown` of the PMS life cycle needs special care in this system: Due to the tight connections between components and their plug-ins, each component has to tidy up its references very carefully. Due to the already mentioned restriction of the Java class loader mechanism, we currently cannot truly remove a plug-in from memory when it is unloaded. But in most cases the cleanup process is sufficient to emulate the desired behavior.

4.4   Related Architectures

Plug-ins are more and more used for software architectures. We mention some well known products and their used concepts. The architectures are classified by their authors either as component or as plug-in systems. However, since the distinction between those systems is ambiguous, we list them here without discussion of details.

The model of JavaBeans (see `http://www.sun.com`) defines a mechanism for reusable components of Java application servers. Enterprise JavaBeans extend this again to allow for the design of standardized server components.

The OpenIDE is an integrated development environment which is based on NetBeans (see `http://www.netbeans.org`). The main advantages are the dynamic mechanism for plug-ins and the openness of the architecture. The disadvantages are the specific overhead and the missing visualization of the mechanisms used.

Gimp (Gnu Image Manipulation Program, see http://www.gimp.org) is a open source product with a static plug-in concept. The concept is strictly oriented towards its main purpose of image manipulation.

Netscape see (see http://www.netscape.com) installs its components on demand. This allows for a lean version. However, the basic mechanism is designed for the presentation of Web pages. It is unclear whether this architecture can be used in general for the design of software architectures.

Poseidon (see http://www.gentleware.com) uses the NetBeans mechanism for its plug-in architecture. The migration from a monolithic to a plug-in-based architecture (like the one presented in this paper) was done in [3]. Poseidon demonstrates the potential of NetBeans.

Eclipse is an open source product that mainly provides an architecture for static plug-ins, which are loaded once at startup time. Usually a restart of Eclipse is required when installing new plug-ins. Dynamic plug-ins are not fully realized in the current version of 3.1.

5   Conclusion

Nets-within-nets is an expressive modeling technique that is capable of modeling dynamic system architectures. Models that are built with these nets can profit from their ability to construct arbitrary and dynamic structures. The reference semantics that is applied in reference nets allows to express extensibility and dependency relationships of system components. Furthermore, the possibility to concretize the model by refinement leading to a functional model is of great advantage when designing, discussing and redesigning a system.

Our generic concept model for a dynamic architecture proves to be an approach that is both, sufficiently abstract for expressive modeling and sufficiently concrete to be able to transfer it to a real-world application. Moreover, it is the only modeling technique—to our knowledge—that is able to represent a flexible, adaptable and dynamic architecture design. The level of abstractness is a benefit to the general design decisions. The level of concreteness helps the architect and developer to experiment and evaluate the model prior to the implementation.

The concept model comes with an explicit top-level net, the PMS. The similarity of structures on the top level and all other levels allows for the introduction of independent service and extension management units on every level. Our model is capable of describing a pluggable plug-in mechanism. Such a model is useful to merge multiple systems with independent management architectures.

The Petri net IDE RENEW has undergone major refactorings and this process is still in progress. However, the preliminary results are promising. It is safe to say that the decision to refactor the system was the right way to go. We achieved a lean and flexible plug-in mechanism that permits arbitrarily nested plug-ins. The IDE has become more flexible and it can now be configured according to the needs of the users—even individually within a multi-user setting. Extending the functionality as a developer has become much easier, due to the fact that extensibility is a first order concept in the system.

Beside just another plug-in mechanism with specific features that are very valuable in the context of our research and development, a visual modeling concept for plug-ins has been presented. In fact, currently well-established modeling techniques are highly elaborated and powerful but also oriented towards static architecture design and very resistant against paradigm shifts. In order to improve modern architecture design many dynamic aspects have to be included as first-order concepts. Extensibility is one of them.

We believe that this approach can be transferred to other application areas and applied as a general concept for various kinds of domains. Moreover, it is possible to generalize the implementation of the concept model as done in RENEW to achieve a generalized core application that together with a conceptual approach can form a base for component-based application design of any kind.

We are looking forward to unleashing the full power of our architecture model by supporting an interleaved multi-formalism simulation support. Thereby, several advantages of different formalisms can be combined to the advantage of the designed model. Such an approach would be very difficult to handle in a monolithic system.

References

1. Lawrence Cabac, Daniel Moldt, and Heiko Rölke. A proposal for structuring Petri net-based agent interaction protocols. In W.M.P. van der Aalst and E. Best, editors, *Lecture Notes in Computer Science: 24th International Conference on Application and Theory of Petri Nets, ICATPN 2003, Netherlands, Eindhoven*, volume 2679, pages 102–120, Berlin Heidelberg: Springer, June 2003.
2. Søren Christensen and Niels Damgaard Hansen. Coloured Petri nets extended with channels for synchronous communication. Technical Report DAIMI PB–390, Computer Science Department, Aarhus University, DK-8000 Aarhus C, Denmark, April 1992.
3. Clemens Eichler. Entwicklung einer Plug-In-Architektur für dynamische Komponenten. Diplomarbeit, University of Hamburg, Department of Computer Science, Vogt-Kölln Str. 30, 22527 Hamburg, Germany, 2002.
4. Thomas Jacob. Implementation einer sicheren und rollenbasierten Workflow-Managementkomponente für ein Petrinetzwerkzeug. Diplomarbeit, University of Hamburg, Department of Computer Science, 2002.
5. Olaf Kummer. *Referenznetze*. Logos-Verlag, Berlin, 2002.
6. Olaf Kummer, Frank Wienberg, and Michael Duvigneau. Renew – The Reference Net Workshop. http://www.renew.de, October 2004. Release 2.0.1.
7. Olaf Kummer, Frank Wienberg, Michael Duvigneau, Jörn Schumacher, Michael Köhler, Daniel Moldt, Heiko Rölke, and Rüdiger Valk. An extensible editor and simulation engine for Petri nets: Renew. In Jordi Cortadella and Wolfgang Reisig, editors, *Applications and Theory of Petri Nets 2004: 25th International Conference, ICATPN 2004, Bologna, Italy, June 2004. Proceedings*, number 3099 in Lecture Notes in Computer Science, pages 484–493. Springer, 2004.
8. Neno Medvidovic and Richard N. Taylor. A classification and comparison framework for software architecture description languages. *IEEE Transaction on Software Engineering*, 26(1):70–93, January 2000.

9. Object Management Group (OMG). *Unified Modeling Language (UML)*, 2004. http://www.uml.org.
10. Wolfgang Reisig. *Elements of Distributed Algorithms: Modeling and Analysis with Petri Nets*. Springer-Verlag New York, October 1997.
11. Heiko Rölke. *Modellierung von Agenten und Multiagentensystemen – Grundlagen und Anwendungen*, volume 2 of *Agent Technology – Theory and Applications*. Logos Verlag, Berlin, 2004.
12. J. Sametinger. *Software Engineering with Reusable Components*. Springer, Berlin, 1997.
13. Jörn Schumacher. Eine Plug-in-Architektur für Renew: Konzepte, Methoden, Umsetzung. Diplomarbeit, University of Hamburg, Department of Computer Science, October 2003.
14. Clemens Szyperski. *Component software: beyond object-oriented programming*. ACM Press books. Addison-Wesley, 2. edition, 2002.
15. Richard Torkar. Dynamic software architectures. In I. Crnkovic and M. Larsson, editors, *Building Reliable Component-based Systems*, chapter 3, pages 21–28. Artech House, 2002.
16. Rüdiger Valk. Petri nets as dynamical objects. In Gul Agha and Fiorella De Cindio, editors, *Workshop Proc. 16th International Conf. on Application and Theory of Petri Nets, Torino, Italy*, June 1995.
17. Frank Wienberg. *Informations- und prozeßorientierte Modellierung verteilter Systeme auf der Basis von Feature-Structure-Netzen*. Dissertation, University of Hamburg, Department of Computer Science, Vogt-Kölln Str. 30, 22527 Hamburg, Germany, 2001.

A High Level Language for Structural Relations in Well-Formed Nets

Lorenzo Capra[1], Massimiliano De Pierro[2], and Giuliana Franceschinis[3]

[1] Università di Milano, Dip. di Informatica e Comunicazione, Milano, Italia
`capra@dico.unimi.it`
[2] Università di Torino*, Dip. di Informatica, Italia
`depierro@di.unito.it`
[3] Università del Piemonte Orientale, Dip. di Informatica, Alessandria, Italia
`giuliana@mfn.unipmn.it`

Abstract. Well-formed Nets (WN) structural analysis techniques allow to study interesting system properties without requiring the state space generation. In order to avoid the net unfolding, which would reduce significantly the effectiveness of the analysis, a symbolic calculus allowing to directly work on the WN colour structure is needed. The algorithms for high level Petri nets structural analysis most often require a common subset of operators on symbols annotating the net elements, in particular the arc functions. These operators are the function difference, the function transpose and the function composition. This paper focuses on the first two, it introduces a language to denote structural relations in WN and proves that it is actually closed under the difference and transpose.

1 Introduction

Several flavours of High Level Petri Nets (HLPN) [1] have been introduced to allow more parametric and compact Petri Net (PN) based representations of complex systems. Although HLPN models can be translated into PN ones by applying an unfolding procedure, usually the unfolding is avoided, and ad-hoc analysis techniques are applied directly to the high level representation: this approach usually leads to improved efficiency and sometimes allows also to obtain parametric results.

Several types of analysis techniques can be applied to HLPN models, for example a reachability graph (RG) can be computed, allowing to perform reachability analysis, or structural analysis methods can be applied, which allow to derive marking independent properties of PN models. Structural analysis results are often used to check some properties before proceeding to the RG construction, and they can be exploited for early detection of modelling errors, or for improving the efficiency of the RG construction process.

---

* The work of Massimiliano De Pierro was done when he was at the Dipartimento d'Informatica, Università del Piemonte Orientale.

G. Ciardo and P. Darondeau (Eds.): ICATPN 2005, LNCS 3536, pp. 168–187, 2005.

This paper is about structural analysis of Well-formed Nets (WN) [2], an HLPN formalism similar to Coloured PNs (CPN) [3], characterised by a structured syntax that has been exploited for developing efficient RG generation algorithms [4]. In particular the paper develops the basis for a symbolic calculus of different structural properties and relations which can be used for a number of different applications. Examples of such relations are the structural conflict and the causal connection relations between transitions, which can be used to improve the RG construction process efficiency (through partial order techniques [5]); they are also useful in the construction of Stochastic WN models (SWN formalism includes timed transitions with exponentially distributed delays, and immediate transitions enriched with priorities and probabilities for implementing different conflict resolution policies), to help the modeller in correctly specifying the priorities and weights of *immediate transitions* (as proposed in [6] for Generalised Stochastic PNs). Another example of useful HLPN structural analysis technique which has been studied in the literature concerns deadlock detection [7]. Finally structural considerations can speed up the computation of enabled transition instances in HLPN discrete event simulation or RG construction (for example in [8] some structural ad-hoc rules are applied to WN models to this purpose).

Some results on HLPN models structural relation computation have been presented in [9, 10], however they apply to a rather restricted subclass of WN, namely Unary Regular Nets: the present paper extends such results to the whole WN class. The core result of this paper is the formalisation of a calculus for the computation of *symbolic structural properties* and of *symbolic structural relations* between WN nodes: more details can be found in [11]. A language for representing such properties and relations in symbolic form is proposed, which is based on the same syntax used to specify WN arc functions, extended with an intersection operator and an output guard (called *filter*): this representation has been chosen with the aim of being easily interpreted by the modeller. The derivation of the symbolic structural properties and relations of interest is based on the application of the following set of operators on the expressions of the language: transpose, difference, support and composition. Only the first two operators are developed in details in this paper (due to space constraints), the other operators are developed in [11] and [12]. A practical example of application of the calculus and a discussion about the results interpretation has been proposed in [13].

The rest of the paper is organised as follows: Section 2 introduces the basic WN's definitions and the notation used in the paper. Section 3 proposes the language for expressing WN structural properties and defines the operators on the elements of such language. Section 4 discusses in details the transpose and difference operators: it proves that the proposed language is closed with respect to these two operators, and illustrates the algorithms to symbolically compute the result of the application of such operators. Section 5 discusses the presented results and the planned future work.

2   Well-Formed Nets Basics

This section highlights the aspects of the WN definition which are useful to describe the topic of this paper, namely the arc function syntax and the colour domain structure. The complete WN formalism definition is not included for the sake of space, and the interested reader can find it in [2].

WN belongs to the Coloured Petri Nets (CPN) family. In WN anyway particular constraints are imposed on the colour specification syntax, they are discussed in detail next. To help introducing the WN basics, the net of Fig. 1(a) is utilised, it shows a modified WN version of the well-known database system CPN model treated in [3]. Figure 1(b) shows a variant of the colour structure of the same example, used to illustrate more complex colour definitions. Symbols F_1 :, F_2 :, F_3 : are not part of the WN definition, but are labels used only to introduce the variant in (b).

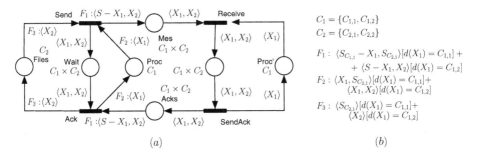

(a)   (b)

Fig. 1. Case (a) and case (b) of the database system WN model

The Colour Domains Structure. In WN, each place and transition has an associated *colour domain* which is defined as the Cartesian product of finite sets called *basic colour classes* and belonging to a finite collection; formally the symbols $C_1, \ldots, C_n$ are used to denote such sets and the mapping $\mathcal{C}(s)$ is used to denote the colour domain of a WN node s, which may be either a transition or a place. Thus $\mathcal{C}(s) = C_1^{m_{1,s}} \times C_2^{m_{2,s}} \times \ldots \times C_n^{m_{n,s}}$, where the superscript $m_{i,s} \in \mathbb{N}$ depends upon s and indicates the multiplicity (i.e. the number of occurrences) of the basic colour class C_i in $\mathcal{C}(s)$. Basic colour classes are not primitive sets meaning that each one may be partitioned into further elements called *static subclasses*: the number of elements of the partition of C_i is denoted $||C_i||$ and the j-th static subclass of C_i is denoted $C_{i,j}$, so that $C_i = \bigcup_{j=1}^{||C_i||} C_{i,j}$ and $\sum_{j=1}^{||C_i||} |C_{i,j}| = |C_i|$. A colour class C_i may be cyclically ordered.

A place colour domain defines the type of objects the place can contain: generally a marking of a place is not simply a set but is a multiset on $\mathcal{C}(p)$. The set of all multisets on $\mathcal{C}(p)$ is denoted $Bag[\mathcal{C}(p)]$, whereas the sum notation $\sum_{j=1}^{|C_i|} \lambda_j . c_j$, $\lambda_j \in \mathbb{N}$, is utilised to denote a multiset on $C_i = \{c_1, c_2, \ldots, c_{|C_i|}\}$.

The net in Fig. 1 has got two basic colour classes $\{C_1, C_2\}$ which respectively model the *database managers* and the *files*; let us look at some place: place Proc has colour domain C_1, whereas places Wait and Mes have colour domain $C_1 \times C_2$. A token in place Mes models a sent message, which is formed by a sender *datatbase manager* and a modified *file*. Version (b) of the database-system partitions both the *database managers* and the *files* into two static subclasses. This partition will allow to model a different system behaviour for elements belonging to different subclasses.

The Arc Function Syntax. As in CPN, functions on colour domains label arcs connecting places and transitions. By the CPN definition a function F labelling an arc between transition t and place p is a mapping which assigns to each colour in $\mathcal{C}(t)$, a multiset of colours in $\mathcal{C}(p)$. The following notation is used to indicate the domain and codomain of a function $F : \mathcal{C}(t) \rightarrow Bag[\mathcal{C}(p)]$.

The WN formalism prescribes a syntax to express such functions, built on a limited set of primitive symbols:

$$F = \sum_i \lambda_i T_i[guard_i], \quad \lambda_i \in \mathbb{N}$$

The innermost term T_i of the definition is a function *tuple* and represents a mapping $\mathcal{C}(t) \rightarrow Bag[\mathcal{C}(p)]$. To denote the colour domain of a function tuple, symbol $\mathcal{C}$ is extended to functions, so that $\mathcal{C}(T_i)$ is the colour domain of the function tuple T_i. Symbols T, T_i are reserved to indicate function tuples. The name function-tuple given to T derives from its syntax which has the form $\langle f_1, \ldots, f_k \rangle$, where each component f_i is a function mapping the elements of $\mathcal{C}(T)$ into elements of $Bag[C_j]$, $C_j \in \{C_1, \ldots, C_n\}$. The application of tuple T to colour $c \in \mathcal{C}(T)$ is defined as $T(c) = \otimes_{i=1}^{k} f_i(c)$ where $\otimes$ is the multiset Cartesian product operator.

The components f_i of a tuple T are called *class-functions* and are expressed as linear combinations of *elementary* functions, formally a class-function $f_i : Bag[\mathcal{C}(T)] \rightarrow Bag[C_j]$ is so defined:

$$f_i = \sum_{k=s_j}^{e_j} \alpha_k X_k + \sum_{q=1}^{||C_j||} \beta_q S_q + \sum_{k=s_j}^{e_j} \gamma_k !X_k, \quad \text{where } \alpha_k, \beta_k, \gamma_k \in \mathbb{Z}$$

The symbols s_i and e_i will be used to denote the position of the first and last occurrence of basic class C_i in a colour domain. If C_i does not appear in the colour domain, then $m_i = e_i = s_i = 0$ (by definition), otherwise $m_i = e_i - s_i + 1$. $X_k, !X_k, S, S_q$ are the elementary functions and represent the limited set of symbols upon which class-functions are defined. Obviously, they have the same domain and codomain as f_i. Namely the projection X_k maps a colour $c \in \mathcal{C}(T)$, $c = \langle c_1, \ldots, c_m \rangle$, into its k-th component $c_k \in C_j$, the successor $!X_k$ maps a colour $c \in \mathcal{C}(T)$ into the successor of its k-th component and may be used only when basic class C_j is ordered, the constant function S maps a colour $c \in \mathcal{C}(T)$ into the constant multiset $\sum_{c \in C_j} c$, the constant S_q on the q-th static subclass of C_j maps its argument into the constant multiset $\sum_{c \in C_{j,q}} c$.

In WN the class-functions must always map into multisets with non negative coefficients, so $\alpha_i, \beta_i, \gamma_i$ must be defined consistently with such assumption. A class-function expressed by means of a linear combination with *positive integer* coefficients of the functions $X_i, S - X_i, S_q - X_i, S, S_q$ and $!X_i$, surely satisfies the above constraints. Hereafter the class functions are assumed to be expressed in this *simple form*: this is not restrictive since any class function expression that satisfies the above constraint (for any possible argument) can always be rewritten in *simple form*, for example $2S - X_i = S + (S - X_i)$. In Fig. 1(a), transition Send defines the system behaviour when a database file is being modified; the functions on the arcs surrounding transition Send highlight this situation: when a database manager m modifies a file f then the messages $\{(m', f) : m' \in C_1 \wedge m' \neq m\}$ are sent over the network. Function F2, selecting the identity of the database manager m of the transition instance Send(m, f) is expressed as X_1 (projection on first component of (m, f) in WN syntax, similarly function F1:$\langle X_2 \rangle$ selects the file identity f, finally the set of messages to be sent over the network is given by F1: $\langle S - X_1, X_2 \rangle$ that returns the result of the Cartesian product $C_1 - m \times f$.

[$guard_i$] : $\mathcal{C}(T_i) \rightarrow Bag[\mathcal{C}(T_i)]$ in the expression of F represents a guard. Guards are functions which are right composed to function tuples, that is they are applied before the function tuple, and act as the identity function for those elements of the domain which satisfy a given condition, represented by the guard, otherwise map into the empty multiset. Guards in WN can also be associated with transitions: in this case they are used to restrict the transition colour domain to those elements that satisfy the guard. WN guards are Boolean expressions whose terms are the predicates $X_i = X_j$, $X_i \neq X_j$, $d(X_i) = d(X_j), d(X_i) \neq d(X_j)$, $d(X_i) = C_{j,q}$ and $d(X_i) \neq C_{j,q}$. The term $X_i = X_j$ ($X_i \neq X_j$) is true if the i-th and j-th component of the (guard) argument are equal (not equal), $d(X_i) = d(X_j)$ ($d(X_i) \neq d(X_j)$) is true if the static subclass to which the i-th and j-th component of the argument belong are equal (not equal), finally $d(X_i) = C_{j,q}$ ($d(X_i) \neq C_{j,q}$) is true if the i-th component of the argument belongs to (does not belong to) static subclass $C_{j,q}$. The truth value of the whole Boolean expression can then be established following the standard evaluation rules for Boolean expressions.

In version (b) of the database-system of Fig. 1 the guards are used on the arc functions to specify different behaviours for managers belonging to different groups: managers in $C_{1,2}$ behave like managers of version (a) for any file in C_2; instead, managers in $C_{1,1}$ access only files in $C_{2,1}$ and in an exclusive way, that is all files in $C_{2,1}$ must be locked prior to modify them and thus transition Send requires $\langle S_{C_{2,1}} \rangle$ from place Files (see function F_3).

The calculus presented in this paper shall often need the application of the functions, so far discussed, to multisets of colours rather than to single colours. The linear extension of the WN functions is considered: the linear extension F* of function F, (where F may be either a function tuple or a class-function or an elementary function) is assumed by definition: $F^*(a + b) = F(a) + F(b)$, $F^*(\lambda.a) = \lambda.F(a)$ where a and b are colours in $\mathcal{C}(F)$, λ is any integer and the domain of F^* is $Bag[\mathcal{C}(F)]$. For convenience, in the rest of the paper F will be used instead of F^* abusing notation.

Although the function tuples in the arc functions may have elements that are sums of class functions, sometimes it is convenient to deal only with function tuples whose elements are elementary functions: the *tuple expansion* operation can be used in this cases. Using the distributive property of the multiset Cartesian product with respect to the multiset sums it is possible to rewrite any WN function tuple into a sum of tuples whose components are (weighted) elementary functions only:

Property 1 (Tuple expansion). Let $T = \langle \sum_{i_1=1}^{j_1} \lambda_{i_1} f_{i_1}^1, \ldots, \sum_{i_k=1}^{j_k} \lambda_{i_k} f_{i_k}^k \rangle$ a WN tuple where f_i^j are elementary WN functions and $\lambda_i \in \mathbb{N}$, then T is rewriteable as a sum of tuples $T = \sum_{i_1=1}^{j_1} \sum_{i_2=1}^{j_2} \cdots \sum_{i_k=1}^{j_k} \langle \lambda_{i_1} f_{i_1}^1, \lambda_{i_2} f_{i_2}^2 \ldots, \lambda_{i_k} f_{i_k}^k \rangle$.

Moreover the weights of the components may be factorised outside the tuple: $\langle \lambda_1 f_1, \ldots, \lambda_k f_k \rangle = (\prod_{i=1}^{k} \lambda_i) \langle f_1, \ldots, f_k \rangle$. This property is useful to simplify the presentation of the algorithms in next sections.

3   Structural Relations in WN

3.1   Motivation

The structured syntax of WN's arc functions, recalled in the previous section, allows to implicitly encode the system's symmetries into its model specification; in this way efficient methods can be applied that exploit such symmetries and build an aggregated state space.

This paper deals with WN syntax from a different point of view: is it appropriate to symbolically express *structural relations* between the coloured nodes of a WN? What type of extension is required to this purpose? Many analysis tools base their algorithms on the ability to derive structural relationships on the net, and sometimes even state-space based methods require some structural consideration; an example of method belonging to this latter group is the *stubborn-set* one ([14, 15, 16]). The method tries to reduce the number of RG states to be generated by reducing the number of considered transition interleavings to be fired from a given marking. This of course must be done in a way that preserves the properties to be studied. A *stubborn set* is a subset of enabled transitions in a given marking m to be fired for generating the (subset of) states reachable from m in this method. Since the set have to be inferred without looking at the future, that is without generate the next states, both structural and marking considerations are used. Structural relations between the nodes of a PN which are often used by algorithms dealing with the stubborn sets computation are the structural conflict (SC), the structural causal connection (SCC) and the structural mutual exclusion (SME), that are all binary relations between transitions. Structural analysis is thus a relevant framework to develop and consider in PN qualitative and quantitative analysis.

The first question is how to express the known PN structural relations, such as the ones mentioned above, in WN. Once this has been done, the second question is how to efficiently manipulate them to extend the PN algorithms to work on WN models: for instance a *symbolic* version of the stubborn-set algorithm might be developed if relations such as SC, SCC and SME could be computed without unfolding the net.

In [9, 10] a convenient way to express coloured structural relations in CPN is introduced. A structural relation $\mathcal{R}$ between two coloured nodes s, s' is defined by means of a function $F(s, s') : Bag[\mathcal{C}(s')] \rightarrow Bag[\mathcal{C}(s)]$ mapping any colour instance c' of s' into the instances c of s that are in relation $\mathcal{R}$ with instance c' of s'. According to the definition, for example, coloured version of the structural conflict and causal connection between transitions t, t' are functions denoted respectively $SC(t, t')$ and $SCC(t, t')$.

To illustrate how the above definition may be exploited in case of WN, let us introduce an example using the modified database-system model (b) of Fig. 1. Assume to represent the predecessor relationship between place Files and transition Send, denoted $\Gamma^-(\text{Files}, \text{Send})$. Such relation expresses for a given coloured instance c of Send the coloured instances of Files connected to it in the graph of the unfolded net. $\Gamma^-(\text{Files}, \text{Send})$ is easy to derive since it is given by the function $\langle S_{C_{2,1}} \rangle[d(X_1) = C_{1,1}] + \langle X_2 \rangle[d(X_1) = C_{1,2}]$ on the arc connecting Files to Send. Thus, evaluating this function for a given colour of Send returns the instances of Files that are connected to it through an output arc in the unfolded net.

Assume now to represent the predecessor between transition Send and place Mes, denoted $\Gamma^-(\text{Send}, \text{Mes})$: although it is related to the function $\langle S_{C_{1,1}} - X_1, S_{C_{2,1}} \rangle[d(X_1) = C_{1,1}] + \langle S - X_1, X_2 \rangle[d(X_1) = C_{1,2}]$ labelling the arc, in this case some computation and reasoning on the unfolded net must be done. With a little extension to the WN's syntax the function $[d(X_1) = C_{1,1}]\langle S - X_1, S \rangle[d(X_1) = C_{1,1} \wedge d(X_2) = C_{2,1}] + [d(X_1) = C_{1,2}]\langle S - X_1, X_2 \rangle$ may be computed, which is the correct result. The interpretation of this expression is the following: given a token (m, f) in place Mes, denoting a pair of a database-manager and a file, the database modification events (transition Send instances) producing it could be composed either by all managers in $C_{1,2}$, except m, that are modifying file f (rightmost addend) or by all managers in $C_{1,1}$, except m, that are modifying any file in class $C_{2,1}$ only (leftmost addend).

The keypoint is the ability to determine algorithmically, and without unfolding the net, such expressions. In [9, 10] it is shown that the predecessor may be computed by means of an operator applied to the involved arc function, and called the *transpose*. Moreover, also the other relations such as the conflict and causal connection may be computed by means of formulae expressed in terms of the arc functions and appropriate operators. Figure 2 illustrates the case of $SCC_p(t, t')$ and $SC_p(t, t')$ formulae.

The remainder of this section introduces the language to express structural relations in WN, whereas the next section formally treats the manipulation of such expressions under a limited set of operators. This paper focuses on two operators, namely the difference $\ominus$ and the transpose $(.)^t$. Although the transpose

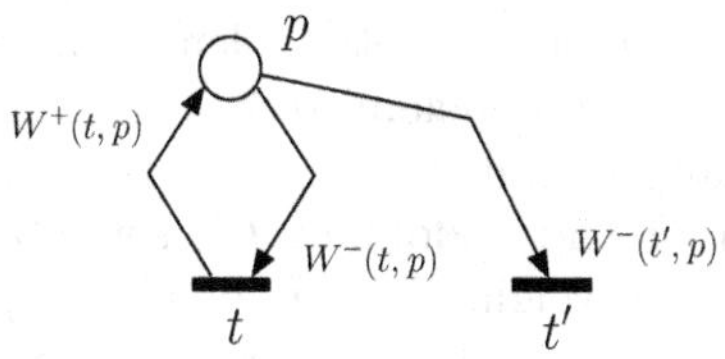

$$SCC_p(t,t') = \overline{(W^+(t,p) \ominus W^-(t,p))^t \circ W^+(t',p)}$$

$$SC_p(t,t') = \overline{(W^-(t,p) \ominus W^+(t,p))^t \circ W^+(t',p)}$$

Fig. 2. Formulae and structural conditions for causal connection and conflict

and difference without the composition operator are not enough to compute the SC and SCC relations, they can be used to derive expressions that can be useful in optimising the computation of the enabled transition instances in a given marking, during RG construction or in simulation. The trivial way of computing the enabled instances of t in marking m consists of considering each possible colour $c \in \mathcal{C}(t)$, and for each input and inhibitor place apply the corresponding arc function, and check whether the place marking contains the required number of coloured tokens. Since $\mathcal{C}(t)$ can be quite large, a method to reduce the number of colour instances to be checked is needed (for example in [8] some heuristics devised to this purpose for speeding up SWN simulation are described). Using simply the transpose and difference operators, it is possible to derive some expressions from which the exact set of enabled transition instances in a given marking can be directly derived, as a function of the marking. In alternative the expressions can be used to derive a superset of the enabled transition instances (usually much smaller than the complete colour domain) to which the normal enabling test can then be applied.

3.2   The Language to Denote WN Structural Relations

The language syntax is an extension of the WN arc function syntax and introduces the following new elements: (1) the successor operator ! is extended to the k-th successor $!^k X_i$ where k ranges in $1, \ldots, |C_j|$ (2) class-function terms may be intersection of the elementary functions, (3) guards can appear on the left of a tuple (i.e. they can be left composed to a tuple), in this case they are called *filters*. The definition of the *language* follows.

Definition 1. *Given a domain* $D = C_1^{m_1} \times \ldots \times C_n^{m_n}$ *and a codomain* $D' = Bag[C_1^{m'_1} \times \ldots \times C_n^{m'_n}]$, $s = \sum_{\forall i} m_i$, $k = \sum_{\forall i} m'_i$, *then*

- $\mathcal{S} = \{X_i, S, S - X_i, S_q, !^k X_i : i \in \{1 \ldots s\} \wedge q \in \{1 \ldots max_{\forall j} ||C_j||\}\}$
- $\mathcal{R}[D, D'] = \sum_j [filter]_j \lambda_j T_j [guard]_j$

  where $T_j = \langle \sum_{i_1=1}^{j_1} \lambda_{i_1} f_{i_1}^{j,1}, \ldots, \sum_{i_k=1}^{j_k} \lambda_{i_k} f_{i_k}^{j,k} \rangle$ *and*
  $f_{i_j}^r = \bigcap_l \Gamma_l : \Gamma_l \in \mathcal{S}$ *where* l *depends on* i_j, r

- $\mathcal{K}[D, D'] = \{F_i : \lambda \langle f_{i,1}, \ldots, f_{i,k} \rangle \wedge f_{i,j} = \bigcap_{i,j}^r \Gamma_{i,j}^r : \Gamma_{i,j}^r \in \mathcal{S}\}$

the language is the union of all the possible D, D' *of* $\mathcal{R}[D, D']$.

Set S is the collection of all the primitive symbols denoting elementary functions. It is observable that symbols $S_q - X_i$ do not belong to S although they satisfy the WN syntax, that is because an equivalent form is utilised, precisely $(S - X_i) \cap S_q$. For a given domain D and codomain D', set $\mathcal{R}[D, D']$ is the collection of the expressions which are weighted sums of tuples possibly preceded by a filter and followed by a guard. The tuples are such that their components are sums of a finite number of terms, and each term is an intersection of a finite number of symbols in S. $\mathcal{K}[D, D']$ denotes the tuples in $\mathcal{R}[D, D']$ (so $\mathcal{K}[D, D'] \subset \mathcal{R}[D, D']$) whose class-functions are constituted by single addends. As an example the following element belongs to the language $\mathcal{R}[C_1 \times C_1 \times C_2; C_1 \times C_2]$: $\langle X_1 \cap (S - X_2), S \rangle + \langle X_1, X_3 \rangle$ and both the tuples appearing in the expression singularly belong to $\mathcal{K}[C_1 \times C_1 \times C_2; C_1 \times C_2]$. Expression $\langle X_1 \cap (S - X_2) + X_1, S - X_2 \rangle$ is also an element of the same language but it does not belong to $\mathcal{K}[C_1 \times C_1 \times C_2; C_1 \times C_2]$.

Note that syntactically different tuples in $\mathcal{R}[D, D']$ may represent the same function, for instance it is easy to verify that $\langle X_1 \cap (S - X_2), S \rangle$ and $\langle X_1, S \rangle[X_1 \neq X_2]$ identify the same function. The language of symbolic expressions used in this paper is the union on any D, D' of $\mathcal{R}[D, D']$.

The operators defined on the language are the *transpose*, the *difference*, the *support* and the *composition*, plus some auxiliary ones. Let $F, G \in \mathcal{R}[D, D']$ be two expressions of the language: the transpose of F is denoted F^t, its support is denoted $\overline{F}$, the difference of F and G is denoted $F \ominus G$, the composition of F and G is denoted $F \circ G$, finally the intersection and the the componentwise multiplication are respectively denoted $F \cap G$ and $F * G$. The following definitions introduce the operators above mentioned.

Definition 2. *(Transpose) Let f be a linear function from $Bag[C]$ to $Bag[D]$ then the transpose of f denoted f^t is a function from $Bag[D]$ to $Bag[C]$ so defined: $f^t(x)(y) = f(y)(x) \quad \forall x \in D, \forall y \in C$.*

Definition 3. *(Composition) Let $F : Bag[D''] \to Bag[D']$ and $G : Bag[D] \to Bag[D'']$ be linear functions, then their composition $F \circ G$ is defined as follows: $(F \circ G)(c) = F[G(c)] \quad \forall c \in D$.*

Definition 4. *(Difference) Let $a = \sum_{i=1}^{|C|} \lambda_i.c_i$ and $b = \sum_{i=1}^{|C|} \gamma_i.c_i$ be multisets on C then $a \ominus b$ is so defined: $a \ominus b = \sum_{i=1}^{|C|} sup[0, \lambda_i - \gamma_i].c$.*
Let F and F' be linear functions from $Bag[C]$ to $Bag[D]$ and $c \in C$ then $(F \ominus F')(c) = F(c) \ominus F'(c)$. The following linear extension is assumed for $F \ominus F'$: let F and F' be linear functions from $Bag[C]$ to $Bag[D]$ and $c, c' \in C$ then the $(F \ominus F')(c + c') = [F(c) \ominus F'(c)] + [F(c') \ominus F'(c')]$.

Observe that the last part of the definition is needed because generally it holds $(F \ominus F')(c + c') \neq F(c + c') \ominus F(c + c')$: the reason is that the second member is equal to $(F(c) + F'(c')) \ominus (F(c) + F(c'))$ and since $\ominus$ and $+$ are not associative (due to the presence of the *sup* in the definition of $\ominus$), it is not equal to the first member.

Definition 5. *(Support) Let $a \in Bag[C]$ then the support of a denoted $\overline{a}$ is a function from $Bag[C]$ to $Bag[C]$ such that $\overline{a} = \sum_{c \in C : a(c) > 0} c$.*
Let F be a linear function from $Bag[C]$ to $Bag[D]$, then its support is defined as $\overline{F}(c) = \overline{F(c)}, \ \forall c \in C$.

Definition 6. *(Multisets Operators) Let $a = \sum_{i=1}^{|C|} \alpha_i.c_i$ and $b = \sum_{i=1}^{|C|} \beta_i.c_i$ be multisets on C then the* Componentwise Multiplication $*$ *is defined as $a * b = \sum_{i=1}^{|C|} (\alpha_i \beta_i).c_i$, and the* Intersection $\cap$ *is defined as $a \cap b = \sum_{i=1}^{|C|} inf[\alpha_i \beta_i].c_i$*

Let E_1 and E_2 be two expressions of the language, which represent two linear functions: the problem to solve now is how to compute *in a completely symbolic way* the expressions of the language representing the function $E_1 \ominus E_2$, E_1^t, $\overline{E}_1$, etc., Section 4 provides a set of *rewriting rules* leading to symbolic manipulation algorithms that solve this problem. The *symbolic* here is relative to the fact that *no function unfolding is done*, that is no evaluation for any value of the argument is done. Each *rewriting rule* is justified by a set of algebraic properties which the operands satisfy. In the rest of the paper the term *rewriting rule* always denotes symbolic manipulation that can be implemented algorithmically. Properties are instead used to justify the *rewriting rule*.

4   Symbolic Manipulation

This section introduces the algebraic properties and the *rewriting rules* needed to deal with the *transpose* and the *difference* operators. It also treats a group of secondary operators needed in the development of the transpose and the difference, namely the intersection and the componentwise multiplication.

4.1   The Transpose Operator

This section deals with the computation of the transpose for the elements of the language. For this purpose an incremental approach is followed. The discussion is organised as follow: in the next two subsections the case of WN arc functions is first illustrated. Two properties are introduced, which respectively give the rules to transpose WN *class-functions* in terms of the *elementary function* transposes and the rules to transpose WN function tuples. The general case in which the function to transpose is a sum of guarded tuples and includes the language extensions, that is the filters and the elementary functions intersection, are treated in the final subsection where a practical algorithm is also illustrated.

WN Elementary and Class-Function Transpose. The first step towards the transpose computation of a tuple consists in formalising the class-function transpose. This requires to consider general linear combinations of elementary functions. Let us consider the following example, which allows us to introduce in an intuitive way some problems and the corresponding solution.

Example 1. Let us consider the class-function $!X_1 + X_2$ so defined: $F : Bag[C_1 \times C_1] \rightarrow Bag[C_1]$ such that $F(c_1, c_2) = !c_1 + c_2$. In this case set C_1 is ordered

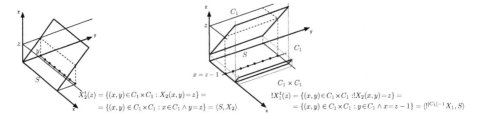

$$X_2^t(z) = \{(x,y) \in C_1 \times C_1 : X_2(x,y) = z\} =$$
$$= \{(x,y) \in C_1 \times C_1 : x \in C_1 \wedge y = z\} = \langle S, X_2 \rangle$$

$$!X_1^t(z) = \{(x,y) \in C_1 \times C_1 : !X_2(x,y) = z\} =$$
$$= \{(x,y) \in C_1 \times C_1 : y \in C_1 \wedge x = z - 1\} = \langle !^{|C_1|-1} X_1, S \rangle$$

Fig. 3. Graphical representation respectively of the transpose of functions X_2 and $!X_1$

because the successor function $!X_1$ is used. Observe that class C_1 occurs twice in the function domain. Let us represent in the three dimensional Cartesian space the mappings of function F. Natural numbers are associated with colour of C_1 and in a way to satisfy the ordering in C_1, v will denote such mapping from C_1 to $\mathbb{N}$ and is defined as $v(c_i) = i$, $\forall c_i \in C_1$, where $i = 1 \ldots |C_1|$. The image $F(c_i, c_j) \in Bag[C_1]$ of the colour $(c_i, c_j) \in Bag[C_1 \times C_1]$, which corresponds to the point $(v(c_i), v(c_j))$ in the plane xy, is represented along the z dimension depicting the whole set of points corresponding to the colours in the multiset $F(c_i, c_j)$, an integer number should also be associated to each image's point and represents the respective colour multiplicity in the multiset, hence there should be a fourth dimension, here not represented for simplicity.

Both X_2 and $!X_1$ may be visualised as planes in the Cartesian space, precisely the planes $z = y$ and $z = x + 1$. The image of F is represented by the points (x, y, z) with discrete coordinates in $1 \ldots |C_1|$ laying on the union of such planes, let us denote Ω such set. Observe that points at the intersection sum up their multiplicity.

Fixing $z = v(c_k)$, the transpose $F^t(c_k)$ is represented by the set of points at the intersection of Ω with the plane $z = v(c_k)$, projected on plane xy. Observe that the multiplicity of such points (x, y) corresponds to that of (x, y, z). Fig. 3 shows a graphical representation of the two components of F.

Informally what above said may be translated into symbolic computation as:

$$F^t = (X_2 + !X_1)^t = X_2^t + !X_1^t = \langle S, X_1 \rangle + \langle !^{|C_1|-1} X_1, S \rangle$$

where $!^{|C_1|-1}$ is equivalent to the predecessor function. $\qquad\square$

The previous example is interesting because it shows how the transpose of general class-functions can be obtained by combining the elementary functions transposes: this observation is generalised and formalised in Property 2. The example suggests that in order to generalise the approach the transposes of all the *elementary functions* are needed. Table 1 shows the transposes of all the WN elementary colour functions.

For the projection and successor functions, the derivation is a straightforward extension of the previous example (see Fig. 3). The transpose of the constant function on the static subclass $C_{i,j}$, at the last line of the table, has been expressed using a guard that maps colours not belonging to the static subclass $C_{i,j}$

Table 1. WN elementary colour function transposes, $D = C_1^{m_1} \times \ldots \times C_n^{m_n}$

$f : Bag[D] \rightarrow Bag[C_i]$	$f^t : Bag[C_i] \rightarrow Bag[D]$	Notes
S	$\langle S, S, \ldots, S \rangle$	the tuple is composed by all S
X_i	$\langle S, \ldots, S, X, S, \ldots, S \rangle$	all S but X at the i-th position
$!^k X_i$	$\langle S, \ldots, S, !^{|C_i|-k} X, S, \ldots, S \rangle$	all S but $!^{|C_i|-k} X$ at the i-th position
$S - X_i$	$\langle S, \ldots, S, S - X, S, \ldots, S \rangle$	all S but $S - X$ at the i-th position
$S_{C_{i,j}}$	$\langle S, S, \ldots, S \rangle \circ [d(X) = C_{i,j}]$	all S and a guard

onto the empty set. The $S_{C_{i,j}}^t$ is thus equal to the empty set for such colours not in $C_{i,j}$ due to the fact that, for a property of linear functions, the image of the empty set is the empty set itself. Table 1 does not show an entry for $(S_q - X_i)^t$, in fact, as discussed previously, it can be rewritten into an expression of the language using symbols in $\mathcal{S}$ only. For this reason its transpose is not provided here.

The next property states how to compute the transpose of a linear combination of linear functions starting from the transposes of the functions themselves. This property is not only useful in this case, while calculating the transpose of the class-functions in terms of the elementary ones, but will be also useful later for calculating the transpose of a sum of tuples $\sum_i [filter_i] \lambda_i T_i [guard_i]$.

Property 2. Let $\{F_i : i = 1 \ldots n\}$ be a collection of functions from $Bag[C]$ to $Bag[D]$ and let $F = \sum_i \lambda_i \cdot F_i$ be a linear combination of such functions, whose coefficients λ_i are in $\mathbb{N}$ and where $+$ and $\cdot$ are the multi-set addition and the scalar multiplication, then the transpose F^t of F is defined by

$$F^t = \sum_i \lambda_i \cdot F_i^t$$

Property 2 appears as an algebraic property of linear functions, however it is combined with the elementary function transposes of Table 1, it also provides a rewriting rule for obtaining the language (sub)expression for the transpose of the linear combination of language (sub)expressions. This is the basic building block on which the tuple transpose computation can be constructed.

WN Tuple Transpose. The algorithm for general WN arc function transposing is based on the ability to compute the transpose of a tuple. The next property expresses the transpose of a tuple in terms of the transposes of its components. In turn the transpose of each component can be computed using Property 2.

Property 3 (Transpose of a function tuple). Let f_j, $j = 1 \ldots k$, be linear functions defined from $Bag[D]$ to $Bag[C_{i_j}]$, where $D = C_1^{m_1} \times \ldots \times C_n^{m_n}$ and $i_j \in \{1, \ldots, n\}$, then

$$\left(\bigotimes_{j=1}^k f_j \right)^t = \prod_{j=1}^k \left(f_j^t \circ \pi_j \right)$$

where the product $\prod$ is the componentwise product on multi-sets, and π_j is defined as $\pi_j : Bag[\bigotimes_{s=1}^{k} C_{i_s}] \rightarrow Bag[C_{i_j}]$ such that $\pi_j(c_1, c_2, \ldots, c_k) = c_j$.

Example 2. An application of the above property is illustrated on a practical case. Let us compute the transpose of the tuple $\langle X_1, S - X_2, S \rangle : C_1 \times C_1 \rightarrow C_1 \times C_1 \times C_2$. Given that by Table 1 the transposes of the components of F are in order $X_1^t : Bag[C_1] \rightarrow Bag[C_1 \times C_1] = \langle X_1, S \rangle, (S - X_2)^t : Bag[C_1] \rightarrow Bag[C_1 \times C_1] = \langle S, S - X_1 \rangle, S : Bag[C_2] \rightarrow Bag[C_1 \times C_1] = \langle S, S \rangle$, the transpose of the tuple can be computed applying property 3. It is noticeable that there is no correlation between symbols X_i appearing in the first step of the computation and symbols X_i appearing in the second, in fact the indices depend on the domain to which they apply. :

$$\langle X_1, S - X_2, S \rangle^t : Bag[C_1 \times C_1 \times C_2] \rightarrow Bag[C_1 \times C_1]$$
$$= \langle X_1, S \rangle \circ \pi_1 * \langle S, S - X_1 \rangle \circ \pi_2 * \langle S, S \rangle \circ \pi_3 =$$
$$= \langle X_1, S \rangle * \langle S, S - X_2 \rangle * \langle S, S \rangle =$$
$$= \langle X_1, S - X_2 \rangle$$

$\square$

To completely define the above calculation by means of an algorithm it is necessary to be able to deal with $*$ (the component-wise multiplication) operation between Cartesian products of multi-sets. The task is straightforward if the following property is considered:

$$\langle a_1, \ldots, a_n \rangle * \langle b_1, \ldots, b_n \rangle = \langle a_1 * b_1, \ldots, a_n * b_n \rangle$$

where $a_j, b_j \in Bag(C_{i_j})$ and $i_j = 1, \ldots, n$.

Observation: when class-functions map into multisets whose coefficients may be in $\{0,1\}$ only, then the product $*$ is equivalent to the *intersection*. Moreover the following property $\lambda f_1 * \gamma f_2 = \lambda \gamma (f_1 \cap f_2)$ holds, where f_1 and f_2 are functions satisfying to the assumption above. The elementary symbols in $\mathcal{S}$ map into multisets with coefficients 0 or 1, then actually it is always possible to replace $*$ with the intersection $\cap$ operator. In the algorithm proposed in the following subsection such feature is utilised. The intersection between elementary symbols may be stored as a mapping assigning to each pair of symbols in $\mathcal{S}$ the expression corresponding to the intersection result. Table 2 shows such mapping. The table shows only the cases leading to simplifications, in all the others situations the rewriting rules simply leave the operation indicated, for instance tuple $\langle X_1 \cap X_2, S \cap X_2 \rangle$ is rewritten as $\langle X_1 \cap X_2, X_2 \rangle$.

Language Expression Transpose. The rules so far discussed allow to operate on expressions of the language composed by a single tuple and which satisfy the WN syntax only. This section completes the transpose computation extending the calculus to any element of the language $\mathcal{R}[D, D']$. The following properties are the basis of the computation:

Table 2. Intersection among elementary functions

Symbolic Operation	Rewriting Rule	Symbolic Operation	Rewriting Rule
$S \cap X_i$	X_i	$X_i \cap !^k X_i$	X_i if $k = |C_{i_j}|$ else ϕ
$X_i \cap X_i$	X_i	$X_i \cap (S - X_i)$	ϕ
$!^k X_i \cap (S - X_i)$	ϕ if $k = |C_{i_j}|$ else $!^k X_i$	$X_i \cap (S - !^k X_i)$	ϕ if $k = |C_{i_j}|$ else X_i
$S \cap !^k X_i$	$!^k X_i$		

Property 4 (Transpose properties).

i. *[Transpose of an intersection] Let f and f' be functions from $Bag[C]$ to $Bag[D]$, if $\forall c \in C$ both f and f' maps into multisets with coefficients in $\{0,1\}$ then $(f \cap f')^t = f^t \cap f'^t$.*

ii. *[Transpose of a guard] the transpose of a guard (filter) is the guard (filter) itself.*

iii. *[Transpose of a composition] Let $f : Bag[D'] \to Bag[D]$ and $f' : Bag[C] \to Bag[D']$ be functions on multiset, the transpose of a composition is the composition of the transposes exchanged: $(f \circ f')^t = f'^t \circ f^t$.*

Properties 4(ii) and 4(iii) complete the elements needed to transpose an expression of the language $\mathcal{R}[D, D']$. In fact, property 2 allowed to rewrite the transpose of a sum $(\sum_i [filter_i] \lambda_i T_i [guard_i])^t$ in terms of its addend in the following way $\sum_i ([filter_i] \lambda_i T_i [guard_i])^t$, Property 4($iii$) about the transpose of a composition of linear functions says that the previous expression is equivalent to $\sum_i [guard_i]^t (\lambda_i T_i)^t [filter_i]^t$, moreover by property 4(ii) filters and guards expressions are not modified by the transpose operator. Thus, so far the calculus needs to do simply manipulation of the expression elements to compute the transpose. The innermost transpose $(\lambda_i T_i)^t$ is by Property 2 simply $\lambda_i (T_i^t)$. The core point is thus the tuple transpose. Property $4i$ allows to extend the transpose computation to class-functions whose addends are not simply symbols in $\mathcal{S}$ but intersection between them. Its application to Property 3, assuming the f_i as intersections, will result in an intersection of as many tuples as the components of the intersection in f_i.

Let us consider again the example of structural relation $\Gamma^-(\text{Send}, \text{Mes})$ for the database-system and show how the algorithm behaves.

$$\left(\langle S_{C_{1,1}} - X_1, S_{C_{2,1}} \rangle [d(X_1) = C_{1,1}] + \langle S - X_1, X_2 \rangle [d(X_1) = C_{1,2}] \right)^t \overset{Prop.2}{=}$$

$$= \left(\langle S_{C_{1,1}} - X_1, S_{C_{2,1}} \rangle [d(X_1) = C_{1,1}] \right)^t + \left(\langle S - X_1, X_2 \rangle [d(X_1) = C_{1,2}] \right)^t \overset{Prop.4iii}{=}$$

$$= [d(X_1) = C_{1,1}]^t \langle S_{C_{1,1}} - X_1, S_{C_{2,1}} \rangle^t + [d(X_1) = C_{1,2}]^t \langle S - X_1, X_2 \rangle^t \overset{Prop.4ii}{=}$$

$$= [d(X_1) = C_{1,1}] \langle S_{C_{1,1}} - X_1, S_{C_{2,1}} \rangle^t + [d(X_1) = C_{1,2}] \langle S - X_1, X_2 \rangle^t$$

Let us compute the transpose of the second tuple in the above expression. Using Table 1 and simplification rules of Table 2 results in the following steps:
$$\langle S - X_1, X_2 \rangle^t = \langle S - X_1, S \rangle \cap \langle S, X_2 \rangle = \langle (S - X_1) \cap S, S \cap X_2 \rangle = \langle S - X_1, X_2 \rangle.$$

Avoiding waste of rewriting can be obtained exploiting that the intersections with S's, shown above, are known a priori. So, some intermediate step may be skipped. In general, if T is the starting tuple and T' its transpose, the algorithm may proceed in this way: T' is initialised with all S and a guard always true. Then, each element s of tuple T that is a basic symbol of set S is considered separately. If $s = S$ then it is ignored. If $s = S_{C_{i,j}}$ then s^t produces a basic predicate which is composed through a logical AND with the guard of tuple T' (see Table 1). If $s \in \{X_i, S - X_i, !^k X_i, S - !^k X_i\}$, the index i of the projection appearing in s defines the position j in tuple T' where the transposed symbol s^t should end up (combined through intersection with the current j-th component of T'), whereas the index of the projection that will appear in s^t is the position of s in T. For the sake of clarity, Fig. 4 shows an example of the application of these rules to a more complex tuple than that of the example.

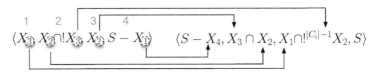

Fig. 4. An example of the application of the rules to transpose a tuple

Let us complete the transpose computation of the above example. The transpose of the first tuple is computed using the optimised steps. Prior to transpose $\langle S_{C_{1,1}} - X_1, S_{C_{2,1}} \rangle^t$ the tuple must be rewritten in simple form $\langle S_{C_{1,1}} \cap (S - X_1), S_{C_{2,1}} \rangle^t$. Thus, let us start with a tuple $T' = \langle S, S \rangle$ of all S and analyse how the transposes of the single components modify it:

- $S_{C_{1,1}}^t$ produces a guard $[d(X_1) = C_{1,1}]$ applied to tuple T', which becomes $\langle S, S \rangle [d(X_1) = C_{1,1}]$ (see Table 1 for the rules about the guard's form);
- $(S - X_1)^t$ produces $S - X_1$ at the first component of tuple T': at this step tuple T' becomes $\langle S - X_1, S \rangle [d(X_1) = C_{1,1}]$;
- $S_{C_{2,1}}^t$ produces a guard $[d(X_2) = C_{2,1}]$ applied to tuple T': the resulting tuple is $\langle S - X_1, S \rangle [d(X_1) = C_{1,1} \wedge d(X_2) = C_{2,1}]$ (see Table 1).

Concluding, the final result is: $[d(X_1) = C_{1,1}]\langle S - X_1, S \rangle [d(X_1) = C_{1,1} \wedge d(X_2) = C_{2,1}] + [d(X_1) = C_{1,2}]\langle S - X_1, X_2 \rangle$.

4.2   The Difference Operation

This section works out the problem of computing the difference $F \ominus F'$ between two expressions F, F' of the language. Situations involving difference $F \ominus F'$ may arise in several applications of the structural analysis. As concrete example, assume the necessity to evaluate the actual multiset of tokens a given transition t firing instance puts into an input place p, this can be obtained by evaluating the expression $W^+(t, p) \ominus W^-(t, p)$ on that instance of t, where $W^+(t, p)$ and

$W^-(t,p)$ are respectively functions on the input and output arcs between t and p. This operation has been already illustrated in Sec. 3 Fig. 2 for the calculus of both the structural relations $SC(t,t')$ and $SCC(t,t')$.

The topic of this section is to represent $F \ominus F'$ by means of an expression of the language, that is to *solve* the $\ominus$ operator. The problem is actually more complex than it may seem, in fact F and F' are generally sums of tuples and due to the $\ominus$ semantics the correct procedure must be followed in doing the subtraction. Initially, in the development of the solution, guards and filters are not consider because of simplicity, so that the problem is finding out an expression $\nu_1 T_1 + \ldots + \nu_m T_m$ satisfying the equation:

$$\nu_1 T_1 + \ldots + \nu_m T_m = (\lambda_1 T_1' + \ldots + \lambda_{m'} T_{m'}') \ominus (\gamma_1 T_1'' + \ldots + \gamma_{m''} T_{m''}'') \quad (1)$$

where $\nu_i, \lambda_i, \gamma_i \in \mathbb{N}$. Once equation (1) has been worked out the more general situation involving filters and guards is treated. The algorithm solving problem in (1) involves several sub-tasks which are discussed next:

a) the most elementary sub-task to be considered is the difference between forms of type $\bigcap_i \Gamma_i$ where $\Gamma_i \in \mathcal{S}$. Exploiting the fact that actually the elements of set $\mathcal{S}$ represent functions mapping into multiset with coefficient 0 or 1, the rules in Table 3 are used in dealing with such difference, such rules are closed on the form given above.

Table 3. (a) rules used to deal with subtraction $\ominus$; (b) rules to rewrite the complement

(a) Class-functions subtraction rules		**(b)** Complement	
a,b are functions mapping into sets,		Symbolic	Rewriting
^{c} is the complement		Operation	Rule
$a \ominus b \;\; \rightarrow a \cap {}^c b$	$a \cap {}^c a \rightarrow \phi$	${}^c!^k X_i$	$S -!^k X_i$
${}^c(a \cap b) \rightarrow {}^c a + {}^c b$	${}^c({}^c a) \;\; \rightarrow a$	${}^c(S -!^k X_i)$	$!^k X_i$
$a \cap a \;\; \rightarrow a$		${}^c S_q$	$\sum_{i \neq q} S_i$
		${}^c S$	ϕ

b) the second sub-task the algorithm dealing with equation (1) requires is the ability to solve single tuples difference $\lambda T \ominus \gamma T'$. When T and T' are elements of $\mathcal{K}[D, D']$, the following property allows to represent the difference $\lambda T \ominus \gamma T'$ into an expression of the language:

Property 5. Let $T = \langle u_1, \ldots, u_n \rangle$ and $T' = \langle v_1, \ldots, v_n \rangle$ be tuples whose components are functions mapping to sets then it holds:

$$\lambda T \ominus \gamma T' = \lambda \langle u_1 \ominus v_1, U' \rangle + \lambda \langle u_1 \cap v_1, U' \ominus V' \rangle + (\lambda \ominus \gamma)(T \cap T')$$

where $U' = \langle u_2, \ldots, u_n \rangle$ and $V' = \langle v_2, \ldots, v_n \rangle$.

The first term of the right member is a tuple, the latter also represents a tuple because $T \cap T' = \langle u_1 \cap v_1, u_2 \cap v_2, \ldots, u_n \cap v_n \rangle$. The second term may be recursively derived applying the property itself. Observing that after the first step the third term never appears, since U' and V' always have coefficient 1, it is easy to verify that the result is a sum of $n+1$ terms. A feature of such rewriting is that the tuples on the conclusion of the Property 5 are disjoint, this will simplify the solution of the general problem of difference (1) as discussed later;

c) the third sub-task considered is the expansion of a tuple $T \in \mathcal{R}[D, D']$, that is its translation into a sum of elements belonging to $\mathcal{K}[D, D']$. By Property 1 such rewriting is easily performable, however, for a motivation that will be clearer later, a pre-transformation is applied to T first, this pre-transformation guarantees that after the *expansion* the resulting tuples will be disjoint. Such rewriting operates on the class-functions of T and is based on the following property.

Property 6. Let F and G functions mapping into sets and $\lambda, \gamma \in \mathbb{N}$ then it holds:

$$\lambda f + \gamma g = \lambda(f \ominus g) + (\lambda + \gamma)(f \cap g) + \gamma(g \ominus f)$$

and the addends $\lambda(f \ominus g), (\lambda + \gamma)(f \cap g), \gamma(g \ominus f)$ are disjoint.

As an example, assume to expand tuple $\langle 2S + X_1, X_2 \rangle$, instead of simply applying $\langle 2S + X_1, X_2 \rangle \rightarrow 2\langle S, X_2 \rangle + \langle X_1, X_2 \rangle$, Property 6 is considered, giving $\langle 2(S - X_1) + (X_1 - S) + 3S \cap X_1, X_2 \rangle$, then using Table 2 and 3 the tuple becomes $\langle 2(S - X_1) + 3X_1, X_2 \rangle$. Expansion finally gives $2\langle S - X_1, X_2 \rangle + 3\langle X_1, X_2 \rangle$.

The remainder of the section illustrates how to utilise the basic steps, introduced above, in a general algorithm to solve (1) $\rightarrow E \in \mathcal{R}[D, D']$. For simplicity of notation the rules next provided consider expressions involving tuples whose multiplicity is 1, later on it will result clear that this is not a restriction. So let us consider the following simplification of problem (1):

$$(T_1 + \ldots + T_k) \ominus (T'_1 + \ldots + T'_{k'}) \tag{2}$$

Assume that expression (2) is composed by tuples mapping into multiset with coefficients 0 and 1, this is not a restriction since, by Property 1, each expression is rewriteable into a form satisfying such assumption. This allows to use sub-task b) and operate on basic differences in the following way: $T \ominus T' \rightarrow E \in \mathcal{R}[D, D']$.

To explain the difference algorithm let us start considering the case where the subtrahend is composed by one term only, namely let us consider $(T_1 + T_2 + \ldots + T_k) \ominus T'_1$, and see how such expression may be translated into an expression of the language using sub-task b), that is how to solve the problem

$$(T_1 + T_2 + \ldots + T_k) \ominus T'_1 \rightarrow E' \in \mathcal{R}[D, D'] \tag{3}$$

observe that, once (3) is solved, and more tuples constitute the subtrahend as in (2), it is sufficient to consider one tuple T'_i of (2) at the time and subtract it from the partial result E'_i: actually, Property 1 is applied to E' and the computation reiterated: $E' \ominus T'_2 \rightarrow E'_2 \in \mathcal{R}[D, D']$ and so on.

Let us thus consider problem (3), using subtask b), it could be rewritten as

$$(T_1 \ominus T_1') + [T_2 \ominus (T_1' \ominus T_1)] + [T_3 \ominus ((T_1' \ominus T_1) \ominus T_2)] + \ldots + \{T_k \ominus [(T_1' \ominus T_1) \ominus \ldots) \ominus T_{k-1}]\}$$

That is, first T_1' is subtracted to T_1, then the part of T_1' not subtracted to T_1, is subtracted to T_2, and so on. It is observable that in this expression only the innermost operations involve simple tuples, while the others involve a tuple and an expression $E \in \mathcal{R}[D, D']$, for this reason a sort of recursion appears in dealing with the formula above. In [11,12] a recursive algorithm is illustrated, which uses the technique above. This paper provides another approach that exploits a feature the expressions of the language often satisfy to improve the solving of (3). In fact, if it is assumed by hypothesis that $T_1, T_2, \ldots, T_k$ in (3) are disjoint for any colour they are applied to, then the formula (3) may be rewritten as $(T_1 \ominus T_1') + (T_2 \ominus T_1') + \ldots + (T_k \ominus T_1')$, which is less complex to manage, in fact each difference can be handled with subtask b) (actually the complexity is moved into the step needed to rewrite the tuples into a disjoint form, anyway it is required to be done only the first time).

To obtain disjoint tuples a two step procedure is followed: first, each time there is the necessity to apply Property 1 to expand the tuples the technique provided in case c) is followed, this assures that in (3) tuples T_i of the minuend eventually coming from the expansion of the same tuple are disjoint; second, tuples coming from different expansions must be disjoint in turn. This latter step may be practically obtained using Property 6 on each pair of tuples and rewriting the expression.

Dealing with Guards. Difference between two guarded tuples may be correctly managed encoding the guards into the respective tuples and utilising the rules of the difference above described. Formally the following rewriting rule is performed on both the minuend and the subtrahend: $T [guard] \rightarrow \langle T; T' \rangle$. The extension T' has as many components as the number of classes appearing in the guard's codomain and has the same semantics of the guard. Here is an example: in the difference $\langle S \rangle \ominus \langle X_1 \rangle [X_2 \neq X_3]$ the tuples, whose domain is assumed to be $C_1 \times C_1 \times C_1$, are rewritten as $\langle S; X_1, X_2, X_3 \rangle \ominus \langle X_1; X_1, X_2 \cap (S - X_3), X_3 \cap (S - X_2) \rangle$. Once the difference is computed the codomain of the result can be reverted to the original codomain translating the extended part of the tuples in guards right after them. Turning back to the above example the difference would result in $\langle S - X_1; X_1, X_2, X_3 \rangle + \langle S \cap X_1; X_1, X_2 \cap X_3, X_3 \rangle$ which reverted to the original codomain becomes $\langle S - X_1 \rangle + \langle X_1 \rangle [X_2 \neq X_3]$. Observe that extended tuples belong to the language.

Table 4. Tuple representation of the basic predicates (assume $i < j$)

Predicate	Tuple
$[X_i = X_j]$	$\langle X_1, \ldots, X_{i-1}, X_i \cap X_j, X_{i+1}, \ldots, X_{j-1}, X_i \cap X_j, X_{j+1}, \ldots, X_k \rangle$
$[X_i \neq X_j]$	$\langle X_1, \ldots, X_{i-1}, X_i \cap (S - X_j), X_{i+1}, \ldots, X_{j-1}, X_j \cap (S - X_i), X_{j+1}, \ldots, X_k \rangle$
$[X_i \in S_q]$	$\langle X_1, \ldots, X_{i-1}, X_i \cap S_q, X_{i+1}, \ldots, X_k \rangle$
$[X_i \notin S_q]$	$\langle X_1, \ldots, X_{i-1}, X_i \cap (S - S_q), X_{i+1}, \ldots, X_k \rangle$

Table 4 shows the rewriting rules of basic predicates into a tuple notation. Basic predicates $d(X_i) = d(X_j)$ and $d(X_i) \neq d(X_j)$ are not shown in table since they can rewritten respectively as $[(X_i \in C_1 \wedge X_j \in C_1) \vee \ldots \vee (X_i \in C_n \wedge X_j \in C_n)]$ and $[(X_i \in C_1 \wedge X_j \notin C_1) \vee \ldots \vee (X_i \in C_n \wedge X_j \in C_n)]$. The translation of general guards in form of Boolean expression built on the basic predicates is described in [11].

5   Conclusions and Future Work

The contribution of this paper is the definition of a high-level language that may be used to represents structural relations between nodes of WN models. The language syntax is very close to the WN arc function syntax, with some extensions, namely the *filters* and the *elementary functions intersection*. Filters are syntactically similar to WN guards but they appear on the left (i.e. they are left composed) of both tuples and functions. The proposed language has two advantages: the first is that it allows to give a compact representation of high-level net structural relations which, being expressed in a language similar to that used by the modeller to describe WN models, may be relatively easy to interpret by the user; the second is that several properties have been proven on the symbols of the language and some relevant operators, leading to the possibility of developing symbolic manipulation algorithms implementing efficient structural analysis, and avoiding the need to unfold the model.

This paper focuses on two main operators, the *transpose* and the *difference*, commonly utilised in many structural analysis methods, providing both the theoretical background to implement them algorithmically and possible algorithms.

The symbolic calculus dealing with the composition operator is described in [11, 12]: in these reports the closure of the language with respect to the composition operator it is not yet proven, but only conjectured. The authors are confident that the conjecture is true: the complete proof is currently under development.

Together with the *composition* operator, the transpose and the difference provide a uniform framework for the implementation of several structural analysis algorithms. A concrete implementation of the symbolic computation algorithms is planned for a future integration in the *GreatSPN* tool. Possible applications of the results presented, which the authors plan to develop, are the computation of enabling functions from the input and inhibitor arc functions of transitions, and the computation of structural conflict and causal connection: the results of these structural analysis algorithms can be used to speed up state space based analysis methods, for example providing a basis for enabled transition instances computation, and for partial order techniques.

Acknowledgments

This work has been partially funded by the MIUR Firb project PERF (RBNE019N8N) and MIUR 60% local research funding.

References

1. Jensen, K., Rozenberg, G., eds.: High-Level Petri Nets. Theory and Application. Springer Verlag (1991)
2. Chiola, G., Dutheillet, C., Franceschinis, G., Haddad, S.: Stochastic well-formed coloured nets for symmetric modelling applications. IEEE TC **42(11)** (1993) 1343–1360
3. Jensen, K.: Coloured Petri Nets. Basic Concepts, Analysis Methods and Practical Use. Volume 1, Basic Concepts. Monographs in Theoretical Computer Science, Springer-Verlag, 2nd corrected printing 1997. ISBN: 3-540-60943-1. (1997)
4. Chiola, G., Dutheillet, C., Franceschinis, G., Haddad, S.: A Symbolic Reachability Graph for Coloured Petri Nets. Theoretical Computer Science B (Logic, semantics and theory of programming) **176** (1997) 39–65
5. Brgan, R., Poitrenaud, D.: An efficient algorithm for the computation of stubborn sets of well formed Petri nets. In: Application and Theory of Petri Nets, LNCS 935, Springer-Verlag (1995) 121–140
6. Teruel, E., Franceschinis, G., De Pierro, M.: Well-defined generalized stochastic Petri nets: A net-level method to specify priorities. IEEE TSE **29(11)** (2003) 962–973
7. Barkaoui, K., Dutheillet, C., Haddad, S.: An efficient algorithm for finding structural deadlocks in colored Petri nets. In: Application and Theory of Petri Nets 1993, LNCS. Volume 691., Springer-Verlag (1993) 69–88
8. Gaeta, R.: Efficient discrete-event simulation of colored petri nets. IEEE TSE **22(9)** (1996) 629–639
9. Dutheillet, C., Haddad, S.: Structural analysis of coloured nets. application to the detection of confusion. In: technical report in MASI 92.16. (1992)
10. Dutheillet, C., Haddad, S.: Conflict sets in colored Petri nets. In: proc. of Petri Nets and Performance Models. (1993) 76–85
11. De Pierro, M.: Ph.D. thesis: Structural analysis of conflicts and causality in GSPN and SWN. Università di Torino - Italia. (2004 - http://www.di.unito.it/~depierro/public/)
12. Capra, L., De Pierro, M., Franceschinis, G.: Symbolic calculus for structural relations in well-formed nets. In: Technical report RT 03-04; http://www.di.unito.it/~depierro/public/, Dipartimento di Informatica e Comunicazione, Università di Milano - Italia (2004)
13. Capra, L., De Pierro, M., Franceschinis, G.: An application example of symbolic calculus for SWN structural relations. In: Proceedings of the 7th International Workshop on Discrete Event Systems, 2004, Reims, France, Elsevier-Oxford (2005)
14. Valmari, A.: Stubborn sets of coloured Petri nets. In: Proceedings of the 12th International Conference on Application and Theory of Petri Nets, 1991, Gjern, Denmark. (1991) 102–121 NewsletterInfo: 39.
15. Valmari, A.: State of the art report: Stubborn sets. Petri Net Newsletter (1994) 6–14
16. Kristensen, L.M., Valmari, A.: sloopy Finding stubborn sets of coloured Petri nets without unfolding. Lecture Notes in Computer Science: 19th Int. Conf. on Application and Theory of Petri Nets, ICATPN'98, Lisbon, Portugal, June 1998 **1420** (1998) 104–123

Derivation of Non-structural Invariants of Petri Nets Using Abstract Interpretation

Robert Clarisó, Enric Rodríguez-Carbonell, and Jordi Cortadella

Universitat Politècnica de Catalunya,
Barcelona, Spain

Abstract. Abstract interpretation is a paradigm that has been successfully used in the verification and optimization of programs. This paper presents a new approach for the analysis of Petri Nets based on abstract interpretation. The main contribution is the capability of deriving non-structural invariants that can increase the accuracy of structural methods in calculating approximations of the reachability space. This new approach is illustrated with the verification of two examples from the literature.

1 Introduction

The analysis of the state space of a Petri Net can be done by using different methods. Traditionally, three types of methods have been proposed [24]:

- Enumeration techniques, which provide an exact characterization for bounded systems and partial approximations for unbounded systems. These techniques suffer from the state explosion problem that often appears in highly concurrent systems.
- Transformation techniques, which alleviate the previous problem by reducing the system into a smaller one that still preserves the properties under analysis.
- Structural techniques, which provide information of the system based on the underlying graph structure of the net. Structural techniques typically compute upper approximations of the state space that can be effectively used for the verification of safety properties.

Structural techniques provide linear descriptions of the state space by exploiting the information given by the state equation [19]. As an example, the following invariant characterizes the markings that fulfill the state equation for the Petri Net in Fig. 1(a):

$$2p_1 + p_2 + p_3 + p_4 + p_5 = 2. \tag{1}$$

The reachability graph is depicted in Fig. 1(b), in which the shadowed states represent spurious (unreachable) markings. The presence of spurious markings is the cost that must be paid when using approximation techniques to calculate the state space.

This paper presents an attempt to explore a different analysis approach that lives between the accuracy of the enumeration methods and the efficiency of structural techniques. The goal is to reduce the set of spurious markings by generating more accurate

G. Ciardo and P. Darondeau (Eds.): ICATPN 2005, LNCS 3536, pp. 188–207, 2005.

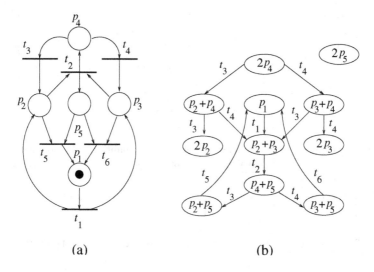

(a)   (b)

Fig. 1. (a) Petri Net (from [24]), (b) reachability graph

Table 1. Non-structural invariants for the Petri Net in Fig. 1(a)

Linear inequalities		Polynomial equalities	
$p_2 + p_4 \leq p_3 + p_5$	(2)	$p_5^2 - p_5 = 0$	(5)
$p_3 + p_4 \leq p_2 + p_5$	(3)	$p_4 p_5 - p_4 = 0$	(6)
$p_5 \leq p_2 + p_3 + p_4$	(4)	$2p_3 p_5 + p_2 + p_4 = p_3 + p_5$	(7)

characterizations of the state space taking into account both the initial marking and the structure of the net.

The approach is based on the paradigm of abstract interpretation [9], successfully used in different areas for the verification and optimization of systems [7]. In this paper we present two abstract domains that are able to derive non-structural invariants for Petri Nets: linear inequalities and polynomial equalities.

As an example, abstract interpretation has been able to obtain the invariants in Table 1 for the previous Petri net.[1]

Some observations on the new invariants:

– The invariants (1)-(4) represent the exact reachability graph.
– The invariants (1) and (5)-(7) also represent the exact reachability graph.

Even though in this case abstract interpretation can characterize the reachability graph exactly, in general it provides an upper approximation, which may be more accurate than the one defined by the structural invariants. Nevertheless, these conservative invariants can be used to prove safety properties of the Petri Net, like boundedness and

---

[1] The invariant (1) is also obtained in both abstract domains.

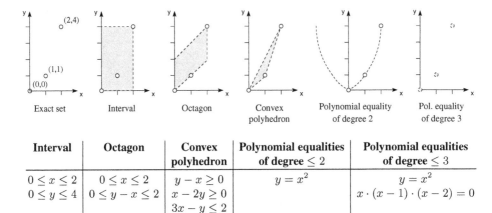

Interval	Octagon	Convex polyhedron	Polynomial equalities of degree ≤ 2	Polynomial equalities of degree ≤ 3
$0 \leq x \leq 2$ $0 \leq y \leq 4$	$0 \leq x \leq 2$ $0 \leq y - x \leq 2$	$y - x \geq 0$ $x - 2y \geq 0$ $3x - y \leq 2$	$y = x^2$	$y = x^2$ $x \cdot (x - 1) \cdot (x - 2) = 0$

Fig. 2. Approximating a set of values (left) with several abstract domains

deadlock freedom (by studying whether the conjunction of the disabling conditions for all the transitions and the invariants is feasible).

Some previous works have also studied the generation of linear inequality invariants. Many of them have been based on the analysis of structural properties and objects [16], including siphons and traps [5]. Another approach, presented in [23], uses Farkas' lemma to generate inductive linear invariants. Finally, Presburger arithmetics [13] and real arithmetics [2] can be used to represent the state space of Petri Nets, providing again linear inequality invariants.

Abstract interpretation offers new chances for the analysis of concurrent systems and, in particular, of Petri Nets. The possibility of deriving non-structural invariants can open the door to a new family of strategies that can better explore the trade-off between accuracy and efficiency in the modeling of concurrent systems. In this paper, the applicability of abstract interpretation is illustrated with the verification of an automated manufacturing system and the alternating bit protocol.

2   Abstract Interpretation

2.1   Fundamentals

Abstract interpretation [9] is a generic approach for the static analysis of complex systems. The underlying notion in abstract interpretation is that of *upper approximation*: to provide an abstraction of a complex behavior with less details. Upper approximations are conservative in the sense that they can be used to prove safety properties, e.g. "no errors in the abstraction" means "no errors in the system". A property about a system such as an invariant is in some way an abstraction: it represents all the states of the system that satisfy the property.

Intuitively, abstract interpretation defines a procedure to compute an upper approximation for a given behavior of a system. This definition guarantees (a) the termination of the procedure and (b) that the result is conservative. An important decision is the choice of the kind of upper approximation to be used, which is called the *abstract domain*. For a given problem, there are typically several abstract domains available. Each abstract domain provides a different trade-off between precision (proximity to the exact result) and efficiency.

There are many problems where abstract interpretation can be applied, several of them oriented towards the compile-time detection of run-time errors in software [6]. For example, some analysis based on abstract interpretation can discover numeric invariants among the variables of a program. Several abstract domains can be used to describe the invariants: intervals [8], octagons [17], convex polyhedra [10] or polynomial equalities [22]. These abstract domains provide different ways to approximate sets of values of numeric variables. For example, Figure 2 shows how these abstract domains can represent the set of values of a pair of variables x and y.

2.2   Application to the Reachability Problem

The reachable markings of a Petri Net can be studied using abstract interpretation. We will consider the classic model of Petri Nets, extended with inhibitor arcs and parameters in the initial marking. A reachable marking of a Petri Net N can be seen as an assignment to k non-negative integer variables, where k is the number of places in N. Therefore, a set of markings can be approximated using the abstract domains from Figure 2. This approach has several benefits. First, the abstract domains can represent large sets of markings compactly. Even infinite sets of markings can be represented efficiently. Moreover, the analysis can work with *parametric* markings where the number of tokens in a place is defined by a parameter. Finally, the approximation leads to a faster generation of the reachable state space. The result may contain some unreachable markings, but all reachable markings will be included in the solution. Thus, all invariants discovered by abstract interpretation hold in all the reachable markings.

We will show the computation of the reachable markings using the convex polyhedra abstract domain, even though other abstract domains could be used. The abstract interpretation procedure applied to the reachability analysis is shown in Figure 3. Intuitively, the algorithm behaves as follows. Initially, only the initial marking is reachable. There may be several enabled transitions, which will discover new reachable markings when they are fired. The algorithm keeps on firing transitions until no more reachable states be found. The enabling condition can be expressed using linear inequalities, while the effect of firing a transition can be expressed as linear assignments. Both operations are available in the abstract domain of convex polyhedra. Notice that each step deals with sets of reachable markings instead of individual markings.

The algorithm consists in computing the following sequence:

$$\text{reachable}^0 = M_0$$
$$\text{reachable}^{i+1} = \text{reachable}^i \cup \text{next}(\text{reachable}^i, T) \,.$$

In this recurrence, M_0 is the initial marking of the Petri Net. The set $\text{next}(M, T)$ represents the markings reached by firing once any transition in $t \in T$ from any marking $m \in M$ such that t is enabled in m. The union operator ($\cup$) is not exact for convex

---

Input: A Petri Net $N = \langle P, T, F, M_0 \rangle$ with a set of places P, a set of transitions T, a flow-relation F and an initial marking M_0.

Output: An upper approximation of the set of reachable markings of N, described in the abstract domain used in this analysis (e.g. a set of linear inequalities or polynomial equalities).

reachable = { M_0 }   # *Start from the initial marking*
do {
  old := reachable
  for each transition $t \in T$ {
    enabling := enablingCondition(t, F)   # *Enabling condition of t defined by F*
    enabled := reachable ∩ enabling   # *Reachable markings where t is enabled*
    if (enabled = ∅) continue   # *Check if t is not enabled yet*
    next := fire(t, enabled, F)   # *Fire t from the enabled markings*
    reachable := reachable ▽ (reachable ∪ next) # *Accumulate the new markings*
  }
} while (reachable ≠ old);

Fig. 3. Abstract interpretation algorithm used to compute the set of reachable markings

polyhedra, so some degree of approximation is introduced in this way. However, solving this recurrence has a problem: there is no guarantee that the algorithm will terminate. If the Petri Net is unbounded, the algorithm might iterate an infinite number of times. So instead of this recurrence, the algorithm solves another recurrence that relies on a *widening* operator (∇). Widening extrapolates the effect of repeating a computation an unbounded number of times. Its definition ensures the termination of the analysis after a finite number of steps. Given A, the states before the computation, and B the states after the computation, the widening is denoted as $A \nabla B$. A possible high-level definition of widening can be "keep the constraints from A that also hold in B", considering that any property modified during the computation might be further modified in later iterations. Using this operator, the recurrence can be rewritten as:

$$\text{reachable}^0 = M_0$$
$$\text{reachable}^{i+1} = \text{reachable}^i \nabla (\text{reachable}^i \cup \text{next}(\text{reachable}^i, T)) .$$

Figure 4 illustrates the effect of the widening on the computation. On the top, we show the computation of the reachable markings if the widening is not used. This computation does not terminate. On the bottom, we see the same computation using widening. In this case, the computation terminates quickly. Notice that, between ($p = 0$) and ($0 \leq p \leq 1$), the only common property is ($p \geq 0$).

2.3   An Example

The example used to present the abstract interpretation algorithm is shown in Figure 5. This Petri Net is modeling a producer-consumer system that communicates through a lossy channel. The left subnet, the producer, generates tokens while the right subnet, the consumer, removes these tokens. The place p_4 models the communication channel

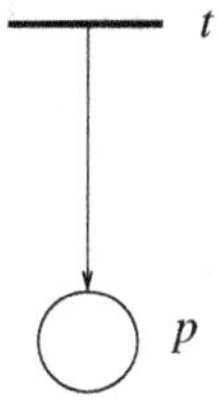

$$(p = 0)$$
$$\downarrow t$$
$$(p = 1) \cup (p = 0) = (0 \leq p \leq 1)$$
$$\downarrow t$$
$$(1 \leq p \leq 2) \cup (0 \leq p \leq 1) = (0 \leq p \leq 2)$$
$$\downarrow t$$
$$\cdots$$

$$(p = 0)$$
$$\downarrow t$$
$$(p = 1) \cup (p = 0) = (0 \leq p \leq 1)$$
$$(p = 0) \nabla (0 \leq p \leq 1) = (p \geq 0)$$
$$\downarrow t$$
$$(p \geq 1) \cup (p \geq 0) = (p \geq 0)$$
$$(p \geq 0) \nabla (p \geq 0) = (p \geq 0) \quad \text{Fixpoint}$$

Fig. 4. Justification of the necessity of a widening operator. On the top, computation of the reachable markings without any widening. On the bottom, computation of the reachable markings using widening

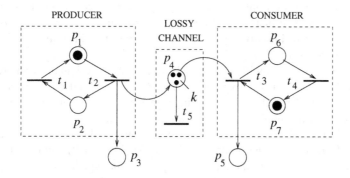

Fig. 5. Petri Net model of a producer-consumer system

between the producer and the consumer, while places p_3 and p_5 count the number of executions of the producer and the consumer respectively. The transition t_5 models the possible loss of data in the channel: some tokens generated by the producer will disappear before reaching the consumer. Initially, we assume that there are already k tokens in the channel, where k is a parameter of the system.

Several interesting invariants hold in this system. For example, the consumer cannot process more elements than those created by the producer or initially in the channel, i.e. $(p_5 \leq p_3 + k)$. It should also be noted that the places p_3, p_4 and p_5 are not bounded, so the set of reachable markings is infinite.

Let us discuss a part of the execution of the algorithm in this example. For the sake of brevity, the constraints of the form $(p_i \geq 0)$ and $(k \geq 0)$ will not be shown. The initial markings, parametrized by the value of k, are the following:

$$(p_1 = 1) \wedge (p_2 = 0) \wedge (p_3 = 0) \wedge (p_4 = k) \wedge (p_5 = 0) \wedge (p_6 = 0) \wedge (p_7 = 1).$$

In these markings, there are three transitions enabled: t_2, t_3 and t_5. When t_2 is fired, the following markings are reached:

$$(p_1 = 0) \wedge (p_2 = 1) \wedge (p_3 = 1) \wedge (p_4 = k + 1) \wedge (p_5 = 0) \wedge (p_6 = 0) \wedge (p_7 = 1).$$

These new markings can be combined with the initial markings using union and widening, producing the invariant:

$$(p_1 + p_4 = k + 1) \wedge (p_2 + k = p_4) \wedge (p_3 + k = p_4) \wedge (p_5 = 0) \wedge$$
$$(p_6 = 0) \wedge (p_7 = 1) \wedge (p_4 \geq k) \wedge (p_4 \leq k + 1).$$

The algorithm does the same computation for the enabled transitions t_3 and t_5. After this step, new transitions become enabled, which once they are fired, increase again the set of reachable markings. The procedure is repeated until the set of markings does not change. When the fixpoint is found in this example, the set of reachable markings is defined by:

$$(p_1 + p_2 = 1) \wedge (p_6 + p_7 = 1) \wedge (p_2 \leq p_3) \wedge (p_5 \geq p_6) \wedge (p_3 + k \geq p_4 + p_5).$$

The most interesting invariant in this result is $(p_3 + k \geq p_4 + p_5)$. This property is stating that any element that is consumed (p_5) or remains in the channel (p_4) was either produced (p_3) or initially available (k). Note that this invariant implies the one presented previously, $(p_5 \leq p_3 + k)$.

The following sections will present in detail two abstract domains that are suitable for the discovery of Petri Net invariants. *Convex polyhedra* and *polynomial equalities* can both represent a large class of interesting invariants. On one hand, convex polyhedra describe the set of reachable markings as a system of linear inequalities, so properties like $(p = 1)$ or $(p_1 \leq p_2)$ are easy to represent. The weakness of convex polyhedra is the loss of precision in the union, e.g. $(p_1 = 3) \cup (p_2 = 3)$ can only be approximated as $(p_1 + p_2 \geq 3)$. On the other hand, polynomial equalities are very precise in terms of describing disjunctions, e.g. $(p_1 = 3) \cup (p_2 = 3)$ can be represented exactly as $((p_1 - 3) \cdot (p_2 - 3) = 0)$. However, the description of inequality properties is more difficult: it is only possible when an upper bound is known, e.g. $(p_1 \leq 2)$ can be represented exactly as $(p_1 \cdot (p_1 - 1) \cdot (p_1 - 2) = 0)$; but for instance, that is not possible with $(p_1 \leq p_2)$.

3   Linear Inequality Invariants

3.1   Convex Polyhedra

Convex polyhedra can be described as the set of solutions of a conjunction of *linear inequality constraints* with rational ($\mathbb{Q}$) coefficients. Let P be a polyhedron over $\mathbb{Q}^n$, then it can be represented as the solution to the system of m inequalities $P = \{X | AX \geq B\}$ where $A \in \mathbb{Q}^{m \times n}$ and $B \in \mathbb{Q}^m$. The set of variables X contains the counters of the number of tokens in each place and the parameters of the initial marking. Convex polyhedra can also be characterized in a *polar* representation by means of a *system of generators*, i.e. as a linear combination of a set of vertices V (points) and a set

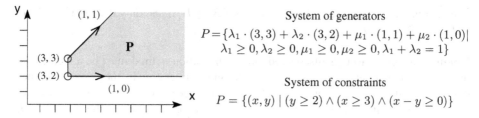

System of generators

$$P = \{\lambda_1 \cdot (3,3) + \lambda_2 \cdot (3,2) + \mu_1 \cdot (1,1) + \mu_2 \cdot (1,0) \mid$$
$$\lambda_1 \geq 0, \lambda_2 \geq 0, \mu_1 \geq 0, \mu_2 \geq 0, \lambda_1 + \lambda_2 = 1\}$$

System of constraints

$$P = \{(x,y) \mid (y \geq 2) \wedge (x \geq 3) \wedge (x - y \geq 0)\}$$

Fig. 6. An example of a convex polyhedron (shaded area) and its double description.

of rays R (vectors). Formally, the convex polyhedron P can also be represented as $P = \{\sum_{v_i \in V} \lambda_i \cdot v_i + \sum_{r_j \in R} \mu_j \cdot r_j \mid \lambda_i \geq 0, \mu_j \geq 0, \sum_i \lambda_i = 1\}$. Figure 6 shows an example of a convex polyhedron and its double description.

The fact that there are two representations is important, because several of the operations for convex polyhedra are computed very efficiently when the proper representation is available. There are efficient algorithms [3, 10] that translate one representation into the other. Also, the dual representations can be used to keep a *minimal* description, removing redundant constraints and generators.

In the remaining of the paper, we will denote the number of tokens in a place p_i as x_i.

3.2   Abstract Semantics

The abstract domain of convex polyhedra provides all the operations required in abstract interpretation. This section will describe the implementation of the operations used in our problem: guards, assignments, test for inclusion ($\subseteq$), union ($\cup$) and widening (∇).

Initial Marking. The initial marking defines the number of tokens in each place, and therefore, the value of all token counter variables. Therefore, the convex polyhedron that represents the initial marking has the following system of constraints: $\wedge_i (x_i = m_i)$, where m_i is the initial number of tokens in place i.

Guards. There are two kinds of guards that arise in our analysis: guards testing the presence of tokens in the input nodes, of the form $(x_1 \geq 1)$; and guards describing inhibitor arcs, of the form $(x_2 = 0)$. In any case, these guards are linear inequalities, so the resulting convex polyhedron only adds these guards to its system of constraints.

Assignments. The assignments that appear in our analysis increase or decrease counter variables by a constant value, e.g. $(x_i := x_i \pm c)$. These assignments can be applied to a convex polyhedron by changing its system of generators: each vertex is modified by increasing variable x_i by the constant c.

Test for Inclusion. In order to decide whether a fixpoint has been reached, the convex polyhedra approximating the reachability set must be compared with the one computed in the previous iteration. This comparison is made using the test for inclusion $(P \subseteq Q)$, which requires both representations of polyhedra. A convex polyhedron P, whose system of generators is the set V of vertices and the set R of rays, is included in a polyhedron Q, whose system of constraints is $AX \geq B$, if and only if $\forall v \in V, Av \geq B$ and $\forall r \in R, Ar \geq 0$.

Union. The new markings discovered when a transition is fired must be added to the previously known set of markings using the union operator. In the convex polyhedra domain, the union of convex polyhedra is not necessarily a convex polyhedron. Therefore, the union of two convex polyhedra is approximated by the *convex hull*, the smallest convex polyhedron that includes both operands. The system of generators of the convex hull can be computed by joining the systems of generators of the operands.

Widening. The extrapolation operator on convex polyhedra works on the system of constraints. The widening $P\nabla Q$ can be simply defined as the inequalities from P that are also satisfied by Q. More complex definitions of widening may provide a better precision in the analysis [1, 15].

The firing of a transition can be modeled as a sequence of these operations. First, testing if the transition is enabled can be performed by guard operations that check the number of tokens in the input places, e.g. $(x_1 \geq 1)$? or $(x_2 = 0)$? for inhibitor arcs. Then, the changes in the number of tokens in a place are modeled as linear assignments, e.g. $(x_1 := x_1 + 1)$. The new reachable markings will be added to the current reachable set using the union and widening operator, as it was described in Section 2.2. Figure 8 shows an example of this computation.

Figure 7 shows several examples of the operations described in this section. Notice that the intersection and linear assignment are exact, while the union and the widening operations are approximate.

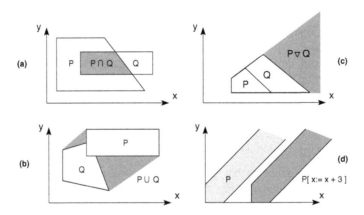

Fig. 7. Example of the operations on convex polyhedra: (a) intersection, (b) union, (c) widening and (d) linear assignment

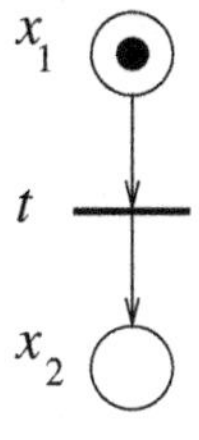

$$x_1 \quad \bullet$$

$$t$$

$$x_2$$

$$(x_1 = 1) \wedge (x_2 = 0) \rightsquigarrow (x_1 + x_2 = 1) \wedge (0 \le x_1 \le 1) \wedge (0 \le x_2 \le 1)$$

(a)

$$(x_1 = 1) \wedge (x_2 = 0) \rightsquigarrow (x_1 + x_2 = 1) \wedge (x_2(x_2 - 1) = 0)$$

(b)

Fig. 8. Computation of the invariants when a transition is fired, in the case of (a) linear inequalities and (b) polynomial equalities

4   Polynomial Equality Invariants

4.1   Ideals of Polynomials

In the abstract domain of convex polyhedra (Section 3), we have represented states as solutions of a system of *linear inequalities*; now, in the domain of ideals of polynomials, we will consider states as solutions of a system of *polynomial equalities*.

Namely, we abstract as follows: given a set of states S, regarded as points in $\mathbb{Q}^n$, the corresponding abstraction is a set of polynomials P with rational coefficients such that $P(\sigma) = 0 \; \forall \sigma \in S$, i.e. all points in S are zeroes of P. This set of polynomials has the algebraic structure of an *ideal*: by definition, an ideal I is a set of polynomials such that it a) contains 0, b) is closed under addition, and c) for any polynomial P, if $Q \in I$ then $P \cdot Q \in I$. Thus, we take ideals as our abstract values: an ideal I is an abstraction of the common zeroes of its polynomials, $\{\sigma \in \mathbb{Q}^n | \; P(\sigma) = 0 \; \forall P \in I\}$, which we call the *variety* of I and denote by $\mathbf{V}(I)$.

The set of polynomials with rational coefficients is denoted as $\mathbb{Q}[X]$. Given a subset $S \subseteq \mathbb{Q}[X]$, the *ideal generated by* S is

$$\langle S \rangle = \{f \in \mathbb{Q}[X] \mid \exists k \ge 1 \; f = \sum_{j=1}^{k} P_j Q_j \text{ with } P_j \in \mathbb{Q}[X], Q_j \in S\}.$$

For an ideal I, a set of polynomials S such that $I = \langle S \rangle$ is called a *basis* of I. By Hilbert's basis theorem, all ideals of polynomials admit a finite basis. Therefore any ideal is associated to a *finite* system of polynomial equality constraints: the ideal $I = \langle P_1(X), ..., P_k(X) \rangle$ corresponds to the system $\{P_1(X) = 0, ..., P_k(X) = 0\}$, or equivalently to the formula $\bigwedge_{j=1}^{k} P_j(X) = 0$.

For example, the ideal $\langle x(x^2 + y^2 - 16), y(x^2 + y^2 - 16) \rangle$ is associated to the system $\{x(x^2 + y^2 - 16) = 0, y(x^2 + y^2 - 16) = 0\}$. Its solutions, which form the variety $\mathbf{V}(\langle x(x^2 + y^2 - 16), y(x^2 + y^2 - 16) \rangle)$, are the union of a circle and a point, pictured in Figure 9. Notice that this set, unlike convex polyhedra, is not convex or even connected.

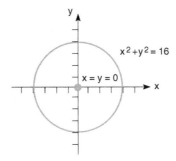

Fig. 9. An example of variety of an ideal

4.2   Abstract Semantics

This section shows the operations required to perform the abstract interpretation of Petri Nets using ideals of polynomials. For a more detailed description, see [22].[2]

Initial Marking. If we are given an initial marking $(m_1, m_2, ..., m_n)$ for the Petri Net (where each m_i may be a constant or a parameter), at first we know that $(x_i = m_i)$ for every place of the net, and so we take the ideal

$$\langle x_1 - m_1, x_2 - m_2, ..., x_n - m_n \rangle$$

as initial ideal.

Guards: Inhibitor Arcs. A guard describing an inhibitor arc, i.e. checking that there are no tokens in the inhibitor place p_i, is of the form $(x_i = 0)$. Similarly as we did with convex polyhedra, we have to add the guard to the system of constraints. In this case, we just need to add the polynomial x_i to the list of generators of the input ideal.

Guards: Presence of Tokens. Testing the presence of tokens is conservatively translated into polynomial disequality guards: checking that there are at least C tokens at place p_i is expressed as $(x_i \neq 0 \wedge \cdots \wedge x_i \neq C - 1)$. Given an input ideal I, for each of these disequalities $(x_i \neq c)$ we want to represent the points that belong to $\mathbf{V}(I)$ but not to $\mathbf{V}(\langle x_i - c \rangle)$, in other words $\mathbf{V}(I) \backslash \mathbf{V}(\langle x_i - c \rangle)$. The polynomials in the *quotient ideal $I : \langle x_i - c \rangle$* [12] evaluate to 0 at this difference of sets, and therefore abstract the states we are interested in; so we take this quotient as output ideal.

Assignments. The assignments that appear in our analysis are of the form $x_i := x_i \pm c$, as they express the change in the number of tokens at place p_i after firing a transition. Given an ideal $I = \langle P_1, ..., P_k \rangle$, we want to compute the effect of applying the assignment $x_i := x_i + c$ on I (in case of a subtraction, we may take c as a negative value). In terms of formulas, we need to express the following assertion using ideals:

$$\exists x_i'(x_i = x_i' + c \wedge (\bigwedge_{j=1}^{k} P_j(x_i \leftarrow x_i') = 0)),$$

---

[2] The abstract domain of ideals of polynomials and its semantics have been simplified in this paper with respect to [22] for the sake of clearness and efficiency.

where x_i' stands for the value of the assigned variable previous to the assignment, and $\leftarrow$ denotes substitution of variables. In this case, the auxiliary variable x_i' can be easily eliminated by substitution, as $x_i' = x_i - c$. So we get the formula $\wedge_{j=1}^k P_j(x_i \leftarrow x_i - c) = 0$, which translated into ideals yields $I(x_i \leftarrow x_i - c)$.

Test for Inclusion. In order to check whether a fixpoint has been reached, we need to test if the newly computed reachable states, represented by the ideal I, are already included in our previous approximation given by I_{prev}. So we need to see if $\mathbf{V}(I) \subseteq \mathbf{V}(I_{prev})$, which can be done by duality by checking that $I \supseteq I_{prev}$.

Union. Unlike with convex polyhedra, we can perform exact unions of states in the domain of ideals of polynomials. Let I, J be ideals corresponding to the sets of states $\mathbf{V}(I)$ and $\mathbf{V}(J)$ respectively, and assume that we want to represent $\mathbf{V}(I) \cup \mathbf{V}(J)$. In this case the abstraction is given by the *intersection ideal* $I \cap J$, which satisfies that $\mathbf{V}(I \cap J) = \mathbf{V}(I) \cup \mathbf{V}(J)$.

Widening. In order to get termination if the initial marking has parameters or the Petri Net is not bounded, we need to introduce a widening operator. Similarly as we did with convex polyhedra, given the ideals I and J we have to perform an upper approximation of $\mathbf{V}(I) \cup \mathbf{V}(J)$. By duality, we have to compute a lower approximation of $I \cap J$; i.e., we need to sieve the polynomials in the intersection so that the result is still sound and also the analysis terminates in a finite number of steps without much loss of precision.

Given a degree bound $d \in \mathbb{N}$ and a graded term ordering $\succ$, our widening operator $I \nabla_d J$ is defined as the ideal generated by the polynomials of a Gröbner basis of $I \cap J$ (with respect to $\succ$) of degree at most d; more formally,

$$I \nabla_d J = \langle \{P \in GB(I \cap J, \succ) \mid \text{degree}(P) \leq d\} \rangle,$$

where $GB(\cdot, \succ)$ stands for a Gröbner basis of an ideal with respect to the term ordering $\succ$. For definitions of Gröbner basis, graded term ordering and related concepts, we refer the reader to [12].

Figure 10 shows several examples of the operations described in this section. Contrary to convex polyhedra, the union operator is exact. Widening can be seen as a parametrized union, where any polynomial in the basis with a degree higher than the bound is abstracted. Varying this bound achieves different levels of precision in the result. For instance, Figures 10(b) and (c) show two widenings with degree bounds 2 and 3 respectively; notice that the latter represents exactly the union of states.

5   Examples

The techniques presented in the previous sections have been implemented and applied to several Petri Net examples from the literature. The linear inequality analysis has been implemented as a C program using the New Polka convex polyhedra library [21]. On the other hand, the polynomial equality analysis is performed by means of the algebraic geometry tool Macaulay 2 [14]. When it is not computationally feasible to work with polynomials over the rationals, we heuristically employ coefficients in a finite field instead. As the invariants obtained in the finite field might not necessarily be invariants in $\mathbb{Q}$, the polynomials thus generated are finally checked to be truly invariants of the system.

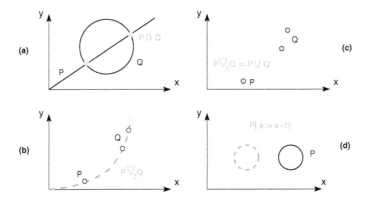

Fig. 10. Example of the operations on ideals of polynomials: (a) intersection, (b) widening with degree bound 2, (c) widening with degree bound 3 (union) and (d) linear assignment

5.1   Comparison with Structural Invariants

Structural invariants [20] are properties of the Petri Net structure, so they are independent of the initial marking. For instance, a Petri net may be *bounded* for a given initial marking, i.e. the number of tokens in each place is bounded in all reachable markings. Moreover, a net may be bounded for any initial marking, i.e. *structurally bounded.* Any structurally bounded net is also bounded, but the reverse does not necessarily hold. The approaches presented in this paper can be used to detect structural properties of a Petri Net: a parameter defines the initial marking of each place, hence the invariants obtained describe properties that are independent on the initial marking. For example, Figure 11(a) shows a Petri Net whose initial marking is defined by the parameters p, q and r: $x_1 = p$, $x_2 = q$, $x_3 = r$. In this Petri Net, the linear inequality invariants that can be computed with our approach appear in Figure 11(b). Notice that these invariants are structural as they are satisfied by any initial marking; for instance, $(r + p \geq x_3)$ meansthat place x_3 is structurally bounded. Invariant polynomial equalities can be similarly obtained using the same concept.

If a net is bounded, the analysis with polynomial invariants discovers the exact state space, for a sufficiently large degree. Thus, if a net is bounded but not structurally bounded, the analysis with polynomial invariants obtains a description of the state space which is more precise than structural invariants.

Regarding the analysis with convex polyhedra, or Petri nets with an infinite state space, using a widening adds some approximation. This approximation may make us fail to discover some structural invariants. In practice, in all the examples that we have studied, the state space computed by abstract interpretation satisfies all the structural invariants.

Furthermore, invariants describing properties which depend on the initial marking can also be computed with our approach. For example, Figure 12 shows a Petri Net where a place p is bounded for the initial marking depicted in the figure, while it is not structurally bounded. Abstract interpretation analysis discovers this property, encoded as the linear inequalities $(0 \leq p \leq 1)$ or the polynomial equality $p \cdot (p - 1) = 0$.

Another example of a non-structural property discovered by these invariants appears in Figure 13. This Petri Net can have a deadlock depending on the initial marking. For

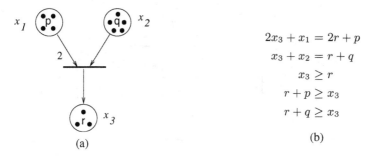

$$2x_3 + x_1 = 2r + p$$
$$x_3 + x_2 = r + q$$
$$x_3 \geq r$$
$$r + p \geq x_3$$
$$r + q \geq x_3$$

(a)   (b)

Fig. 11. (a) Petri Net with a parametric initial marking and (b) the computed linear inequality invariants

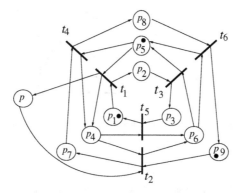

Fig. 12. Petri Net with a bounded place p which is not structurally bounded

instance, for the initial marking $p_1 p_8$ there is no deadlock, while for the initial marking p_1 the deadlocks $p_3 p_8$ and $p_5 p_8$ are reachable. The invariants produced by both linear inequalities and polynomial equalities are sufficient to prove the absence of deadlocks for the former initial marking, as the conjunction of these invariants with the disabling conditions of all the transitions is unfeasible.

5.2   Automated Manufacturing System

Figure 14 shows a Petri Net model of an automated manufacturing system [25]. This manufacturing system consists of several elements: four machines ($M_1 - M_4$), two robots ($R_1 - R_2$), two buffers with capacity 3 ($B_1 - B_2$) and an assembly cell. The place x_1 models the entry point for raw material, while the place x_{10} (x_{15}) represents the availability of the buffer B_1 (B_2). The place x_{12} (x_{13}) models the availability of the robot R_1 (R_2), whereas the place x_{25} represents the delivery point for the final product. Finally, the places x_4, x_7, x_{16} and x_{19} model the availability of the machines M_1 to M_4.

The initial marking of this Petri Net is as follows. The entry point x_1 has an undetermined number of tokens p, as we want to study the behavior of the system depending on the quantity of available raw materials. The capacities of the buffers, x_{10} and x_{15}, have 3 tokens as the buffers have size 3. Finally, places x_2, x_4, x_7, x_{12}, x_{13}, x_{16}, x_{19} and x_{24} have one token, and the rest of places have no tokens in the initial marking.

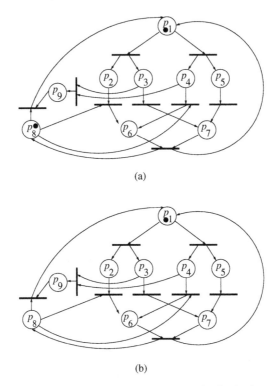

(a)

(b)

Fig. 13. Petri Net with a non-structural deadlock: (a) no deadlock, (b) potential deadlock

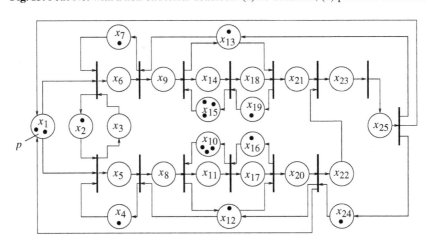

Fig. 14. Petri Net model of an automated manufacturing system

Some relevant properties in this system are *boundedness* and *liveness*, i.e. deadlock freedom. In previous work, these properties have been studied in detail. In [25], it was proven that the system is bounded, and that it is live only for some values of p, namely $2 \le p \le 4$. A different approach based on integer programming [4] managed to prove liveness for a wider interval of values, $1 \le p \le 8$. Also, a sequence of firings leading to a deadlock when $p > 8$ was shown. Later work has revisited these results using other techniques such as Presburger arithmetics [13], real arithmetics [2] and inductive linear inequalities based on Farkas' lemma [23].

We have analyzed the manufacturing system using both linear inequality invariants and polynomial equality invariants. Using linear inequalities, the computation takes 2.2 seconds and 68 Mb of memory. The resulting invariants are the following (we show the equalities and the inequalities separatedly):

$$x_{18} + x_{19} = 1 \qquad\qquad\qquad\qquad\qquad\qquad\qquad x_2 + x_3 = 1$$
$$x_8 + x_{12} + x_{20} = 1 \qquad\qquad\qquad\qquad\qquad\quad x_4 + x_5 = 1$$
$$x_{22} + x_{23} + x_{24} + x_{25} = 1 \qquad\qquad\qquad\qquad x_6 + x_7 = 1$$
$$x_9 + x_{13} + x_{21} + x_{23} + x_{25} = 1 \qquad\qquad\qquad x_{10} + x_{11} = 3$$
$$x_{19} + x_{17} + x_{15} + x_{13} + x_{11} + x_7 + x_5 + x_2 - x_{24} - x_{12} = 5 \qquad x_{14} + x_{15} = 3$$
$$x_{19} + x_{17} + x_{15} + x_{13} + x_{11} + x_7 + x_5 - x_{24} - x_{12} - x_3 = 4 \qquad x_{16} + x_{17} = 1$$
$$x_{19} + x_{15} + x_{13} + x_{12} + x_7 + p - x_{17} - x_{11} - x_5 - x_1 = 7$$

$$x_{25} + x_{24} + x_{23} \leq 1 \qquad\qquad\qquad\qquad\qquad x_7 \leq 1$$
$$x_{25} + x_{23} + x_{21} + x_{13} \leq 1 \qquad\qquad\qquad\quad x_{15} \leq 3$$
$$x_{19} + x_{15} + x_{13} + x_7 - x_{24} - x_1 \leq 5 \qquad\quad x_{19} \leq 1$$
$$x_{19} + x_{17} + x_{15} + x_{13} + x_{11} + x_7 + x_5 - x_{24} - x_{12} \geq 4 \qquad x_{20} + x_{12} \leq 1$$

This set of constraints suffices to prove that the manufacturing system is bounded.

On the other hand, we have applied our analysis with ideals so as to discover polynomial equality invariants of degree at most 2. In order to speed up the computation, we have employed a finite field $\mathbb{Z}_p$ instead of $\mathbb{Q}$, with p a relatively big prime number (in particular, we have taken $p = 32749$, the largest prime allowed in **Macaulay 2**); the generation of the candidate invariants takes 16 minutes and 304 Mb of memory. After checking that the polynomials obtained are invariant, which requires 6 minutes and 490 Mb, we get (we separate the linear invariants from the quadratic ones):

$$x_8 + x_{12} + x_{20} = 1 \qquad\qquad\qquad\qquad\qquad\qquad\qquad\quad x_2 + x_3 = 1$$
$$x_9 + x_{13} + x_{21} + x_{23} + x_{25} = 1 \qquad\qquad\qquad\qquad\quad x_4 + x_5 = 1$$
$$x_{24} + 2x_{16} + 2x_{12} + 2x_{10} + 2x_4 - x_2 - x_1 + p = 12 \qquad x_6 + x_7 = 1$$
$$x_{16} - x_{18} - x_{14} + x_{13} + x_{12} + x_{10} - x_6 + x_4 - x_1 + p = 7 \qquad x_{10} + x_{11} = 3$$
$$x_{19} + x_{16} - x_{14} + x_{13} + x_{12} + x_{10} - x_6 + x_4 - x_1 + p = 8 \qquad x_{14} + x_{15} = 3$$
$$x_{21} - x_{22} + 2x_{16} + x_{13} + 2x_{12} + 2x_{10} + x_9 + 2x_4 - x_2 - x_1 + p = 12 \quad x_{16} + x_{17} = 1$$

$$x_2^2 = x_2 \qquad\qquad\qquad\qquad\qquad\qquad\qquad\qquad x_8 x_{12} = 0$$
$$x_4^2 = x_4 \qquad\qquad\qquad\qquad\qquad\qquad\qquad\qquad x_9 x_{13} = 0$$
$$x_6^2 = x_6 \qquad\qquad\qquad\qquad\qquad\qquad\qquad\quad x_{13} x_{21} = 0$$
$$x_8^2 = x_8 \qquad\qquad\qquad\qquad\qquad\qquad\qquad\qquad x_9 x_{21} = 0$$
$$x_9^2 = x_9 \qquad\qquad\qquad\qquad\qquad\qquad\qquad\qquad x_9 x_{23} = 0$$
$$x_{12}^2 = x_{12} \qquad\qquad\qquad\qquad\qquad\qquad\qquad x_{21} x_{23} = 0$$
$$x_{13}^2 = x_{13} \qquad\qquad\qquad\qquad\qquad\qquad\qquad x_{13} x_{23} = 0$$
$$x_{16}^2 = x_{16} \qquad\qquad\qquad\qquad\qquad\qquad\qquad x_{23}^2 = x_{23}$$
$$x_{21}^2 = x_{21}$$

and four other more complex polynomial constraints.

Unlike with linear inequalities, these invariants are not enough to prove that *all* places in the net are bounded; the reason is that polynomial equalities cannot express relations such as $x_1 \leq p$, where the bounds are parametric. However, *some of the* places can be shown to be bounded: for instance, $x_2 = x_2^2$ means that either $x_2 = 0$ or $x_2 = 1$. Notice that each family of constraints uses a different approach to represent boundedness: linear inequalities give an upper bound on the number of tokens (for example, the linear invariant $x_{20} + x_{12} \leq 1$ implies $x_{12} \leq 1$), whereas polynomial equalities encode an exact disjunction of the possible values.

Further, by means of these quadratic constraints, together with the implicit invariant that all variables are non-negative, it is also possible to prove that the system is deadlock-free for $1 \leq p \leq 8$; in order to do that, we show that the conjunction of the invariants and the disabling conditions of all the transitions is not satisfiable. Moreover, as in [23], for $p = 9$ we are able to isolate four potential deadlocks, one of which is the deadlock corresponding to the sequence of firings in [4] mentioned above.

5.3   Alternating Bit Protocol

The Petri Net in Figure 15 models the alternating bit protocol for retransmitting lost or corrupted messages. The correctness of the protocol has already been shown in previous work: while in [11] all the proofs were done by hand, in [13] Fribourg and Olsén employed Presburguer arithmetics to automatically characterize the reachable states and thus prove that the system behaves properly.

The initial marking is $x_1 = 1$, $x_{13} = 1$ and $x_i = 0$ for $1 \leq i \leq 16$, $i \neq 1, 13$. Notice that the net has eight inhibitor arcs, linked to the places x_j, $j = 5..12$, which can be proved to be unbounded; these inhibitor arcs are pictured as circle-headed arrows on the figure. Thus, this example cannot be handled by other techniques for generating invariants that do not deal with equality guards, such as [18].

For this Petri Net, by means of convex polyhedra the following linear constraints:

$$x_4 + x_3 + x_2 + x_1 = 1 \qquad x_{16} + x_{15} + x_{14} + x_{13} = 1$$

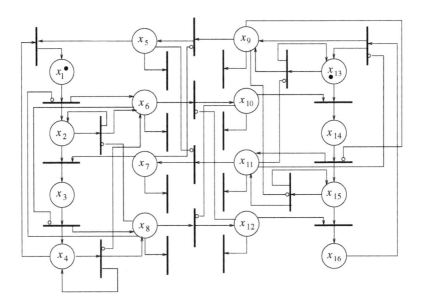

Fig. 15. Petri Net model of the alternating bit protocol

are obtained as invariants in 1.1 seconds using 65 Mb of memory. In this case, these linear invariants are not enough to show correctness, due to the fact that convex polyhedra cannot represent disjunctions in general.

As regards the analysis with polynomials, in 74.7 seconds and using 21 Mb of memory we get the following linear and quadratic constraints:

$$x_4 + x_3 + x_2 + x_1 = 1 \qquad\qquad x_{16} + x_{15} + x_{14} + x_{13} = 1$$

$x_1 x_2 = 0$	$x_2 x_3 = 0$	$x_1^2 = x_1$
$x_1 x_3 = 0$	$x_1 x_6 = 0$	$x_2^2 = x_2$
$x_2 x_8 = 0$	$x_3 x_8 = 0$	$x_3^2 = x_3$
$x_5 x_7 = 0$	$x_6 x_8 = 0$	$x_{13}^2 = x_{13}$
$x_9 x_{11} = 0$	$x_{10} x_{12} = 0$	$x_{14}^2 = x_{14}$
$x_9 x_{15} = 0$	$x_{11} x_{13} = 0$	$x_{15}^2 = x_{15}$
$x_{11} x_{14} = 0$	$x_{13} x_{14} = 0$	$x_6 x_3 + x_6 x_2 = x_6$
$x_{13} x_{15} = 0$	$x_{14} x_{15} = 0$	$x_{14} x_9 + x_{13} x_9 = x_9$

Note that, out of the 26 computed constraints, just the first two are linear and coincide with the linear equalities obtained above using convex polyhedra; the rest of the polynomial constraints are implicitly defining disjunctions. This explains why the linear inequality analysis does not yield enough information to verify the system.

Unfortunately, the quadratic invariants do not suffice to prove the correctness of the Petri Net either. This leads us to generate *cubic* invariant polynomials, i.e. of degree 3; the computation takes 102.9 seconds and 34 Mb of memory and gives

$x_2 x_7 x_9 = 0$	$x_7 x_9 x_{10} = 0$	$x_2 x_{12} x_{13} = x_2 x_{12}$
$x_2 x_7 x_{12} = 0$	$x_7 x_{10} x_{13} = 0$	$x_5 x_8 x_{13} = x_5 x_8$
$x_2 x_7 x_{13} = 0$	$x_8 x_9 x_{10} = 0$	$x_5 x_{12} x_{13} = x_5 x_{12}$
$x_2 x_{11} x_{12} = 0$	$x_2 x_5 x_6 = x_5 x_6$	$x_6 x_{12} x_{13} = x_6 x_{12}$
$x_5 x_8 x_{10} = 0$	$x_2 x_5 x_{10} = x_5 x_{10}$	$x_7 x_9 x_{13} = x_7 x_9$
$x_5 x_8 x_{11} = 0$	$x_2 x_5 x_{11} = x_5 x_{11}$	$x_7 x_9 x_{13} = x_7 x_9$
$x_5 x_{11} x_{12} = 0$	$x_2 x_6 x_9 = x_6 x_9$	$x_8 x_9 x_{13} = x_8 x_9$
$x_6 x_7 x_9 = 0$	$x_2 x_6 x_{12} = x_6 x_{12}$	$x_9 x_{12} x_{13} = x_9 x_{12}$
$x_6 x_7 x_{12} = 0$	$x_2 x_6 x_{13} = x_6 x_{13}$	$x_2 x_5 x_{13} + x_5 = x_5 x_{13} + x_2 x_5$
$x_6 x_7 x_{13} = 0$	$x_2 x_9 x_{10} = x_9 x_{10}$	$x_2 x_9 x_{13} + x_9 = x_9 x_{13} + x_2 x_9$
$x_6 x_{11} x_{12} = 0$	$x_2 x_{10} x_{13} = x_{10} x_{13}$	

in addition to the quadratic constraints above. Unlike with the quadratic case, the cubic invariants allow us to prove that

$$\left\{ \begin{array}{l} x_1 = 1 \implies (x_2 = x_3 = x_4 = x_6 = x_7 = x_{10} = x_{11} = x_{14} = x_{15} = x_{16} = 0) \wedge (x_{13} = 1) \\ x_3 = 1 \implies (x_1 = x_2 = x_4 = x_5 = x_8 = x_9 = x_{12} = x_{13} = x_{14} = x_{16} = 0) \wedge (x_{15} = 1) \\ x_{14} = 1 \implies (x_1 = x_3 = x_4 = x_7 = x_8 = x_{11} = x_{12} = x_{13} = x_{15} = x_{16} = 0) \wedge (x_2 = 1) \\ x_{16} = 1 \implies (x_1 = x_2 = x_3 = x_5 = x_6 = x_9 = x_{10} = x_{13} = x_{14} = x_{15} = 0) \wedge (x_4 = 1) \end{array} \right.$$

which implies that the system is correct (see [11]).

6   Conclusions

The applicability of abstract interpretation can be extended to the analysis of Petri Nets. This paper has presented an approach that can generate a rich set of invariants using this paradigm.

Abstract interpretation is a general approach that accepts different algorithmic techniques to calculate approximations. We believe that different strategies can be studied to explore the trade-off between efficiency and accuracy in analyzing concurrent systems.

Acknowledgements. This work has been partially funded by CICYT TIC2001-2476-C03-01, CICYT TIN 2004-07925, the LogicTools project (CICYT TIN 2004-03382), and the FPU grants AP2002-3862 and AP2002-3693 from the Spanish MECD. We would also like to thank M. Müller-Olm for insightful discussions.

References

1. R. Bagnara, P. M. Hill, E. Ricci, and E. Zaffanella. Precise widening operators for convex polyhedra. In R. Cousot, editor, *Int. Symp. on Static Analysis*, volume 2694 of *Lecture Notes in Computer Science*, pages 337–354, San Diego, California, USA, 2003. Springer-Verlag.
2. B. Bérard and L. Fribourg. Reachability analysis of (timed) Petri nets using real arithmetic. In *Proc.Int. Conf. on Concurrency Theory*, volume 1664 of *LNCS*, pages 178–193, 1999.
3. N. Chernikova. Algoritm for discovering the set of all solutions of a linear programming problem. *USSR Computational Mathematics and Mathematical Physics*, 6(8):282–293, 1964.
4. F. Chu and X.-L. Xie. Deadlock analysis of Petri nets using siphons and mathematical programming. *IEEE Transactions on Robotics and Automation*, 13(6):793–804, Dec. 1997.
5. F. Commoner. *Deadlocks in Petri Nets*. Wakefield: Applied Data Research, Inc., CA-7206–2311, 1972.
6. P. Cousot. Abstract interpretation: Achievements and perspectives. In *Proc. of the SSGRR 2000 Computer & eBusiness Int. Conf.* Scuola Superiore G. Reiss Romoli, July 2000.
7. P. Cousot. Abstract interpretation based formal methods and future challenges, invited paper. In R. Wilhelm, editor, *Informatics — 10 Years Back, 10 Years Ahead*, volume 2000 of *Lecture Notes in Computer Science*, pages 138–156. Springer-Verlag, Berlin, Germany, 2001.
8. P. Cousot and R. Cousot. Static determination of dynamic properties of programs. In *Proc. of the 2nd Int. Symposium on Programming*, pages 106–130. Dunod, Paris, France, 1976.
9. P. Cousot and R. Cousot. Abstract interpretation: a unified lattice model for static analysis of programs by construction or approximation of fixpoints. In *Proc. ACM SIGPLAN-SIGACT Symp. on Principles of Programming Languages*, pages 238–252. ACM Press, 1977.
10. P. Cousot and N. Halbwachs. Automatic discovery of linear restraints among variables of a program. In *Proc. ACM SIGPLAN-SIGACT Symp. on Principles of Programming Languages*, pages 84–97. ACM Press, New York, 1978.
11. J. M. Couvreur and E. Paviot-Adet. New structural invariants for petri nets analysis. In Valette, R., editor, *Proc. Int. Conf. on Application and Theory of Petri Nets*, volume 815 of *LNCS*, pages 199–218. Springer-Verlag, 1994.
12. D. Cox, J. Little, and D. O'Shea. *Ideals, Varieties and Algorithms. An Introduction to Computational Algebraic Geometry and Commutative Algebra*. Springer-Verlag, 1998.
13. L. Fribourg and H. Olsén. Proving safety properties of infinite state systems by compilation into presburger arithmetics. In *Proc.Int. Conf. on Concurrency Theory*, volume 1243, pages 213–227. Lecture Notes in Computer Science, July 1997.
14. D. R. Grayson and M. E. Stillman. Macaulay 2, a Software System for Research in Algebraic Geometry. Available at http://www.math.uiuc.edu/Macaulay2/.
15. N. Halbwachs, Y.-E. Proy, and P. Roumanoff. Verification of real-time systems using linear relation analysis. *Formal Methods in System Design*, 11(2):157–185, 1997.

16. G. Memmi and J. Vautherin. Computation of flows for unary-predicates/transition nets. *Lecture Notes in Computer Science: Advances in Petri Nets 1984*, 188:455–467, 1985.
17. A. Miné. The octagon abstract domain. In *Analysis, Slicing and Tranformation (in Working Conference on Reverse Engineering)*, IEEE, pages 310–319. IEEE CS Press, Oct. 2001.
18. M. Müller-Olm and H. Seidl. Computing Polynomial Program Invariants. *Information Processing Letters (IPL)*, 91(5):233–244, 2004.
19. T. Murata. State equation, controllability, and maximal matchings of petri nets. *IEEE Trans. Autom. Contr.*, 22(3):412–416, June 1977.
20. T. Murata. Petri nets: Properties, analysis and applications. *Proc. of the IEEE*, 77(4), 1989.
21. New Polka: Convex Polyhedra Library. http://www.irisa.fr/prive/bjeannet/newpolka.html.
22. E. Rodríguez-Carbonell and D. Kapur. An Abstract Interpretation Approach for Automatic Generation of Polynomial Invariants. In *Int. Symp. on Static Analysis (SAS 2004)*, volume 3148 of *Lecture Notes in Computer Science*, pages 280–295. Springer-Verlag, 2004.
23. S. Sankaranarayanan, H. Sipma, and Z. Manna. Petri net analysis using invariant generation. In *Verification: Theory and Practice*, pages 682–701. Springer-Verlag, 2003.
24. M. Silva, E. Teruel, and J. M. Colom. Linear algebraic and linear programming techniques for the analysis of place/transition net systems. *Lecture Notes in Computer Science: Lectures on Petri Nets I: Basic Models*, 1491:309–373, 1998.
25. M. Zhou, F. DiCesare, and A. Desrochers. A hybrid methodology for synthesis of Petri net models for manufacturing systems. *IEEE Transactions on Robotics and Automation*, 8(3):350–361, June 1992.

Modeling Multi-valued Genetic Regulatory Networks Using High-Level Petri Nets

Jean-Paul Comet, Hanna Klaudel, and Stéphane Liauzu

LaMI, UMR CNRS 8042, Université d'Evry-Val d'Essonne,
Boulevard François Mitterrand, 91025 Evry Cedex, France
{comet, klaudel}@lami.univ-evry.fr

Abstract. Regulatory networks are at the core of all biological functions from bio-chemical pathways to gene regulation and cell communication processes. Because of the complexity of the interweaving retroactions, the overall behavior is difficult to grasp and the development of formal methods is needed in order to confront the supposed properties of the biological system to the model. We revisit here the tremendous work of R. Thomas and show that its binary and also its multi-valued approach can be expressed in a unified way with high-level Petri nets.

A compact modeling of genetic networks is proposed in which the tokens represent gene's expression levels and their dynamical behavior depends on a certain number of biological parameters. This allows us to take advantage of techniques and tools in the field of high-level Petri nets. A developed prototype allows a biologist to verify systematically the coherence of the system under various hypotheses. These hypotheses are translated into temporal logic formulae and the model-checking techniques are used to retain only the models whose behavior is coherent with the biological knowledge.

1   Introduction

To elucidate the principles that govern biological complexity, computer modeling has to overcome *ad hoc* explanations in order to make emerge novel and abstract concepts[1]. Computational *system biology*[2] tries to establish methods and techniques that enable us to understand biological systems as systems, including their robustness, design and manipulation[3, 4]. It means to understand: the structure of the system, such as gene/metabolic/signal transduction networks, the dynamics of such systems, methods to control, design and modify systems in order to cope with desired properties[5].

Biological regulatory networks place the discussion at a biological level instead of a biochemical one, that allows one to study behaviors more abstractly. They model interactions between biological entities, often macromolecules or genes. They are statically represented by oriented graphs, where vertices abstract the biological entities and arcs their interactions. Moreover, at a given stage, each vertex has a numerical value to describe the level of concentration of the corre-

G. Ciardo and P. Darondeau (Eds.): ICATPN 2005, LNCS 3536, pp. 208–227, 2005.

sponding entity. The dynamics correspond to the evolutions of these concentration levels and can be represented, for instance, by differential equation systems.

R. Thomas introduced in the 70's a boolean approach for regulatory networks to capture the qualitative nature of the dynamics. He proved its usefulness in the context of immunity in bacteriophages[6, 7]. Later on, he generalized it to multi-valued levels of concentration, so called "generalized logical" approach. Moreover, the vertices of R. Thomas' regulatory networks are abstracted into "variables" allowing the cohabitation of heterogeneous information (e.g., adding environmental variables to genetic ones).

The R. Thomas boolean approach has been justified as a discretization of the continuous differential equation system[8], then has been confronted to the more classical analysis in terms of differential equations[9]. Taking into account "singular states", Thomas and Snoussi showed that all steady states can be found *via* the discrete approach[10]. More recently Thomas and Kaufman have shown that the discrete description provides a qualitative fit of the differential equations with a small number of possible combinations of values for the parameters[11].

A direct or indirect influence of a gene on itself corresponds to a closed oriented path which constitutes a feedback circuit. Feedback circuits are fundamental because they decide the existence of steady states of the dynamics: it has been stated and proved [12, 13, 14, 15] that at least one positive regulatory circuit is necessary to generate multistationarity whereas at least one negative circuit is necessary to obtain a homeostasis or a stable oscillatory behavior[16].

These static properties (number of stationary states) can be reinforced by introducing some properties on the dynamics of the system extracted from the biological knowledge or hypotheses. It becomes necessary to construct models which are coherent not only with the previous static conditions but also with the dynamical ones. Formal methods from computer science should be able to help modeler to automatically perform this verification. In [17, 18] the machinery of formal methods is used to revisit R. Thomas' regulatory networks: all possible state graphs are generated and model checkers help to select those which satisfy the temporal properties. All this approach is based on the semantics of the regulatory graph, *i.e.*, its dynamics, which has to be computed before. The state explosion phenomenon in the transition graph limits the readability of these modelings and the possible extensions like, for instance, the introduction of delays for transitions. These observations motivated our interest for applying in this context the Petri net theory.

In this article we present a modeling of the R. Thomas' regulatory networks in terms of high-level Petri nets. To ensure the appropriateness between both formalisms, we first present formally the biological regulatory graphs which describe the interactions between biological entities, the parameters which pilot the behaviors of the system and the associated dynamics (section 2). Then, after a brief introduction to the high-level Petri nets, a modeling of regulatory graphs is introduced in section 3. In section 4 we show how model-checking can be used to determine which models have to be considered. Sections 5 and 6 illustrate our approach with the LTL model-checker Maria, and describe our prototype for computer aided modeling in the context of genetic regulatory networks. The last section 7 discusses the results and the possible extensions.

2   Genetic Regulatory Graphs and Associated Semantics

In this section we formally define the biological regulatory networks. We first introduce the biological regulatory graph which represents interactions between biological entities. A vertex represents a variable (which can abstract a gene or its protein for instance). The interactions between genes are represented by arcs and each arc is labelled with the sign of the interaction: "−" for an inhibition and "+" for an activation.

The interactions have almost always a sigmoid nature (see Fig. 1): for each positive interaction of i on j, if variable i has a concentration below a certain threshold (defined by the inflection point of the sigmoid), then the variable j is not influenced by i, and for any concentration of i above this threshold, j is activated by i (and symmetrically for negative interactions). Figure 1 assumes that the variable i acts positively on j and negatively on j'; each curve represents the concentration of the target after a sufficient delay for the regulator i to act on it. Three regions are relevant: the first one corresponds to the situation where i does neither activate j nor inhibit j', the second to the situation where i activates j and does not inhibit j', and the last one corresponds to the situation where i activates j and inhibits j'. This justifies the discretization of the concentration of i into three abstract levels (0, 1 and 2) corresponding to the previous regions and constituting the only relevant information from a qualitative point of view.

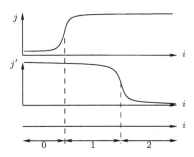

Fig. 1. The discretization is supervised by the thresholds of actions on targets

For a variable which has τ targets (itself possibly included), $\tau + 1$ abstract levels have to be considered if all thresholds are distinct, but possibly less in the case where two or more thresholds are equal.

Definition 1. *A biological regulatory graph is a labelled directed graph* $G = (V, E)$ *where each vertex* i *of* V, *called a variable, is provided with a boundary* $\beta_i \in \mathbb{N}$ *less or equal to the out-degree (the number of out-going arcs) of* i, *except*

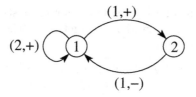

Fig. 2. Regulatory graph for mucus production in *pseudomonas aeruginosa*

if the out-degree is 0 in which case $\beta_i = 1$. *Each edge* $(i \longrightarrow j)$ *is labelled with a pair* (t_{ij}, ϵ_{ij}) *where* t_{ij}, *called the threshold of the interaction, is a natural number between 1 and* β_i *and* $\epsilon_{ij} \in \{-, +\}$ *is its sign.*

The threshold t_{ij} of a positive interaction $(i \longrightarrow j)$ determines the conditions which allow the variable i to stimulate j: if variable i has an abstract level below t_{ij}, the interaction is not active and j is not stimulated, otherwise, it is. For negative interactions, the conditions are symmetrical.

At a given stage, each variable of a regulatory graph has a unique abstract concentration level. So, the *state* of the system may be represented as a vector of concentration levels n_i of each variable i.

Definition 2. *A state of a biological graph is a tuple* $n = (n_1, n_2, ..., n_p)$, *where* p *is the number of variables and* n_i *is the abstract concentration level of the variable* i *with* $n_i \in \mathbb{N}$ *and* $n_i \leq \beta_i$.

Running example: We take as running example the mucus production in *Pseudomonas aeruginosa*. These bacteria are commonly present in the environment and secrete mucus only in lungs affected by cystic fibrosis. As this mucus increases the respiratory deficiency of the patient, it is the major cause of mortality. The regulatory network which controls the mucus production has been elucidated [19]. The main regulator for the mucus production, AlgU, supervises an operon which is made of 4 genes among which one codes for a protein that is an inhibitor of AlgU. Moreover AlgU favors its own synthesis. The regulatory network can then be simplified into the regulatory graph of Fig. 2, where variable 1 represents AlgU, and variable 2 its inhibitor. The order of thresholds t_{12} and t_{11} is not deductible from biological knowledge and in fact both orderings have to be considered [1]. Figure 2 assumes that $t_{12} < t_{11}$. Variable 1 can take three different abstract concentration levels: 0,1 or 2. Similarly, variable 2 which is an inhibitor of variable 1, can take two levels: 0 and 1. Consequently, there are 6 possible states $(0, 0)$, $(0, 1)$, $(1, 0)$, $(1, 1)$, $(2, 0)$ and $(2, 1)$.

Up to now, the discretization of continuous concentrations into the abstract levels allows us to define when a regulator has an influence on its targets, but we need to determine towards which abstract levels the targets are attracted.

---

[1] Two regulatory graphs should be considered; for simplicity we explain the concepts on one of them.

To answer this question, one has to know for each state n which regulators are actually effective on the considered target i, in other words, which are the "*resources*" of i in the state n.

Definition 3. *Given a biological regulatory graph* $G = (V, E)$ *and a possible abstract state* $n = (n_1, n_2, ..., n_p)$, *the set of* resources *of a variable* i *is the set*

$$R_i(n) =$$
$$\left\{ j \in V \mid (j \overset{(t_{ji}, \epsilon_{ji})}{\longrightarrow} i) \in E, (((n_j \geq t_{ji}) \wedge (\epsilon_{ji} = +)) \vee ((n_j < t_{ji}) \wedge (\epsilon_{ji} = -))) \right\}.$$

$R_i(n)$ contains the activators of i whose abstract level is above the threshold and the inhibitors of i whose abstract level is below the threshold. A resource is either the presence of an activator or the *absence* of an inhibitor.

It remains to define towards which abstract level a variable i is attracted when its resources are ω. We call this level the *attractor* of i for resources ω and denote it by $k_{i,\omega}$.

Having the values of attractors, it is straightforward to define the *synchronous state graph* of the biological network. For a state $n = (n_1, n_2, ..., n_p)$ where p is the number of variables, we take for each variable i, the attractor $k_{i,R_i(n)}$, where $R_i(n)$ is the set of resources of variable i when the system is in the state n. The synchronous state graph is obtained by setting for the unique possible next state $n' = (k_{1,R_1(n)}, k_{2,R_2(n)}, ...k_{p,R_p(n)})$, the state towards which the system is attracted. However, this definition has at least two drawbacks:

- first, it allows two or more variables to change simultaneously, while the probability that several variables pass through their respective thresholds at the same time is negligible *in vivo*. But we do not know which one will pass through its threshold first;
- and second, it does not prevent that a variable passes directly two or more thresholds, which is not realistic because an abstract concentration level should evolve gradually.

Then, an improved semantics is defined in terms of an *asynchronous state graph* which:

- replaces each diagonal transition of the synchronous state graph (transition with 2 or more variables changing their concentration levels) by the collection of transitions each of them modifying only one of the involved variables,
- replaces a transition of length greater or equal to 2 (which passes two or more thresholds at once) by a transition of length 1 in the same direction.

We then introduce the evolution operator which allows us to define formally the asynchronous state graph.

Definition 4. *Let* $x, k \in \mathbb{N}$. *The evolution operator* $\blacktriangleright$ *is defined as follows:*

$$x \blacktriangleright k = \begin{cases} x - 1 & \text{iff } x > k \\ x + 1 & \text{iff } x < k \\ x & \text{otherwise.} \end{cases}$$

Definition 5. *Let* $G = (V, E)$ *a regulatory graph with* p *variables. Its asynchronous state graph is defined as follows :*

- *the set of vertices is the set of states* $\Pi_{i \in V}[0, \beta_i] = \{(n_1, ..., n_p) \in \mathbb{N}^p \mid \forall i \in [1, p], n_i \leq \beta_i\}$
- *there is a transition from the state* $n = (n_1, ..., n_p)$ *to* $m = (m_1, ..., m_p)$ *iff*

$$\begin{cases} \exists \ unique \ i \in [1, p] \ such \ that \ m_i \neq n_i \\ m_i = (n_i \blacktriangleright k_{i, R_i(n)}) \end{cases} or \begin{cases} m = n \\ \forall i \in [1, p], \ n_i = (n_i \blacktriangleright k_{i, R_i(n)}). \end{cases}$$

From the R. Thomas modeling towards a modeling with Petri nets. A natural way to define an equivalent Petri net, *i.e.*, whose dynamics is exactly the same as the asynchronous approach of R. Thomas, consists in introducing inhibitor arcs [20]. A place is associated to each gene, and a transition to each attractor. The input places of the transition corresponding to $k_{i, \omega}$ are all the predecessors of i, with the places of ω connected to the transition by standard arcs and the predecessors of i not included in ω connected to the transition by inhibitor arcs.

In such a modeling, there are as many transitions as attractors. For a non trivial regulatory graph, it leads to a Petri net which is difficult to interpret because of its size. Moreover, in general, using inhibitor arcs changes the complexity of the Petri net class and may lead to introduce difficulties in proofs of some properties.

3 Modeling with High-Level Petri Nets

3.1 Introduction to High-Level Petri Nets

Definition 6. *A (low-level) Petri net is a triple* $L = (S, T, W)$, *where* S *is a set of places,* T *is a set of transitions, such that* $S \cap T = \emptyset$ *and* $W : (S \times T) \cup (T \times S) \to \mathbb{N}$ *is a weight function.*

A *marking* of a Petri net (S, T, W) is a mapping $M : S \to \mathbb{N}$, which associates to each place a natural number of *tokens*. The behavior of such a net, starting from an arbitrary *initial marking*, is determined by the usual definitions for place/transition Petri nets.

High-level nets that we consider can be viewed as simple abbreviations of the low-level ones.

Let *Val* and *Var* be fixed but suitably large disjoint sets of *values* and *variables*, respectively. A *multiset* over a set E is a function $\mu : E \to \mathbb{N}$; μ is finite if $\{e \in E \mid \mu(e) > 0\}$ is finite. We denote by $\mathcal{M}_f(E)$ the set of finite multisets over E. The set of all well-formed *predicates* built from the sets *Val*, *Var* and a suitable set of operators is denoted by Pr.

Definition 7. *A high-level Petri net, HLPN for short, is a triple* (S, T, ι), *where* S *and* T *are disjoint sets of* places *and* transitions, *and* ι *is an inscription function with domain* $S \cup (S \times T) \cup (T \times S) \cup T$ *such that:*

- *for every place* $s \in S$, $\iota(s) \subseteq Val$, *is the* type *of* s, *i.e., the set of possible values the place may carry;*
- *for every transition* $t \in T$, $\iota(t)$ *is the* guard *of* t, *i.e., a predicate from* Pr;
- *for every arc* $(s, t) \in (S \times T) : \iota((s, t)) \in \mathcal{M}_f(Val \cup Var)$ *is a multi-set of variables or values (analogously for arcs* $(t, s) \in (T \times S)$). *The inscriptions* $\iota((s, t))$ *and* $\iota((t, s))$ *will generally be abbreviated as* $\iota(s, t)$ *and* $\iota(t, s)$, *respectively. The arcs with empty inscriptions are omitted.*

A *marking* of a high-level Petri net (S, T, ι) is a mapping $M : S \to \mathcal{M}_f(Val)$ which associates to each place $s \in S$ a multi-set of values from its type $\iota(s)$. A *binding* is a mapping $\sigma : Var \to Val$ and an *evaluation* of an entity η (which can be a variable, a vector or a (multi-)set of variables, etc.) through σ is defined as usual and denoted by $\eta[\sigma]$.

The transition rule specifies the circumstances under which a marking M' is reachable from a marking M. A transition t is *activated* at a marking M if there is an *enabling binding* σ for variables in the inscription of t (making the guard true) and in inscriptions of arcs around t such that $\forall s \in S : \iota(s, t)[\sigma] \leq M(s)$, *i.e.*, there are enough tokens of each type to satisfy the required flow. The effect of an occurrence of t, under an enabling binding σ, is to remove tokens from its input places and to add tokens to its output places, according to the evaluation of arcs' annotations under σ.

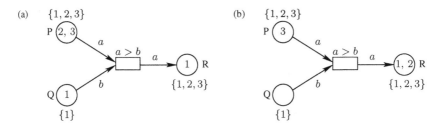

Fig. 3. A simple marked high-level Petri net before (a) and after (b) the firing of the transition

For the example of Fig. 3-(a), the marking is given by: $M(P) = \{2, 3\}$, $M(Q) = \{1\}$, $M(R) = \{1\}$. Bindings are $\sigma_1 = \begin{cases} a \to 1 \\ b \to 1 \end{cases}$, $\sigma_2 = \begin{cases} a \to 2 \\ b \to 1 \end{cases}$ and $\sigma_3 = \begin{cases} a \to 3 \\ b \to 1 \end{cases}$. Only σ_2 and σ_3 are enabling (σ_1 does not make the guard true). At the marking M, the transition t is activated for both σ_2 and σ_3. Figure 3-(b) shows the new marking if σ_2 is chosen.

An important property of high-level Petri nets is that they may be unfolded to low-level ones, which may be helpful when using various verification tools.

The unfolding operation associates a low-level net $\mathcal{U}(N)$ with every high-level net N, as well as a marking $\mathcal{U}(M)$ of $\mathcal{U}(N)$ with every marking M of N.

Definition 8. *Let* $N = (S, T, \iota)$; *then* $\mathcal{U}(N) = (\mathcal{U}(S), \mathcal{U}(T), W)$ *is defined as follows:*

- $\mathcal{U}(S) = \{s_v \mid s \in S \text{ and } v \in \iota(s)\}$;
- $\mathcal{U}(T) = \{t_\sigma \mid t \in T \text{ and } \sigma \text{ is an enabling binding of } t\}$;
- $W(s_v, t_\sigma) = \displaystyle\sum_{\substack{a \in \iota(s,t) \\ a[\sigma] = v}} \iota(s,t)(a)$, *where* $\iota(s,t)(a)$ *is the number of occurrences*

of a *in the multiset* $\iota(s,t)$, *and analogously for* $W(t_\sigma, s_v)$.

Let M be a marking of N. The unfolding of a marking $\mathcal{U}(M)$ is defined as follows: for every place $s_v \in \mathcal{U}(S)$, $(\mathcal{U}(M))(s_v) = (M(s))(v)$ where $(M(s))(v)$ is the number of occurrences of v in the marking $M(s)$ of s. Thus, each elementary place $s_v \in \mathcal{U}(S)$ contains as many tokens as the number of occurrences of v in the marking $M(s)$. Figure 4 presents the Petri net obtained by unfolding of the high-level Petri net of Fig. 3-(a).

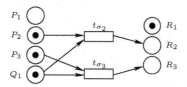

Fig. 4. Unfolded Petri net of high-level Petri net of Fig. 3-(a)

3.2   Modeling of Genetic Regulatory Networks

We can represent a regulatory network by a high-level Petri net which has a unique transition and as many places as genes in the regulatory graph. Each place corresponds to a gene i and carries one token: its abstract concentration level n_i. The marking of this net corresponds thus to an abstract state $n = \{n_1, \ldots, n_p\}$. The transition can fire at a marking n leading to the marking n' if its guard

$$\text{asyn\_guard}(n, n') =$$

$$\left\{ \begin{array}{c} \left(\exists i \in [1, p], \ (n_i \neq k_{i, R_i(n)}) \wedge (n'_i = n_i \blacktriangleright k_{i, R_i(n)}) \wedge (\forall j \neq i, n'_j = n_j) \right) \\ \vee \\ \left(\forall i \in [1, p], \ (n_i = k_{i, R_i(n)}) \wedge (n'_i = n_i) \right) \end{array} \right\}$$

is true. This guard translates directly the asynchronous semantics of R. Thomas. Indeed, the marking represents a stable steady state for the asynchronous semantics, when the attractors $k_{i, R_i(n)}$ equal the current concentrations n_i for all

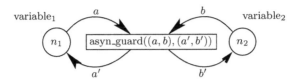

Fig. 5. HLPN modeling a genetic regulatory network with 2 genes. Each place abstracts a gene and carries one token: its abstract concentration level. The guard expresses the relationship between the current marking $(n_{\text{variable}_1}, n_{\text{variable}_2})$ and the possible updated marking $(n'_{\text{variable}_1}, n'_{\text{variable}_2})$

variables $i \in V$. The guard gives in this case the same marking for the next one[2]. If the marking does not correspond to any stable steady state, then some variables are not equal to their attractors. Following Thomas' semantics, a possible next marking is a marking for which the variables do not change unless one of them which changes (plus or minus one) in the direction of its attractor.

Figure 5 represents the high-level Petri net for the running example. Let us consider furthermore that the attractors of the running example are given: $k_{1,\{\}} = 0$, $k_{1,\{1\}} = 2$, $k_{1,\{2\}} = 2$, $k_{1,\{1,2\}} = 2$, $k_{2,\{\}} = 0$ and $k_{2,\{1\}} = 1$. Let us choose the state $(1,0)$ as the initial marking. The resources of variable 1 are $R_1((1,0)) = \{2\}$ since both inhibitor and activator are absent, and the resources of variable 2 are $R_2((1,0)) = \{1\}$ since the activator is present. Both variables are attracted towards values different from their current values. Two possible new markings are possible: $(2,0)$ if the variable 1 evolves first, and $(1,1)$ otherwise. Globally, the sequence of markings during an execution corresponds to a particular possible path.

Thus, for implementation reasons, we propose a more compact modeling which is in fact a folding of the previous one. It consists in a unique place called *cell*, which abstracts the cell in which each token represents a specific gene and its expression level. A particular structured type *gene* is needed: it represents a couple $(gene, level)$ where *level* is the abstract level of the variable *gene*. The arc inscriptions are also modified: the input one becomes $\{(1, a_1), \ldots, (p, a_p)\}$, with $a_i \neq a_j$ for $i \neq j$, and the output one becomes $\{(1, a'_1), \ldots, (p, a'_p)\}$, with $a'_i \neq a'_j$ for $i \neq j$. The state of the system is now represented by the set of tokens present in the unique place. Let be $\nu = \{(i, n_i), i \in [1, p]\}$ and $\nu' = \{(i, n'_i), i \in [1, p]\}$. The guard of the unique transition can then be written:

$$
\text{asyn\_guard}(\nu, \nu') =
\left\{
\begin{array}{c}
\left(\exists i \in [1, p], \ (n_i \neq k_{i, R_i(n)}) \wedge (n'_i = n_i \blacktriangleright k_{i, R_i(n)}) \wedge (\forall j \neq i, n'_j = n_j) \right) \\
\vee \\
\left(\forall i \in [1, p], \ (n_i = k_{i, R_i(n)}) \wedge (n'_i = n_i) \right)
\end{array}
\right\}
$$

---

[2] The self loop corresponds in this case to the stability of the system which is quite different from a deadlock, especially when verifying temporal properties.

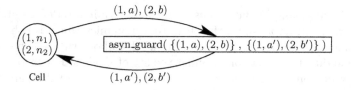

Fig. 6. Second HLPN modeling of a genetic regulatory network with 2 genes

Then, for every biological/genetic regulatory network, the high-level Petri net has a unique place and a unique transition. The only things that distinguish different nets are the number and the type of tokens, and the attractors $k_{i,R_i(n)}$ which are present in the guard. This property is very useful in practice because it allows us to have a generic description (and so a generic source file) and to generate automatically the high-level Petri net for an arbitrary genetic regulatory network. As an illustration, Fig. 6 presents the high-level Petri net for the regulatory graph of our running example.

4   Determination of Valuable Models

The execution of regulatory networks depends on the initial state which is often unknown, so the analysis of its behavior has to be performed a priori from every possible one. Parameters $\{k_{i,\omega}, i \in V$ and $\omega \subseteq G^{-1}(i)\}$, where $G^{-1}(i)$ denotes the set of predecessors of variable i in the regulatory graph, play also a major role on the dynamics of the regulatory networks. In fact, they are sufficient for entirely defining their dynamics. For this reason, a set of particular attractor values is called a *model* of the system. Unfortunately, most often they cannot be deduced from experiments and the modeler has to consider the different possible values of attractors. For a given regulatory graph, the number of different sets of attractor values, *i.e.*, the number of models, is exponential with the number of predecessors of each variable. More precisely, this number is equal to $\prod_{i \in V} 2^{|G^{-1}(i)|}$. This enormous number prevents us to construct all possible dynamics of the regulatory network and to let biologist select only interesting ones. Some interesting results can therefore be used for reducing the number of models to be considered. In [10] the modeling of Thomas is seen as a discretization of a particular class of continuous differential equation systems, and attractors $k_{i,\omega}$ reflect a discretization of sums of ratios of positive constants. In such a case, the attractors in the set $\{k_{i,\omega} \mid i \in V$ and $\omega \subseteq G^{-1}(i)\}$ have to satisfy the following constraints

$$k_{i,\emptyset} = 0 \text{ and } \omega \subset \omega' \implies k_{i,\omega} \leq k_{i,\omega'}.$$

Nevertheless, it is possible to enlarge the set of models which can be described by this discrete formalism leading to slacken these previous constraints which cannot be added arbitrarily.

Then, the modeling activity focuses on the determination of a suitable class of models, *i.e.*, attractor values that lead to a dynamics which is coherent with the experimental knowledge. Biological knowledge about the behavior can then be used as indirect criteria constraining the set of models. For instance, multistationarity or homeostasis which are experimentally observable, are useful to reduce the set of attractor values. This relies on notions of positive/negative functional circuits and of their characteristic states [21].

But, it is possible to take into account not only such conditions as homeostasis and multistationarity, but also particular temporal properties extracted from biological knowledge or hypotheses[18]. These knowledge or hypotheses may be translated into a formal temporal language as LTL (Linear Temporal Logic) or CTL (Computational Tree Logic) in order to be manipulated automatically by computer. The coherence of the model may then be verified automatically by model checking.

For the running example, the presence of a positive circuit in the regulatory graph of *Pseudomonas aeruginosa* makes possible a dynamics with two stable steady states which would correspond, from a biological point of view, to an epigenetic switch (stable change of phenotype without mutation) from the non-mucoid state (the bacterium does not produce mucus) to the mucoid state (it does). In other words, the question is to find at least one model of the bacteria, which is compatible with the known biological results and which has a multistationarity where one stable steady state produces mucus and the other one does not. It turns out that the mucus production is triggered by a high level of variable 1. Then, a recurrent production of mucus is equivalent to the fact that the concentration level n_1 of variable 1 is repeatedly equal to 2. So, the stationarity of the mucoid state can be expressed as:

$$(n_1 = 2) \implies XF(n_1 = 2) \tag{1}$$

where $XF\varphi$ means that φ will be satisfied in the future. Moreover, we know that the bacteria never produce mucus by them-selves when starting from a basal state (second stable steady state):

$$(n_1 = 0) \implies G(\neg(n_1 = 2)). \tag{2}$$

However, even if it may be easy to express in this way particular properties of a given system, proposing a general method allowing to express formally a biological hypothesis remains a difficult open problem.

Nevertheless the formal properties being given, it becomes possible to design a general approach for selecting suitable models. We are interested in models which lead to a dynamics coherent with the considered temporal properties. This approach can be summarized as follows (see Fig. 7):

1. Design the regulatory graph corresponding to the biological system. Because of the partial information on the system, the biological regulatory network can be represented by several regulatory graphs. For this step it is not necessary to describe all details of the system but only the key concepts. In

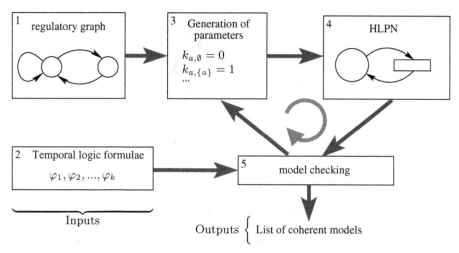

Fig. 7. Computer aided modeling approach. The first step consists in designing the regulatory graph (1) and the temporal logic formulae expressing temporal properties of the underlying biological system (2). Generation of a potential model (3) and construction of the HLPN (4). The model checker is then called for verifying the temporal properties (5). If the model satisfies the properties, the model is stored. Back to (3) for generating another potential model

particular, positive and negative circuits have to be present as well as their intertwined interactions.

2. Design the temporal logic formulae which express formally dynamical knowledge or hypotheses that biologist want to take into account. This step should be performed with care in order not to forget important information.
3. Generate, from the regulatory graphs, all potential models (set of attractor values).
4. Construct the high-level Petri net for each of them.
5. For each Petri net, call the model checker for verifying if the temporal properties are satisfied. Return only the models and associated state graphs which satisfy the formulae.

5   Implementation with Maria

For implementing, we chose the model checker MARIA [22] (Modular Reachability Analyzer for Algebraic System Nets) which takes as input a high-level Petri net described in a particular language (see Fig. 13 in the appendix for an example), a LTL formula and performs the checking.

As said above, the execution of regulatory networks depends on the initial state and so the checkings have to be done from each possible initial state unless the formulae specify the opposite. Maria makes possible to specify a unique initial marking. Then, it is necessary to include a mechanism allowing to take

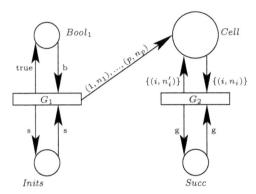

Fig. 8. Implementation in MARIA of the HLPN. The guard $G_1 = (b = false) \wedge (s = ((1, n_1), (2, n_2), ..., (p, n_p)))$ is fired only once for the initialization step. G_2 is the asynchronous guard completed with the number of the gene which evolves (see text)

into account all interesting initial markings. In order to do this, we add two places and a new transition (see Fig. 8).

- The first place, *Inits*, contains all initial states that have to be considered for the current checking. If no restriction has to be taken into consideration, all states are added in this place.
- The second place, $Bool_1$, contains only a boolean token initially set to false, which means that the initial state has to be chosen. Then the token becomes true.
- The transition G_1 takes an initial marking from *Inits* and the value false from $Bool_1$ and generates the corresponding tokens in the place *Cell*, the value true in $Bool_1$ and gives back the initial marking to *Inits*.

At the beginning, the place *Cell* is empty. The transition G_2 is not enabled but the transition G_1 can fire. It chooses a possible initial state and generates the corresponding tokens in the place *Cell*. The token of the place $Bool_1$ becomes true and the transition G_1 is disabled. The fact that G_1 gives back to *Inits* the chosen initial state prevents to generate supplementary states of the Petri net which do not correspond to anything in the regulatory network.

The high-level Petri net which models an asynchronous state graph contains some non-determinism due to the fact that several successor states may be reachable from a given state. Representing this in Maria assumes the definition of a supplementary place *Succ* initialized with all natural numbers from 0 to p where p is the number of genes present in the regulatory graph. The transition G_2 reads the current state, chooses a particular token from the place *Succ* and generates the next state according to the *Succ'* token. For example, if the token 3 is chosen, the next marking corresponds to the state where only gene 3 has changed. If 0 is chosen, it means that no gene evolves.

The guard of G_2 is almost as asyn_guard seen before, the only change concerns the token g read from the place *Succ*. It can be written as follows:

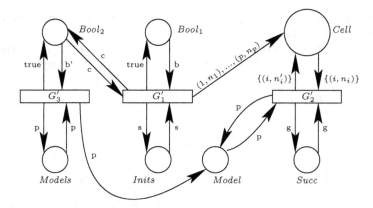

Fig. 9. Implementation a HLPN manipulating a set of models for a regulatory network

$$G_2(\nu, \nu', g) =$$
$$\left\{ \begin{array}{c} \left((g \neq 0) \wedge (n_g \neq k_{g,R_g(n)}) \wedge (n'_g = n_g \blacktriangleright k_{g,R_g(n)}) \wedge (\forall j \neq g, n'_j = n_j) \right) \\ \vee \\ (g = 0) \wedge \left(\forall i \in [1,p], \ (n_i = k_{i,R_i(n)}) \wedge (n'_i = n_i) \right) \end{array} \right\}$$

With this modeling it is possible to implement the global approach described in section 4. In such an approach the majority of models are rejected. It could be interesting to test directly a set of models which are considered as non suitable in order to reject them in only one call to the model checker. We construct now a new modeling which is able to manage a set of models.

Three supplementary places and a transition are added (see Fig. 9). The place *Models* is initialized with all models considered as suitable, the place *Model* is initially empty and carries the current model to check, and *Bool₂* contains a boolean token which is false iff *Model* is empty. The transition G'_3 takes a possible model from *Models* and the value false from *Bool₂* and generates the corresponding token in the place *Model* as well as the value true in *Bool₂*. The transition G'_2 which simulates the evolution of the regulatory network reads now the attractors from the place *Model*, and the guard is modified accordingly. The size of this Petri net does not depend on the number of genes nor on the number of models.

This modeling does not replace the previous one but completes it. It becomes possible to verify the temporal properties on a set of models similarly as it was possible to do it for a set of initial markings.

This approach could also be used for finding a first model that is compatible with the temporal properties. Let us assume that we are looking for a model that satisfies a set of formulae $\varphi_1, \varphi_2, ..., \varphi_n$. The model checker MARIA tries to validate the formulae for all possible paths and for all possible models. If no path contradicts the formulae, Maria answers that the Petri net satisfies them. In our current modeling, it would mean that all possible models are coherent with the biological temporal properties. If one path refutes the conjunction of the formulae, then the computation stops.

Let us initialize the Petri net with all possible models and give to the model checker the formula $XX(\neg(\varphi_1 \wedge \varphi_2 \wedge ... \wedge \varphi_n))$. The double X expresses that the formulae are checked after the initialization of attractors and after the initialization of markings. The model checker is looking for a model which contradicts the negation of the conjunctions of formulae. If there exists such a path, Maria stops and gives the associated attractors. This direct way does not allow us to exhibit all models that we are looking for but only the "first" one.

Unfortunately, the used temporal logic limits this approach. Let us assume that we are looking for models satisfying the temporal property φ. If we give the model checker the negation of the formula, Maria tries to exhibit a path that contradicts the negation, $i.e.$, that satisfies the formula. This path corresponds to a model, but for this parameterization, there could exist another path that refutes the formula. In fact, in order to develop such an approach, one needs to express in the formulae that something is possible or that something is true for all possible choices, in other words, the suitable temporal logic would be rather CTL. For example, if we have to check the formula on all paths and for all initial markings, we are looking for a model that satisfies the formula $EX\ AX\ \varphi$. It means that there exists a model for which all initial markings lead to a state where φ is satisfied. The model checker is given the negation of the previous formula, $AX\ EX\ \neg\varphi$, and it will answer, if any, a path that refutes the previous formula, $i.e.$, a model such that for all initial markings, we have φ.

As mentioned before this modeling permits the user to exhibit, if any, a model which satisfies the temporal properties. The existence of such a model proves the coherence between the regulatory graph and the temporal properties, but it is not sure that the biological system works in the same way. From a biological point of view it is more useful to exhibit all possible models which are coherent with the temporal properties since it permits the biologist to explain the behaviors by various models. It would be interesting to develop a model checker which enumerate all counterexamples.

6   Prototype for Computer Aided Modeling

We have designed and developed a prototype for computer aided modeling which implements our general approach described above. It contains principally 3 modules (see Fig. 10):

- The first module permits the user to design the regulatory graph with a user friendly interface (see Fig. 12 in the appendix). NetworkEditor is able to generate a XML file to represent the regulatory graph according to the GINML document type definition [23].
- FormulaEditor helps the user to write the LTL formula which describes the biological knowledge or hypothesis on the dynamics of the system. The editor translates the LTL formula into the Maria LTL format.

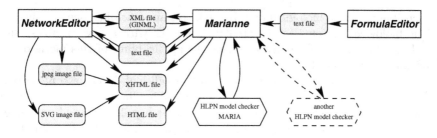

Fig. 10. A prototype for computer aided modeling

- `Marianne` takes as input a GINML file or a text file representing a biological regulatory graph and an LTL formula. Marianne computes then for each potential model the corresponding HLPN. It selects the models which satisfy the LTL formula using MARIA. It offers also the possibility to generate the corresponding asynchronous state graph (see Fig. 11 in the appendix).

7   Conclusion and Future Work

We have defined a modeling of regulatory networks in terms of high-level Petri nets. Applied to the *Pseudomonas aeruginosa*, this modeling approach selects 4 models leading to 4 different asynchronous state graphs for each regulatory graph corresponding to two possible orderings of t_{11} and t_{12} (see footnote 1). These 8 models prove that the proposed regulatory graphs of *Pseudomonas aeruginosa* are coherent with the hypothesis of epigenetic switch. Figure 13 presents one model which satisfies the temporal specifications and Fig. 11 the corresponding state graph obtained directly from Maria. If *Pseudomonas aeruginosa* is actually compatible with one of these models (no matter which model, because there are observationally equivalent), it could open new therapeutics in prospects. Since the formula 2 is known to be satisfied, one has just to confirm *in vivo* the formula 1. An experiment schema may be suggested by the structure of the formula: it consists in pulsing variable$_1$ up to saturation by an external signal, and in checking, after a transitory phase due to the pulse, if the mucus production persists [19].

Besides this biological case study, the contribution of the paper is also on a more abstract level. Indeed, our approach overpasses the pure application context and allows a computer aided manipulation of the semantics of the discrete modeling of R. Thomas. It consists in defining an automatic translation from regulatory graphs to high-level Petri nets and to provide the means to express and check some behavioral properties. In particular, temporal properties expressed in temporal logics can be checked in order to confirm or refute some biological hypotheses. These analyses may be performed using various existent Petri net methods and tools. Moreover, the method based on an intensive use of model checking has been shown both useful from the computer science point of view and sensible from the biological one. Indeed, this approach has been applied for various biological systems represented by networks with a few variables (supported also by our tools), which confirms the applicability of this method to a significant class of problems.

Our compact and generic representation through high-level Petri nets opens up some extensions. First, recent extensions [24] taking into account non sigmoid character of the interaction function may easily be handled. Second, resources, like time or energy, may be introduced in the net model, for instance, using high-level buffers, as in [25, 26]. Moreover, Petri net representation naturally leads to various kinds of semantics. In particular, one may consider non sequential ones, which allow on one hand combating the state explosion and on the other hand using more efficient verification techniques [27, 28].

The paper presents also a user friendly environment we developed, helping biologists in modeling and analyzing regulatory networks and to express desired properties. Our experiments showed that it would be interesting to extend the model checker MARIA with CTL logics. Also, since Maria allows the user to unfold the model into the native input formats of PEP [28], LoLA [29] or Prod [30], it gives a possibility to use different analysis techniques offered by these tools. In particular, the problem of using a CTL model checker may be resolved with Prod.

This approach has been compared in terms of efficiency with another environment for regulatory networks [18], which uses the classical CTL model checker NuSMV [31] and the execution times were similar. It was not surprising because the semantics was explicitly sequential. We hope that for some extensions accepting truly concurrent behaviors, the verification could be more efficient if partial order representation and dedicated tools are used.

Acknowledgement. The authors thank genopole®-research in Evry (H. Pollard and P. Tambourin) for constant supports. We gratefully acknowledge the members of the genopole® working groups *observability* and G^3 for stimulating interactions.

References

1. Huang, S.: Genomics, complexity and drug discovery: insights from boolean network models of cellular regulation. Pharmacogenomics. **2** (2001) 203–22
2. Wolkenhauer, O.: Systems biology: the reincarnation of systems theory applied in biology? Brief Bioinform. **2** (2001) 258–70
3. Kitano, H.: Computational systems biology. Nature **420** (2002) 206–10
4. Hasty, J., McMillen, D., Collins, J.: Engineered gene circuits. Nature **420** (2002) 224–30
5. Kitano, H.: Looking beyond the details: a rise in system-oriented approaches in genetics and molecular biology. Curr. Genet. **41** (2002) 1–10
6. Thomas, R., Gathoye, A., Lambert, L.: A complex control circuit. regulation of immunity in temperate bacteriophages. Eur. J. Biochem. **71** (1976) 211–27

7. Thomas, R.: Logical analysis of systems comprising feedback loops. J. Theor. Biol. **73** (1978) 631–56
8. Snoussi, E.: Qualitative dynamics of a piecewise-linear differential equations : a discrete mapping approach. Dynamics and stability of Systems **4** (1989) 189–207
9. Kaufman, M., Thomas, R.: Model analysis of the bases of multistationarity in the humoral immune response. J. Theor. Biol. **129** (1987) 141–62
10. Snoussi, E., Thomas, R.: Logical identification of all steady states : the concept of feedback loop caracteristic states. Bull. Math. Biol. **55** (1993) 973–991
11. Thomas, R., Kaufman, M.: Multistationarity, the basis of cell differentiation and memory. I. & II. Chaos **11** (2001) 170–195
12. Plathe, E., Mestl, T., Omholt, S.: Feedback loops, stability and multistationarity in dynamical systems. J. Biol. Syst. **3** (1995) 569–577
13. Snoussi, E.: Necessary conditions for multistationarity and stable periodicity. J. Biol. Syst. **6** (1998) 3–9
14. Cinquin, O., Demongeot, J.: Positive and negative feedback: striking a balance between necessary antagonists. J. Theor. Biol. **216** (2002) 229–41
15. Soulé, C.: Graphical requirements for multistationarity. ComPlexUs **1** (2003) 123–133
16. Thomas, R., Thieffry, D., Kaufman, M.: Dynamical behaviour of biological regulatory networks - I. Bull. Math. Biol. **57** (1995) 247–76
17. Pérès, S., Comet, J.P.: Contribution of computational tree logic to biological regulatory networks: example from pseudomonas aeruginosa. In: CMSB'03. Volume 2602 of LNCS. (2003) 47–56
18. Bernot, G., Comet, J.P., Richard, A., Guespin, J.: Application of formal methods to biological regulatory networks: Extending Thomas' asynchronous logical approach with temporal logic. J. Theor. Biol. **229** (2004) 339–347
19. Guespin-Michel, J., Kaufman, M.: Positive feedback circuits and adaptive regulations in bacteria. Acta Biotheor. **49** (2001) 207–18
20. Chaouiya, C., Remy, E., Ruet, P., Thieffry, D.: Qualitative modelling of genetic networks: From logical regulatory graphs to standard petri nets. In: ICATPN 2004. LNCS 3099, Springer-Verlag (2004) 137–156
21. Thomas, R., Thieffry, D., Kaufman, M.: Dynamical behaviour of biological regulatory networks - I. biological role of feedback loops an practical use of the concept of the loop-characteristic state. Bull. Math. Biol. **57** (1995) 247–76
22. Mäkelä, M.: Maria: Modular reachability analyser for algebraic system nets. In: ICATPN 2002. Number 2360 in LNCS, Springer-Verlag (2002) 434–444
23. Chaouiya, C., Remy, E., Mossé, B., Thieffry, D.: GINML: towards a GXL based format for logical regulatory networks and dynamical graphs. http://www.esil.univ-evry.fr/~chaouiya/GINsim/ginml.html (2003)
24. Bernot, G., Cassez, F., Comet, J.P., Delaplace, F., Müller, Roux, O., Roux, O.: Semantics of biological regulatory networks. In: Biology BioConcur'2003. (2003)
25. Pommereau, F.: Modèles composables et concurrents pour le temps-réel. PhD thesis, Université Paris 12 (2002)
26. Klaudel, H., Pommereau, F.: Asynchronous links in the pbc and m-nets. In: ACSC'99. Volume 1742 of LNCS., Springer-Verlag (1999) 190 – 200
27. Khomenko, V., Koutny, M., Vogler, W.: Canonical prefixes of petri net unfoldings. Acta Informatica **40** (2003) 95–118
28. Grahlman, B.: The state of PEP. In: AMAST'98. Number 1548 in LNCS, Springer-Verlag (1999) 522–526
29. Schmidt, K.: LoLA: a low level analyser. In Nielsen, M., Simpson, D., eds.: ICTPN 2000. Volume 1825 of LNCS., Springer-Verlag (2000) 465–474

30. Varpaaniemi, K., Halme, J., Hiekkanen, K., Pyssysalo, T.: Prod: reference manual. Technical Report B13, Helsinki University of Technology, Finland (1995)
31. Cimatti, A., Clarke, E., Giunchiglia, F., Roveri, M.: Nusmv: a reimplementation of smv. In: STTT'98. BRICS Notes Series, NS-98-4 (1998) 25–31

Appendix

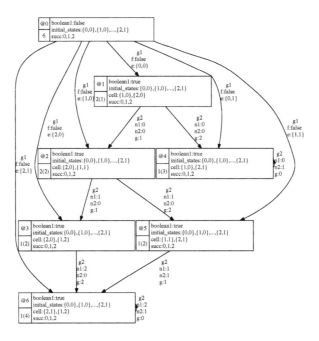

Fig. 11. State graph obtained directly from Maria

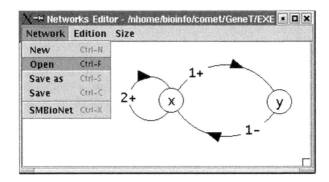

Fig. 12. The interface allows the user to specify the regulatory network in term of regulatory graph. The thresholds and the sign of the interactions are added on vertices

```
/////////////////////////
// DEFINITION OF TYPES //
/////////////////////////
typedef unsigned (1..2) numero_of_gene;
typedef unsigned (0..2) level;
typedef struct{ numero_of_gene g;
               level n;            } gene;
typedef unsigned (0..2) level_gene1;
typedef unsigned (0..1) level_gene2;
typedef struct{ level_gene1 n1;
               level_gene2 n2;   } state;
typedef unsigned (0..2) succt;
typedef bool flag;


/////////////////////////////
// DEFINITION OF FUNCTIONS //
/////////////////////////////
state vector(gene g1,gene g2)
    is state({is level_gene1 g1.n,
             is level_gene2 g2.n});

state synchronous(state e) is state(e?
    {2,1}:  // attractor for the state {2,1}
    {2,1}:  // attractor for the state {1,1}
    {0,1}:  // attractor for the state {0,1}
    {2,1}:  // attractor for the state {2,0}
    {2,1}:  // attractor for the state {1,0}
    {2,1}   // attractor for the state {0,0});

state step1(state e) is state(
  (e.n1<synchronous(e).n1)  ?{+e.n1,e.n2}:
  ( (e.n1>synchronous(e).n1)?{|e.n1,e.n2}:e));

state step2(state e) is state(
  (e.n2<synchronous(e).n2)  ?{e.n1,+e.n2}:
  ( (e.n2>synchronous(e).n2)?{e.n1,|e.n2}:e));

gene gene1(state e) is gene({1,e.n1});
gene gene2(state e) is gene({2,e.n2});
```

```
/////////////////////////////
// DEFINITION OF PLACES //
/////////////////////////////
place boolean1 flag: false;
place initial_states state: state e  : e;
place cell gene;
place succ succt: succt g: g;


///////////////////////////////////
// DEFINITION OF TRANSITIONS  //
///////////////////////////////////

trans g1
in  {  place boolean1: f;
       place initial_states : e;          }
out {  place boolean1: true;
       place initial_states : e;
       place cell : (gene1(e),gene2(e)); }
gate (f==false);


trans g2
   {  state v=vector({1,n1},{2,n2});
      state v1=step1(v);
      state v2=step2(v);                 }
in  {  place succ : g;
       place cell : {1,n1},{2,n2};       }
out {  place succ : g;
       place cell :   (b ?
                   (gene1(v2),gene2(v2)): // b==2
                   (gene1(v1),gene2(v1)): // b==1
                   ({1,n1},{2,n2})    ); // b==0
   }
gate (
  (g==0 && v1.n1==v.n1 && v2.n2==v.n2 ) ||
  (g==1 && v1.n1!=v.n1) ||
  (g==2 && v2.n2!=v.n2) );

deadlock fatal;
```

Fig. 13. File describing the high-level Petri net corresponding to the regulatory network of Fig. 12 for model checker Maria. The values of attractors are the following: $k_{1,\emptyset} = 0$, $k_{1,\{1\}} = 2$, $k_{1,\{2\}} = 2$, $k_{1,\{1,2\}} = 2$, $k_{2,\emptyset} = 1$ and $k_{2,\{1\}} = 1$

Termination Properties of TCP's Connection Management Procedures

Bing Han and Jonathan Billington

Computer Systems Engineering Centre,
University of South Australia,
Mawson Lakes, SA 5095, Australia
Bing.Han@postgrads.unisa.edu.au
Jonathan.Billington@unisa.edu.au

Abstract. The Transmission Control Protocol (TCP) is the most widely used transport protocol in the Internet, providing a reliable data transfer service to many applications. This paper analyses TCP's Connection Management procedures for correct termination and absence of deadlocks. The protocol is assumed to be operating over a reordering lossless channel and is modelled using Coloured Petri nets. The following connection management scenarios are examined using state space analysis: client-server and simultaneous opening; orderly release; and abortion. The results demonstrate that TCP terminates correctly for client-server and simultaneous connection establishment, orderly release after the connection is established and aborting of connections. However, we discover a deadlock when connection release is initiated before the connection has been fully established when operating over a reordering lossless channel.

1   Introduction

The Transmission Control Protocol (TCP) [3,32] provides a reliable data transfer service to Internet applications such as the web and email, which ensures that data will be delivered in order and without loss or duplication. TCP is also the basis for the development of two new protocols for the Internet: SCTP [39] and DCCP [19]. TCP is a complex protocol originally specified in RFC 793 [32] using narrative descriptions, message sequence diagrams, and a finite state machine (FSM) diagram. It was then improved and modified in [1,3,8,9,15–17,23,29]. The number of bugs reported in TCP implementations [29] spans 60 pages. This and other experience led us to believe that a more formal approach to TCP specification and analysis may prove beneficial.

TCP comprises a connection management protocol for establishing and terminating connections and a data transfer protocol for reliable data transfer. Before data transfer begins, a connection needs to be set up between two end points, each of which is identified by a socket comprising an IP address and a *port* number [32]. TCP uses a *three-way handshake* [40] to open a connection, that is, three messages are exchanged by the two communicating entities. A TCP connection is full duplex allowing independent data flow in both directions. The

G. Ciardo and P. Darondeau (Eds.): ICATPN 2005, LNCS 3536, pp. 228–249, 2005.

connection is fully released when both ends close in an orderly manner and the procedure is known as *orderly release*. The connection establishment, release and abort procedures are known as *TCP Connection Management*, which is critical for reliable data delivery. Research on verifying TCP's correctness usually falls into two camps: those that verify the connection management protocol and those that verify the data transfer protocol. This paper focuses on verifying TCP Connection Management.

Using Coloured Petri nets (CPNs) [18], we model TCP Connection Management according to the pseudo code provided in Section 3.9 of RFC 793. We use a realistic model of the Internet's service that can lose, delay and re-order packets. In this paper, we analyse TCP Connection Management operating over a reordering channel with no loss. This ensures [2] that errors such as unspecified receptions will be discovered as terminal states and not removed by loss. Two important functional properties of the protocol are examined: correct termination and absence of deadlocks. The analysis is undertaken by checking the properties over the state spaces of the CPN model and is conducted incrementally [2]. We present detailed results to back up every conclusion reached.

This paper is organised as follows. Section 2 provides an introduction to TCP Connection Management. Section 3 reviews related work. Section 4 briefly describes the CPN model. Section 5 defines the desired properties of TCP: termination and absence of deadlocks. Section 6 describes the different configurations of the CPN model. Sections 7 and 8 summarise and discuss the state space analysis results. Finally, section 9 concludes this paper.

2   TCP Connection Management

The messages exchanged in a TCP connection are known as *segments*. A segment is a sequence of 32-bit words, comprising header fields and a data field. At the beginning of the header fields are the 16 bit source and destination port fields. They are used to identify the two communicating application processes. Every octet of data sent by a TCP entity is assigned a sequence number. The 32-bit sequence number field contains the sequence number of the first data octet in the segment, known as the sequence number of the segment. The acknowledgement number field (also 32-bit) contains the next sequence number that the sender of the segment is expecting to receive.

There are also six 1-bit control flags in the header: URG (urgent), ACK (acknowledgement), PSH (push), RST (reset), SYN (synchronisation), and FIN (finish). Control bits URG and PSH are concerned with data segments, which we do not describe here. A segment which has the SYN flag set is the first segment sent in a connection and its sequence number is the initial sequence number for the connection. A segment with the FIN flag set indicates that the sender of the segment has no more data to send. The ACK flag indicates that the acknowledgement number field of the segment is valid. When set, the RST flag informs the receiver of the segment to abort the connection.

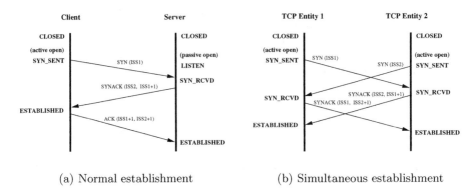

(a) Normal establishment   (b) Simultaneous establishment

Fig. 1. Message sequences for TCP connection establishment

TCP maintains the state of a connection by storing a set of variables in a data structure known as the *transmission control block* (TCB) [32]. The important variables for TCP connection management relate to sequence numbers. They are: send oldest unacknowledged (SND_UNA), send next (SND_NXT), initial send sequence number (ISS) and receive next (RCV_NXT). When TCP transmits a segment, it increments SND_NXT. When TCP accepts a segment, it increases RCV_NXT and sends an acknowledgement. Upon the receipt of an acknowledgement, TCP advances SND_UNA. The amount by which each of the three variables is increased is given by the length of the segment in octets, i.e., the sequence space for both data and control bits SYN and FIN. The SYN and FIN bits each require one sequence number [32]. If the SYN or FIN segment does not carry data, the amount by which each variable is advanced is 1.

Connection Establishment. A connection is initiated by the TCP entity (TCP client) that sends a SYN segment, and is responded to by the peer TCP entity (TCP server). The TCP server receiving a SYN segment has no way of telling whether it is a new SYN to open a connection or an old duplicate SYN from an earlier incarnation. Therefore it must ask the other side (through the exchange of segments) to verify the SYN. This process is illustrated in Fig. 1 (a). TCP states (e.g., CLOSED, SYN_SENT and ESTABLISHED) are written next to the vertical lines. User commands (i.e., active open and passive open) are written in parentheses. The sequence number and the acknowledgement number (when relevant) are included with the segment name. TCP also allows both sides to initiate a connection *simultaneously*, as illustrated in Fig. 1(b).

Connection Release. Orderly release ensures that all data is received before the connection is fully closed. As shown in Fig. 2(a), the procedure comprises an exchange of FIN and ACK segments by each party. Assume the FIN segment has sequence number x and acknowledgement number y. If no data is transferred by TCP entity 1 after the connection is established, then x=ISS1+1 and y=ISS2+1. The ACK has an acknowledgement number x+1, that is, the sequence number of

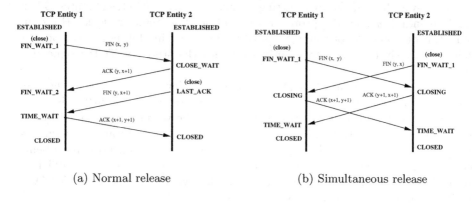

(a) Normal release   (b) Simultaneous release

Fig. 2. Message sequences for TCP connection release

the FIN segment plus 1. The sequence number of the ACK depends on whether data has been sent before the FIN is received. If no data is sent, then the ACK has sequence number y, as shown in the figure.

After initiating orderly release, TCP entity 1 receives the ACK from TCP entity 2 and changes state from FIN_WAIT_1 to FIN_WAIT_2, waiting for a connection release request from TCP entity 2. Meanwhile, data can still be transmitted from TCP entity 2 to 1 (but not vice versa) until the user of TCP entity 2 issues a *close* command. TCP entity 2 then sends a FIN segment. If no data is sent by TCP entity 2 after it receives the FIN from TCP entity 1, then the sequence number of the FIN sent to TCP entity 1 is the same as that of its preceding ACK segment. The acknowledgement number of the FIN is also x+1 as no other segment with a higher sequence number will have been received.

Upon receipt of the FIN segment, TCP entity 1 enters the TIME_WAIT state and responds to the FIN with an ACK segment that has sequence number x+1 and acknowledgement number y+1. When TCP entity 2 receives the ACK from TCP entity 1, it enters CLOSED from LAST_ACK. TCP entity 1 remains in TIME_WAIT for two maximum segment lifetimes (MSL) before entering CLOSED. The MSL is the longest time that a segment can exist in the Internet (about 2 minutes) [32]. Fig. 2(b) illustrates the simultaneous release procedure.

3   Related Work

In their pioneering work on verifying TCP Connection Management, Sunshine and Dalal [40] address the design issues of early versions [4–6] of TCP, which was still evolving at that time. They conduct informal case studies (manual walk-through of the sequences) to investigate the functional behaviour of TCP's connection establishment procedures and their result corroborates that of Tomlinson [42] on the necessity of employing the three-way handshake in connection establishment. In addition, they construct a simplified reachability graph of the three-way handshake by omitting details such as symmetrical states. Their analysis indicates that the protocol is free of deadlocks and livelocks. However, simultaneous opening of connections is not analysed in [40], nor is the orderly release procedure. This significantly simplifies the study of TCP Connection Management.

Based on another early version [31] of TCP, Schwabe [33] specifies the connection establishment protocol using the SPEX language [34]. SPEX is based on a non-deterministic state transition system. The verification is aided by the theorem prover AFFIRM [28]. Schwabe finds that in a simultaneous open scenario, the presence of an old duplicate SYN prevents the connection from being established. However, the simultaneous open Schwabe investigated is different from that in the current TCP specification [32], where TCP sends a SYNACK rather than an ACK upon receiving a SYN.

Kurose and Yemini [20] specify and verify the connection establishment protocol based on another version [30] of TCP. They only consider the client-server situation and specify the protocol using a PASCAL-like language. They formalise two desired properties of the protocol using temporal logic. One property is that if the connection is established, then the send next sequence number (SND_NXT) should equal the receive next sequence number (RCV_NXT). The other property is that eventually the connection will be established. They prove the two properties are correct through deductive reasoning [10].

Lin [21] specifies a connection establishment protocol [40,42] using finite state machines and proves that the connection will be eventually established. Again, only the client-server connection is examined. In addition, the behaviour of the TCP server is not correctly specified as it does not go through LISTEN.

Mehrpour and Karbowiak [24] model and analyse an early and simplified version [38] of TCP using Numerical Petri nets [41]. They model TCP segments by their names without sequence and acknowledgement numbers. Hence, the model is inadequate. It is also incomplete in that no arc inscriptions are given.

Sidhu and Blumer [35] spotted a few problems in the military standard TCP [25], while they were studying the protocol with a view to verifying connection management. According to [25], the military standard TCP is the same as TCP [32] except that it is specified with action statements and tables. The problems they find are syntax errors of the specification and the consequences of the errors are that: (1) TCP can not receive data sent with a SYN segment; (2) TCP can not establish a connection if the first SYN is lost; and (3) TCP can not receive data while in the ESTABLISHED state.

Murphy and Shankar [26,27] specify a transport protocol using a state transition model and invariant and progress assertions. The protocol is similar to TCP only with respect to the connection establishment procedures. They reported some correctness problems of TCP connection establishment when duplicate SYNs are present in the channel. The problems are described under three scenarios. In the first two scenarios, the connection is initiated in a client-server mode. After the TCP client sends a SYN, an old duplicate SYN arrives and the connection finishes with one TCP entity being in CLOSED and the other in

LISTEN. So the old duplicate SYN prevents the connection from being established by the desired SYN. A third scenario shows that in a connection where each TCP entity is in LISTEN, and when two old duplicate SYNs arrive at each end, each TCP entity sends ACKs indefinitely, resulting in a livelock.

Smith [36,37] specifies TCP Connection Management using the *general timed automaton* [22]. He demonstrates that the *quiet time* of two maximum segment lifetime (MSL) given in [32] is insufficient to prevent old duplicate data from being delivered, which was previously pointed out by Murphy and Shankar [27]. Smith uses a phase-based approach to specify the protocol, which is different from the way TCP is specified in [32]. This makes it difficult to validate this model against the official TCP specification. As well as excluding simultaneous opening of connections and user aborts, Smith's specification does not address the following details that are part of TCP's functional behaviour and can have an impact on its logical correctness: (1) the release of a connection in any state including SYN_RCVD, (2) entering TIME_WAIT from FIN_WAIT_1 upon receiving a FINACK segment, as specified in RFC 1122 [3], and (3) state variable SND_UNA (send oldest unacknowledged number) that is used to check whether or not an ACK is a duplicate segment.

In [12], we provided a quite simple model of TCP Connection Management based on the FSM diagram [32]. The model does not include protocol details such as sequence numbers and state variables. A more detailed model enhanced with these features is given in [2, 13], which is structured according to a *state-based* approach. Using an *event processing* approach, we re-structured the model to comply with the way TCP is specified in [32] and also incorporated the retransmission mechanism and lossy channel [14]. The contribution of this paper is two-fold: (1) investigating TCP Connection Management's correctness when including retransmissions; and (2) providing some insights into TCP's operation by discussing a deadlock that can occur, even when the channel is lossless.

4   TCP Connection Management CPN

This section briefly describes our TCP Connection Management model. The full model and its description is provided in [11]. The basic idea of modelling TCP is that we consider two peer TCP entities, communicating over the Internet Protocol (IP) as well as interacting with their application processes. Figure 3 shows the TCP_Overview page. Places User_1 and User_2 model TCP user commands, such as active (A_Open) and passive (P_Open) open. The TCB places model TCP states and the transmission control block state variables. Places H1_H2 and H2_H1 model TCP buffers and all network storage (e.g., router buffers). H1_H2 indicates the data flow direction is from host 1 to host 2, whereas H2_H1 indicates data flow in the opposite direction. Transitions Lossy_Channel1 and Lossy_Channel2 can be switched on and off by their guards to model lossy and non-lossy channels respectively. The TCP entities are modelled by two substitution transitions: TCP'1 and TCP'2.

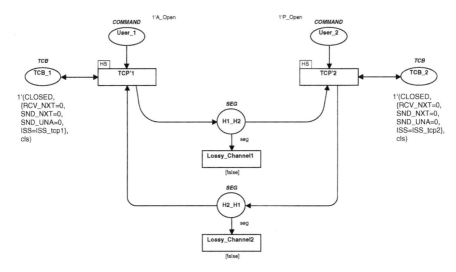

Fig. 3. Top level CPN page: TCP_Overview

The CPN model contains another 18 CPN pages and a declarations page, as shown in Fig. 4. The TCP_Overview page is at the first level of the hierarchy. The second level contains the Event_Processing page, which is known as a page instance for TCP'1 and TCP'2. It models TCP's responses to user commands, segment arrivals and retransmissions. The Segment_Processing page models the processing of segments for each of TCP's 11 states.

In modelling TCP Connection Management, we consider that security and precedence are always met and only consider a single connection between users, which allows us to omit port numbers. A segment in the CPN model contains a sequence number, an acknowledgement number and four control bits: SYN, ACK, FIN and RST.

In contrast to data transfer, TCP Connection Management only consumes a small portion of the sequence number space. If we omit data transfer and choose a small value for the initial sequence number for each TCP entity, sequence numbers will not wrap. In this case, we don't need to implement modulo arithmetic. Given that connection management segments are always small (when there is no data), it is reasonable to assume that the receive window is always big enough to accept incoming segments. We can thus omit modelling the window field in segments and implementing checks associated with window size. This simplifies the model, but means that our results may not be applicable when sequence numbers do wrap. We also assume that segments can be lost, delayed, and reordered while traversing the network.

The official TCP specification [3, 32] is incomplete in terms of specifying the retransmission mechanism for the connection management protocol. In particular, it does not provide an adequate description of which type of segments need to be retransmitted. It only mentions the retransmission of data segments on page 10 of RFC 793 [32] and the retransmission of SYNs on page 95 of RFC 1122 [3].

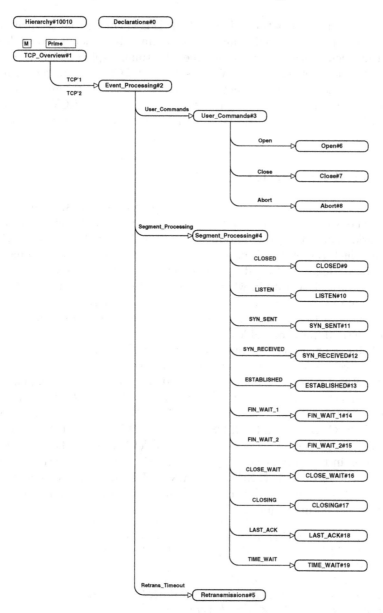

Fig. 4. The Hierarchy page of TCP Connection Management CPN

According to Wright and Stevens [44], TCP retransmits a SYN, SYNACK and FIN to recover from loss in the channel. It is also not clear from [3, 32] when TCP sets and turns off timers. From our reading of RFC 793 (pages 10 and 77) and our understanding of the protocol, we believe that retransmissions related to connection management occur as follows. When sending a segment (SYN, SYNACK or FIN), the TCP entity puts a copy of the segment into its retransmission queue and starts a timer. If an acknowledgement for that segment arrives in time, the copy is removed from the queue and the timer is stopped. If the acknowledgement does not arrive in time, the timer expires, TCP retransmits the segment and resets the timer. We consider that TCP sets four retransmission timers for connection management: (1) when it sends a SYN and enters state SYN_SENT; (2) when it sends a SYNACK and enters SYN_RCVD; (3) when it sends a FIN and enters FIN_WAIT_1; and (4) when it sends a FIN and enters LAST_ACK.

5 Desired Properties

To define the termination and absence of deadlocks properties, we firstly introduce some notation. Let 'empty' represent an empty multiset. Let M_0 denote the initial marking and $M(p)$ the marking of place p. A desired terminal marking is a state in which the protocol terminates correctly under normal circumstances. We use M_{dt} to denote a desired terminal marking in Definition 3. An acceptable terminal marking is a state in which the protocol terminates successfully under special situations, as explained later. We use M_{at} to denote an acceptable terminal marking in Definition 4. For TCP, a deadlock is a terminal marking that is neither desired nor acceptable.

We define four projection functions that project the transmission control block, effectively a 6 tuple, to a TCP state and three state variables. The projection functions are used to define the desired and acceptable terminal markings of TCP Connection Management. In Definition 1, the set $STATE$ is the set of TCP states: CLOSED, LISTEN, SYN_SENT, SYN_RCVD, EST, FIN_WAIT_1, FIN_WAIT_2, CLOSE_WAIT, CLOSING, LAST_ACK and TIME_WAIT. The set $LISTENstat$ contains two values, 'lis' and 'cls', which indicate a TCP entity has been in LISTEN or not respectively. This is used to determine the next state TCP enters from SYN_SENT or SYN_RCVD upon receiving a RST segment. The set $\mathbb{N}^4$ is the set of values of the four TCP state variables, where each component is a natural number.

Definition 1. *The projection functions are given by*
  $State : STATE \times \mathbb{N}^4 \times LISTENstat \to STATE$
  where $State(s, sv, i) = s$
  $SndNxt : STATE \times \mathbb{N}^4 \times LISTENstat \to \mathbb{N}$
  where $SndNxt(s, (RCV\_NXT, SND\_NXT, SND\_UNA, ISS), i) = SND\_NXT$
  $SndUna : STATE \times \mathbb{N}^4 \times LISTENstat \to \mathbb{N}$
  where $SndUna(s, (RCV\_NXT, SND\_NXT, SND\_UNA, ISS), i) = SND\_UNA$
  $RcvNxt : STATE \times \mathbb{N}^4 \times LISTENstat \to \mathbb{N}$
  where $RcvNxt(s, (RCV\_NXT, SND\_NXT, SND\_UNA, ISS), i) = RCV\_NXT$

Each projection function takes a tuple as its argument and hence can not be applied directly to the marking of place TCB, which is a singleton multiset. To solve this, we need a function that converts a singleton multiset into its basis element, as defined below.

Definition 2. *Let S_{MS_1} be the set of all singleton multisets over a basis set S:* $S_{MS_1} = \{\{(s^c 1)\} \mid s \in S\}$. *A function that converts a singleton multiset to its basis element is given by* $f_c : S_{MS_1} \to S$, *where* $f_c(\{(s^c 1)\}) = s$.

The termination property is concerned with whether a protocol can terminate in a desired state. For TCP Connection Management, this means that a connection should be able to be established, released and aborted successfully. We formulate the termination property as follows.

Property 1. *The termination property for TCP Connection Management comprises:*
*(I) **Successful Establishment** — If one host requests that a connection be opened and the other host is willing to accept the connection or to request that the same connection be opened at the same time, then the connection should be established.*
*(II) **Proper Establishment** — When a connection has been established at both ends, the variable SND_NXT must equal SND_UNA at the same end of the connection, which must also equal RCV_NXT at the other end of the connection.*
*(III) **Successful Release** — If one host requests that the connection be closed and the other host is willing to close the connection or also requests that the connection be closed at the same time, then the connection should be able to be released.*
*(IV) **Successful Abort** — A connection should be able to be aborted in the case that either or both users issue an abort command.*

The successful and proper establishment properties can be verified by examining the terminal markings of the state space of TCP connection establishment. We define the desired terminal state for connection establishment as follows.

Definition 3. *The desired terminal state, M_{dt}, for connection establishment is a marking where*

$$M_{dt}(H1\_H2) = empty \tag{1}$$
$$M_{dt}(H2\_H1) = empty \tag{2}$$
$$State(f_c(M_{dt}(TCB\_1))) = EST \tag{3}$$
$$State(f_c(M_{dt}(TCB\_2))) = EST \tag{4}$$
$$SndNxt(f_c(M_{dt}(TCB\_1))) = SndUna(f_c(M_{dt}(TCB\_1)))$$
$$= RcvNxt(f_c(M_{dt}(TCB\_2))) \tag{5}$$
$$SndNxt(f_c(M_{dt}(TCB\_2))) = SndUna(f_c(M_{dt}(TCB\_2)))$$
$$= RcvNxt(f_c(M_{dt}(TCB\_1))) \tag{6}$$

The desired terminal state for connection establishment has no messages left in the channel and both TCP entities in state ESTABLISHED with state variables related as described in (II) of Property 1. The user places are not included in the definition as their purpose is to allow different scenarios to be analysed using an incremental methodology [2]. Thus the details of the markings of the user places are not important for termination.

Below we formalise two acceptable terminal states for connection establishment.

Definition 4. *An acceptable terminal state, M_{at}, for connection establishment satisfies:*

$$M_{at}(H1\_H2) = empty \tag{7}$$

$$M_{at}(H2\_H1) = empty \tag{8}$$

$$M_{at}(TCB\_1) = 1`(CLOSED, (0,0,0,0), cls) \tag{9}$$

$$M_{at}(TCB\_2) = 1`(LISTEN, (0,0,0,ISS), lis) \ or \ 1`(CLOSED, (0,0,0,0), cls) \tag{10}$$

Equation 9 indicates that when a TCP entity enters CLOSED, its state variables are reset to 0, implying that the TCB for that connection is deleted. The LISTEN flag is reset to the initial value 'cls'. Equation 10 indicates that when a TCP entity returns to LISTEN, its state variables are reset to 0 except for the initial send sequence number, which is set for the next connection instance. The LISTEN flag has a value of 'lis', indicating it has been in LISTEN.

In client-server connection establishment, the protocol can also terminate in a state where the TCP client is in CLOSED and the TCP server is in LISTEN. This is acceptable because when the TCP server is down and therefore in CLOSED and receives a SYN, it sends a RST which changes the state of the TCP client from SYN_SENT to CLOSED. The TCP server then enters LISTEN after it has come up on receiving a passive open.

Another acceptable state where both TCP entities are CLOSED can happen in both client-server and simultaneous connection establishment. In client-server connection establishment, it can result when the TCP server retransmits the SYNACK after sending a SYNACK in response to a SYN, and then aborts the connection (due to the maximum number of retransmissions being reached). On aborting the connection, the TCP server sends a RST and enters CLOSED. The RST can terminate the connection at the client side. In simultaneous connection establishment, this acceptable state can result from when one TCP entity opens the connection while the other TCP entity is CLOSED.

The desired terminal state for connection release and abort is given in Definition 5.

Definition 5. *A desired terminal state for connection release and abort is a marking where*

$$M_{dt}(H1\_H2) = empty \tag{11}$$

$$M_{dt}(H2\_H1) = empty \tag{12}$$

$$M_{dt}(TCB\_1) = 1`(CLOSED, (0,0,0,0), cls) \tag{13}$$

$$M_{dt}(TCB\_2) = 1`(CLOSED, (0,0,0,0), cls) \tag{14}$$

TCP should be able to terminate in a desired or acceptable terminal marking, and be absent from deadlocks.

6   Analysis Approach

The analysis is undertaken for two models for a reordering but lossless channel: Model 1 (without retransmissions) and Model 2 (with retransmissions). To ensure the detection of errors in each phase of the protocol, we take an incremental approach [2] which considers different connection scenarios. Each scenario is defined by configuring, for each model, the initial markings of places User_1 and User_2 and place TCB. As shown in Table 1, each model has 11 configurations. Configurations A and B are used to analyse the client-server and simultaneous connection establishment procedures respectively. Configurations C, D and E examine the connection release procedures. The other configurations investigate

Table 1. The initial marking for each configuration

Config.	Initial Marking
A	$M_0(\text{User\_1}) = 1\text{'A\_Open}$ $M_0(\text{User\_2}) = 1\text{'P\_Open}$ $M_0(\text{TCB\_1}) = 1\text{'(CLOSED}, (0,0,0,10), \text{cls})$ $M_0(\text{TCB\_2}) = 1\text{'(CLOSED}, (0,0,0,20), \text{cls})$
B	$M_0(\text{User\_1}) = 1\text{'A\_Open}$ $M_0(\text{User\_2}) = 1\text{'A\_Open}$ $M_0(\text{TCB\_1}) = 1\text{'(CLOSED}, (0,0,0,10), \text{cls})$ $M_0(\text{TCB\_2}) = 1\text{'(CLOSED}, (0,0,0,20), \text{cls})$
C	$M_0(\text{User\_1}) = 1\text{'Close}$ $M_0(\text{User\_2}) = 1\text{'Close}$ $M_0(\text{TCB\_1}) = 1\text{'(EST}, (21,11,11,10), \text{cls})$ $M_0(\text{TCB\_2}) = 1\text{'(EST}, (11,21,21,20), \text{cls})$
D	$M_0(\text{User\_1}) = 1\text{'A\_Open} + +1\text{'Close}$ $M_0(\text{User\_2}) = 1\text{'P\_Open} + +1\text{'Close}$ $M_0(\text{TCB\_1}) = 1\text{'(CLOSED}, (0,0,0,10), \text{cls})$ $M_0(\text{TCB\_2}) = 1\text{'(CLOSED}, (0,0,0,20), \text{cls})$
E	$M_0(\text{User\_1}) = 1\text{'A\_Open} + +1\text{'Close}$ $M_0(\text{User\_2}) = 1\text{'A\_Open} + +1\text{'Close}$ $M_0(\text{TCB\_1}) = 1\text{'(CLOSED}, (0,0,0,10), \text{cls})$ $M_0(\text{TCB\_2}) = 1\text{'(CLOSED}, (0,0,0,20), \text{cls})$
F	$M_0(\text{User\_1}) = 1\text{'A\_Open} + +1\text{'Abort}$ $M_0(\text{User\_2}) = 1\text{'P\_Open} + +1\text{'Abort}$ $M_0(\text{TCB\_1}) = 1\text{'(CLOSED}, (0,0,0,10), \text{cls})$ $M_0(\text{TCB\_2}) = 1\text{'(CLOSED}, (0,0,0,20), \text{cls})$
G	$M_0(\text{User\_1}) = 1\text{'A\_Open} + +1\text{'Abort}$ $M_0(\text{User\_2}) = 1\text{'A\_Open} + +1\text{'Abort}$ $M_0(\text{TCB\_1}) = 1\text{'(CLOSED}, (0,0,0,10), \text{cls})$ $M_0(\text{TCB\_2}) = 1\text{'(CLOSED}, (0,0,0,20), \text{cls})$
H	$M_0(\text{User\_1}) = 1\text{'Close} + +1\text{'Abort}$ $M_0(\text{User\_2}) = 1\text{'Close} + +1\text{'Abort}$ $M_0(\text{TCB\_1}) = 1\text{'(EST}, (21,11,11,10), \text{cls})$ $M_0(\text{TCB\_2}) = 1\text{'(EST}, (11,21,21,20), \text{cls})$
I	$M_0(\text{User\_1}) = 1\text{'A\_Open} + +1\text{'Close}$ $M_0(\text{User\_2}) = 1\text{'P\_Open} + +1\text{'Abort}$ $M_0(\text{TCB\_1}) = 1\text{'(CLOSED}, (0,0,0,10), \text{cls})$ $M_0(\text{TCB\_2}) = 1\text{'(CLOSED}, (0,0,0,20), \text{cls})$
J	$M_0(\text{User\_1}) = 1\text{'A\_Open} + +1\text{'Abort}$ $M_0(\text{User\_2}) = 1\text{'P\_Open} + +1\text{'Close}$ $M_0(\text{TCB\_1}) = 1\text{'(CLOSED}, (0,0,0,10), \text{cls})$ $M_0(\text{TCB\_2}) = 1\text{'(CLOSED}, (0,0,0,20), \text{cls})$
K	$M_0(\text{User\_1}) = 1\text{'A\_Open} + +1\text{'Close}$ $M_0(\text{User\_2}) = 1\text{'A\_Open} + +1\text{'Abort}$ $M_0(\text{TCB\_1}) = 1\text{'(CLOSED}, (0,0,0,10), \text{cls})$ $M_0(\text{TCB\_2}) = 1\text{'(CLOSED}, (0,0,0,20), \text{cls})$

the connection management procedures involving aborts. The markings of places H1_H2 and H2_H1 are 'empty' for each configuration and are not included in the table. The initial send sequence number (ISS) is set to the same value for each configuration, that is, for one side of the connection ISS is 10 and for the other side ISS is 20. A_Open and P_Open represent active open and passive open commands respectively.

As part of an incremental approach, we consider Configuration C that models the releasing of a connection after it has been established. In Configuration C, each user place is initialised to have a Close token, with both TCP entities in their established states and with their state variables satisfying the *proper establishment* property (i.e., item (II) of Property 1). In addition to releasing a connection after it has been established, TCP allows the user to close a connection even before it has been established, for example, closing from the SYN_RCVD state. Configuration D models the situation where closing a connection can occur as soon as either side makes a connection request, that is, after the client user issues an active open or after the server user issues a passive open. Configuration E models closing a connection once a simultaneous open is attempted at either side. Configurations F – K consider various connection scenarios involving user aborts.

7   Termination and Absence of Deadlocks (Model 1)

The CPN model is analysed using Design/CPN [43] on a machine with an Intel Pentium 2.6GHz CPU and 1GB RAM. Table 2 shows the statistics of the state space for each configuration of Model 1, which include calculation time (in seconds), the number of markings ($|V|$), arcs ($|A|$), terminal markings (TMs), and deadlocks (DLs). Except for Configurations D and E, each configuration analysed satisfies the termination property and is free from deadlocks. Configuration D contains one deadlock, which is explained in Section 7.1. Configuration E contains two deadlocks which are similar to that of Configuration D.

7.1   Failure to Terminate the Connection

A path leading to the deadlock of Configuration D is shown in Fig. 5 with its corresponding time sequence diagram in Fig. 6. The scenario begins with a

Table 2. State space statistics of the TCP CM configurations (Model 1)

Config.	Time	$|V|$	$|A|$	TMs	DLs
1-A	0	11	12	2	0
1-B	0	42	60	2	0
1-C	0	57	92	1	0
1-D	0	225	455	3	1
1-E	3	2850	8260	6	2
1-F	0	51	91	3	0
1-G	1	355	870	4	0
1-H	1	356	792	13	0
1-I	0	79	141	2	0
1-J	0	73	129	3	0
1-K	1	742	1896	4	0

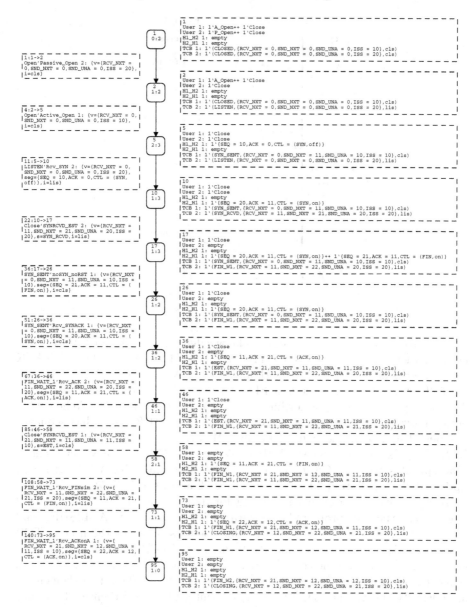

Fig. 5. A path in the occurrence graph for Configuration D (Model 1)

passive open from TCP entity 2 and then TCP entity 1 sends a SYN segment
with sequence number 10. TCP entity 2 responds with a SYNACK which has
sequence number 20 and acknowledgement number 11. It then sets SND_UNA
to 20 and enters the SYN_RCVD state. Now the user of TCP entity 2 issues
a close command, which results in a FIN segment being sent with sequence
number 21 and acknowledgment number 11. Note SND_UNA is still 20, since

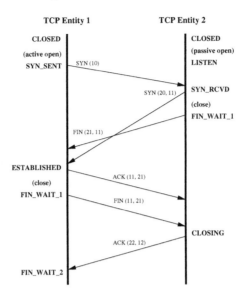

Fig. 6. Connection release fails

TCP entity 2 has yet to receive an acknowledgement for its SYNACK. TCP entity 2 then changes state from SYN_RCVD to FIN_WAIT_1. The SYNACK sent earlier is delayed in the network and is overtaken by the FIN and received by TCP entity 1. Because it is a FIN segment received in state SYN_SENT, TCP entity 1 ignores it, according to page 75 of RFC 793 [32]. When TCP entity 1 receives the delayed SYNACK, it sends an ACK which has sequence number 11 and acknowledgement number 21 (acknowledging the receipt of the SYNACK) and goes to ESTABLISHED. When the ACK is received, TCP entity 2 accepts it since its ACK number (21) is greater than SND_UNA (20), and updates SND_UNA from 20 to 21 (see pages 72 and 73 of RFC 793).

If TCP entity 1 (a client) is waiting for data to be sent by TCP entity 2 (the server), then the protocol hangs with TCP entity 1 in ESTABLISHED and TCP entity 2 in FIN_WAIT_1, and no data being sent (since the user of TCP entity 2 has issued a *close* command). If the user of TCP entity 1 issues a close command to terminate the connection, a FIN is sent to TCP entity 2, which has sequence number 11 and acknowledgement number 21. TCP entity 1 then enters state FIN_WAIT_1. Since TCP entity 2 has sent a FIN to TCP entity 1 and has yet to receive an acknowledgment of its FIN, it treats this FIN from TCP entity 1 as a simultaneous release request. TCP entity 2 sends an acknowledgement for it, which has sequence number 22 and acknowledgement number 12, and enters CLOSING from FIN_WAIT_1. When receiving this ACK, TCP entity 1 changes state from FIN_WAIT_1 to FIN_WAIT_2, and is waiting for TCP entity 2 to send a FIN to complete the graceful release. TCP entity 1 is oblivious to the FIN it received while it was in SYN_SENT. On the other hand, TCP entity 2 is expecting TCP entity 1 to send an acknowledgement to its first FIN, which will never come. So the protocol is deadlocked.

7.2   Resolution of the Deadlock Using Retransmission

In this section, we discuss the impact of TCP's retransmission mechanism on removing this deadlock. Referring to Fig. 6, when TCP sends a FIN in state FIN_WAIT_1, it sets a retransmission timer for the FIN. The timer keeps running until it receives an acknowledgment to its FIN or until it expires. We firstly consider the situation where the timer expires while TCP entity 2 is in FIN_WAIT_1. When the timer expires, the FIN is retransmitted. We consider two important cases: (1) when all the retransmitted FINs arrive before the SYNACK; and (2) when one of the retransmitted FINs arrives after the SYNACK. If the FIN arrives before the SYNACK, then TCP entity 1 will discard it the same as before. This can happen a number of times until TCP entity 2 reaches its maximum number of retransmissions, if the SYNACK is greatly delayed. If TCP entity 2 reaches its maximum number of retransmissions, then TCP aborts the connection. That is, it sends a RST and enters CLOSED. On arrival the RST changes the state of TCP entity 1 to CLOSED. On the other hand, if one of the retransmitted FINs arrives after the SYNACK, then TCP entity 1 can close the connection in the usual way. Hence the deadlock problem will not occur in the situation where the timer expires in FIN_WAIT_1.

If the timer is still running when TCP entity 2 receives a FIN and enters CLOSING (see Fig. 6), then the deadlock can also be removed. Consider that TCP entity 2 retransmits a FIN in state CLOSING, which has sequence number 21 and acknowledgement number 12. The FIN will be received either before TCP entity 1 receives ACK(22,12) or after it receives ACK(22,12). In the first case, TCP entity 1 enters TIME_WAIT due to receiving this FIN (which is a FINACK) and sends an ACK. When TCP entity 2 receives this ACK, it enters TIME_WAIT. Segment ACK(22,12) will be discarded by TCP entity 1 as a duplicate ACK. Both TCP entities will then enter CLOSED from TIME_WAIT. In the second case, TCP entity 1 enters TIME_WAIT after receiving the retransmitted FIN and sends an ACK. When TCP entity 2 receives this ACK, it enters TIME_WAIT. The connection will then be CLOSED at both ends after the 2MSL timers expire. Hence the deadlock will not occur in this case also.

To conclude, FIN retransmission can remove the deadlock. In the case where out of order segments are buffered and reordered in the receiver before delivery, TCP retransmissions are required to recover from loss. However, in this case, the medium does not lose segments, so one would not expect retransmission to be necessary. The conclusion we draw from this informal analysis is corroborated by the analysis in Section 8.

8   Termination and Absence of Deadlocks (Model 2)

Due to space limitations, we only present the state space statistics of Configurations A – H of Model 2 (see Table 3). Results for Configurations I – K can be found in [11]. In the Configuration column of Table 3, a case is represented by three elements: (1) the model number; (2) the configuration class; and (3) a four tuple of natural numbers. The four tuple represents the maximum number of retransmissions set for a SYN, SYNACK, FIN retransmitted in FIN_WAIT_1 and/or CLOSING, and FIN retransmitted in LAST_ACK. For example, in case 2-D-(0,1,0,1), the tuple indicates that the maximum number of retransmissions for SYN is 0, for SYNACK is 1, for FIN retransmitted in FIN_WAIT_1 and/or CLOSING is 0, and for FIN retransmitted in LAST_ACK is 1. The maximum number of retransmissions for each segment is identical for each end.

Except for Configurations D and E where each maximum number of retransmissions is limited to 1, the other configurations have the maximum number of retransmissions limited to 2. Since Configurations A and B are only concerned with connection establishment, the last two numbers of the tuple are 0 since there are no FINs retransmitted. This rule also applies to Configurations F and G, where no close command is issued. Configuration C is concerned with connection release from the ESTABLISHED state, hence the first two numbers of the tuple concerning the retransmissions of the SYN and SYNACK are always 0. This also applies to Configuration H. For Configurations I – K, the last number in the tuple is 0. This is because neither of the TCP entities go through state LAST_ACK (there is one close token in one of the user places), so no FINs are retransmitted in LAST_ACK.

We limit the maximum computation time to 6 hours for configurations other than D and E, and to 24 hours for D and E. If a case takes more than these limits to compute, we classify it as having a large state space requiring advanced state space techniques, such as the sweep-line method [7], which we would like to apply in the future.

Except for Configurations D and E, each case analysed satisfies the termination property and contains no deadlocks. Referring to Table 1, Configuration D has an A_Open token and a Close token in place User_1, and a P_Open and a Close token in place User_2. If the first element in the four-tuple of Configuration D is set to 1 (which means that a SYN can be retransmitted once by user 1), then the deadlock discovered in Model 1 can still exist. This is because the SYN can be delayed in the network and arrive at TCP entity 2 when it is in CLOSING. TCP entity 2 drops the SYN and sends an ACK in response due to its unacceptable sequence number, and TCP entity 1, in FIN_WAIT_2, accepts the ACK and remains in the same state. Hence there are two deadlocks, one has the SYN retransmitted and the other does not. If the second element in the four-tuple of Configuration D is set to 1 (which means that a SYNACK can be retransmitted once by user 2), then the deadlock can exist for the same reason. If the third element of the tuple is set to 1 (which means that a FIN can be retransmitted in FIN_WAIT_1 or CLOSING), then the number of deadlocks is 0. This corroborates the discussion in Section 7.2. Since the scenarios leading to the deadlock do not involve a TCP entity being in LAST_ACK, the retransmission of FIN in state LAST_ACK does not affect the deadlock.

Cases 2-D-(1,1,0,0) and 2-D-(1,1,0,1) each have both the first and second elements in the tuple set to 1, so there exist four combinations of the value of the retransmission counter for SYN and SYNACK, and hence four deadlocks.

Table 3. State space statistics of Configuration A – H of Model 2

Configuration	hh:mm:ss	|V|	|A|	TMs	DLs
2-A-(0,1,0,0)	00:00:00	41	60	4	0
2-A-(1,0,0,0)	00:00:00	99	176	6	0
2-A-(1,1,0,0)	00:00:00	438	1006	12	0
2-A-(0,2,0,0)	00:00:00	78	133	5	0
2-A-(2,0,0,0)	00:00:00	279	575	9	0
2-A-(1,2,0,0)	00:00:01	823	2012	16	0
2-A-(2,1,0,0)	00:00:01	1480	4067	18	0
2-A-(2,2,0,0)	00:00:03	2880	8388	24	0
2-B-(0,1,0,0)	00:00:01	1191	3334	7	0
2-B-(1,0,0,0)	00:00:02	1465	3938	8	0
2-B-(1,1,0,0)	00:01:01	29254	112180	28	0
2-B-(0,2,0,0)	00:00:06	5083	16320	14	0
2-B-(2,0,0,0)	00:00:12	8837	27614	18	0
2-B-(1,2,0,0)	00:11:59	116420	489726	56	0
2-B-(2,1,0,0)	03:40:24	247977	1123602	63	0
2-C-(0,0,0,1)	00:00:00	311	744	3	0
2-C-(0,0,1,0)	00:00:06	5138	16294	11	0
2-C-(0,0,1,1)	00:00:09	6742	21384	15	0
2-C-(0,0,0,2)	00:00:00	691	1802	5	0
2-C-(0,0,2,0)	00:06:13	73707	282914	38	0
2-C-(0,0,1,2)	00:00:13	9394	30122	19	0
2-C-(0,0,2,1)	00:07:03	78815	301236	44	0
2-C-(0,0,2,2)	00:09:25	87291	332366	50	0
2-D-(0,0,0,1)	00:00:01	1309	3899	5	1
2-D-(0,1,0,0)	00:00:01	1362	3498	7	2
2-D-(1,0,0,0)	00:00:02	1810	4810	6	2
2-D-(0,1,0,1)	00:00:11	8743	31122	11	2
2-D-(0,0,1,0)	00:00:15	10156	34825	11	0
2-D-(1,1,0,0)	00:00:15	10381	33056	16	4
2-D-(1,0,0,1)	00:00:16	10481	38071	10	2
2-D-(0,0,1,1)	00:00:28	16612	59184	15	0
2-D-(0,1,1,0)	00:04:26	64871	258399	24	0
2-D-(1,1,0,1)	00:05:07	65381	273981	24	4
2-D-(1,0,1,0)	00:07:16	77940	317337	24	0
2-D-(0,1,1,1)	00:14:34	104046	426872	32	0
2-D-(1,0,1,1)	00:48:35	126098	530381	32	0
2-E-(0,0,0,1)	00:00:48	19354	75158	8	2
2-E-(0,1,0,0)	00:06:26	45293	163984	21	8
2-E-(1,0,0,0)	00:22:50	98627	392610	24	8
2-E-(0,0,1,0)	00:25:25	122654	516530	12	0
2-E-(0,1,0,1)	09:10:37	328023	1524604	29	8
2-F-(0,1,0,0)	00:00:00	202	420	7	0
2-F-(1,0,0,0)	00:00:00	371	890	6	0
2-F-(1,1,0,0)	00:00:02	1611	4401	14	0
2-F-(0,2,0,0)	00:00:00	479	1098	10	0
2-F-(2,0,0,0)	00:00:01	1236	3461	9	0
2-F-(1,2,0,0)	00:00:04	3792	11162	20	0
2-F-(2,1,0,0)	00:00:07	5978	19196	21	0
2-F-(2,2,0,0)	00:00:23	14563	50327	30	0
2-G-(0,1,0,0)	00:00:05	4524	13810	15	0
2-G-(1,0,0,0)	00:00:12	8495	29146	16	0
2-G-(1,1,0,0)	00:10:59	103670	422326	60	0
2-G-(0,2,0,0)	00:00:35	21612	73264	32	0
2-G-(2,0,0,0)	00:04:47	66355	271050	36	0
2-H-(0,0,0,1)	00:00:01	1078	2698	21	0
2-H-(0,0,1,0)	00:00:21	13544	44478	53	0
2-H-(0,0,1,1)	00:00:30	17874	58326	69	0
2-H-(0,0,0,2)	00:00:02	2278	6072	29	0
2-H-(0,0,1,2)	00:00:49	25740	84568	85	0
2-H-(0,0,2,0)	02:09:17	188793	743234	167	0
2-H-(0,0,2,1)	02:34:09	203217	795326	191	0
2-H-(0,0,2,2)	03:36:12	229587	893332	215	0

Cases 2-D-(1,1,1,0) and 2-D-(1,1,1,1) could not be generated in 24 hours, but we conjecture that they will not have deadlocks given that the third element of the tuple is 1.

Also referring to Table 1, Configuration E has an A_Open token and a Close token in place User_1, and an A_Open and a Close token in place User_2. If the first element in the four-tuple of a case of Configuration E is set to 1 (which means that a SYN can be retransmitted by either end), then there are four combinations of the retransmission values. Because Configuration E of Model 1 has two deadlocks, we expect these cases of Configuration E of Model 2 to have eight deadlocks, differing only in the marking of the Retransmission Counter place for each TCP entity. Similar results hold for the cases where the second element in the four-tuple is set to 1, which means that a SYNACK can be retransmitted by either end. If the third element of the tuple is set to 1, then the number of deadlocks is 0, again corroborating the discussion in Section 7.2. Finally, the number of the fourth element indicates the maximum number of FINs retransmitted in state LAST_ACK, which does not affect the deadlock.

9   Conclusions

Although TCP has been used in the Internet for over twenty years, its protocol mechanisms are still not fully understood, due to its complexity. This is demonstrated in [29] where some 60 pages of errors were reported in TCP implementations in 1999. Moreover, TCP is being used as the basis for the development of new Internet protocols [19, 39]. Thus it is important to have a more fundamental understanding of TCP's operations. To address this situation we have developed a detailed CPN model of TCP's Connection Management procedures with a view to verifying their correctness. We have also provided insights into TCP's operation through in-depth functional analysis.

In this paper, we have investigated the termination properties of TCP Connection Management. We define the notions of desired and acceptable terminal states, and then check that these are the only possible dead markings that the TCP CPN model can enter. We check these properties for 11 different connection management scenarios using state spaces. Our analysis indicates that, over a reordering channel without loss, TCP terminates correctly for client-server and simultaneous connection establishment, orderly release after the connection is established and aborting of connections. We have discovered, however, that termination problems can occur when the user releases the connection while the TCP entity is in the SYN_RCVD state, i.e. before the connection has been fully set up. In this case TCP can deadlock: the client can be waiting for data from the server, while the server is waiting for an acknowledgement of its FIN that never comes. Further, if the client decides to close, TCP also deadlocks where the client waits for the server to close its half connection, while the server is still waiting for an acknowledgement of its FIN.

Loss of TCP segments can occur either due to routers dropping packets or due to the detection of transmission errors. When there is no loss and out of order segments are buffered and reordered in the receiver before delivery, retransmissions should not be needed. However, we have also shown that retransmission of the FIN in state FIN_WAIT_1 or CLOSING removes these deadlocks, for the cases investigated. It is therefore recommended that TCP implementations always include a retransmission timer for FIN segments, even over media that do not lose or corrupt packets and have negligible transmission errors.

References

1. M. Allman, V. Paxson, and W. Stevens. TCP Congestion Control. Request for Comments 2581, IETF, April 1999.
2. J. Billington, G.Gallasch, and B. Han. A Coloured Petri Net Approach to Protocol Verification. In *Lectures on Concurrency and Petri Nets: Advances in Petri Nets*, volume 3098 of *Lecture Notes in Computer Science*, pages 210–290. Springer-Verlag, 2004.
3. R. Braden. Requirements for Internet Host — Communication Layers. RFC 1122, IETF, October 1989.
4. V. G. Cerf. Specification of TCP Internet Transmission Control Program, TCP (Version 2), March 1977. available from DARPA/IPTO.
5. V. G. Cerf, Y. K. Dalal, and C. A. Sunshine. Specification of Internet Transmission Control Program. INWG Note 72, December 1974.
6. V. G. Cerf and J. B. Postel. Specification of Internet Transmission Control Program, TCP (Version 3), January 1978. available from USC/ISI.
7. S. Christensen, L. M. Kristensen, and T. Mailund. A Sweep-line Method for State Space Exploration. In *Lecture Notes in Computer Science*, volume 2031, pages 450–464, Berlin, Germany, 2001. Springer-Verlag.
8. S. Floyd. HighSpeed TCP for Large Congestion Windows. RFC 3649, IETF, 2003.
9. S. Floyd and T. Henderson. The NewReno Modification to TCP's Fast Recovery Algorithm. RFC 2582, IETF, April 1999.
10. B. Hailpern. Verifying Concurrent Processes Using Temporal Logic. Technical Report 195, Computer Systems Lab, Stanford University, U.S.A, 1980.
11. B. Han. Formal Specification of the TCP Service and Verification of TCP Connection Management. PhD Thesis, University of South Australia, Australia, December 2004.
12. B. Han and J. Billington. An Analysis of TCP Connection Management Using Coloured Petri nets. In *Proceedings of the 5th World Multi-Conference on Systemics, Cybernetics and Informatics (SCI'2001)*, pages 590–595, Orlando, Florida, July 2001.
13. B. Han and J. Billington. Validating TCP Connection Management. In *Proceedings of the Workshop on Software Engineering and Formal Methods, Adelaide, Australia*, volume 12 of *Conferences in Research and Practice in Information Technology*, pages 47–55, June 2002.
14. B. Han and J. Billington. Experience using Coloured Petri Nets to Model TCP's Connection Management Procedures. In *Proceedings of the 5th Workshop and Tutorial on Practical Use of Coloured Petri Nets and the CPN tools, Aarhus, Denmark*, pages 57–76, October 2004.
15. V. Jacobson and R. Braden. TCP Extensions for Long-Delay Paths. RFC 1072, IETF, October 1988.

16. V. Jacobson, R. Braden, and D. Borman. TCP Extensions for High Performance. RFC 1323, IETF, May 1992.

17. V. Jacobson, R. Braden, and L. Zhang. TCP Extension for High-Speed Paths. RFC 1185, IETF, October 1990.

18. K. Jensen. *Coloured Petri Nets: Basic Concepts, Analysis Methods and Practical Use, Volume 1, Basic Concepts*. Springer-Verlag, Berlin, 1997.

19. E. Kohler, M. Handley, and S. Floyd. Datagram Congestion Control Protocol. Internet Draft draft-ietf-dccp-spec-08.txt, IETF, October 2004.

20. J. F. Kurose and Y. Yemini. The Specification and Verification of a Connection Establishment Protocol Using Temporal Logic. In *Protocol Specification, Testing, and Verification*, pages 43–62. North-Holland Publishing Company, 1982.

21. H. P. Lin. Modelling a Transport Layer Protocol using First-order Logic. In *Proc. of the ACM SIGCOMM Conference on Communications Architecture and Protocols*, pages 92–100, September 1986.

22. N. Lynch. *Distributed Algorithms*. Morgan Kaufmann Publishers, Inc., 1996.

23. J. Martin, A. Nilsson, and I. Rhee. Delay Based Congestion Avoidance for TCP. *IEEE/ACM Transactions on Networking*, 11(3):356–369, 2003.

24. H. Mehrpour and A. E. Karbowiak. Modelling and Analysis of DOD TCP/IP Protocol Using Numerical Petri Nets. In *Proc. IEEE Region 10 Conf. on Computer Communication Systems*, pages 617–622, Hong Kong, September 1990.

25. Military Standard Transmission Control Protocol. MIL-STD-1778, August 1983.

26. S. L. Murphy. Service Specification and Protocol Construction for a Layered Architecture. PhD Thesis, University of Maryland, USA, May 1990.

27. S. L. Murphy and A. U. Shankar. Connection Management for the Transport Layer: Service Specification and Protocol Verification. *IEEE Transactions on Communications*, 39(12):1762–1775, December 1991.

28. D. R. Musser. Abstract Data Type Specifications in the Affirm System. In *Proc. Specifications of Reliable Software Conference*, pages 47–57. IEEE, March 1979.

29. V. Paxson. Known TCP Implementation Problems. RFC 2525, IETF, March 1999.

30. J. Postel. Transmission Control Protocol Version 4, February 1979.

31. J. Postel. DoD Standard Transmission Control Protocol. RFC 761, IETF, January 1980.

32. J. Postel. Transmission Control Protocol. RFC 793, IETF, September 1981.

33. D. Schwabe. Formal Specification and Verification of a Connection Establishment Protocol. In *Proc. of the Seventh Symposium on Data Communications*, pages 11–26, New York, NY, USA, 1981. ACM Press.

34. D. Schwabe. Formal Techniques for Specification and Verification of Protocols. PhD Thesis, Report CSD 810401, Computer Science Department, University of California at Los Angeles, 1981.

35. D. P. Sidhu and T. P. Blumer. Some Problems with the Specification of the Military Standard Transmission Control Protocol. RFC 964, IETF, November 1985.

36. M. A. Smith. Formal Verification of Communication Protocols. In *Formal Description Techniques IX: Theory, Applications and Tools*, pages 129–144. Chapman & Hall, London, UK, October 1996.

37. M. A. Smith. Formal Verification of TCP and T/TCP. PhD Thesis, M.I.T., USA, September 1997.

38. W. Stallings. *Department of Defense (DOD) Protocol Standards*, volume 3. Howard W. Sams & Company, 1978.

39. R. Stewart, etc. Stream Control Transmission Protocol. RFC 2960, IETF, October 2000.

40. C. A. Sunshine and Y. K. Dalal. Connection Management in Transport Protocols. *Computer Networks*, 2(6):454–473, December 1978.
41. F. J. W. Symons. Modelling and Analysis of Communication Protocols Using Numerical Petri Nets. PhD Thesis, University of Essex, May 1978.
42. R. S. Tomlinson. Selecting Sequence Numbers. In *Proc. of ACM SIG-COM/SIGOPS Interprocess Communications Workshop*, pages 11–23, Santa Monica, California, March 1975.
43. University of Aarhus. *Design/CPN Online*. Web site: `http://www.daimi.au.dk/designCPN`, 2003.
44. G. R. Wright and W. R. Stevens. *TCP/IP Illustrated, Vol.2 : The Implementation*. Addison-Wesley, Reading, MA, 1995.

Soundness of Resource-Constrained Workflow Nets

Kees van Hee, Alexander Serebrenik, Natalia Sidorova,
and Marc Voorhoeve

Department of Mathematics and Computer Science,
Eindhoven University of Technology,
P.O. Box 513, 5600 MB Eindhoven, The Netherlands
{k.m.v.hee, a.serebrenik, n.sidorova, m.voorhoeve}@tue.nl

Abstract. We study concurrent processes modelled as workflow Petri nets extended with resource constraints. We define a behavioural correctness criterion called *soundness*: given a sufficient initial number of resources, all cases in the net are guaranteed to terminate successfully, no matter which schedule is used. We give a necessary and sufficient condition for soundness and an algorithm that checks it.

Keywords: Petri nets; concurrency; workflow; resources; verification.

1 Introduction

In systems engineering, coordination plays an important role on various levels. Workflow management systems coordinate the activities of human workers; the principles underlying them can also be applied to other software systems, like middleware and web services. Petri nets are well suited for modelling and verification of concurrent systems; for that reason they have proven to be a successful formalism for Workflow systems (see e.g. [1, 2, 3, 4]).

Workflow systems can be modelled by so-called *Workflow Nets (WF-nets)* [1], i.e. Petri nets with one initial and one final place and every place or transition being on a directed path from the initial to the final place. The execution of a *case* is represented as a firing sequence that starts from the initial marking consisting of a single token on the initial place. The token on the final place with no garbage (tokens) left on the other places indicates the *proper termination* of the case execution. A model is called *sound* iff every reachable marking can terminate properly.

WF-nets are models emphasising the partial ordering of activities in the process and abstracting from *resources*, such as machines, manpower or money, which may further restrict the occurrence of activities. In this paper we consider the influence of *resources* on the processing of cases in Workflow Nets. We consider here only durable resources, i.e. resources that are claimed and released during the execution, but not created or destroyed. We introduce the notion of the *Resource-Constrained Workflow net (RCWF-net)*, which is a workflow net

G. Ciardo and P. Darondeau (Eds.): ICATPN 2005, LNCS 3536, pp. 250–267, 2005.

consisting of a production (sub)net — a workflow net where resources are abstracted away, and a number of resource places restricting the functionality of the production net.

We adapt the notion of generalised soundness introduced for WF-nets [11] to the nature of RCWF-nets: We say that an RCWF-net with k case tokens (tokens on the initial place of the production net) and a resource marking R is (k, R)-sound iff all cases can terminate properly, whatever choices are made during the execution, and all resources are returned to their places. We will say that an RCWF-net is *sound* iff there exists a resource marking R_0 such that the RCWF-net is (k, R)-sound for any number of cases k and any resource marking $R \geq R_0$. This definition is very natural, especially in the area of business processes, since we would like to have a system specification such that any number of orders could be processed correctly, and buying new machines or obtaining additional financial resources would not require to reconsider the specification, lest the system become unreliable.

In many practical applications, cases processed in the Workflow net are independent of each other, which can be modelled by introducing simple colours for tokens going through the production net. We build a transition system corresponding to the work of the production net with a single initial token, extending this transition system with the information about consumptions and releases of resources for every transition in it. Then we represent this transition system as a state machine, which can be considered as a model of the production net where the colours of tokens can be removed without influencing the system behaviour, and finally we extend this state machine up to the RCWF-net by adding resource places according to information about resource consumptions/releases that we have for every transition of the net. Thus our task of checking the correctness of arbitrary RCWF-nets is reduced to checking the correctness of RCWF-nets whose production nets are state machines.

In this paper we consider only RCWF-nets with one resource type, which is sufficient for many practical applications (memory and money are typical examples of such resources). We give a necessary and sufficient condition of soundness for the nets of this class and give a decision algorithm with a polynomial complexity w.r.t. the number of states of the state machine describing the behaviour of the production net.

Related Work. The problem of the correct functioning of parallel processes that share resources is not new at all. The famous banker's algorithm of Dijkstra (cf. [8]) is one of the oldest papers on this topic. The problem of the banker's algorithm is different from ours, because in the bankers algorithm a schedule (i.e. an ordering of processes for granting their resource claims) is designed. It is a pessimistic approach because it assumes that each process might eventually claim its maximal need for resources, a number that has to be known in advance. In our situation the pessimistic scheduling is too restrictive. Another important difference is that we do not consider a scheduling strategy at all: we look for conditions such that a workflow engine can execute tasks (i.e. fire transitions) as soon as all preliminary work has been done, if there are enough resources available. So the workflow engine may assign resources considering the local state only. This means that if the processes are designed properly, a standard workflow engine can be used to execute the process in a sound way.

The problem of resource sharing in flexible manufacturing systems has been studied extensively, specifically by modelling them as Petri nets (see [14, 13, 10, 6, 9] for an overview of works in this field). In these works the authors focus on extending a model that represents the production process with a scheduler in order to avoid deadlocks and to use resources in the most efficient way. As mentioned above, our goal is to allow the workflow engine to execute processes without further scheduling. Therefore we concentrate here on fundamental correctness requirements for RCWF-nets: resource conservation laws (every claimed resource is freed before the case terminates and no resource is created) and the absence of deadlocks and livelocks that occur due to the lack of resources.

In [5] the authors consider structural analysis of Workflow nets with shared resources. Their definition of structural soundness corresponds approximately to the existence of k cases and R resource tokens such that the net is sound for this k and R. We consider systems where a number of cases with id's are going through the net and the number of available resources can vary; so we require that the system should work correctly for any number of cases and resources. Therefore the results of [5] are not applicable to our case.

The rest of the paper is organised as follows. In Section 2, we sketch the basic definitions related to Petri nets and Workflow nets. In Section 3 we introduce the notion of Resource-Constrained Workflow Nets and define the notion of soundness for RCWF-nets. In Section 4 we give a necessary and sufficient condition of soundness and in Section 5 we give a decision algorithm for soundness. We conclude in Section 6 with discussing the obtained results and indicating directions for future work.

2 Preliminaries

$\mathbb{N}$ denotes the set of natural numbers and $\mathbb{Q}$ the set of rational numbers.

Let P be a set. A *bag (multiset)* m over P is a mapping $m : P \to \mathbb{N}$. The set of all bags over P is $\mathbb{N}^P$. We use $+$ and $-$ for the sum and the difference of two bags and $=, <, >, \leq, \geq$ for comparison of bags, which are defined in a standard way. We overload the set notation, writing $\emptyset$ for the empty bag and $\in$ for the element inclusion. We write e.g. $m = 2[p] + [q]$ for a bag m with $m(p) = 2$, $m(q) = 1$, and $m(x) = 0$ for all $x \notin \{p, q\}$. As usual, $|m|$ stands for the number of elements in bag m.

For (finite) *sequences* of elements over a set T we use the following notation: The empty sequence is denoted with ϵ; a non-empty sequence can be given by listing its elements. A concatenation of sequences σ_1 and σ_2 is denoted with $\sigma_1\sigma_2$, $t\sigma$ and σt stand for the concatenation of t and sequence σ and vice versa, and σ^n for the concatenation of n sequences σ.

Transition Systems. A *transition system* is a tuple $E = \langle S, Act, T \rangle$ where S is a set of *states*, Act is a finite set of *action names* and $T \subseteq S \times Act \times S$ is a *transition relation*. A *process* is a pair (E, s_0) where E is a transition system and $s_0 \in S$ an initial state. We denote $(s_1, a, s_2) \in T$ as $s_1 \xrightarrow{a}_E s_2$, and we say that a leads from s_1 to s_2 in E. We omit E and write $s \xrightarrow{a} s'$ whenever no ambiguity can arise. For a sequence of transitions $\sigma = t_1 \ldots t_n$ we write $s_1 \xrightarrow{\sigma} s_2$ when $s_1 = s^0 \xrightarrow{t_1} s^1 \xrightarrow{t_2} \ldots \xrightarrow{t_n} s^n = s_2$. In this case we say that σ is a trace of E. Finally, $s_1 \xrightarrow{*} s_2$ means that there exists a sequence $\sigma \in T^*$ such that $s_1 \xrightarrow{\sigma} s_2$. We say that s_2 is *reachable* from s_1 iff $s_1 \xrightarrow{*} s_2$.

Petri Nets. A *Petri net* is a tuple $N = \langle P, T, F^+, F^- \rangle$, where:

- P and T are two disjoint non-empty finite sets of *places* and *transitions* respectively; we call the elements of the set $P \cup T$ *nodes* of N;
- F^+ and F^- are mappings $(P \times T) \to \mathbb{N}$ that are *flow functions* from transitions to places and from places to transitions respectively.

$F = F^+ - F^-$ is the *incidence matrix* of net N.

We present nets with the usual graphical notation.

Given a transition $t \in T$, the *preset* $^\bullet t$ and the *postset* $t^\bullet$ of t are the *bags* of places where every $p \in P$ occurs $F^-(p, t)$ times in $^\bullet t$ and $F^+(p, t)$ times in $t^\bullet$. Analogously we write $^\bullet p, p^\bullet$ for pre- and postsets of places. We will say that a node n is a *source* node iff $^\bullet n = \emptyset$ and n is a *sink* node iff $n^\bullet = \emptyset$.

A marking m of N is a bag over P; markings are states (configurations) of a net. A pair (N, m) is called a *marked* Petri net. A transition $t \in T$ is *enabled* in marking m iff $^\bullet t \leq m$. An enabled transition t may *fire*. This results in a new marking m' defined by $m' \stackrel{\text{def}}{=} m - {}^\bullet t + t^\bullet$. We interpret a Petri net N as a transition system/process where markings play the role of states and firings of the enabled transitions define the transition relation, namely $m + {}^\bullet t \xrightarrow{t} m + t^\bullet$, for any $m \in \mathbb{N}^P$. The notion of reachability for Petri nets is inherited from the transition systems. We denote the set of all markings reachable in net N from marking m as $\mathcal{R}(N, m)$. We will drop N and write $\mathcal{R}(m)$ when no ambiguity can arise.

Place Invariants. (see [12]) A *place invariant* is a row vector $I : P \to \mathbb{Q}$ such that $I \cdot F = 0$. When talking about invariants, we consider markings as *vectors*.

State Machines. A subclass of Petri nets that we will heavily use further on is *state machines*. State machines can represent conflicts by a place with several output transitions, but they cannot represent concurrency and synchronisation. Formally: Let $N = \langle S, T, F \rangle$ be a Petri net. N is a *state machine* (SM) iff $\forall t \in T : |^\bullet t| = 1 \wedge |t^\bullet| = 1$.

Workflow Petri Nets. In this paper we primarily focus upon the *Workflow Petri nets (WF-nets)* [1]. As the name suggests, WF-nets are used to model the processing of tasks in workflow processes. The initial and final nodes indicate respectively the initial and final states of processed cases.

Definition 1 (WF-net). *A Petri net N is a* Workflow net (WF-net) *iff:*

1. *N has two special places: i and f. The initial place i is a source place, i.e. $^\bullet i = \emptyset$, and the final place f is a sink place, i.e. $f^\bullet = \emptyset$.*
2. *For any node $n \in (P \cup T)$ there exists a path from i to n and a path from n to f.*

We consider the processing of multiple tasks in Workflow nets, meaning that the initial place of a Workflow net may contain an arbitrary number of tokens. Our goal is to provide correctness criteria for the design of these nets. One natural correctness requirement is *proper termination*, which is called *soundness* in the WF-net theory. We will use the generalised notion of soundness for WF-nets introduced in [11]:

Definition 2 (soundness of WF-nets).
N is k-sound for some $k \in \mathbb{N}$ iff for all $m \in \mathcal{R}(k[i])$, $m \overset{*}{\longrightarrow} k[f]$.
N is sound *iff it is k-sound for all $k \in \mathbb{N}$.*

3   Resource-Constrained Workflow Nets

Workflow nets specify the handling of tasks within the organisation, factory, etc. without taking into account resources available there for the execution. We extend here the notion of WF-nets in order to include information about the use of resources into the model.

A resource belongs to a type; we have one place per resource type in the net where the resources are located when they are free. We assume that resources are durable, i.e. they can neither be created nor destroyed, they are claimed during the handling procedure and then released again. Typical examples of resources are money, memory, manpower, machinery. By abstracting from the resource places we obtain the WF-net that we call *production net*.

Definition 3 (RCWF-net). *A WF-net $N = \langle P_p \cup P_r, T, F_p^+ \cup F_r^+, F_p^- \cup F_r^- \rangle$ with initial and final places $i, f \in P_p$ is a* Resource-Constrained Workflow net (RCWF-net) *with the set P_p of production places and the set P_r of resource places iff*

- *$P_p \cap P_r = \emptyset$,*
- *F_p^+ and F_p^- are mappings $(P_p \times T) \to \mathbb{N}$,*
- *F_r^+ and F_r^- are mappings $(P_r \times T) \to \mathbb{N}$, and*
- *$N_p = \langle P_p, T, F_p^+, F_p^- \rangle$ is a WF-net, which we call the production net of N.*

Workflow Nets with Id-Tokens. Cases processed in the Workflow net are often independent of each other, i.e. tokens related to different cases cannot interfere with each other. This can be modelled by assigning a *unique id-colour* to each case, and allowing firings only on the tokens of the same colour. Colouring does not concern the resource tokens: resources are shared by all cases processed in the net and are colourless.

Therefore, we extend the semantics of Petri nets by introducing *id-tokens*. Our RCWF-nets will have tokens of two types: coloured tokens on production places, which are pairs (p, a), where p is a place and $a \in Id$ is an identifier, and uncoloured tokens on resource places. We assume Id to be a countable set. We will write x_p for the projection of $x \in \mathbb{N}^P$ on production places (coloured part of the marking) and x_r for the projection of x on resource places (uncoloured part). A transition $t \in T$ is *enabled* in m iff $(^{\bullet}t)_r \leq m$ and there exists $a \in Id$ such that m_p contains tokens on $(^{\bullet}t)_p$ with identifier a. A firing of t results in consuming these tokens and producing tokens with identifier a to $(t^{\bullet})_p$ and uncoloured tokens to $(t^{\bullet})_r$. Later on, we will use the extended semantics when working with id-tokens, and the standard semantics for classical tokens.

Though being a very simple sort of coloured nets, WF-nets with id-tokens are often expressive enough to reflect the essence of a modelled process, separating different cases which are processed in the net concurrently.

Soundness of RCWF-Nets. Soundness in WF-nets is the property that says that every marking reachable from an initial marking with k tokens on the initial place terminates properly, i.e. it can reach a marking with k tokens on the final place, for an arbitrary natural number k. In the RCWF-net, the initial marking of the net is a marking with some tokens on the initial place and a number of resource tokens on the resource places. With the proper termination for RCWF-nets we mean that the resource tokens are back to their resource places and all tasks are processed correctly, i.e. all the places of N_p except for f are empty. Moreover, we want the net to work properly not only with some fixed amount of resources but also with any greater amount: we want the verified system to work correctly also when more money, memory, manpower, or machinery is available. On the other hand, it is clear that there is some minimal amount of resources needed to guarantee that the system can work at all.

Another correctness requirement that should be reflected by the definition of soundness is that resource tokens cannot be created during the processing, i.e. at any moment of time the number of available resources does not exceed the number of initially given resources. The extended definition of soundness reads thus as follows:

Definition 4 (soundness of RCWF-nets). *Let N be an RCWF-net.*
N is (k, R)-sound for some $k \in \mathbb{N}, R \in \mathbb{N}^{P_r}$ iff for all $m \in \mathcal{R}(\sum_{a \in Id}[(i, a)] + R)$ with $|Id| = k$ holds: $m_r \leq R$ and $m \xrightarrow{*} (\sum_{a \in Id}[(f, a)] + R)$.
N is k-sound iff there exists $R \in \mathbb{N}^{P_r}$ such that N is (k, R')-sound for all $R' \geq R$.
N is sound iff there exists $R \in \mathbb{N}^{P_r}$ such that N is (k, R')-sound for all $k \in \mathbb{N}, R' \geq R$.

The soundness problem is a *parameterised* problem formulated on a *coloured* Petri net. We will first use the nature of the colouring to reduce this problem to a problem on an *uncoloured* net.

Lemma 5. *The production net of a sound RCWF-net is 1-sound.*

Proof. Since we want to prove 1-soundness, we only have to consider the processing of a single case in the net, and therefore all production tokens have the same colour, which we abstract from. Let N be a sound RCWF-net and assume that N_p is not 1-sound. Then there exist a firing sequence σ and a production marking m_p such that $[i] \xrightarrow{\sigma}_{N_p} m_p$ and $m_p \not\xrightarrow{}_{N_p} [f]$. Take enough resources $m_0 \in \mathbb{N}^{P_r}$ to enable σ in N, then $m_p + m_r$ is reachable in $(N, [i] + m_0)$ but $m_p + m_r \not\xrightarrow{}_N [f] + m_0$, which contradicts the soundness of the RCWF-net.   □

1-soundness of the production net is thus a necessary condition of the soundness of the RCWF-net. 1-soundness of a WF-net can be checked by checking that the closure[1] of the WF-net is live and bounded [1]. In the rest of the paper we assume that the check of 1-soundness of the production net has been done and its result is positive.

Corollary 6. *For any sound RCWF-net $N, \mathcal{R}(N_p, [(i, a)])$ is finite for any $a \in Id$.*

Proof. All production tokens in $(N_p, [i, a])$ will have colour a and thus the colour does not influence the behaviour of the net and we can abstract from it. Assume $\mathcal{R}(N_p, [i])$ is infinite. Then there are $m_1, m_2 \in \mathcal{R}(N_p, [i])$ such that $m_2 = m_1 + \delta$ for some $\delta > \emptyset$. Since N is sound, N_p is 1-sound and $m_1 \xrightarrow{*}_{N_p} [f]$. Thus $m_1 + \delta \xrightarrow{*}_{N_p} [f] + \delta$. Hence $[f] + \delta \in \mathcal{R}(N_p, m_2) \subseteq \mathcal{R}(N_p, [i])$ and $[f] + \delta \xrightarrow{*}_{N_p} [f]$, which is impossible since f is a sink place and any transition of N_p has at least one output place.   □

Given an RCWF-net N with one resource type we construct a resource-constrained state machine WF-net with the same behaviour as N as follows. First, let T be a transition system corresponding to $(N_p, [i])$ extended with the information about resource consumption and production for every transition of T. Then we build a resource-constrained state machine workflow net N' by creating a place for every state of T and a transition with the corresponding resource consumption/production for every transition of T. Observe that due to the use of id-tokens, N' is sound iff N is. Hence, we can check soundness of an RCWF-net by checking soundness of the corresponding state machine workflow net.

In this paper we restrict our attention to Resource-Constrained Workflow nets with one type of resources. This is a typical situation in various practical applications with memory, money or manpower being the considered resource. Therefore, in the remainder of the paper we consider only state machine workflow nets with one resource type (SM1WF-nets):

Definition 7. *An RCWF-net $N = \langle P_p \cup P_r, T, F_p^+ \cup F_r^+, F_p^- \cup F_r^- \rangle$ is called a state machine workflow net with one resource type (SM1WF-net) if $P_r = \{r\}$ and the production net N_p of N is a state machine.*

---

[1] The closure of a WF-net N is the net obtained by adding to N a transition with f as the input place and i as the output place.

Note that a production token in the SM1WF-net represents a part of a production marking of the original RCWF-net related to one case (one id-colour). Thus all production tokens in the SM1WF-net have different id-colours. Note that every firing in an SM1WF-net requires only one production token (and a number of resource tokens) and results in the production of a single production token (and a number of resource tokens). Therefore we can abstract from colours when considering soundness of SM1WF-nets.

For SM1WF-nets we write $°t$ and $t°$ for the input/output place of t in the production net.

4   Soundness Check for SM1WF-Nets

In this section we will give a necessary and sufficient condition for the soundness of SM1WF-nets. We start by introducing a notion of *path* that we will use here. Unlike a trace, a path does not deal with the processing of multiple production tokens. Formally, given an SM1WF-net N, a *path* is a sequence $t_1 \ldots t_n$ of transitions in T such that $\forall k : 1 \leq k < n : t_k° = °t_{k+1}$. We write $°\sigma$ and $\sigma°$ for the input and the output place of a nonempty path $\sigma = t_1 \ldots t_n$, i.e. for $°t_1, t_n°$ respectively. A path σ is called a *successor* of a path ρ (and ρ a *predecessor* of σ) if $\rho° = °\sigma$. Their juxtaposition $\rho\sigma$ then is again a path of N.

With every path we associate three numbers: its resource production, consumption and effect.

Definition 8. *Let N be an SM1WF-net. The resource effect $\mathcal{E}$, production $\mathcal{P}$ and consumption $\mathcal{C}$ are defined as follows:*

- *for the empty path ϵ, $\mathcal{E}(\epsilon) = \mathcal{P}(\epsilon) = \mathcal{C}(\epsilon) = 0$;*
- *for a path t, $t \in T$, $\mathcal{E}(t) = t^\bullet(r) - {}^\bullet t(r)$, $\mathcal{P}(t) = t^\bullet(r)$, and $\mathcal{C}(t) = {}^\bullet t(r)$;*
- *for a path σt, $\mathcal{E}(\sigma t) = \mathcal{E}(\sigma) + \mathcal{E}(t)$, $\mathcal{P}(\sigma t) = \max(\mathcal{P}(t), \mathcal{P}(\sigma) + \mathcal{E}(t))$ and for a path $t\sigma$, $\mathcal{C}(t\sigma) = \max(\mathcal{C}(t), \mathcal{C}(\sigma) - \mathcal{E}(t))$.*

The notion of effect allows us to distinguish three kinds of paths. A path σ is called a *C-path (consumption path)* if $\mathcal{E}(\sigma) < 0$, an *E-path (equality path)* if $\mathcal{E}(\sigma) = 0$, and a *P-path (production path)* if $\mathcal{E}(\sigma) > 0$.

Example 9. Now we will illustrate the intuitive meaning of $\mathcal{E}, \mathcal{P}$ and $\mathcal{C}$ on an example and in the rest of the section we will prove that $\mathcal{E}, \mathcal{P}$ and $\mathcal{C}$ confirm this intuition indeed. Consider paths tu and vx of SM1WF-net N in Fig. 1.[2] The resource effect of these paths $\mathcal{E}(tu) = 1 - 4 + 5 - 2 = 0$ and $\mathcal{E}(vx) = 3 - 1 + 3 - 2 = 3$, which corresponds to the change of the number of resource tokens due to the firing of the transitions of the corresponding path. $\mathcal{P}(tu) = \max(\mathcal{P}(u), \mathcal{P}(t) + \mathcal{E}(u)) = \max(5, 1 + 3) = 5$ and $\mathcal{P}(vx) = \max(\mathcal{P}(x), \mathcal{P}(v) +$

---

[2] Instead of drawing a resource place and its in- and outgoing arcs, we put the weights of the arcs from and to the resource place under the corresponding transitions. So $(4, 1)$ for transition t means that t consumes 4 resource tokens and then releases 1 resource token.

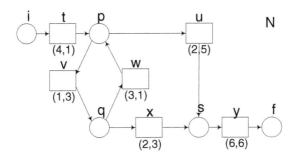

Fig. 1. Example of an SM1WF-net

$\mathcal{E}(x)) = \max(3, 3 + 1) = 4$. Note that $\mathcal{P}(tu), \mathcal{P}(vx)$ correspond to the minimal number of resource tokens we are guaranteed to have immediately after the firing of tu/vx respectively. $\mathcal{C}(tu) = \max(\mathcal{C}(t), \mathcal{C}(u) - \mathcal{E}(t)) = \max(4, 2 + 3) = 5$ and $\mathcal{C}(vx) = \max(\mathcal{C}(v), \mathcal{C}(x) - \mathcal{E}(v)) = \max(1, 2 - 2) = 1$. $\mathcal{C}(tu)$ and $\mathcal{C}(vx)$ correspond to the minimal number of resource tokens needed to make the firings of tu/vx possible.

4.1   Properties of the Resource-Effect Function

Lemma 10. *Let N be a sound SM1WF-net. Then for any place $p \in P_p$ and any two paths σ and ρ such that $^\circ\sigma = {}^\circ\rho = i$ and $\sigma^\circ = \rho^\circ = p$ holds $\mathcal{E}(\sigma) = \mathcal{E}(\rho) \leq 0$.*

Proof. Since N is sound, N_p is sound as well and there exists a firing sequence γ such that $[p] \xrightarrow{\gamma} [f]$. Take R large enough to make both $\sigma\gamma$ and $\rho\gamma$ firable from $[i] + R[r]$. Thus $[i] + R[r] \xrightarrow{\sigma} [p] + (R + \mathcal{E}(\sigma))[r] \xrightarrow{\gamma} [f] + R[r]$ and $[i] + R[r] \xrightarrow{\rho} [p] + (R + \mathcal{E}(\rho))[r] \xrightarrow{\gamma} [f] + R[r]$, which implies that $\mathcal{E}(\sigma) = \mathcal{E}(\rho)$. Moreover, since N is sound and thus no resource creation happens, $R + \mathcal{E}(\sigma) \leq R$, i.e. $\mathcal{E}(\sigma) \leq 0$.   □

Thus, in a sound SM1WF-net, each production place p has a unique weight defined as $-\mathcal{E}(\sigma)$ for some σ such that $^\circ\sigma = i$ and $\sigma^\circ = p$, showing how many resources a production token on place p possesses. (Clearly, the weight can be equivalently defined as $\mathcal{E}(\rho)$ where ρ is some sequence with $^\circ\rho = p$ and $\rho^\circ = f$.) This observation leads to the following place invariant property for sound SM1WF-nets:

Lemma 11. *Let N be a sound SM1WF-net with the initial place i, the final place f, and the resource place r. Then there exists a unique place invariant W such that $W(i) = W(f) = 0$, $W(r) = 1$. Moreover, for every place $p \in P_p$, $W(p) = -\mathcal{E}(\sigma)$ for any σ with $^\circ\sigma = i$ and $\sigma^\circ = p$, and hence $W(p) \geq 0$ for all $p \in P_p$.*

Proof. The proof is done in a constructive way. Since N is sound, we have a unique mapping $W : P \to \mathbb{N}$ such that for every place $p \in P_p$ $W(p) = -\mathcal{E}(\sigma)$

where σ is some path with $^\circ\sigma = i$ and $\sigma^\circ = p$, and $W(r) = 1$. By construction, for any sound net $W(i) = W(f) = 0$ and $W(p) \geq 0$, for all $p \in P_p$.

Now we will show that W is a place invariant, i.e. $W \cdot F = 0$. Since N_p is a state machine, a column of F corresponding to a transition t has -1 in the cell $^\circ t$, 1 in t° and $t^\bullet(r) - {}^\bullet t(r)$ in the resource place r. Hence, the product of W and the t-column of F can be written as $-W(^\circ t) + W(t^\circ) + (t^\bullet(r) - {}^\bullet t(r)) \cdot W(r) = \mathcal{E}(\sigma) - \mathcal{E}(\sigma t) + t^\bullet(r) - {}^\bullet t(r) = 0$ (σ is some path with $^\circ\sigma = i$ and $\sigma^\circ = {}^\circ t$). Since the same reasoning can be applied to any transition t, we have $W \cdot F = 0$.

By induction on the length of σ with $^\circ\sigma = i, \sigma^\circ = p$, it is easy to show that W is unique, i.e. for any invariant W' such that $W'(i) = W'(f) = 0$ and $W'(r) = 1$ we have $W(p) = -\mathcal{E}(\sigma)$. $\qquad\square$

Thus the existence of such an invariant is a *necessary condition* of soundness. This condition can be easily checked by standard algebraic techniques. For net N from Fig. 1 the invariant is $r + 3p + q$, i.e. the weights of places are $W(p) = 3$, $W(q) = 1$ and $W(s) = 0$. We assume further on that N is an SM1WF-net with a unique place invariant W satisfying $W(i) = W(f) = 0$ and $W(r) = 1$, and moreover, we have $W(p) \geq 0$.

4.2   Properties of the Consumption and Production Functions

The following lemma states that at least $\mathcal{C}(\sigma)$ resources are needed to execute σ and at least $\mathcal{P}(\sigma)$ resources become available after the execution of σ.

Lemma 12. *Let σ be a path in N. Then*

1. *If $M \xrightarrow{\sigma} M'$ for some markings M, M', then $M'(r) \geq \mathcal{P}(\sigma)$ and $M(r) \geq \mathcal{C}(\sigma)$.*
2. *$[^\circ\sigma] + \mathcal{C}(\sigma)[r] \xrightarrow{\sigma} [\sigma^\circ] + \mathcal{P}(\sigma)[r]$ if $\sigma \neq \epsilon$.*

Proof. We prove Part 1 by induction on the length of σ. If $\sigma = \epsilon$, the lemma holds. We prove the $\mathcal{P}$-part by setting $\sigma = \rho t$. Let M'' be such that $M \xrightarrow{\rho} M'' \xrightarrow{t} M'$. By the induction hypothesis, $M''(r) \geq \mathcal{P}(\rho)$ and thus $M'(r) \geq \max(\mathcal{P}(t), \mathcal{P}(\rho) + \mathcal{E}(t))$, i.e., $M'(r) \geq \mathcal{P}(\sigma)$, completing the proof of the $\mathcal{P}$-part in Part 1. We omit the proof of the $\mathcal{C}$-part since it can be obtained analogously by taking $\sigma = t\rho$.

Part 2 follows from the existence of markings M and M' such that

$$M \xrightarrow{\sigma} [\sigma^\circ] + \mathcal{P}(\sigma)[r] \quad \text{and} \quad [^\circ\sigma] + \mathcal{C}(\sigma)[r] \xrightarrow{\sigma} M'. \tag{1}$$

We prove (1) by induction on the length of σ. The case $\sigma = t$, where $t \in T$, is clear. For the $\mathcal{P}$-part, let $\sigma = \rho t$, with $\rho \neq \epsilon$. By the induction hypothesis, there exists M'' such that $M'' \xrightarrow{\rho} [\rho^\circ] + \mathcal{P}(\rho)[r]$. Note that $\mathcal{P}(\sigma) = \mathcal{P}(\rho t) = \max(\mathcal{P}(t), \mathcal{P}(\rho) + \mathcal{E}(t))$. We distinguish between two cases:

- If $\mathcal{P}(\sigma) = \mathcal{P}(\rho) + \mathcal{E}(t)$, then $\mathcal{P}(\rho) + \mathcal{E}(t) \geq \mathcal{P}(t)$, i.e., $\mathcal{P}(\rho) \geq \mathcal{P}(t) - \mathcal{E}(t) = {}^\bullet t(r)$. Hence, $\mathcal{P}(\rho) \geq \mathcal{C}(t)$ and $[\rho^\circ] + \mathcal{P}(\rho)[r] \xrightarrow{t} [\sigma^\circ] + (\mathcal{P}(\rho) + \mathcal{E}(t))[r]$. Recall that $\mathcal{P}(\rho) + \mathcal{E}(t) = \mathcal{P}(\sigma)$, i.e., $[\rho^\circ] + \mathcal{P}(\rho)[r] \xrightarrow{t} [\sigma^\circ] + \mathcal{P}(\sigma)$, so we take $M = M''$ and have $M \xrightarrow{\sigma} [\sigma^\circ] + \mathcal{P}(\sigma)[r]$.

- If $\mathcal{P}(\sigma) = \mathcal{P}(t)$, then $\mathcal{P}(\rho) + \mathcal{E}(t) \leq \mathcal{P}(t)$, i.e., $\mathcal{P}(\rho) \leq \mathcal{P}(t) - \mathcal{E}(t)$. Therefore, $\mathcal{P}(\rho) \leq \mathcal{C}(t)$ and we take $M = M'' + (\mathcal{C}(t) - \mathcal{P}(\rho))[r]$. Thus, $M \xrightarrow{\rho} [\rho^\circ] + \mathcal{C}(t)[t] \xrightarrow{t} [\sigma^\circ] + \mathcal{P}(t)[t]$. Since $\sigma = \rho t$, $M \xrightarrow{\sigma} [\sigma^\circ] + \mathcal{P}(\sigma)[r]$.

The $\mathcal{C}$-part is analogous, using $\sigma = t\rho$. Due to Part 1 of the lemma, M and M' in (1) satisfy $M \geq [^\circ\sigma] + \mathcal{C}(\sigma)[r]$ and $M' \geq [\sigma^\circ] + \mathcal{P}(\sigma)[r]$. Hence $M \xrightarrow{\sigma} M' + \delta$ where $\delta = M - ([^\circ\sigma] + \mathcal{C}(\sigma)[r])$ and $M + \delta' \xrightarrow{\sigma} M'$ where $\delta' = M' - ([\sigma^\circ] + \mathcal{P}(\sigma)[r])$. Thus we conclude that $\delta = \delta' = \emptyset$ and $[^\circ\sigma] + \mathcal{C}(\sigma)[r] \xrightarrow{\sigma} [\sigma^\circ] + \mathcal{P}(\sigma)[r]$.
□

Corollary 13. $\mathcal{E}(\sigma) = \mathcal{P}(\sigma) - \mathcal{C}(\sigma)$ and $\mathcal{E}(\sigma) = W(^\circ\sigma) - W(\sigma^\circ)$ for all σ.

Proof. Follows directly from Lemma 12.(2) and the definition of W.   □

Corollary 14. Let $k > 0$ and σ be a path such that $\mathcal{E}(\sigma) \leq 0$. Then,

$$k[^\circ\sigma] + (\mathcal{C}(\sigma) - (k-1) * \mathcal{E}(\sigma))[r] \xrightarrow{\sigma^k} k[\sigma^\circ] + \mathcal{P}(\sigma)[r]$$

Proof. The proof is done by induction on k with the use of Lemma 12(2) and Corollary 13.   □

Next we show that under certain conditions two paths can be swapped.

Lemma 15 (Interchange Lemma). Let M, M' be markings and σ, ρ be paths such that $\mathcal{E}(\sigma) \leq 0 \leq \mathcal{E}(\rho)$, and ρ is not a successor of σ. If $M \xrightarrow{\sigma\rho} M'$ then $M \xrightarrow{\rho\sigma} M'$.

Proof. Let M_1 be a marking such that $M \xrightarrow{\sigma} M_1 \xrightarrow{\rho} M'$. Since $\sigma^\circ \neq {}^\circ\rho$, $M_1 \geq [\sigma^\circ] + [^\circ\rho] + \max(\mathcal{C}(\rho), \mathcal{P}(\sigma))[r]$. Hence, $M \geq [^\circ\sigma] + [^\circ\rho] + \max(\mathcal{C}(\rho) - \mathcal{E}(\sigma), \mathcal{C}(\sigma))[r]$. Since $\mathcal{E}(\sigma) \leq 0$, there exists a marking M_2 such that $M \xrightarrow{\sigma} M_2$ and $M_2 \geq [^\circ\sigma] + [\rho^\circ] + \max(\mathcal{P}(\rho) - \mathcal{E}(\sigma), \mathcal{C}(\sigma) + \mathcal{E}(\rho))[r]$. Therefore, $M_2 \geq [^\circ\sigma] + [\rho^\circ] + (\mathcal{C}(\sigma) + \mathcal{E}(\rho))[r]$. Recall that $\mathcal{E}(\rho) \geq 0$, so $M_2 \geq [^\circ\sigma] + \mathcal{C}(\sigma)[r]$ and thus $M \xrightarrow{\rho\sigma} M'$.   □

The next lemma gives implicit lower bounds for the number of resources in states reachable from the initial marking and states that reach the final marking.

Lemma 16. Let $M, M' \in \mathbb{N}^P$ with $M(r) < M'(r)$.
If $M' \xrightarrow{*} M$, there exists a C-path ρ such that $M \geq [\rho^\circ] + \mathcal{P}(\rho)[r]$.
If $M \xrightarrow{*} M'$, there exists a P-path σ such that $M \geq [^\circ\sigma] + \mathcal{C}(\sigma)[r]$.

Proof. Let $M' \xrightarrow{\alpha} M$. We normalise the trace α as follows. We write α as the concatenation of paths $\sigma_1 \ldots \sigma_n$, where no σ_{k+1} is a successor of σ_k. If α contains a C-path σ_k succeeded by a P-path or by an E-path σ_{k+1}, we swap them in α, obtaining α'. By the interchange lemma, $M' \xrightarrow{\alpha'} M$. We continue with normalizing α' further by using the same procedure. The normalisation

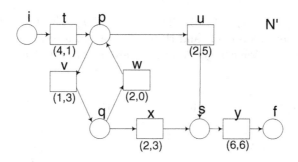

Fig. 2. Example of an unsound SM1WF-net

process terminates since every swap decreases the number of P- and E-paths following a C-path.

Thus, there exists a trace β such that $M' \xrightarrow{\beta} M$ and the division of β into paths consists of a number of P- and/or E-paths followed by C-paths. Since $M(r) < M'(r)$, β contains at least one C-path. Let ρ be the last path of β. Then ρ is a C-path, $M(\rho^\circ) > 0$ and by statement (1) of Lemma 12, $M(r) \geq \mathcal{P}(\rho)$.

Similarly, if $M \xrightarrow{\gamma} M'$, there exists a trace δ containing P-paths followed by C- and/or E-paths such that $M \xrightarrow{\delta} M'$. Since $M(r) < R$, δ contains at least one P-path. Let σ be the first P-path. Then by Lemma 121, $M(r) \geq \mathcal{C}(\sigma)$.   □

We will show that the $\mathcal{C}$-bound in Lemma 16 is sharp. (Sharpness of the $\mathcal{P}$-bound can be proved but is not needed here.)

Lemma 17. *Let $k_0 > 0$ and let σ be a C-path. Then there exist $k > k_0$ and $R \in \mathbb{N}$ such that $k[i] + R[r] \xrightarrow{*} k[\sigma^\circ] + \mathcal{P}(\sigma)[r]$.*

Proof. Let $p = {}^\circ\sigma, q = \sigma^\circ$. There exists a path ρ with ${}^\circ\rho = i, \rho^\circ = p$. Since we assume the existence of the place invariant as described in Lemma 11, $\mathcal{E}(\rho) \leq 0$. So by Corollary 14, $k[i] + (\mathcal{C}(\rho) - (k - 1) * \mathcal{E}(\rho))[r] \xrightarrow{\rho^k} k[p] + \mathcal{P}(\rho)[r]$ for all $k > 0$. Since $\mathcal{E}(\sigma) < 0$, there exists $k > k_0$ such that $\mathcal{C}(\sigma) - (k - 1)\mathcal{E}(\sigma) \geq \mathcal{P}(\rho)$. By taking $R = \mathcal{C}(\rho) - (k - 1) * \mathcal{E}(\rho) + (\mathcal{C}(\sigma) - (k - 1)\mathcal{E}(\sigma) - \mathcal{P}(\rho))$, we obtain by Corollary 14: $k[i] + R(k)[r] \xrightarrow{\rho^k} k[p] + (\mathcal{C}(\sigma) - (k - 1)\mathcal{E}(\sigma))[r] \xrightarrow{\sigma^k} k[q] + \mathcal{P}(\sigma)[r]$.
    □

The construction described in the proof of Lemma 17 will be later on used for giving a meaningful verification feedback on unsound nets, namely we will construct an example of a deadlock/livelock in an unsound net.

Example 18. Consider the consumption path $\sigma = w$ in net N' from Fig. 2 (which differs from net N from Fig. 1 only in the resource consumption/production of transition w); ${}^\circ\sigma = q, \sigma^\circ = p, \mathcal{E}(\sigma) = -2, \mathcal{C}(\sigma) = 2$. Take tv as ρ; $\mathcal{E}(\rho) = -1, \mathcal{C}(\rho) = 4, \mathcal{P}(\rho) = 3$. Pick some $k \in \mathbb{N}$ satisfying $\mathcal{C}(\sigma) - (k - 1)\mathcal{E}(\sigma) \geq \mathcal{P}(\rho)$, i.e. $k \geq 1.5$, and choose R as $\mathcal{C}(\rho) - (k-1) * \mathcal{E}(\rho) + (\mathcal{C}(\sigma) - (k-1)\mathcal{E}(\sigma) - \mathcal{P}(\rho)) =$

$4 + (k-1) + 2 + 2(k-1) - 3 = 3k$. Then $k[i] + 3k[r] \xrightarrow{(tv)^k} k[q] + 2k[r] \xrightarrow{w^k} k[p]$. Note that no resources are left and thus we obtained a deadlock since we need resources to proceed. We can get R larger than any given number just by taking a larger k.

Finally, we are ready to state the main theorem, giving a necessary and sufficient condition for the soundness of SM1WF nets.

Theorem 19. *An SM1WF net N is sound iff there exists a unique place invariant W such that $W(i) = W(f) = 0$, $W(r) = 1$, and moreover $W(p) \geq 0$ for all $p \in P_p$, and for each C-path ρ there is a successor P-path σ such that $\mathcal{P}(\rho) \geq \mathcal{C}(\sigma)$.*

Proof. ($\Rightarrow$): Assume there exists a C-path ρ such that all succeeding P-paths σ satisfy $\mathcal{P}(\rho) < \mathcal{C}(\sigma)$. By Lemma 17, there exist k and $R > \mathcal{P}(\rho)$ such that $k[i] + R[r] \xrightarrow{*} M = k[\rho^\circ] + \mathcal{P}(\rho)[r]$. If $M \xrightarrow{*} k[f] + R[r]$, by Lemma 16 there exists a P-path σ with $M(r) \geq \mathcal{C}(\sigma)$, contradicting the assumption. So $M \not\xrightarrow{*} k[f] + R[r]$ and the net is not sound.
($\Leftarrow$): Let R_0 be a maximal $\mathcal{C}(\rho)$ over all paths ρ of N with $\rho^\circ = f$. The choice of R_0 ensures that if at least R_0 resources are present, one token in the production net can be successfully transferred from any place to f.

Suppose that $R \geq R_0$ and $k[i] + R[r] \xrightarrow{*} M$. We prove by induction on $R - M(r)$ that there exists a marking M' with $M'(r) = R$ such that $M \xrightarrow{*} M'$, i.e., that for any reachable marking there is a way to continue and to return all the resources consumed so far. Note that $M(r) \leq R$ since the number of the resources consumed is always non-negative, i.e. no resources are created (due to the existence of the place invariant W). If $M(r) = R$, the statement clearly holds. If $M(r) < R$, by applying Lemma 16 to $k[i] + R[r] \xrightarrow{*} M$, we conclude that there exists a C-path ρ such that $M \geq [\rho^\circ] + \mathcal{P}(\rho)[r]$. By the condition of the theorem, there exists a P-path σ and a marking $M'' = M - [^\circ\sigma] + [\sigma^\circ] + \mathcal{E}(\sigma)[r]$ such that $M \xrightarrow{\sigma} M''$, so $M \xrightarrow{*} M''$. Since $M''(r) > M(r)$, the induction hypothesis is applicable to M'', i.e. finally we obtain that $M'' \xrightarrow{*} M'$ and $M'(r) = R$.

Let $p \in P_p$ be such that $M'(p) > 0$. Then since $R \geq R_0$ and by the choice of R_0, we have $[p] + R[r] \xrightarrow{*} [f] + R[r]$. So $M' \xrightarrow{*} M' - [p] + [f]$. We can repeat this procedure for all $p \neq f$ with $M(p) > 0$, reaching $k[f] + R[r]$. $\square$

Note that the net may be unsound if it contains a deadlock (a nonterminal marking where there are not enough resources to proceed any further even with one single step) or a livelock (there are always enough resources to make a following step, but all possible steps are not "progress"-steps, i.e. we cannot leave the cycle in order to terminate properly). With a slight modification of the condition in Theorem 19 we can diagnose whether the net has no deadlock: along with the invariant requirement we require that for each C-path ρ there is a successor path σ (no matter whether σ is a P-path or a C-path) such that $\mathcal{P}(\rho) \geq \mathcal{C}(\sigma)$. This reflects the requirement that there is always some next step possible. If the net has no deadlock but does not meet the requirements of Theorem 19, this net has a livelock.

5   Decision Algorithm

The necessary and sufficient condition formulated in Theorem 19 allows to characterise soundness of SM1WF-nets. The condition as stated is however not directly verifiable, since infinitely many different paths should be taken into account. In this subsection we show that checking finitely many paths is sufficient. The decision algorithm we give here is polynomial in the size of the SM1WF-net.

We start by the following simple observation.

Lemma 20. *Let σ be a cyclic path (i.e. $°\sigma = \sigma°$). Then for any ρ_1, ρ_2 such that $°\sigma = \rho_1°$ and $\sigma° = °\rho_2$ we have $\mathcal{E}(\rho_1\sigma\rho_2) = \mathcal{E}(\rho_1\rho_2)$, $\mathcal{P}(\rho_1\sigma\rho_2) = \mathcal{P}(\rho_1\rho_2)$, and $\mathcal{C}(\rho_1\sigma\rho_2) = \mathcal{C}(\rho_1\rho_2)$.*

Proof. For $\mathcal{E}$ the lemma follows from Lemma 11. Results for $\mathcal{P}$ and $\mathcal{C}$ can be obtained analogously.   □

Hence, to check the condition of Theorem 19 it is sufficient to consider acyclic paths only. Since there are finitely many acyclic paths, soundness of SM1WF-nets is decidable. As we showed in Section 3, the soundness of RCWF-nets can be reduced to the soundness of SM1WF-nets, and thus we can conclude the following:

Corollary 21. *Soundness of RCWF-nets with one resource is decidable.*

Next we give an efficient decision algorithm for SM1WF-nets. The algorithm is based on the following property of paths.

Lemma 22. *Let ρ, σ be paths such that $\rho° = °\sigma$. Then, $\mathcal{P}(\rho\sigma) = \max(\mathcal{P}(\rho) + \mathcal{E}(\sigma), \mathcal{P}(\sigma))$ and $\mathcal{C}(\rho\sigma) = \max(\mathcal{C}(\rho), \mathcal{C}(\sigma) - \mathcal{E}(\rho))$.*

Proof. Suppose $[°\rho] + A[r] \xrightarrow{\rho} [\rho°] + B[r] = [°\sigma] + B[r] \xrightarrow{\sigma} [\sigma°] + C[r]$. Then by Lemma 12.(1) (applied both to ρ and to σ) $A \geq \mathcal{C}(\rho)$ and $C \geq \mathcal{P}(\sigma)$. Since $B = A + \mathcal{E}(\rho) \geq \mathcal{C}(\rho) + \mathcal{E}(\rho) = \mathcal{P}(\rho)$ and $C = B + \mathcal{E}(\sigma)$, i.e. $B = C - \mathcal{E}(\sigma) \geq \mathcal{P}(\sigma) - \mathcal{E}(\sigma) = \mathcal{C}(\sigma)$, we deduce that $B \geq \max(\mathcal{P}(\rho), \mathcal{C}(\sigma))$. Thus $A \geq \max(\mathcal{C}(\rho), \mathcal{C}(\sigma) - \mathcal{E}(\rho))$ and $C \geq \max(\mathcal{P}(\sigma), \mathcal{P}(\rho) + \mathcal{E}(\sigma))$. By applying Lemma 12.(2) to ρ and σ, we deduce that $[°\rho] + \max(\mathcal{C}(\rho), \mathcal{C}(\sigma) - \mathcal{E}(\rho))[r] \xrightarrow{\rho} [\rho°] + \max(\mathcal{P}(\rho), \mathcal{C}(\sigma))[r] \xrightarrow{\sigma} [\sigma°] + \max(\mathcal{P}(\rho) + \mathcal{E}(\sigma), \mathcal{P}(\sigma))[r]$ indeed. Finally, using Lemma 12.(1) and Lemma 12.(2) on $\rho\sigma$, we conclude that $\mathcal{P}(\rho\sigma) = \max(\mathcal{P}(\rho) + \mathcal{E}(\sigma), \mathcal{P}(\sigma))$ and $\mathcal{C}(\rho\sigma) = \max(\mathcal{C}(\rho), \mathcal{C}(\sigma) - \mathcal{E}(\rho))$.   □

For $X = \emptyset$ we assume $\min X = \omega$. For $p, q \in P_p$, we define $\mu(p, q)$ as $\min\{\mathcal{P}(\sigma) + W(q) \mid °\sigma = p \wedge \sigma° = q\}$. If $°\sigma = p$ and $\sigma° = q$, then $\mathcal{C}(\sigma) + W(p) = \mathcal{P}(\sigma) + W(q)$, so $\mu(p, q)$ can alternatively be defined as $\min\{\mathcal{C}(\sigma) + W(p) \mid °\sigma = p \wedge \sigma° = q\}$. Then, the condition from Theorem 19 can be now reformulated in the following way, assuming the existence of the place invariant W:

Corollary 23. *N is sound if and only if*

$$\forall x \in P_p : \min\{\mu(y, x) \mid W(y) < W(x)\} \geq \min\{\mu(x, y) \mid W(y) < W(x)\}.$$

Analogously to Corollary 23, we can show that SM1WF-net has no deadlock iff

$$\forall x \in P_p : \min \{\mu(y, x) \mid W(y) < W(x)\} \geq \min \{\mu(x, y)\}.$$

With these conditions we can diagnose SM1WF-nets as sound, non-sound due to deadlock, or non-sound due to livelock.

Function μ has the following important property:

Lemma 24. *For all p and q in P_p we have $\mu(p, q) = \min \{\max(\mu(p, x), \mu(x, q)) \mid x \in P_p\}$.*

Proof. Recall that $\mu(p, q)$ is defined as $\min \{\mathcal{P}(\sigma) + W(q) \mid {}^\circ\sigma = p \wedge \sigma^\circ = q\}$. Every path from p to q can be seen as $\rho_1\rho_2$ for some paths ρ_1 from p to some x and ρ_2 from x to q. Hence, $\mathcal{P}(\sigma) + W(q) = \mathcal{P}(\rho_1\rho_2) + W(q)$, and by Lemma 22, $\mathcal{P}(\rho_1\rho_2) + W(q) = \max(\mathcal{P}(\rho_1) + \mathcal{E}(\rho_2), \mathcal{P}(\rho_2)) + W(q) = \max(\mathcal{P}(\rho_1) + \mathcal{E}(\rho_2) + W(q), \mathcal{P}(\rho_2) + W(q))$. Since ρ_2 is one of the possible paths from x to q, $\mathcal{P}(\rho_2) + W(q) \geq \mu(x, q)$. By Corollary 13, $\mathcal{E}(\rho_2) + W(q) = W(x)$. Therefore, $\mathcal{P}(\rho_1) + \mathcal{E}(\rho_2) + W(q) = \mathcal{P}(\rho_1) + W(x)$ and $\mathcal{P}(\rho_1) + \mathcal{E}(\rho_2) + W(q) \geq \mu(p, x)$. Summarizing these two parts we obtain $\mathcal{P}(\sigma) + W(q) = \max(\mathcal{P}(\rho_2) + W(q), \mathcal{P}(\rho_1) + \mathcal{E}(\rho_2) + W(q)) \geq \max(\mu(x, q), \mu(p, x))$. Thus, $\mu(p, q) \geq \min \{\max(\mu(p, x), \mu(x, q)) \mid x \in P_p\}$.

Let s be such that $\min \{\max(\mu(p, x), \mu(x, q)) \mid x \in P_p\} = \max(\mu(p, s), \mu(s, q))$, i.e. the minimum is reached on s, and let $\mu(p, s) = \mathcal{P}(\sigma) + W(s)$ for some σ with ${}^\circ\sigma = p$ and $\sigma^\circ = s$ and $\mu(s, q) = \mathcal{P}(\gamma) + W(q)$ for some γ with ${}^\circ\gamma = s$ and $\gamma^\circ = q$. Then, $\sigma\gamma$ is a path from p to q and it should be taken into account while computing the minimum for $\mu(p, q)$. Hence, $\mu(p, q) \leq \mathcal{P}(\sigma\gamma) + W(q) = \max(\mathcal{P}(\sigma) + \mathcal{E}(\gamma), \mathcal{P}(\gamma)) + W(q) = \max(\mathcal{P}(\sigma) + \mathcal{E}(\gamma) + W(q), \mathcal{P}(\gamma) + W(q)) = \max(\mathcal{P}(\sigma) + W(s), \mu(s, q)) = \max(\mu(p, s), \mu(s, q)$. It implies that $\mu(p, q) \leq \min \{\max(\mu(p, x), \mu(x, q)) \mid x \in P_p\}$.

Therefore, $\mu(p, q) = \min \{\max(\mu(p, x), \mu(x, q)) \mid x \in P_p\}$.   □

Lemma 24 leads to the following efficient algorithm for computing μ. For two matrices $A, B : P_p \times P_p \to \mathbb{N}$, $A = (a(p, q)), B = (b(p, q))$, we define $A \circ B = (c(p, q))$ where $c(p, q) = \min \{\max(a(p, x), b(x, q)) \mid x \in P_p\}$. The matrix $\mu(p, q)$ is computed by initializing the matrix $A = (a(p, q))$ by $a(p, p) = 0$ and $a(p, q) = \min \{\mathcal{P}(t) + W(q) \mid t \in T \wedge {}^\circ t = p \wedge t^\circ = q\}$. We then compute the subsequent powers of A with respect to $\circ$. The computation eventually reaches the fixpoint since the values in the matrix can be changed only to strictly smaller ones with respect to a well-founded ordering on $\mathbb{N} \cup \{\omega\}$. Moreover, A^k takes into account all paths of length up to k. Therefore, the process terminates after no more steps than the length of the longest acyclic path in the net. Upon termination the matrix becomes $(\mu(p, q))$.

Example 25. In our example net from Fig. 1, we have only one transition t leading from i to p and $W(i) = 0$, $W(p) = 3$, $\mathcal{C}(t) = 4$, $\mathcal{P}(t) = 1$, giving $a(i, p) = 4$ initially. Our full initial matrix A and its iterations become

$$A = \begin{array}{c|ccccc} & i & p & q & s & f \\ \hline i & 0 & 4 & \omega & \omega & \omega \\ p & \omega & 0 & 4 & 5 & \omega \\ q & \omega & 4 & 0 & 3 & \omega \\ s & \omega & \omega & \omega & 0 & 6 \\ f & \omega & \omega & \omega & \omega & 0 \end{array} \quad A^2 = \begin{array}{c|ccccc} & i & p & q & s & f \\ \hline i & 0 & 4 & 4 & 5 & \omega \\ p & \omega & 0 & 4 & 4 & 6 \\ q & \omega & 4 & 0 & 3 & 6 \\ s & \omega & \omega & \omega & 0 & 6 \\ f & \omega & \omega & \omega & \omega & 0 \end{array} \quad A^3 = \begin{array}{c|ccccc} & i & p & q & s & f \\ \hline i & 0 & 4 & 4 & 4 & 6 \\ p & \omega & 0 & 4 & 4 & 6 \\ q & \omega & 4 & 0 & 3 & 6 \\ s & \omega & \omega & \omega & 0 & 6 \\ f & \omega & \omega & \omega & \omega & 0 \end{array}$$

We find $A^4 = A^3$, so A^3 gives the desired $\mu(x, y)$. We now check our condition: $\forall x \in P_p : \min \{\mu(y, x) \mid W(y) < W(x)\} \geq \min \{\mu(x, y) \mid W(y) < W(x)\}$.

Now $\min \{\mu(y, x) \mid W(y) < W(x)\} = \min \{\mu(x, y) \mid W(y) < W(x)\} = \omega$ for $x \in \{i, r, f\}$, since $W(i) = W(r) = W(f) = 0$ and no place has a smaller weight. Since $W(p) = 3$ and all other places have smaller weight, we have $\min \{\mu(y, p) \mid W(y) < W(p)\} = 4$ and $\min \{\mu(x, y) \mid W(y) < W(p)\} = 4$. Finally, for $x = q$ we have $\min \{\mu(y, q) \mid W(y) < W(q)\} = 4$ and $\min \{\mu(x, y) \mid W(y) < W(q)\} = 3$. Our condition holds, so the net is sound.

Now consider net N' from Fig. 2. Then the $(\mu(x, y))$ is computed as follows:

$$A = \begin{array}{c|ccccc} & i & p & q & s & f \\ \hline i & 0 & 4 & \omega & \omega & \omega \\ p & \omega & 0 & 4 & 5 & \omega \\ q & \omega & 3 & 0 & 3 & \omega \\ s & \omega & \omega & \omega & 0 & 6 \\ f & \omega & \omega & \omega & \omega & 0 \end{array} \quad A^2 = \begin{array}{c|ccccc} & i & p & q & s & f \\ \hline i & 0 & 4 & 4 & 5 & \omega \\ p & \omega & 0 & 4 & 4 & 6 \\ q & \omega & 3 & 0 & 3 & 6 \\ s & \omega & \omega & \omega & 0 & 6 \\ f & \omega & \omega & \omega & \omega & 0 \end{array} \quad A^3 = \begin{array}{c|ccccc} & i & p & q & s & f \\ \hline i & 0 & 4 & 4 & 4 & 6 \\ p & \omega & 0 & 4 & 4 & 6 \\ q & \omega & 3 & 0 & 3 & 6 \\ s & \omega & \omega & \omega & 0 & 6 \\ f & \omega & \omega & \omega & \omega & 0 \end{array}$$

A^3 is the fixpoint. Now, $\min \{\mu(y, p) \mid W(y) < W(p)\} = 3$ and $\min \{\mu(x, y) \mid W(y) < W(p)\} = 4$, so the net is not sound. Moreover, $\min \{\mu(x, y)\} = 4$, and thus we can use the construction from the proof of Lemma 17 to reproduce a deadlock from this net (see Example 18).

Observe that the computation proposed strongly resembles the All-Pairs Shortest Paths problem (Floyd-Warshal algorithm, see [7]; for a more efficient algorithm see [15]; also see [16] for a survey). Hence, our computation can benefit from efficient matrix multiplication algorithms. Moreover, to decrease the number of multiplication steps we can repeatedly square the result of the previous step, i.e., instead of $A, A^2, A^3, \ldots$ we compute $A, A^2, A^4, \ldots$. The number of multiplication steps is logarithmic in the length of the longest acyclic path of the net.

Corollary 26. *For an SM1WF-net with P places and T transitions the soundness decision algorithm presented above has complexity of $O(P^3 \log P + T)$.*

6 Conclusion

We have introduced an extension of Workflow nets: *Resource-Constrained Workflow nets* and defined a notion of soundness on this class of nets, which is an extension of the soundness notion for WF-nets. In addition to the soundness requirements for WF-nets, soundness for RCWF-nets states that no resources are created during the processing and all resources are returned to their resource place when the processing is completed; moreover, no deadlock or livelock can arise due to the lack of resources. We showed how to reduce the problem of soundness for a general class of RCWF-nets with one resource type to the problem of soundness for SM1WF-nets and gave a necessary and sufficient condition of soundness for SM1WF-nets. The decision algorithm we described has a polynomial complexity w.r.t. the number of states of the production net marked with a single initial token.

Future Work. We have considered here the problem of soundness for RCWF-nets with one resource type. Finding a necessary and sufficient condition of soundness for RCWF-nets with multiple resource types is left for future work. Another direction for future research is to find a method to transform a given unsound RCWF-net into a sound one by applying modifications of one type only: transitions may claim and release more resources than in the original situation.

Future work includes also the integration of our algorithm into tools working with this class of nets.

References

1. W. M. P. van der Aalst. Verification of workflow nets. In P. Azéma and G. Balbo, editors, *Application and Theory of Petri Nets 1997, ICATPN'1997*, volume 1248 of *Lecture Notes in Computer Science*. Springer-Verlag, 1997.
2. W. M. P. van der Aalst. The Application of Petri Nets to Workflow Management. *The Journal of Circuits, Systems and Computers*, 8(1):21–66, 1998.
3. W. M. P. van der Aalst. Workflow verification: Finding control-flow errors using Petri-net-based techniques. In W. M. P. van der Aalst, J. Desel, and A. Oberweis, editors, *Business Process Management: Models, Techniques, and Empirical Studies*, volume 1806 of *Lecture Notes in Computer Science*, pages 161–183. Springer-Verlag, 1999.
4. W. M. P. van der Aalst and K. M. van Hee. *Workflow Management: Models, Methods, and Systems*. MIT Press, 2002.
5. K. Barkaoui and L. Petrucci. Structural analysis of workflow nets with shared resources. In *Workflow management: Net-based Concepts, Models, Techniques and Tools (WFM'98)*, volume 98/7 of *Computing science reports*, pages 82–95. Eindhoven University of Technology, 1998.
6. J. Colom. The resource allocation problem in flexible manufacturing systems. In W. van der Aalst and E. Best, editors, *Application and Theory of Petri Nets 2003, ICATPN'2003*, volume 2679 of *Lecture Notes in Computer Science*, pages 23–35. Springer-Verlag, 2003.
7. T. H. Cormen, C. E. Leiserson, and R. L. Rivest. *Introduction to Algorithms*. MIT Press, 1990.
8. E. W. Dijkstra. Ewd 623. *Selected writings on computing: a personal perspective*, 1982.
9. J. Ezpeleta. Flexible manufacturing systems. In C. Girault and R. Valk, editors, *Petri nets for systems engineering*. Springer-Verlag, 2003.

10. J. Ezpeleta, J. M. Colom, and J. Martínez. A Petri net based deadlock prevention policy for flexible manufacturing systems. *IEEE Transactions on Robotics and Automation*, 11(2):173–184, 1995.
11. K. van Hee, N. Sidorova, and M. Voorhoeve. Soundness and separability of workflow nets in the stepwise refinement approach. In W. van der Aalst and E. Best, editors, *Application and Theory of Petri Nets 2003, ICATPN'2003*, volume 2679 of *Lecture Notes in Computer Science*, pages 337–356. Springer-Verlag, 2003.
12. K. Lautenbach. *Liveness in Petri Nets*. Internal Report of the Gesellschaft für Mathematik und Datenverarbeitung, Bonn, Germany, ISF/75-02-1, 1975.
13. M. Silva and E. Teruel. Petri nets for the design and operation of manufacturing systems. *European Journal of Control*, 3(3):182–199, 1997.
14. M. Silva and R. Valette. Petri nets and flexible manufacturing. In G. Rozenberg, editor, *Applications and Theory of Petri Nets*, volume 424 of *Lecture Notes in Computer Science*, pages 374–417. Springer, 1990.
15. T. Takaoka. Subcubic cost algorithms for the all pairs shortest path problem. *Algorithmica*, 3(20):309–318, 1998.
16. U. Zwick. Exact and approximate distances in graphs – a survey. In F. Meyer auf der Heide, editor, *Algorithms – ESA 2001, 9th Annual European Symposium, Aarhus, Denmark, August 28-31, 2001, Proceedings*, Lecture Notes in Computer Science, pages 33–48. Springer Verlag, 2001.

High-Level Nets with Nets and Rules as Tokens

Kathrin Hoffmann[1], Hartmut Ehrig[1], and Till Mossakowski[2]

[1] Institute for Software Technology and Theoretical Computer Science,
Technical University Berlin
[2] BISS, Department of Computer Science,
University of Bremen

Abstract. High-Level net models following the paradigm "nets as tokens" have been studied already in the literature with several interesting applications. In this paper we propose the new paradigm "nets and rules as tokens", where in addition to nets as tokens also rules as tokens are considered. The rules can be used to change the net structure. This leads to the new concept of high-level net and rule systems, which allows to integrate the token game with rule-based transformations of P/T-systems. The new concept is based on algebraic high-level nets and on the main ideas of graph transformation systems. We introduce the new concept with the case study "House of Philosophers", a dynamic extension of the well-known dining philosophers. In the main part we present a basic theory for rule-based transformations of P/T-systems and for high-level nets with nets and rules as tokens leading to the concept of high-level net and rule systems.

Keywords: High-level net models, algebraic high-level nets, nets and rules as tokens, integration of net theory and graph transformations, case study: House of Philosophers, algebraic specifications, graph grammars and Petri net transformations.

1 Introduction

The paradigm "nets as tokens" has been introduced by Valk in order to allow nets as tokens, called object nets, within a net, called a system net (see [Val98, Val01]). This paradigm has been very useful to model interesting applications in the area of workflow, agent-oriented approaches or open system networks. Especially his concept of elementary object systems [Val01] has been used to model the case study of the hurried philosophers proposed in [Sil01]. In elementary object systems object nets can move through a system net and interact with both the system net and with other object nets. This allows to change the marking of the object net, but not their net structure. According to the requirements of the hurried philosophers in [Sil01] the philosophers have the capability to introduce a new guest at the table, which - in the case of low level Petri nets - certainly changes the net structure of the token net representing the philosophers at the table. We use the notion of token net instead of object net in order to avoid confusion with features of object-oriented modeling. Instead our intention

G. Ciardo and P. Darondeau (Eds.): ICATPN 2005, LNCS 3536, pp. 268–288, 2005.

is to consider the change of the net structure as rule-based transformation of Petri nets in the sense of graph transformation systems [Ehr79, Roz97]. In order to integrate the token game of Petri nets with rule-based transformations, we propose in this paper the new paradigm "nets and rules as tokens" leading to the concept of high-level net and rule systems.

In Section 2 we show how this new concept can be used to model the main requirements of the hurried philosophers [Sil01]. Of course, this concept has interesting applications in all areas where dynamic changes of the net structure have to be considered while the system is still running. This applies especially to flexible workflow systems (see [Aal02]) and medical information systems (see [Hof00]).

In Section 3 we introduce the basic theory for rule-based transformations of P/T-systems. This theory is inspired by graph transformation systems [Ehr79, Roz97], which have been generalized already to net transformations systems in [EHK91, EP04], including high-level and low-level nets. The theory in these papers is based on pushouts in the corresponding categories according to the double-pushout approach of graph transformations in [Ehr79]. In order to improve the intuition of our concepts for the Petri net community we give in this paper an explicit approach of rule-based transformations for P/T-systems, which is new and extends the theory of P/T-net transformations taking into account also initial markings, and avoids categorical terminology like pushouts. Moreover, the interaction of the token game and transformation of nets - as considered in this paper - has not been studied up to now.

In Section 4 we introduce high-level nets with nets and rules as tokens leading to our new concept of high-level net and rule (HLNR) systems motivated above. This new concept is based on algebraic high-level (AHL) nets [PER95] using the terminology of [EHP02]. In order to model nets and rules as tokens we present a specific signature together with a corresponding algebra with specific sorts for P/T-systems and rules. Moreover, there are operations corresponding to firing of a transition and applying a rule to a P/T-system respectively. Since AHL-nets are based on classical algebraic specifications (see [EM85]) we are able to give a set theoretic definition of domains and operations. In order to obtain also an algebraic specification we need algebraic higher-order specifications as presented in HASCASL [Hets, SM02], which allows to specify function types with set-theoretic notions of semantics using intensional algebras.

In Section 5 we discuss specification and implementation aspects for our approach. More precisely, we discuss how the concept of algebraic higher-order (AHO) nets based on HASCASL, which has been already introduced in [HM03], can be used to specify the algebra of HLNR-systems. Since tools for HASCASL already have been implemented [Mos05, Hets] this is an important step towards implementation and tool support for HLNR-systems. Unfortunately, this is not possible using CPN tools [RWL03] for Coloured Petri (CP) Nets [Jen92]. Actually, CP-Nets are based on an extension of the functional language Standard ML [MTH97]. As Standard ML does not allow functional equivalence testing, it is not suitable for our purpose where we need a form of functional equivalence. The conclusion in Section 6 includes proposals for future work.

2   Case Study: House of Philosophers

In order to illustrate the concepts described in Section 3 and Section 4 we will present a small system inspired by the case study "the Hurried Philosophers" of C. Sibertin-Blanc proposed in [Sil01] which is a refinement of the well-known classical "Dining Philosophers".

Requirements. In our case study "House of Philosophers" presented below we essentially consider the following requirements:

1. There are three different locations in the house where the philosophers can stay: the library, the entrance-hall, and the restaurant;
2. In the restaurant there are different tables where one or more philosophers can be placed to have dinner;
3. Each philosopher can eat at a table only when he has both forks, i.e. the philosophers at each table follow the rules of the classical "Dining Philosophers";
4. The philosophers in the entrance-hall have the following additional capabilities:
   (a) They are able to invite another philosopher in the entrance-hall to enter the restaurant and to take place at one of the tables;
   (b) They are able to ask a philosopher at one of the tables with at least two philosophers to leave the table and to enter the entrance-hall.

System Level. In Fig. 1 we present the system level of our version of the case study. The system level is given by a high-level net and rule system, short HLNR-system, which will be explained in Section 4. The marking of the HLNR-system shows the distribution of the philosophers at different places in the house and the firing behavior of the HLNR-system describes the mobility of the philosophers. There are three different locations in the house where the philosophers can stay: the library, the entrance-hall, and the restaurant. Each location is represented by its own place in the HLNR-system in Fig. 1. Initially there are two philosophers at the library, one philosopher at the entrance-hall, and four additional philosophers are at table 1 resp. table 2 (see Fig. 5 and Fig. 6) in the restaurant.

Philosophers may move around, which means they might leave and enter the library and they might leave and enter the tables in the restaurant. The mobility aspect of the philosophers is modeled by transitions termed *enter* and *leave library* as well as *enter* and *leave restaurant* in our HLNR-system in Fig. 1. While the philosophers are moving around, the static structure of the philosophers is changed by rule-based transformations. E.g. a philosopher enters the restaurant and arrives at a table. Then the structure and the seating arrangement of the philosophers have to be changed. For this reason, we have tokens of type *Rules*, $rule_1$, ..., $rule_4$, which are used as resources. Because the philosophers have their own internal behavior, there are two transitions, *start/stop reading* and *start/stop activities*, to realize the change of the behavior.

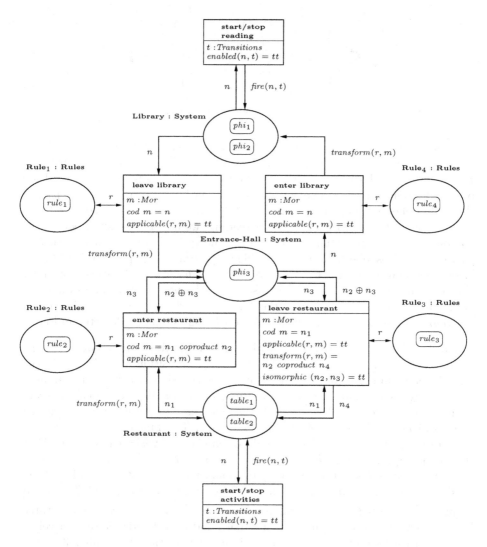

Fig. 1. High-level net and rule system of "House of Philosophers"

Fig. 2. Token net phi_1 of philosopher 1   **Fig. 3.** Token net phi'_1 of philosopher 1

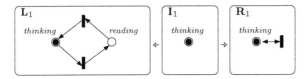

Fig. 4. Token rule of rule $rule_1$

Token Level. The token level consists of two different types of tokens: P/T-systems and rules. They are represented as tokens in the places typed *System* and *Rules* of the HLNR-system in Fig. 1. The tokens on system places are modeled by P/T-systems, i.e. Petri nets with an initial marking. In Fig. 2 the net phi_1 of the philosopher 1 is depicted, which - in the state *thinking* - is used as a token on the place *Library* in Fig. 1. To start reading, we use the transition *start/stop reading* of the HLNR-system in Fig. 1. First the variable n is assigned to the net phi_1 of the philosopher 1 and the variable t to a transition $t_0 \in T_0$ where T_0 is a given vocabulary of transitions. The condition *enabled(n,t)=tt* means that under this assignment t_0 is an enabled transition in the net of phi_1. The evaluation of the term $fire(n, t)$ computes the follower marking of the net (i.e. token $reading_1$) and we obtain the new net phi_1' of the philosopher 1 depicted in Fig. 3.

Mobility of Philosophers by Application of Rules. We assume that the philosopher 1 wants to leave the library, i.e. the transition *leave library* in the HLNR-system in Fig. 1 must fire. For this purpose we have to give an assignment for the variables n, r and m in the net inscriptions of the transition. They are assigned to the net phi_1 (see Fig. 2), the rule $rule_1$ (see Fig. 4), and a match morphism $m_1 : L' \to G$ between P/T-systems. The first condition *cod m=n* requires $G = phi_1$ and the second condition *applicable(r,m)=tt* makes sure that rule $rule_1$ is applicable to phi_1, especially $L' = L_1$, s.t. the evaluation of the term $transform(r,m)$ leads to the new net phi_1'' isomorphic to R_1 of $rule_1$ in Fig. 4. As result of this firing step phi_1 is removed from place *Library* and phi_1'' is added on place *Entrance-Hall*. In general, a rule $r = (L \xleftarrow{i_1} I \xrightarrow{i_2} R)$ is given by three P/T- systems called left-hand side, interface, and right-hand side respectively.

In a further step the philosopher 1 is invited by the philosopher 3 to enter the restaurant in order to take place as a new guest at the table 1. The philosopher 3 accompanies philosopher 1 but returns to the entrance-hall. The token net phi_3 of philosopher 3 is isomorphic to R_1 of $rule_1$ in Fig. 4 where *thinking* in R_1 is replaced by $thinking_3$. Currently the philosophers 4 and 5 are at the table 1 (see Fig. 5). Both philosophers may start eating, but apparently compete for their shared forks, where *left fork$_4$=right fork$_5$* and *left fork$_5$=right fork$_4$*. Analogously table 2 has the same net structure as table 1 but different philosophers are sitting at table 2 (see Fig. 6). To introduce the philosopher 1 at the table 1 the seating arrangement at table 1 has to be changed. In our case the new guest takes place between philosopher 4 and 5. Formally, we apply rule $rule_2 = (L_2 \xleftarrow{i_1} I_2 \xrightarrow{i_2} R_2)$, which is depicted in the upper row of Fig. 7 and used as token on place $Rule_2$. We have to give an assignment v for the variables

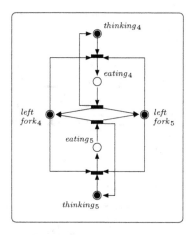

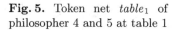

Fig. 5. Token net $table_1$ of philosopher 4 and 5 at table 1

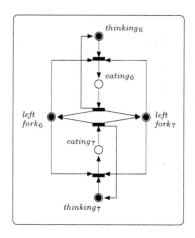

Fig. 6. Token net $table_2$ of philosopher 6 and 7 at table 2

of the transition *enter restaurant*, i.e. variables n_1, n_2, n_3, r, and m. The assignment v is defined by $v(n_1) = table_1$, $v(n_2) = phi_1''$, $v(n_3) = phi_3$, $v(r) = rule_2$, and $v(m) = g$ (see match morphism $g : L_2 \to G$ in Fig. 7). Then we compute the disjoint union of the P/T-system phi_1'' and the P/T-system $table_1$ as denoted by the net inscription n_1 *coproduct* n_2 in the firing condition of the transition *enter restaurant*. The result is the disjoint union of both nets shown as P/T-system G in Fig. 7.

In our case the match g maps $thinking_j$ and $eating_j$ in L_2 to $thinking_4$ and $eating_4$ in G of Fig. 7. The condition $cod\ m = n_1$ *coproduct* n_2 makes sure that the codomain of g is equal to G. The second condition $applicable(r,m)=tt$ checks if $rule_2$ is applicable with match g to G (see "gluing condition" (Def. 4) and "applicability" (Def. 5) in Section 3). In the direct transformation shown in Fig. 7 we delete in a first step $g(L_2 \setminus I_2)$ from G leading to P/T-system C. Note, that a positive check of the "gluing condition' makes sure that C is a well-defined P/T-system (see Prop. 2 in Section 3). In a second step we glue together the P/T-systems C and R_2 along I_2 leading to P/T-system H in Fig. 7. H shows the new version of table 1 given by the net $table_1'$ of table 1, where philosophers 1, 4, and 5 are sitting at the table, all of them in state *thinking*. The effect of firing the transition *enter restaurant* in Fig. 1 with assignments of variables as discussed above is the removal of P/T-systems phi_1'' from place *Entrance Hall* and $table_1$ from place *Restaurant* and adding P/T-System $table_1'$ to the place *Restaurant*.

Philosophers in the entrance-hall have the capability to ask one of the philosophers in the restaurant to leave; this is realized in our system by the transition *leave restaurant* in Fig. 1. We use the rule $rule_3$ defined as inverse of $rule_2$ in Fig. 7, i.e. $rule_3 = (R_2 \xleftarrow{i_2} I_2 \xrightarrow{i_1} L_2)$, which is present as a token on place $Rule_3$. This rule is applied with inverse direct transformation to that depicted in Fig. 7. Finally, the rule $rule_4$ is the inverse of rule $rule_1$ (see Fig. 4), enabling

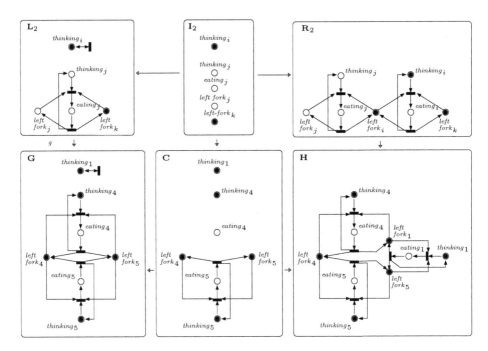

Fig. 7. Direct Transformation

the philosopher to enter the library by firing of the transition *enter library* in Fig. 1. We have to guarantee that after the application of *rule₃* the philosopher who is leaving the restaurant goes into the entrance-hall. In our case one philosopher is asked by philosopher 3 in the entrance-hall to leave the table. Formally this is denoted by the firing condition $isomorphic(n_2, n_3) = tt$ which ensures that the net of the philosophers denoted by n_2 is isomorphic to the net phi_3 of philosopher 3 denoted by n_3.

The execution of philosopher activities at different tables, i.e. the firing of the transition *start/stop activities* in Fig. 1, is analogously defined as the firing of the transition *start/stop reading* described above.

Validation of Requirements. Our case study "House of Philosophers" satisfies the requirements presented in the beginning of this section.

1. The three different locations in the house are represented by places *Library*, *Entrance-Hall*, and *Restaurant* in Fig. 1;
2. In the initial state we have the two tables *table₁* with philosophers 4 and 5 and *table₂* with philosophers 6 and 7 on place *Restaurant*. In a later state also philosopher 1 is sitting at *table₁* as shown by net H of Fig. 7;
3. If there are $n \geq 2$ philosophers sitting at each table, the table with n philosophers is presented by the classical "Dining Philosophers" net;
4. The capability of a philosopher in the entrance-hall to invite another philosopher to enter (leave) the restaurant is given by firing of the transition

enter restaurant (leave restaurant) in Fig. 1. The applicability of the rule *rule*$_3$ ensures that a philosopher only leaves a table with at least two philossphers.

Related Work. In [ADC01] there are several other solutions for the case study "the Hurried Philosophers" modeled by different kinds of (high-level) net classes. Most of these approaches integrate object-oriented modeling and Petri nets, including e.g. inheritance, encapsulation, and dynamic binding, etc. In this paper we do not need features of object-oriented modeling. But it is an interesting aspect to extend our approach by integration of these features.

In the solution of the case study using elementary object systems [Val01], each philosopher has his own place and the exchange of forks between the philosophers is realized by an interaction relation. By contrast in our case each table is modeled by its own P/T-system and describes the states and the seating arrangement of present philosophers. Moreover we use rule-based transformations to change the structure of P/T-systems, especially the states and the seating arrangement. In the sense of object-oriented modeling it might be considered to split up the table with philosophers into a net table with only the table properties and nets for each philosopher at the table. In fact our approach allows to model such self-contained components but this would lead to a much more complex model. The advantage of our approach compared with elementary object systems is a more flexible modeling technique. While the HLNR-system in Fig. 1 is fixed we can add further philosophers and philosophers at tables by adding further tokens of type *System* to our model. Analogously we can add further token rules to realize other kinds of transformations.

Note, that elementary object systems [Val01] allow a simple notion of nets as tokens, such that most principles of elementary net theory are respected and extended. Here on the one hand the system-object interaction relation consists of transitions in the system net and transitions in the object net which have to be fired in parallel, and on the other hand the object-object interaction relation guards the parallel firing of transitions in different object nets. By contrast, we are using different formal frameworks for the token level and the system level. In order to integrate interaction relations into our concept of HLNR-system we can extend the signature and the algebra of the algebraic high-level net by appropriate operations and formulate the dependencies between transitions in the firing conditions of the HLNR-system. In this way we can show that elementary object systems can be translated into semantically equivalent HLNR-systems extended by interaction relations.

The idea of controlled modification of token nets is discussed in the context of linear logic Petri nets [Far99] and feature structure nets [Wie01]. The difference to our approach is that in those approaches, the modification is not carried out by rule tokens, but by transition guards. We are not restricted to define a specific token rule for each transition, but we are able to give a (multi-)set of token rules as resources bound to each transition, which realize the local replacement of subnets.

3   Rule-Based Transformation of P/T-Systems

In this section we present rule-based transformations of nets following the double-pushout (DPO) approach of graph transformations in the sense of [Ehr79, Roz97]. As net formalism we use P/T-systems following the notation of "Petri nets are Monoids" in [MM90]. In this notation a P/T-system is given by $PN = (P, T, pre, post, M^0)$ with pre- and post domain functions $pre, post : T \rightarrow P^\oplus$ and initial marking $M^0 \in P^\oplus$, where $P^\oplus$ is the free commutative monoid over the set P of places with binary operation $\oplus$. Note that M^0 can also be considered as function $M^0 : P \rightarrow \mathbb{N}$ with finite support and the monoid notation $M^0 = 2p_1 \oplus 3p_2$ means that we have two tokens on place p_1 and three tokens on p_2. A transition $t \in T$ is M-enabled for a marking $M \in P^\oplus$ if we have $pre(t) \leq M$ and in this case the follower marking M' is given by $M' = M \ominus pre(t) \oplus post(t)$. Note that the inverse $\ominus$ of $\oplus$ is only defined in $M_1 \ominus M_2$ if we have $M_2 \leq M_1$.

In order to define rules and transformations of P/T-systems we have to introduce P/T-morphisms which are suitable for our purpose.

Definition 1 (P/T-Morphisms).
Given P/T-systems $PN_i = (P_i, T_i, pre_i, post_i, M_i^0)$ for $i = 1, 2$, a P/T-morphism $f : PN_1 \rightarrow PN_2$ is given by $f = (f_P, f_T)$ with functions $f_P : P_1 \rightarrow P_2$ and $f_T : T_1 \rightarrow T_2$ satisfying

> **(1)** $f_P^\oplus \circ pre_1 = pre_2 \circ f_T$ *and* $f_P^\oplus \circ post_1 = post_2 \circ f_T$
> **(2)** $f_P^\oplus(M_{1|p}^0) \leq M_{2|f_P(p)}^0$ *for* $p \in P_1$

*Note that the extension $f_P^\oplus : P_1^\oplus \rightarrow P_2^\oplus$ of $f_P : P_1 \rightarrow P_2$ is defined by $f_P^\oplus(\sum_{i=1}^n k_i \cdot p_i) = \sum_{i=1}^n k_i \cdot f_P(p_i)$ and the restriction $M_{1|p}^0$ by $M_{1|p}^0 = M_1^0(p) \cdot p$ where M_1^0 is considered as function $M_1^0 : P \rightarrow \mathbb{N}$. (1) means that f is compatible with pre- and post domain and (2) that the initial marking of N_1 at place p is smaller or equal to that of N_2 at $f_P(p)$. Moreover the P/T-morphism f is called strict if $f_P^\oplus(M_{1|p}^0) = M_{2|f_P(p)}^0$ and f_P, f_T are injective **(3)**.*

The category defined by P/T-systems and P/T-morphisms is denoted by **PTSys** where the composition of P/T-morphisms is defined componentwise for places and transitions. Examples of P/T-morphisms are given in Fig. 7.

The next step in order to define transformations of P/T-systems is to define the gluing of P/T-systems in analogy to concatenation in the string case.

Definition 2 (Gluing of P/T-Systems).
Given P/T-systems $PN_i = (P_i, T_i, pre_i, post_i, M_i^0)$ for $i = 0, 1, 2$ with strict inclusion $inc : PN_0 \rightarrow PN_1$ and P/T-morphism $f : PN_0 \rightarrow PN_2$. Then the gluing PN_3 of PN_1 and PN_2 via (PN_0, f), written $PN_3 = PN_1 +_{(PN_0, f)} PN_2$, is defined by the following diagram (1), called "gluing diagram", with

> *1. $\forall p \in P_1 = P_0 \uplus (P_1 \setminus P_0) : f_P'(p) = \underline{if}\ p \in P_0\ \underline{then}\ f_P(p)\ \underline{else}\ p$*
> *$\forall t \in T_1 = T_0 \uplus (T_1 \setminus T_0) : f_T'(t) = \underline{if}\ t \in T_0\ \underline{then}\ f_T(t)\ \underline{else}\ t$*

2. $PN_3 = (P_3, T_3, pre_3, post_3, M_3^0)$ *with*
 - $P_3 = P_2 \uplus (P_1 \setminus P_0)$, $T_3 = T_2 \uplus (T_1 \setminus T_0)$,
 - $pre_3(t) = $ *if* $t \in T_2$ *then* $pre_2(t)$
   else $f_P'^{\oplus}(pre_1(t))$,
 - $post_3(t) = $ *if* $t \in T_2$ *then* $post_2(t)$
   else $f_P'^{\oplus}(post_1(t))$ *and*
 - $M_3^0 = M_2^0 \oplus (M_1^0 \ominus M_0^0)$.

$$\begin{array}{ccc} PN_0 & \xrightarrow{\ inc\ } & PN_1 \\ \scriptstyle f \downarrow & (1) & \downarrow \scriptstyle f' \\ PN_2 & \xrightarrow[\ inc'\ ]{} & PN_3 \end{array}$$

Remark 1. The disjoint union in the definition of P_3 and T_3 takes care of the problem that there may be places or transitions in PN_2, which are - by chance - identical to elements in $P_1 \setminus P_0$ or $T_1 \setminus T_0$, but only elements in PN_0 and $f(PN_0)$ should be identified. In this case the elements of $P_1 \setminus P_0$ and $T_1 \setminus T_0$ should be renamed before applying the construction above.

Proposition 1 (Gluing of P/T-Systems).
The gluing $PN_3 = PN_1 +_{(PN_0, f)} PN_2$ is a well-defined P/T-system such that $f' : PN_1 \to PN_3$ is a P/T-morphism, $inc' : PN_2 \to PN_3$ is a strict inclusion and the gluing diagram (1) commutes, i.e. $f' \circ inc = inc' \circ f$.

Proof. 1. PN_3 *is a well-defined P/T-system, because $pre_3, post_3 : T_3 \to P_3^{\oplus}$ are well-defined functions. Now $f' = (f_P', f_T') : PN_1 \to PN_3$ is a P/T-morphism, because we have $pre_3 \circ f_T' = f_P'^{\oplus} \circ pre_1$ (and similar for post) by case distinction:*

Case 1. For $t \in T_0$ we have $pre_3(f_T'(t)) = pre_3(f_T(t)) = pre_2(f_T(t)) = f_P^{\oplus}(pre_0(t)) = f_P'^{\oplus}(pre_0(t)) = f_P'^{\oplus}(pre_1(t))$.

Case 2. For $t \in T_1 \setminus T_0$ we have $pre_3(f_T'(t)) = pre_3(t) = f_P'^{\oplus}(pre_1(t))$.

We have marking compatibility of f' by:

Case 1. For $p \in P_0$ we have
$$f_P'^{\oplus}(M_{1|p}^0) = f_P^{\oplus}(M_{0|p}^0) \leq M_{2|f_P(p)}^0 \leq M_{3|f_P(p)}^0 = M_{3|f_P'(p)}^0.$$

Case 2. For $p \in P_1 \setminus P_0$ we have
$$f_P'^{\oplus}(M_{1|p}^0) = f_P'^{\oplus}((M_1^0 \ominus M_0^0)_{|p}) = (M_1^0 \ominus M_0^0)_{|p} \leq M_{3|f_P'(p)}^0$$

2. *$inc' : PN_2 \to PN_3$ is a P/T-system inclusion by construction. The marking M_3^0 is well-defined because $M_0^0 \leq M_1^0$ and $M_{0|p}^0 = M_{1|p}^0$ for $p \in P_0$ by strict inclusion $inc : PN_0 \to PN_1$. Moreover inc' is strict, because we have $M_1^0 \ominus M_0^0 \in (P_1 \setminus P_0)^{\oplus}$ which implies for $p \in P_2$ $M_{2|p}^0 = M_{3|p}^0$.*

3. *$f' \circ inc = inc' \circ f$ by construction*

Remark 2. The gluing diagram (1) is a pushout diagram in the category **PTSys**. This implies that the transformation of P/T-systems defined below is in the spirit of the double-pushout approach for graph transformations and high-level replacement systems (see [Ehr79, EHK91]).

Two examples of gluing and gluing diagrams are given in Fig. 7, where $G = L_2 +_{I_2} C$ and $H = R_2 +_{I_2} C$ in the left hand and the right hand gluing diagram respectively. Our next goal is to define rules, application of rules and transformations of P/T-systems.

Definition 3 (Rule of P/T-Systems). *A rule $r = (L \xleftarrow{i_1} I \xrightarrow{i_2} R)$ of P/T-systems consists of P/T-systems L, I, and R, called left-hand side, interface, and right-hand side of r respectively, and two strict P/T-morphisms $I \xrightarrow{i_1} L$ and $I \xrightarrow{i_2} R$ which are inclusions.*

Remark 3. The application of a rule r to a P/T-system G is given by a P/T-morphism $L \xrightarrow{m} G$, called match. Now a direct transformation $G \xRightarrow{r} H$ via r can be constructed in two steps. In a first step we construct the context C given by $(G - m(L)) \cup m \circ i_1(I)$ and P/T-morphisms $I \xrightarrow{c} C$ and $C \xrightarrow{c_1} G$, where c_1 is a strict inclusion. This means we remove the match $m(L)$ from G and preserve the interface $m \circ i_1(I)$. In order to make sure that C becomes a subsystem of G we have to require a "gluing condition" (see Def. 4). This makes sure that C is a P/T-system

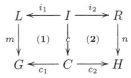

and we have $m \circ i_1 = c_1 \circ c$ in the "context diagram" (1). In the second step we construct H as gluing of C and R along I, this means we obtain the gluing diagram (2) from $I \xrightarrow{c} C$ and $I \xrightarrow{i_2} R$.

Now we define the gluing condition and the context construction.

Definition 4 (Gluing Condition).
Given a strict inclusion morphism $i_1 : I \to L$ and a P/T-morphism $m : L \to G$ the gluing points GP, dangling points DP and the identification points IP of L are defined by
$$GP = P_I \cup T_I$$
$$DP = \{p \in P_L | \exists t \in (T_G \setminus m_T(T_L)) : m_P(p) \in pre_G(t) \oplus post_G(t)\}$$
$$IP = \{p \in P_L | \exists p' \in P_L : p \neq p' \wedge m_P(p) = m_P(p')\}$$
$$\cup \{t \in T_L | \exists t' \in T_L : t \neq t' \wedge m_T(t) = m_T(t')\}$$
where $p \in P_L = \sum_{i=1}^n k_i \cdot p_i$ means $p = p_i$ and $k_i \neq 0$ for some i. Then the gluing condition is satisfied if all dangling and identifications points are gluing points, i.e $DP \cup IP \subseteq GP$.

Proposition 2 (Context P/T-System). *Given a strict inclusion $i_1 : I \to L$ and a P/T-morphism $m : L \to G$ then the following context P/T-system C is well-defined and leads to the following commutative diagram (1), called "context diagram", if the gluing condition $DP \cup IP \subseteq GP$ is satisfied.*

$C = (P_C, T_C, pre_C, post_C, M_C^0)$ *is defined by*

$$P_C = (P_G \setminus m_P(P_L)) \cup m_P(P_I),$$
$$T_C = (T_G \setminus m_T(T_L)) \cup m_T(T_I),$$
$$pre_C = pre_{G|C}, post_C = pre_{G|C} \; and$$
$$M_C^0 = M_{G|C}^0.$$

$$
\begin{array}{ccc}
I & \xrightarrow{\;i_1\;} & L \\
{\scriptstyle c}\downarrow & (1) & \downarrow{\scriptstyle m} \\
C & \xrightarrow[\;c_1\;]{} & G
\end{array}
$$

The morphisms in (1) are defined by $c : I \to C$ *to be the restriction of* $m : L \to G$ *to* I, *and* $c_1 : C \to G$ *to be a strict inclusion.*

Proof. The P/T-system C and $pre_C, post_C : T_C \to P_C^\oplus$ with $pre_C = pre_{G|C}$ and $post_C = pre_{G|C}$ are well-defined if $DP \cup IP \subseteq GP$. For $t \in T_C$ we have to show $pre_C(t) \in P_C^\oplus$ (and similar for $post_C(t)$).

Case 1. For $t \in T_G \setminus m_T(T_L)$ we have $pre_C(t) = pre_G(t) = \sum_{i=1}^n k_i \cdot p_i$. Assume $p_i \notin P_C$ for some $i \leq n$. Then $p_i \in m_P(P_L) \setminus m_P(P_I)$ with $p_i \in pre_G(t)$. Hence there is $p_i' \in P_L \setminus P_I$ with $m_P(p_i') = p_i$. This implies $p_i' \in DP$ and $p_i' \notin GP$ and contradicts the gluing condition $DP \cup IP \subseteq GP$.

Case 2. For $t \in m_T(T_I)$ we have $t' \in T_I$ with $t = m_T(t')$. This implies $pre_C(t) = pre_G(t) = pre_G(m_T(t')) = m_P^\oplus(pre_L(t')) = m_P^\oplus(pre_I(t')) \in m_P^\oplus(P_I^\oplus) = (m_P(P_I))^\oplus \subseteq P_C^\oplus$.

Moreover $c : I \to C$ *satisfies the marking condition (2) in Def. 1, because this is true for* $m : L \to G$ *and* c *is restriction of* m. *Finally* $c_1 : C \to G$ *is a strict inclusion by construction. This leads to the commutative diagram (1) in* **PTSys**.

Remark 4. Note that we have not used the "identification condition" $ID \subseteq GP$, which is part of the gluing condition. But this is needed to show that the context diagram (1) is - up to isomorphism - also a gluing diagram and hence a pushout diagram in the category **PTSys**. This means that C is constructed in such a way that G becomes the gluing of L and C via I, i.e. $G \cong L +_I C$.

An example of a context diagram is the left diagram in Fig. 7, where C is the context P/T-system for $i_2 : I_2 \to L_2$ and $g : L_2 \to G$. Now a direct transformation is given by the combination of a context diagram and a gluing diagram.

Definition 5 (Applicability of Rules and Transformation).

A rule $r = (L \xleftarrow{i_1} I \xrightarrow{i_2} R)$ *is called applicable at match* $L' \xrightarrow{m} G$ *if* $L = L'$ *and the gluing condition is satisfied for* i_1 *and* m. *In this case we obtain a context P/T-system* C *with context diagram (1) and a gluing diagram (2) with* $H = C +_I R$ *leading to a direct transformation* $G \overset{r}{\Longrightarrow} H$ *consisting of the following diagrams (1) and (2). A (rule-based) transformation* $G \overset{*}{\Longrightarrow} H$ *is a sequence of direct transformations* $G = G_0 \overset{r_1}{\Longrightarrow} G_1 \overset{r_2}{\Longrightarrow} \ldots \overset{r_n}{\Longrightarrow} G_n = H$ *with* $G = H$ *for* $n = 0$. *An example for a direct transformation is given in Fig. 7.*

$$
\begin{array}{ccccc}
L & \xleftarrow{\;i_1\;} & I & \xrightarrow{\;i_2\;} & R \\
{\scriptstyle m}\downarrow & (1) & {\scriptstyle c}\downarrow & (2) & \downarrow{\scriptstyle n} \\
G & \xleftarrow[\;c_1\;]{} & C & \xrightarrow[\;c_2\;]{} & H
\end{array}
$$

Remark 5. As pointed out in Remark 2 and Remark 4 already the context diagram (1) and the gluing diagram (2) are pushout diagrams in the category **PTSys**. Hence a direct transformation $G \stackrel{r}{\Longrightarrow} H$ is given by the two pushouts (1) and (2), also called double pushout (DPO). In the DPO-approach of graph transformations (see [Ehr79]), high-level replacement systems [EHK91] and Petri net transformations [EP04] a direct transformation is defined by a DPO-diagram. For P/T-systems our definition is equivalent up to isomorphism to the existence of a DPO in the category **PTSys**.

4   High-Level Nets with Nets and Rules as Tokens

In this section we review the definition of algebraic high-level (AHL) nets in the notation of [EHP02] and [EM85] for algebraic specifications. Moreover we present a specific HLNR-SYSTEM-SIG signature and algebra. Both are essential for our new notion of high-level net and rule (HLNR) systems in order to model high-level nets with nets and rules as tokens.

Definition 6 (Algebraic High-Level Net). *An algebraic high-level (AHL) net* $AN = (\text{SP}, P_{AN}, T_{AN}, pre_{AN}, post_{AN}, cond_{AN}, type_{AN}, A)$ *consists of*

- *an algebraic specification* $\text{SP} = (\Sigma, E; X)$ *with signature* $\Sigma = (S, OP)$, *equations* E, *and additional variables* X;
- *a set of places* P_{AN} *and a set of transitions* T_{AN};
- *pre- and post conditions* $pre_{AN}, post_{AN} : T_{AN} \to (T_\Sigma(X) \otimes P_{AN})^\oplus$;
- *firing conditions* $cond_{AN} : T_{AN} \to \mathcal{P}_{fin}(Eqns(\Sigma; X))$;
- *a type of places* $type_{AN} : P_{AN} \to S$ *and*
- *a* (Σ, E)-*algebra* A

where the signature $\Sigma = (S, OP)$ *consists of sorts* S *and operation symbols* OP, $T_\Sigma(X)$ *is the set of terms with variables over* X, $(T_\Sigma(X) \otimes P_{AN}) = \{(term, p) | term \in T_\Sigma(X)_{type_{AN}(p)}, p \in P_{AN}\}$ *and* $Eqns(\Sigma; X)$ *are all equations over the signature* Σ *with variables* X.

Definition 7 (Firing Behavior of AHL-Nets). *A marking of an AHL-Net* AN *is given by* $M_{AN} \in CP^\oplus$ *where* $CP = (A \otimes P_{AN}) = \{(a, p) | a \in A_{type_{AN}(p)}, p \in P_{AN}\}$.

The set of variables $Var(t) \subseteq X$ *of a transition* $t \in T_{AN}$ *are the variables of the net inscriptions in* $pre_{AN}(t), post_{AN}(t)$ *and* $cond_{AN}(t)$. *Let* $v : Var(t) \to A$ *be a variable assignment with term evaluation* $v^\sharp : T_\Sigma(Var(t)) \to A$, *then* (t, v) *is a consistent transition assignment iff* $cond_{AN}(t)$ *is validated in* A *under* v. *The set* CT *of consistent transition assignments is defined by* $CT = \{(t, v) | (t, v)$ *consistent transition assignment*$\}$.

A transition $t \in T_{AN}$ *is enabled in* M_{AN} *under* v *iff* $(t, v) \in CT$ *and* $pre_A(t, v) \leq M_{AN}$, *where* $pre_A : CT \to CP^\oplus$ *defined by* $pre_A(t, v) = \hat{v}(pre(t)) \in (A \otimes P_{AN})^\oplus$ *and* $\hat{v} : (T_\Sigma(Var(t)) \otimes P_{AN})^\oplus \to (A \otimes P_{AN})^\oplus$ *is the* *obvious extension of $v^\sharp$ to terms and places (similar $post_A : CT \to CP^\oplus$). Then the follower marking is computed by $M'_{AN} = M_{AN} \ominus pre_A(t, v) \oplus post_A(t, v)$.*

The marking graph MG of AN consists of all markings $M \in CP^\oplus$ as nodes and all $M_{AN} \xrightarrow{(t,v)} M'_{AN}$ as edges where M'_{AN} is the follower marking of M_{AN} provided that t is enabled in M_{AN} under v with $(t, v) \in CT$. For an initial marking $INIT$ of AN the reachability graph RG is the subgraph of MG reachable from $INIT$.

In order to allow P/T-systems and rules as tokens of an AHL-net AN we provide a specific specification SP and SP-algebra A based on the construction in the previous section. In fact, it is sufficient to consider as specific SP a signature, called HLNR-System-SIG, together with a suitable HLNR-System-SIG-algebra A, where HLNR-System refers to high-level net and rule systems.

Definition 8 (HLNR-System-SIG Signature and Algebra).
Given vocabularies T_0 and P_0, the signature HLNR-System-SIG is given by

HLNR-System-SIG $=$

sorts: $Transitions, Places, Bool, System, Mor, Rules$
opns: $tt, ff{:}\to Bool$
         $enabled : System \times Transitions \to Bool$
         $fire : System \times Transitions \to System$
         $applicable : Rules \times Mor \to Bool$
         $transform : Rules \times Mor \to System$
         $coproduct : System \times System \to System$
         $isomorphic : System \times System \to Bool$
         $cod : Mor \to System$

and the HLNR-System-SIG-algebra A for P/T-systems and rules as tokens is given by

- $A_{Transitions} = T_0, A_{Places} = P_0, A_{Bool} = \{true, false\}$,
- A_{System} *the set of all P/T-systems over T_0 and P_0, i.e.*
  $A_{System} = \{PN | PN = (P, T, pre, post, M) \text{ P/T-system}, P \subseteq P_0, T \subseteq T_0\}$
        $\cup \{undef\}$,
- A_{Mor} *the set of all P/T-morphisms for A_{System}, i.e.*
  $A_{Mor} = \{f | f : PN \to PN' \text{ P/T-morphism with } PN, PN' \in A_{System}\}$,
- A_{Rules} *the set of all rules of P/T-systems, i.e.*
  $A_{Rules} = \{r | r = (L \xleftarrow{i_1} I \xrightarrow{i_2} R) \text{ rule of P/T-systems with}$
          $\text{strict inclusions } i_1, i_2\}$,
- $tt_A = true, ff_A = false$,
- $enabled_A : A_{System} \times T_0 \to \{true, false\}$ *for $PN = (P, T, pre, post, M)$ with*

$$enabled_A(PN, t) = \begin{cases} true & \text{if } t \in T, pre(t) \leq M \\ false & \text{else} \end{cases}$$

- $fire_A : A_{System} \times T_0 \to A_{System}$ *for $PN = (P, T, pre, post, M)$ with*

$$fire_A(PN, t) = \begin{cases} (P, T, pre, post, M \ominus pre(t) \oplus post(t)) \\ \qquad\qquad\qquad\qquad\qquad if\ enabled_A(PN, t) = tt \\ undef \qquad\qquad\qquad\qquad else \end{cases}$$

- $applicable_A : A_{Rules} \times A_{Mor} \rightarrow \{true, false\}$ with

$$applicable_A(r, m) = \begin{cases} true & if\ r\ is\ applicable\ at\ match\ m \\ false & else \end{cases}$$

- $transform_A : A_{Rules} \times A_{Mor} \rightarrow A_{System}$ with

$$transform_A(r, m) = \begin{cases} H & if\ applicable_A(r, m) \\ undef & else \end{cases}$$

where for $L \xrightarrow{m} G$ and $applicable_A(r, m) = true$ we have a direct transformation $G \xRightarrow{r} H$,
- $coproduct_A : A_{System} \times A_{System} \rightarrow A_{System}$ the disjoint union (i.e. the two P/T-systems are combined without interaction) with

$$coproduct_A(PN_1, PN_2) = \underline{if}\ (PN_1 = undef \vee PN_2 = undef)\ \underline{then}\ undef \\ \underline{else}\ ((P_1 \uplus P_2), (T_1 \uplus T_2), pre_3, post_3, M_1 \oplus M_2)$$

where $pre_3, post_3 : (T_1 \uplus T_2) \rightarrow (P_1 \uplus P_2)^{\oplus}$ are defined by
$pre_3(t) = \underline{if}\ t \in T_1\ \underline{then}\ pre_1(t)\ \underline{else}\ pre_2(t)$
$post_3(t) = \underline{if}\ t \in T_1\ \underline{then}\ post_1(t)\ \underline{else}\ post_2(t)$
- $isomorphic_A : A_{System} \times A_{System} \rightarrow \{true, false\}$ with

$$isomorphic_A(PN_1, PN_2) = \begin{cases} true & if\ PN_1 \cong PN_2 \\ false & else \end{cases}$$

where $PN_1 \cong PN_2$ means that there is a strict P/T-morphism $f = (f_P, f_T) : PN_1 \rightarrow PN_2$ s.t. f_P, f_T are bijective functions,
- $cod_A : A_{Mor} \rightarrow A_{System}$ with $cod_A\ (f : PN_1 \rightarrow PN_2) = PN_2$.

Definition 9 (High-Level Net and Rule Systems).

Given the signature HLNR-SYSTEM-SIG *and the* HLNR-SYSTEM-SIG-*algebra A as above, a high-level net and rule system* $HLNR = (AN, INIT)$ *consists of an AHL-net AN (see Def. 6) with* SP $= ($HLNR-SYSTEM-SIG$; X)$ *where X are variables over* HLNR-SYSTEM-SIG*, and initial marking* $INIT$ *of AN such that*

1. *all places* $p \in P_{AN}$ *are either*
   - *system places i.e.* $p \in P_{Sys} = \{p \in P_{AN}|type_{AN}(p) = System\}$ *or*
   - *rule places i.e.* $p \in P_{Rules} = \{p \in P_{AN}|type_{AN}(p) = Rules\}$,
2. *all rule places* $p \in P_{Rules}$ *are contextual, i.e. for all transitions* $t \in T_{AN}$ *connected with* p *there exists a variable* $r \in X$ *such that* $pre_{AN}(t)_{|p} = post_{AN}(t)_{|p} = r$, *i.e. in the net structure of AN the connection between p and t is given by a double arrow labeled with the variable r.*

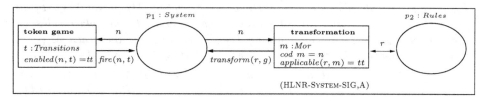

Fig. 8. Basic high-level net and rule system

Remark 6. Our notion of HLNR-systems has static rules. This means that the tokens representing our rules do not move and remain unchanged on the rule places (see Section 6 for extensions). According to our paradigm "nets and rules as tokens" we only allow system and rule places, but no places which are typed by other sorts of HLNR-SYSTEM-SIG. A HLNR-system with only one system place and one rule place is called basic HLNR-system.

Example 1 (Basic HLNR-System). A basic HLNR-system with system place p_1 and rule place p_2 is shown in Fig. 8 where the empty initial marking can be replaced by suitable P/T-systems resp. rules on these places.

Example 2 (House of Philosophers). In Section 2 we have given a detailed discussion of the HLNR-system "House of Philosophers" as given in Fig. 1 with system places *Library*, *Entrance-Hall*, and *Restaurant* and rule places $Rule_1, \ldots, Rule_4$.

The behavior of a HLNR-system $HLNR = (AN, INIT)$ is given by the reachability graph in the sense of AHL-nets (see Def. 7), but it can be represented more explicitly as follows:

Proposition 3 (Reachability Graph of High-Level Net and Rule System). *The reachability graph RG of a HLNR-system $HLNR = (AN, INIT)$ can be characterized as follows:*

1. *Each node of RG is represented by a system family $F \in (A_{System} \times P_{Sys})^{\oplus}$ i.e. $F = \sum_{i=1}^{n}(PN_i, p_i)$ with $PN_i \in A_{System}$ and $p_i \in P_{Sys}$;*

2. *Each edge of RG is represented by $F \xrightarrow{(t_{AN}, v)} F'$, where $(t_{AN}, v) \in CT_{AN}$ is a consistent transition assignment.*

A system family $F = \sum_{i=1}^{n}(PN_i, p_i)$ is well-formed if $PN_i \neq undef$ for all $i = 1, \ldots, n$. If the system family of $INIT$ is well-formed and all $(t_{AN}, v) \in CT_{AN}$ of RG are strongly consistent, i.e. all terms of sort System in $pre_{AN}(t_{AN}), post_{AN}(t_{AN})$ and $cond_{AN}(t_{AN})$ are evaluated under $v^{\sharp}$ to defined elements $PN \neq undef$, then we have:

3. *The reachability graph RG is well-formed, i.e. the system families of all nodes of RG are well-formed.*

Proof. Each node of RG is given by a marking $M_{AN} \in (A \otimes P_{AN})^{\oplus}$, i.e. $M_{AN} = \sum_{i=1}^{n}(a_i, p_i)$ with $p_i \in P_{AN}$ and $a_i \in A_{type(p)}$. Since we have $P_{AN} = P_{Sys} \cup$

P_{Rules} and all rule places are contextual the restriction M_{Rules} of M_{AN} to all $p_i \in P_{Rules}$ is the same for all markings and represents the token rules on the rule places in the initial marking $INIT$. This means that each M_{AN} is uniquely represented by the restriction M_{Sys} of M_{AN} to all $p_i \in P_{Sys}$, w.l.o.g. $M_{Sys} = \sum_{i=1}^{n'}(a_i, p_i)$ with $n' \le n$ and $p_i \in P_{Sys}, a_i \in A_{System}(i = 1, \ldots, n')$. This means $M_{Sys} \in (A_{System} \times P_{Sys})^{\oplus}$. Hence each M_{AN} of RG is represented by the system family M_{Sys} and each edge $M_{AN} \xrightarrow{(t_{AN}, v)} M'_{AN}$ by $M_{Sys} \xrightarrow{(t_{AN}, v)} M'_{Sys}$.

If $INIT_{Sys}$ is well-formed then for each $M_{Sys} \xrightarrow{(t_{AN}, v)} M'_{Sys}$ with well-formed M_{Sys} strong consistency of (t_{AN}, v) implies that also M'_{Sys} is well-formed. This implies that the reachability graph RG is well-formed.

Remark 7. Strong consistency of $(t_{AN}, v) \in CT_{AN}$ can be achieved for a HLNR-system $HLNR$ by including equations of the form $enabled(n, t) = tt$ or $applicable(r, m) = tt$ into $cond_{AN}(t_{AN})$ as shown in Fig. 1 and Fig. 8.

An interesting special case of HLNR-systems are basic HLNR-systems as presented in Fig. 8 of Example 1. Let us assume that the initial marking is given by a P/T-system PN on place p_1 and a set $RULES$ of token rules on place p_2. Then $(PN, RULES)$ can be considered as *reconfigurable P/T-system* in the following sense: on the one hand we can apply the token game and on the other hand rule-based transformations of the net structure of PN. Moreover these activities can be interleaved. This allows to model changes of the net structure while the system is running. This is most important for changes on the fly of large systems, where it is important to keep the system running, while changes of the structure of the system have to be applied. It would be especially important to analyze under which conditions the token game activities are independent of the transformations. This problem is closely related to local Church-Rosser properties for graph resp. net transformations, which are valid in the case of parallel independence of transformations (see [Ehr79, EP04]).

5 Specification and Implementation Aspects

In the previous section we have presented an explicit version of HLNR-systems based on AHL-nets. The main idea was to present a set theoretical version of the HLNR-SYSTEM-SIG-algebra A which defines our concept of "nets and rules as tokens". For various reasons it is also interesting to present an algebraic specification of this algebra. Unfortunately first-order algebraic specifications in the sense of [EM85] or CASL [CAS94] are not suitable for this purpose. Actually we need higher-order features which are provided by HASCASL [SM02], a higher-order extension of the common algebraic specification language CASL.

HASCASL-specifications combine the simplicity of algebraic specifications with higher-order features including function types. It is geared towards specification of functional programs, in particular in Haskell. The semantics of HASCASL is defined by a set-theoretic notion of intensional algebras. The advantage is that in an intensional setting function equivalence testing is possible within some models. Moreover, we can distinguish between different functions that exhibit the same behavior. By contrast extensional equality of functions means that two functions are equal if they always produce the same results for the same arguments. Standard ML, the data type part of Coloured Petri (CP) nets [Jen92], cannot implement equality on function types. This means that it would be difficult to consider P/T-systems and rules as defined in Section 3 as first-class citizens and thus tokens in CP-nets. In our technical report [HM04] we have presented a HASCASL-specification of P/T-systems, P/T-morphism and of rule-based transformations according to the definitions in Section 3. This leads to the formalism of algebraic higher-order (AHO) nets [HM03] where in contrast to AHL-nets higher-order algebraic specifications in HASCASL are used. Since tools for HASCASL already have been implemented this is a first step towards an implementation of our approach presented in this paper.

In fact several aspects of HLNR-systems are supported by tools. The algebraic approach to graph transformations which can also be used for rule-based transformations of nets, is supported by the graph transformation environment AGG (see the homepage of [AGG]). AGG includes an editor for graphs and graph grammars, a graph transformation engine, and a tool for the analysis of graph transformations. On top of the graph transformation system AGG there is the GENGED environment (see the homepage of [Gen]) that supports the generic description of visual modeling languages for the generation of graphical editors and the simulation of the behavior of visual models. Especially, rule-based transformations for P/T-systems can be expressed using GENGED. These transformations can be coupled to other Petri net tools using the Petri Net Kernel [KW01], a tool infrastructure for editing, simulating, and analyzing Petri nets of different net classes and for integration of other Petri net tools. On the level of the data type part the Heterogeneous Tool Set (Hets) (see the homepage of [Hets]) provides a parser and static analysis for CASL and HASCASL-specifications; theorem proving support in form of a translation to the Isabelle/HOL prover is under development. Also, a translation tool from a HASCASL subset to Haskell is provided.

6   Conclusion and Future Work

In this paper we have presented the new concept of high-level nets with rules and nets as tokens and initial marking, short HLNR-systems, which realizes our new paradigm of "nets and rules as tokens". This extends Valk's paradigm "nets as tokens" and also partly his notion of elementary object systems [Val98, Val01]. In Section 2 we have presented a detailed case study of the "House of Philosophers", which allows to give an example driven introduction to HLNR-systems. Moreover we have discussed the relationship to other approaches and pointed out that it seems to be useful and possible to extend our approach by object-oriented features and also to an interaction relation in the sense of Valk.

In Section 3 we have presented the main concepts for our paradigm "nets and rules as tokens". Due to the net inscriptions a firing step in the system level realizes on the one hand the computation of the follower marking of a net (i.e. a P/T-system) and on the other hand the modification of a net by an appropriate rule. Thus transformations become effectively included in the system enabling the system to transform nets as tokens in a formal way by using also rules as tokens. For this purpose we have introduced rule-based transformations for P/T-systems in this paper. In fact we have presented an explicit version of transformations avoiding pushout constructions, but our approach is equal - up to isomorphism - to a double-pushout (DPO) approach in the sense of [Ehr79, EHK91], which will allow to obtain also several other results already known for the DPO-approach [Roz97]. From this point of view the paper presents an interesting integration of concepts in the area of graph transformations and Petri nets.

In HLNR-systems the coupling of a set of rules as tokens to certain transitions is fixed due to the given net topology. In future work we will consider also the migration of rules as tokens. This means the mechanism of mobility and modification presented in our example could be transferred to rules as tokens in order to achieve even more expressive models. The mobility aspect of rules as tokens can be easily introduced by further transitions connecting places of the type *Rules*. However the modification of rules as tokens (see [PP01]) requires an extension of the corresponding algebra in Section 4.

Another interesting aspect for future work is to study transformations of P/T-systems which preserve properties like safety or liveness. Especially in the area of workflow modeling the notion of soundness (which comprises liveness) is of importance (see e.g. [Aal98]). Here we can use the approach of property preserving rules (see [PU03] for an overview). To integrate these kinds of rules into HLNR-systems the set of rules A_{Rules} of the HLNR-SYSTEM-SIG-algebra A (see Section 4) would have to be restricted to property preserving rules.

Finally in Section 5 we have presented several specification and implementation aspects which are useful towards tool-support for our new concepts.

References

[Aal98]   W.M.P. van der Aalst. The Application and Theory of Petri Nets to Workflow Management Systems. *The Journal of Circuits, Systems and Computers*, 8:21-66. 1998.

[Aal02]   W.M.P. van der Aalst. Inheritance of Workflows: An Approach to Tackling Problems Related to Change. *Theoretical Computer Science*, 270(1-2):125–203. 2002.

[AGG]   AGG Homepage. http://tfs.cs.tu-berlin.de/agg.

[ADC01]   G. Agha, F. De Cindio, and G. Rozenberg, editors. *Concurrent Object-Oriented Programming and Petri Nets*, LNCS 2001. Springer, 2001.

[CAS94]   Mosses, Peter D. *CASL Reference Manual - The Complete Documentation of the Common Algebraic Specification Language*, LNCS 2960. Springer, 2004.

[Ehr79]   H. Ehrig. Introduction to the algebraic theory of graph grammars (A survey). In *Graph Grammars and their Application to Computer Science and Biology*, LNCS 73, pages 1–69. Springer, 1979.

[EHK91]   H. Ehrig, A. Habel, H.-J. Kreowski, and F. Parisi-Presicce. Parallelism and concurrency in high-level replacement systems. *Math. Struct. in Comp. Science*, 1:361–404, 1991.

[EM85]   H. Ehrig, B. Mahr. *Fundamentals of Algebraic Specifications 1: Equations and Initial Semantics*. Springer Verlag, EATCS Monographs in Theoretical Computer Science, 1992.

[EHP02]   H. Ehrig, K. Hoffmann, J. Padberg, P. Baldan, and R. Heckel. High Level Net Processes. In *Formal and Natural Computing*, LNCS 2300, pages 191-219. Springer, 2002.

[EP04]   H. Ehrig and J. Padberg. Graph Grammars and Petri Net Transformations. In *Lectures on Concurrency and Petri Nets. Special Issue Advanced Course PNT*, LNCS 3098, pages 496-536. Springer, 2004.

[Far99]   B. Farwer A Linear Logic View of Object Petri Nets. *Fundamenta Informaticae, IOS Press*, 37:225-246, 1999.

[Gen]   GenGED Homepage. http://tfs.cs.tu-berlin.de/genged.

[Hets]   Hets Homepage. http://www.tzi.de/cofi/hets.

[Hof00]   K. Hoffmann. Run Time Modification of Algebraic High Level Nets and Algebraic Higher Order Nets using Folding and Unfolding Constructions. In G. Hommel, editor, *Proc. of 3th Int. Workshop of Communication Based Systems*, pages 55-72. Kluwer, 2000.

[HM03]   K. Hoffmann and T. Mossakowski. Algebraic Higher-Order Nets: Graphs and Petri Nets as Tokens. In M. Wirsing, D. Pattinson, and R. Henicker, editors, *Proc. of 16th Int. Workshop of Algebraic Development Techniques*, LNCS 2755, pages 253–267. Springer, 2003.

[HM04]   K. Hoffmann and T. Mossakowski. Integration of Petri nets and Rule-Based Transformations using Algebraic Higher-Order Nets. Technical Report, Technical University of Berlin, 2004.

[Jen92]   K. Jensen. *Coloured Petri Nets - Basic Concepts, Analysis Methods and Practical Use*, volume 1: Basic Concepts. Springer Verlag, EATCS Monographs in Theoretical Computer Science, 1992.

[KW01]   E. Kindler and M. Weber. The Petri net kernel – an infrastructure for building Petri net tools. *Software Tools for Technology Transfer*, 3(4):486–497, 2001.

[MM90]   J. Meseguer and U. Montanari. Petri Nets are Monoids. *Information and Computation*, 88(2):105–155, 1990.

[MTH97]   R. Milner, M. Tofte, R. Harper, and D. MacQueen. *The Definition of Standard ML - Revised*. MIT Press, 1997.

[Mos05]   T. Mossakowski. Heterogeneous specification and the heterogeneous tool set. Habilitation thesis, University of Bremen, 2004.

[PER95]   J. Padberg, H. Ehrig, and L. Ribeiro. Algebraic high-level net transformation systems. *Mathematical Structures in Computer Science*, 5:217–256, 1995.

[PU03]   J. Padberg, and M. Urbasek. Rule-Based Refinement of Petri Nets: A Survey. In *Advances in Petri Nets: Petri Net Technologies for Modeling Communication Based Systems*, LNCS 2472, pages 161–196. Springer, 2003.

[PP01]   Francesco Parisi-Presicce. On modifying high level replacement systems. In Hartmut Ehrig, Claudia Ermel, and Julia Padberg, editors, *Electronic Notes in Theoretical Computer Science*, volume 44. Elsevier, 2001.

[RWL03]   A. Ratzer, L. Wells, H. Lassen, M. Laursen, J. Qvortrup, M. Stissing, M.
           Westergaard, S. Christensen, K. Jensen. CPN Tools for Editing, Simulat-
           ing, and Analysing Coloured Petri Nets. In *Proc. of the 24th Int. Con-
           ference on Applications and Theory of Petri Nets (ICATPN 2003)*, pages
           450–462, LNCS 2679. Springer, 2003.

[Roz97]   G. Rozenberg, editor. *Handbook of Graph Grammars and Computing by
           Graph Transformations, Volume 1: Foundations*. World Scientific, 1997.

[SM02]   L. Schröder and T. Mossakowski. HasCASL: Towards integrated specifica-
           tion and development of Haskell programs. In H. Kirchner and C. Reingeis-
           sen, editors, *Algebraic Methodology and Software Technology, 2002*, LNCS
           2422, pages 99–116. Springer-Verlag, 2002.

[Sil01]   C. Silbertin-Blanc. The Hurried Philosophers. In G. Agha, F. De Cindio,
           and G. Rozenberg, editors, *Concurrent Object-Oriented Programming and
           Petri Nets*, LNCS 2001, pages 536–537. Springer, 2001.

[Val98]   Rüdiger Valk. Petri Nets as Token Objects: An Introduction to Elementary
           Object Nets. *Proc. of the International Conference on Application and
           Theory of Petri Nets*, LNCS 1420, pages 1–25, 1998.

[Val01]   R. Valk. Concurrency in Communicating Object Petri Nets. In G. Agha,
           F. de Cindio, and G. Rozenberg, editors, *Concurrent Object-Oriented Pro-
           gramming and Petri Nets*, LNCS 2001, pages 164–195. Springer, 2001.

[Wie01]   F. Wienberg. *Informations- und prozeorientierte Modellierung verteilter
           Systeme auf der Basis von Feature-Structure-Netzen*. PhD thesis, Univer-
           sity Hamburg, 2001.

Can I Execute My Scenario in Your Net?

Gabriel Juhás, Robert Lorenz, and Jörg Desel

Lehrstuhl für Angewandte Informatik,
Katholische Universität Eichstätt, 85071 Eichstätt, Germany
{gabriel.juhas, robert.lorenz, joerg.desel}@ku-eichstaett.de

Abstract. In this paper we present a polynomial algorithm to decide whether a scenario (given as a Labelled Partial Order) is executable in a given place/transition Petri net while preserving at least the given amount of concurrency (adding no causality). In the positive case the algorithm computes a process net that respects the concurrency formulated by the scenario. We moreover present a polynomial algorithm to decide whether the amount of concurrency given by a Labelled Partial Order is maximal, i.e. whether the Labelled Partial Order precisely matches a process net w.r.t. causality and concurrency of the events, if this process net represents a minimal causality of events among all process nets.

1 Introduction

Specifications of distributed systems are often formulated in terms of scenarios. In other words, it is often part of the specification that some scenarios should be executable by the system. Given the system, a natural question is whether a scenario can be executed. In this paper we consider Petri net models instead of systems, and we restrict our consideration to place/transition Petri nets. Transforming the above question to this model, we ask whether a given scenario represents a possible execution of a given Petri net. If the answer is positive for all specified scenarios then the Petri net model can be used as a design specification of the system to be implemented. We have not been precise w.r.t. scenarios and executions yet. In general, there are different ways to represent single executions of Petri nets. The most prominent concepts are occurrence sequences, i.e., sequences of transition names that can occur consecutively, and process nets ([4, 5]), i.e., Petri nets representing transition occurrences by events (transitions of process nets) with explicit pre- and post-conditions representing token occurrences of the original net (places of process nets). Playing the token game, it is very easy to check whether a given sequence of transition names is in fact an occurrence sequence of a given net with initial marking. For process nets, we can easily verify the defining conditions of a process net, which reads for marked place/transition Petri nets as follows:

- The underlying net has no cycles, hence the transitive closure of the relation given by arcs is a partial order,
- conditions are not branched,
- no event has an empty pre-set or an empty post-set,
- events are labelled by transitions and conditions by places,

G. Ciardo and P. Darondeau (Eds.): ICATPN 2005, LNCS 3536, pp. 289–308, 2005.
© Springer-Verlag Berlin Heidelberg 2005

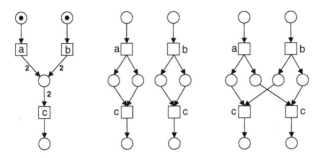

Fig. 1. A p/t-net (left figure) with two of its process nets

- the set of conditions with empty pre-set corresponds to the initial marking (if place p has $m_0(p)$ initial tokens then $m_0(p)$ conditions of this set are labelled by p),
- the pre- and post-sets of events respect the pre- and post-sets of the corresponding transitions (if a transition t consumes k tokens from a place p then each event labelled by t has k pre-conditions labelled by p, and similarly for post-conditions).

Clearly, deciding whether an occurrence sequence can be executed in a place/transition net as well as deciding whether an acyclic labelled net is a process net of a place/transition net can be done in linear time (w.r.t. the size of the occurrence sequence / process net) if the size of the original net is assumed to be constant. For motivation purpose, consider the following example. In Figure 1 a place/transition net is shown. One possible occurrence sequence is $a\,b\,c\,c$. Figure 1 also shows two process nets.

Occurrence sequences lack any information about independence and causality. So it is impossible to specify that events should occur concurrently by an occurrence sequence. Therefore, as soon as concurrency of events has to be specified, occurrence sequences cannot be used for specification of scenarios. Process nets are also not very suitable for specification purposes for two reasons. First, conditions are labelled by names of places of the model specified. So it is not possible to specify that two events have to occur in some order but it is rather also necessary to state which place is responsible for establishing this order. So the specification includes already details of an implementation. The second disadvantage is that a process net determines the precise causality between events. Hence it is not possible to specify a scenario with two events that may either occur (causally) ordered or concurrently. One way to overcome these problems is to specify scenarios in terms of Labelled Partial Orders of events, where the labels refer to the transitions of the specified model. These *LPO*s are called *pomsets* (partially ordered multisets) in [9], emphasizing their close relation to partially ordered sets (we have multisets here because the same transition can occur more than once in a pomset, formally represented by two distinct events labelled by the same transition name). LPOs are called *partial words* in [6], emphasizing their close relation to words or sequences; the total order of elements in a sequence is replaced by a partial order. Actually, pomsets and partial words do not distinguish isomorphic LPOs, because the order of transition occurrences only depends on the labels. So LPOs are somehow in-between occurrence sequences and process nets.

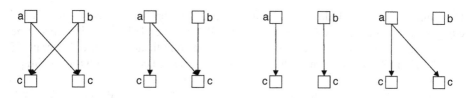

Fig. 2. Four labelled partial orders. All except the most right one are executions of the net in Figure 1. The third one (from the left) is the only strict execution of this net. The most left one refers to both process nets shown in Figure 1

An LPO represents a scenario that can (or cannot) be executed by a marked Petri net. The order defined between events of a so-called *executable* LPO is interpreted as follows: If two events e_1 and e_2, labelled by transitions t_1 and t_2 respectively, are ordered ($e_1 < e_2$) then either t_1 occurs causally before t_2 or both occur concurrently. If e_1 and e_2 are not ordered, then concurrent executions of t_1 and t_2 is demanded. Another interpretation of the order between events defines *strictly executable* LPOs: If e_1 and e_2 are ordered ($e_1 < e_2$) then t_1 is demanded to occur causally before t_2. If e_1 and e_2 are not ordered, then concurrent executions of t_1 and t_2 is demanded.

It is immediate to see that occurrence sequences are special cases of LPOs: A sequence $t_1 t_2 \ldots t_n$ can be viewed as a partially ordered set of events $e_1 < e_2 < \cdots < e_n$ where e_i is labelled by t_i for $1 \leqslant i \leqslant n$. Process nets can be translated to LPOs by removing all conditions and keeping the partial order for the events. We will call such LPOs runs. Formally, an LPO is executable by a given marked place/transition net if it includes (is more sequentialized than) a run of the net. An LPO is strictly executable, if it equals a minimal run (w.r.t. to set inclusion).

Figure 2 shows four LPOs. The first three represent executions of the net in Figure 1. The third one (from the left) is the only strict execution (it is the run given by the first process given in Figure 1). The fourth one is not an execution of this net.

The aim of this paper is to provide an efficient algorithm for deciding whether a given LPO is executable by a given place/transition net. We will provide such an algorithm and prove that its runtime is $O(n^4 |P|)$, where n is the number of events of the LPO and $|P|$ is the number of places of the net. In the positive case, the algorithm computes a run included in the LPO, and thus an underlying process net. Moreover, we provide a (polynomial) characterization which tells us whether the computed run equals the LPO.

The surprising message of this paper might not be the existence of a polynomial algorithm but conversely the fact that this is not a trivial problem. In fact, for elementary Petri nets or 1-safe place/transition nets there exists an immediate algorithm to decide the problem because a unique corresponding process net can be constructed from an LPO – if it exists. The crucial point for place/transition nets is that in general there is not a unique process net corresponding to a given LPO (i.e. an LPO can include different runs). For example, the first LPO given in Figure 2 refers to both process nets of Figure 1.

Astrid Kiehn [8] and Walter Vogler [11] showed that an LPO can be executed if and only if for each cut of the LPO the marking reached by firing all transitions corresponding to events smaller than the cut enables the multiset of transitions given by the cut.

Unfortunately, this result does not lead to an efficient algorithm because the number of cuts grows exponentially with the size of the LPO in general. This result seems to be not very surprising. However, its proof in [8] is quite complicated, and the shorter proof in [10, 11] employs a version of the nontrivial Marriage Theorem known from Graph Theory.

The construction of the algorithm for testing executability and the proof of our result is based on Flow Theory [2]. We will transform a part of our construction to the problem of finding a maximal flow in a flow network associated to the LPO. This maximal flow problem was extensively studied for decades. In [7] an algorithm is presented running in cubic time, and there are several improvements since then (see [3] for an overview). We obtain our complexity result by repeated transformations to flow networks and computations of the maximal flow.

The structure of the remainder of this paper is as follows. In section 2 we provide standard definitions of place/transition nets, occurrence sequences, process nets and LPOs. In Section 3 we establish the so-called flow property of an LPO as a necessary and sufficient condition for its executability. Section 4 is the core of the paper. By applying a maximum flow algorithm we decide whether a given LPO satisfies the flow property. Section 5 introduces the strong flow property which characterizes LPOs corresponding exactly to a run of a process net and briefly presents a polynomial test of strict executability of LPOs.

2   Place/Transition Nets

We use $\mathbb{N}$ to denote the nonnegative integers. Given a function f from A to B and a subset C of A we write $f|_C$ to denote the restriction of f to the set C. Given a finite set A, the symbol $|A|$ denotes the cardinality of A. The set of all multi-sets over a set A is denoted by $\mathbb{N}^A$. Given a binary relation $R \subseteq A \times A$ over a set A, the symbol R^+ denotes the transitive closure of R.

2.1   Place/Transition Net Definitions

Definition 1 (Net).
A net is a triple (P, T, F), where P is a finite set of places, T is a finite set of transitions, satisfying $P \cap T = \emptyset$, and $F \subseteq (P \cup T) \times (T \cup P)$ is a flow relation.

Let (P, T, F) be a net and $x \in P \cup T$ be an element. The *preset* $\bullet x$ is the set $\{y \in P \cup T \mid (y, x) \in F\}$, and the *post-set* $x\bullet$ is the set $\{y \in P \cup T \mid (x, y) \in F\}$. Given a set $X \subseteq P \cup T$, this notation is extended by $\bullet X = \bigcup_{x \in X} \bullet x$ and $X \bullet = \bigcup_{x \in X} x \bullet$. For technical reasons, we consider only nets in which every transition has a nonempty pre-set and a nonempty post-set.

Definition 2 (Place/transition net).
A place/transition net (shortly p/t-net) N is a quadruple (P, T, F, W), where (P, T, F) is a net and $W : F \to \mathbb{N}^+$ is a weight function.

We extend the weight function W to pairs of net elements $(x, y) \in (P \times T) \cup (T \times P)$ satisfying $(x, y) \notin F$ by $W((x, y)) = 0$.

A *marking* of a net $N = (P, T, F, W)$ is a function $m : P \to \mathbb{N}$, i.e. a multi-set over P.

Definition 3 (Occurrence rule).
Let $N = (P, T, F, W)$ be a p/t-net. A transition $t \in T$ is enabled to occur in a marking *m of N iff $m(p) \geq W((p, t))$ for every place $p \in \bullet t$. If a transition t is enabled to occur in a marking m, then its occurrence* leads to the new marking *m' defined by $m'(p) = m(p) - W((p, t)) + W((t, p))$ for every $p \in P$.*

Definition 4 (Marked p/t-net).
A marked p/t-net *is a pair (N, m_0), where N is a p/t-net and m_0 is a marking of N called* initial marking.

2.2 Labelled Partial Orders

In this section we recall the definition of semantics of p/t-nets based on labelled partial orders, also known as partial words [6] or pomsets [9]. For proofs of the presented results see e.g. [10].

Definition 5 (Directed graph, (Labelled) partial order).
A directed graph *is a pair $(V, \to)$, where V is a finite set of nodes and $\to \subseteq V \times V$ is a binary relation over V called the* set of arcs. *As usual, given a binary relation $\to$ we write $a \to b$ to denote $(a, b) \in \to$.*

A partial order *is a directed graph $po = (V, <)$, where $<$ is an irreflexive and transitive binary relation on V.*

For a set $S \subseteq V$ and a node $v \in V \setminus S$ we write $v < S$, if $v < s$ for a node $s \in S$.

Two nodes v, v' of a partial order $(V, <)$ are called independent *if $v \not< v'$ and $v' \not< v$. By $co \subseteq V \times V$ we denote the set of all pairs of independent nodes of V. A* co-set *in a partial order $(V, <)$ is a subset $S \subseteq V$ fulfilling: $\forall x, y \in S : x \, co \, y$. Clearly the relation co is symmetric and reflexive. A* cut *is a maximal co-set.*

Given partial orders $po_1 = (V, <_1)$ and $po_2 = (V, <_2)$, we say that po_2 is a sequentialization *of po_1 if $<_1 \subseteq <_2$.*

A labelled partial order *is a triple $lpo = (V, <, l)$, where $(V, <)$ is a partial order, and l is a* labelling function *on V. If X is a set of labels of lpo, i.e. $l : V \to X$, then for a cut $S \subseteq V$, we define the multi-set $\mu(S) \subseteq \mathbb{N}^X$ by $\mu(S)(x) = |\{v \in V \mid v \in S \wedge l(v) = x\}|$.*

We use the above notation defined for partial orders also for labelled partial orders.

Definition 6 (Enabledness of LPOs).
A labelled partial order $lpo = (V, <, l)$ with $l : V \to T$ is called enabled to occur in *a marking m if the following statement holds: For every cut S of $<$ and every $p \in P$: $m(p) + \sum_{v \in V \wedge v < S}(W((l(v), p)) - W((p, l(v)))) \geq \sum_{v \in S} W((p, l(v)))$. Its occurrence leads to the marking $m'(p)$ given by $m'(p) = m(p) + \sum_{v \in V}(W((l(v), p)) - W((p, l(v))))$.*

A labelled partial order $lpo = (V, <, l)$ enabled in m is said to be minimal *iff there exists no labelled partial order $lpo' = (V, <', l)$ enabled in m with $<' \subset <$.*

Proposition 1. *If a labelled partial order is enabled in* m *and leads to* m', *then every sequentialization is enabled in* m *and leads to* m', *too.*

2.3   Processes, Runs and Executability of LPOs

Definition 7 (Occurrence net).
An occurrence net *is a net* $O = (B, E, G)$ *such that* $| \bullet\, b|, |b\, \bullet\, | \leqslant 1$ *for every* $b \in B$ *(places are* unbranched*) and* O *is* acyclic, *i.e. the transitive closure* G^+ *of* G *is a partial order. Places of an occurrence net are called* conditions *and transitions of an occurrence net are called* events.

The set of conditions of an occurrence net $O = (B, E, G)$ which are minimal (maximal) according to G^+ is denoted by $Min(O)$ $(Max(O))$. Clearly, $Min(O)$ and $Max(O)$ are cuts w.r.t. G^+ (recall that events have nonempty pre- and post-sets by assumption).

Definition 8 (Process).
Let (N, m_0) *be a marked p/t-net,* $N = (P, T, F, W)$. *A* process *of* (N, m_0) *is a pair* $K = (O, \rho)$, *where* $O = (B, E, G)$ *is an occurrence net and* $\rho : B \cup E \to P \cup T$ *is a labelling function, satisfying*

(i) $\rho(B) \subseteq P$ *and* $\rho(E) \subseteq T$.
(ii) $\forall e \in E, \forall p \in P : |\{b \in \bullet e \mid \rho(b) = p\}| = W((p, \rho(e)))$ *and*
    $\forall e \in E, \forall p \in P : |\{b \in e \bullet \mid \rho(b) = p\}| = W((\rho(e), p))$.
(iii) $\forall p \in P : |\{b \in Min(O) \mid \rho(b) = p\}| = m_0(p)$.

Definition 9 (Run).
Let $K = (O, \rho)$ *be a process of a marked p/t-net* (N, m_0). *The labelled partial order* $lpo_K = (E, G^+|_{E \times E}, \rho|_E)$ *is called* run *of* (N, m_0) *representing* K.

A run $lpo = (E, <, l)$ *of* (N, m_0) *is said to be* minimal *iff there exists no other run* $lpo' = (E, <', l)$ *of* (N, m_0) *with* $<' \subset <$.

Definition 10 (Executability of LPOs). *A labelled partial order* $lpo = (V, \prec, l)$ *is called* executable *in a marked p/t-net* (N, m_0) *if there exists a run* $(V, <, l)$ *of* (N, m_0) *with* $< \subseteq \prec$.

A labelled partial order $lpo = (V, \prec, l)$ *is called* strictly executable *in a marked p/t-net* (N, m_0) *if it is a minimal run of* (N, m_0).

Directly from the definition of processes we obtain:

Proposition 2. *Every run of* (N, m_0) *is enabled in* m_0.

From proposition 1 and proposition 2 follows:

Proposition 3. *If a labelled partial order is executable in* (N, m_0), *then it is also enabled in* m_0.

The important result completing the relationship between enabled and executable labelled partial orders was proven in [8, 10].

Theorem 1. *If a labelled partial order is enabled in* m_0 *in a p/t-net* N, *then it is executable in* (N, m_0).

3   Flow Property

As described in the introduction, the definition of enabledness of LPOs is inherently exponential, since an LPO can have exponentially many cuts in the number of nodes. That means, the definition is not appropriate to develop a test of executability.

Instead, we introduce the so called *flow property* of labelled partial orders w.r.t. a marked p/t-net (N, m_0). In this section we show: A labelled partial order fulfills the flow property w.r.t (N, m_0) if and only if it is executable in (N, m_0). In the next section we will give a polynomial test of fulfilling the flow property for a labelled partial order. In the positive case, this test will compute a run included in this labelled partial order.

We fix a marked p/t-net (N, m_0) and a place p of N. Given a labelled partial order $lpo = (V, <, l)$ with $l(V) = T$ we assign non-negative integers to its edges through a so called *flow function*. The aim is to find a flow function assigning values $x(v, v')$ to edges (v, v') in such a way that there is a process with exactly $x(v, v')$ post-conditions of v labelled by p which are also pre-conditions of v'. Thus, such a flow function of lpo abstracts from individuality of conditions of a process and encodes the flow relation of this process by natural numbers. Clearly, finding such a flow function for every place means that lpo includes the run of this process.

In order to simplify the formal definition of the flow property, let us define an extension of $lpo = (V, <, l)$ by adding an initial node which is smaller than all nodes from V and which is labelled by a new label.

Definition 11 (0-extension of a labelled partial order). *Let $lpo = (V, <, l)$ be a labelled partial order. Then a labelled partial order $lpo^0 = (V^0, <^0, l^0)$, where $V^0 = (V \cup \{v_0\})$, $v_0 \notin V$, $<^0 = < \cup(\{v_0\} \times V)$, $l^0(v_0) \notin l(V)$ and $l^0|_V = l$, is called 0-extension of lpo.*

Assigning natural numbers to the arcs of a 0-extension of a labelled partial order we define a so called flow function of this labelled partial order (with v_0 as its unique smallest element).

Definition 12 (Flow function of a labelled partial order). *Let $lpo = (V, <, l)$ be a labelled partial order and $lpo^0 = (V^0, <^0, l^0)$ be a 0-extension of lpo. A function $x :<^0 \to \mathbb{N}$ is called flow function of lpo. For $v \in V$, we denote*

- $\sum_{v' <^0 v} x((v', v))$ *the ingoing flow of v w.r.t. x, and*
- $\sum_{v <^0 v'} x((v, v'))$ *the outgoing flow of v w.r.t. x.*

Let $lpo = (V, <, l)$ be a run representing a process of (N, m_0). For every place p define the *canonical flow function of the run w.r.t. p*, by counting for every $v < v'$ the number of post-conditions of v labelled by p which are pre-conditions of v' in the process. The outgoing flow of the source event v_0 represents the the number of minimal conditions labelled by p which are used by further events.

Definition 13 (Canonical flow function of a run). *Let $K = (O, \rho)$ be a process of (N, m_0) with $O = (B, V, G)$ and let $lpo = (V, <, l)$ be the run representing K. Let $lpo^0 = (V^0, <^0, l^0)$ be a 0-extension of lpo. Define $v_0\bullet = Min(O)$ for the unique* *smallest element* v_0 *of* $(V^0, <^0)$. *We define for every place* $p \in P$ *the* flow function $x_p : <^0 \to \mathbb{N}_0$ *of lpo as follows:*

$$x_p(v, v') = |\{b \in B \mid \rho(b) = p \wedge b \in v \bullet \cap \bullet v'\}|.$$

By definition, this canonical flow function respects the weight function and the initial marking of (N, m_0) in the following sense:

(A) The ingoing flow of an event v equals the number of tokens consumed from place p by the occurrence of transition $l(v)$.

(B) The outgoing flow of an event v (i.e. the number of post-conditions of v labelled by p which are used as pre-conditions of other events) is less than or equal to the number of tokens which are produced by the occurrence of transition $l(v)$ in place p. In particular, the outgoing flow of the source event v_0 is less or equal to the number of tokens in place p of the initial marking m_0.

In general, we say that an arbitrary labelled partial order, whose labels are transitions of (N, m_0), *fulfils the flow property w.r.t.* (N, m_0), if for every place there exists a flow function which fulfils the properties (A) and (B).

Definition 14 (Flow property). *Let* $lpo = (V, <, l)$ *be a labelled partial order with* $l(V) = T$ *and let* $lpo^0 = (V^0, <^0, l^0)$ *be a* 0-extension *of lpo. Denote* $W((l(v_0), p)) = m_0(p)$ *for each place* p. *We say that lpo fulfils the flow property w.r.t.* (N, m_0) *if the following statement holds: For every place* $p \in P$ *there exists a flow* $x_p : <^0 \to \mathbb{N}$ *such that*

(A) *For every* $v' \in V$: $\sum_{v <^0 v'} x_p(v, v') = W((p, l(v')))$.
(B) *For every* $v \in V^0$: $\sum_{v <^0 v'} x_p(v, v') \leqslant W((l(v), p))$.

The ingoing flow *of a node* v *w.r.t.* x_p *is also called* (A)-sum *of* x_p *w.r.t.* v *and the* outgoing flow *of a node* v *w.r.t.* x_p *is also called* (B)-sum *of* x_p *w.r.t.* v.

Given a run lpo of (N, m_0), it follows directly from the definitions of processes and runs that for every $p \in P$ the canonical flow function x_p of lpo fulfils the statements (A) and (B):

Lemma 1. *Every run of* (N, m_0) *fulfils the flow property w.r.t.* (N, m_0).

By the definition of the flow property, given a labelled partial order $lpo = (V, <, l)$ fulfilling (A) and (B) w.r.t. a place p and a flow function x_p and a labelled partial order $lpo' = (V, <', l)$ with $< \subset <'$ we have: lpo' fulfils (A) and (B) w.r.t. to the place p and the flow function x'_p given by $x'_p|_< = x_p$ and $x'_p|_{<' \setminus <} = 0$. Therefore:

Lemma 2. *Every labelled partial order executable in* (N, m_0) *fulfils the flow property w.r.t.* (N, m_0).

The following lemma states the converse:

Lemma 3. *Let* $lpo = (V, \prec, l)$ *be a labelled partial order which fulfils the flow property w.r.t.* (N, m_0). *Then lpo is executable in* (N, m_0), *i.e. there exists a run* $(V, <, l)$ *of* (N, m_0) *such that* $< \subseteq \prec$.

Proof. From the definition of the flow property, for every place $p \in P$ there exists a function x_p which fulfils (A) and (B). We will fix these functions and use them to construct a process $K = (O, \rho)$ of (N, m_0) with $O = (B, V, G)$ and $\rho|_V = l$, satisfying $<= G^+|_{V \times V} \subseteq \prec$. According to the definition of runs, this will conclude the proof.

For convenience, denote $V = \{v_1, \ldots, v_{|V|}\}$ such that $v_i \prec v_j$ implies $i < j$. First define the set of conditions and the labelling of conditions. For every event $v \in V^0$ we define the set of post-conditions of v labelled by $p \in P$:

$$B_p^v = \{p_v^1, \ldots p_v^{W((l(v),p))}\}.$$

Thus, the number of these post-conditions equals the value $W((l(v), p))$. Especially, the number of post-conditions of v_0 labelled by $p \in P$ equals $m_0(p)$. Denote $B_p = \cup_{v \in V^0} B_p^v$ the set of all conditions labelled by p. Define the labelling of conditions by $\rho(b) = p$ for $b \in B_p$. Finally, the set of all conditions of the process is given by $B = \cup_{p \in P} B_p$.

It remains to define the flow relation G. It is the union of all ingoing and outgoing arcs of all events $v \in V$. An event $v \in V$ has an outgoing arc to each of its post-conditions (observe that $v_0 \notin V$). Thus, the set of outgoing arcs of an event $v \in V$ labelled by $p \in P$ is

$$G_p^{v\bullet} = \{v\} \times B_p^v.$$

The ingoing arcs are defined w.r.t. the flow functions. If $x_p(v, v_m) > 0$, then we connect exactly $x_p(v, v_m)$ post-conditions of v labelled by p with v_m. In order to avoid branching of conditions, we connect the post-conditions $p_v^1, \ldots, p_v^{x_p(v,v_m)}$ with the event v_m which has the smallest index m from all events v_m with $x_p(v, v_m) > 0$, and so on. Formally, define the set of ingoing arcs from conditions labelled by $p \in P$ to an event $v_m \in V$ by

$$G_p^{\bullet v_m} = \{(p_v^i, v_m) \mid v \in V_0, \, x_p(v, v_m) > 0,$$
$$\sum_{j < m} x_p(v, v_j) < i \leqslant \sum_{j \leqslant m} x_p(v, v_j)\}.$$

Because x_p fulfils (B), i.e. the number of post-conditions of an event $v \in V^0$ is not less than the outgoing flow of v, by this construction any event $v_m \in V$ is connected with exactly $x_p(v, v_m)$ post-conditions of v labelled by p whenever $x_p(v, v_m) > 0$. Because of this and because x_p also fulfils (A), by this construction every $v_m \in V$ has exactly $W((p, l(v_m)))$ pre-conditions labelled by $p \in P$. Finally denote $G_p = \cup_{v \in V}(G_p^{\bullet v} \cup G_p^{v \bullet})$ for every $p \in P$ and $G = \cup_{p \in P} G_p$.

By construction, the conditions are unbranched and the defined net is acyclic, i.e. $O = (B, V, G)$ is an occurrence net. From the previous reasoning $K = (O, \rho)$ is a process of (N, m_0).

It remains to show that $<= G^+|_{V \times V} \subseteq \prec$. Denote $R = \{(v, v') \in V \times V \mid v \bullet \cap \bullet v' \neq \emptyset\}$. Observe that $G^+|_{V \times V} = R^+$ and (by construction of G) we have $(v, v') \in R \implies (\exists p \in P : x_p(v, v') > 0)$. Because $x_p(v, v') > 0$ implies $v \prec v'$ and $\prec$ is transitive, this gives $<= G^+|_{V \times V} \subseteq \prec$. $\qquad \square$

4   Testing the Flow Property

In this section we give a polynomial algorithm to test whether a labelled partial order $lpo = (V, <, l)$ with $l(V) = T$ fulfils the flow property w.r.t. (N, m_0). In the case that lpo fulfils the flow property, the algorithm constructs flow functions for all places.

4.1   The Algorithm

We describe the algorithm for a fixed place p. Let $lpo^0 = (V^0, <^0, l^0)$ be a 0-extension of lpo. Throughout this section denote $V^0 = \{v_0, v_1, \ldots, v_n\}$ with $v_i < v_j \Rightarrow i < j$. The algorithm starts with a flow function x_0 fulfilling part (A) of the flow property. Such x_0 always exists, e.g. set $x_0(v_0, v') = W((p, l(v')))$ for each $v' \in V$ and $x_0(v, v') = 0$ otherwise. In general this flow function will not fulfil part (B) of the flow property. We denote max_0 the smallest index for which x_0 does not fulfil property (B):

- $\sum_{v_{max_0} <^0 v'} x_0(v_{max_0}, v') > W((l(v_{max_0}), p))$, and
- $\forall j < max_0 \colon \sum_{v_j <^0 v'} x_0(v_j, v') \leqslant W((l(v_j), p))$.

Thus, the aim is to modify x_0, such that the (B)-sum of x_0 w.r.t. the index max_0 is reduced as much as necessary to fulfil (B) w.r.t. max_0, while preserving property (A) for all indexes and property (B) for all indexes smaller than max_0. In Subsection 4.4 we will describe in detail a procedure which modifies x_0 in such a way while minimizing the (B)-sum w.r.t. max_0 in some sense. In the following we will refer to this procedure as the *main procedure* of the algorithm. Repeating the main procedure, we get the algorithm:

1. Set $i = 0$ and compute a flow function x_i fulfilling (A).
2. If x_i does not fulfil (B):
   - Compute the smallest index max_i for which x_i does not fulfil (B).
   - Repeat:
     * Apply the main procedure to modify x_i into a flow function x_{i+1} (in such a way that x_{i+1} still fulfils (A) for all indexes, still fulfils (B) for all indexes smaller than max_i, and the (B)-sum of x_{i+1} w.r.t. max_i is smaller than (or equal to) the (B)-sum of x_i w.r.t. max_i).
     * If x_{i+1} does not fulfil (B), compute the smallest index max_{i+1} for which x_{i+1} does not fulfil (B).
     * Set $i = i + 1$.
     until x_i fulfils (B) or $max_i = max_{i-1}$.

The algorithm terminates, if either x_i fulfils property (B) or $max_i = max_{i-1}$. In the first case x_i is a flow function, for which lpo fulfils the flow property. In the second case we will prove in Subsection 4.5 that lpo is not enabled w.r.t. (N, m_0).

The algorithm has to be applied for every place $p \in P$. Since $max_i \leqslant n$, the main procedure is repeated at most n times. The main procedure itself requires at most $O(n^3)$ time as shown in the Subsection 4.4. Altogether we get that the test of executability takes $O(n^4 |P|)$ time.

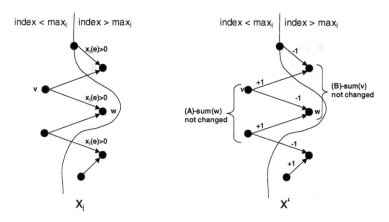

Fig. 3. The left part shows an example of a flow decreasing sequence of the first kind with $k = 3$. The right part shows the modifying operation to get a flow function x' from x_i satisfying $x'(v_{max_i}, w^1) < x_i(v_{max_i}, w^1)$, property (A) for all indexes and property (B) for all indexes smaller than max_i

4.2   Flow Decreasing Sequences

We start with a brief motivation of the main procedure for modifying x_i: Consider the following two possibilities of reducing the (B)-sum of x_i w.r.t. max_i while respecting property (A) for all indexes and property (B) for all indexes smaller than max_i.

The first possibility is to move positive values $x_i(v_{max_i}, v')$ onto edges (v_j, v_l) with $j > max_i$ (see Figure 3): Suppose a sequence of nodes $v^0 = v_{max_i}, w^1, v^1, \ldots, w^k, v^k$ such that

- $w^j \neq w^m$ and $v^j \neq v^m$ for $j \neq m$,
- $x(v^j, w^{j+1}) > 0$ and $v^j < w^j$,
- For each $0 < j < k$ there is an index $m < max_i$ with $v^j = v_m$,
- For $j = k$ there is an index $m > max_i$ with $v^k = v_m$.

Such a sequence allows to modify x_i into a new flow function x' defined as follows:

$$x'(v^j, w^j) = x_i(v^j, w^j) + 1,$$
$$x'(v^j, w^{j+1}) = x_i(v^j, w^{j+1}) - 1.$$

Obviously x' satisfies $x'(v_{max_i}, w^1) < x_i(v_{max_i}, w^1)$, property (A) for all indexes and property (B) for all indexes smaller than max_i. This modification can be applied for each such sequence as long as $x'(v^j, w^{j+1}) > 0$ for all j, thus reducing the (B)-sum of x_i w.r.t. max_i. As a consequence $x'(v^k, w^k) > x_i(v^k, w^k)$, i.e. the (B)-sum w.r.t. v^k is increased. Nevertheless property (B) remains satisfied for all indexes smaller than max_i.

The second possibility is to move positive values $x_i(v_{max_i}, v')$ onto edges (v_j, v_l) with $j < max_i$ and $\sum_{v_j < v'} x_i(v_j, v') < W((l(v_j), p))$ (see Figure 4): Suppose a sequence of nodes $v^0 = v_{max_i}, w^1, v^1, \ldots, w^k, v^k$ such that

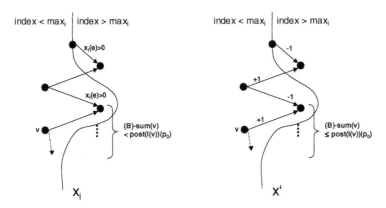

Fig. 4. The left part shows an example of a flow decreasing sequence of the second kind with $k = 2$. The right part shows the modifying operation to get a flow function x' from x_i satisfying $x'(v_{max_i}, w^1) < x_i(v_{max_i}, w^1)$, property (A) for all indexes and property (B) for all indexes smaller than max_i

- $w^j \neq w^m$ and $v^j \neq v^m$ for $j \neq m$,
- $x(v^j, w^{j+1}) > 0$ and $v^j < w^j$,
- For each $0 < j \leqslant k$ there is an index $m < max_i$ with $v^j = v_m$,
- $\sum_{v_k < v'} x_i(v_k, v') < W((l(v_k), p))$.

Such a sequence allows to modify x_i into a new flow function x' defined as follows:

$$x'(v^j, w^j) = x_i(v^j, w^j) + 1,$$
$$x'(v^j, w^{j+1}) = x_i(v^j, w^{j+1}) - 1.$$

Obviously x' satisfies $x'(v_{max_i}, w^1) < x_i(v_{max_i}, w^1)$, property (A) for all indexes and property (B) for all indexes smaller than max_i. This modification can be applied for each such sequence as long as $x'(v^j, w^{j+1}) > 0$ for all j and $\sum_{v_k < v'} x'(v_k, v') < W((l(v_k), p))$, thus reducing the (B)-sum of x_i w.r.t. max_i. As a consequence $x'(v^k, w^k) > x_i(v^k, w^k)$, i.e. the (B)-sum w.r.t. v^k is increased. Nevertheless property (B) remains satisfied for all indexes smaller than max_i.

Such sequences (of *the first or second kind*) will be called *flow decreasing sequences w.r.t. x_i*). We want reduce the (B)-sum of the modified flow w.r.t. max_i by flow decreasing sequences in a maximal way. This can be done by transforming this problem for $(V, <, l)$ w.r.t. the flow function x_i into a *maximum flow problem* for a suitable *flow network*. The maximum flow problem is intensively studied since four decades and several algorithms running in cubic time in the number of nodes ([7]) and faster (see e.g. [3] for an overview) were developed.

4.3   The Associated Flow Network

We briefly introduce the necessary notations:

Definition 15 (Flow network). *A flow network is a directed graph* $(V', \rightarrow)$ *together with a capacity function c assigning nonnegative integers to edges in* $E_{\rightarrow} = \{(v, v') \mid v \rightarrow v' \vee v' \rightarrow v\}$, *satisfying: there is a node* $s \in V$, *called* source, *with no incoming edges w.r.t.* $\rightarrow$, *there is a node t, called* sink, *with no outgoing edges w.r.t.* $\rightarrow$, *and* $c(v, v') = 0$ *for* $v \not\rightarrow v'$.

    A flow f *in* $(V', \rightarrow, c)$ *is a function assigning integers to edges in* $E_{\rightarrow}$ *in such a way that*

- *f does not exceed c:* $f(v, v') \leqslant c(v, v')$.
- $f(v, v') = -f(v', v)$.
- *For each node v except source and sink the flow into (resp. out of) v equals 0:*
  $$\sum_{(v',v) \in E_{\rightarrow}} f(v', v) = 0.$$

The value $|f|$ *of a flow* f *is defined as the outgoing flow of the source (or equivalently the ingoing flow of the sink)* $\sum_{s \rightarrow v'} f(s, v')$. *The* maximal flow *is the flow with maximal value among all flows.*

    Given a flow f, *the* residual capacity c_r *w.r.t.* f *of* $(v, v') \in E_{\rightarrow}$ *is defined by* $c_r(v, v') = c(v, v') - f(v, v')$ *if* $v \rightarrow v'$ *and* $c_r(v, v') = f(v', v)$ *if* $v \not\rightarrow v'$. *The* residual network *of a flow* f *consists of all edges* $e \in E_{\rightarrow}$ *with* $c_r(e) > 0$ *together with the residual capacity. A* flow augmenting path *w.r.t. a flow* f *of a flow network is a simple path from source to sink in the residual network of* f.

One of the first algorithms solving the maximum flow problem was the flow augmenting path method by Ford and Fulkerson ([2]). They proved the following theorem giving a characterization of the maximum flow:

Theorem 2. *Let* f *be a flow in a flow network. If there is no flow augmenting path w.r.t.* f, *then* f *is maximal.*

The flow network $G(x_i) = (V(x_i), \rightarrow, c)$ associated to $lpo = (V, <, l)$ and x_i is defined in such a way that the flow decreasing sequences in $(V, <)$ (w.r.t. x_i) will correspond to flow augmenting paths in $(V(x_i), \rightarrow, c)$ (w.r.t. to the zero flow). The possibility of reducing the (B)-sum w.r.t. max_i through a flow decreasing sequence by a certain amount shall exactly correspond to a flow of the same amount through an associated augmenting path from source to sink. Therefore the capacity restricting the flow on edges (v^j, w^{j+1}) corresponds to the value of the flow function $x_i(v^j, w^{j+1})$. Since a node v_m with $m < max_i$ can serve as a node v^j in a flow decreasing sequence and as a node w^j in the same or another flow decreasing sequence, we split each node $v \in V$ into two nodes in the flow network: (v, out), playing the role of a node v^j, and (v, in), playing the role of a node w^j. Formally, we define $V(x_i) = (V \times \{in, out\}) \cup \{t\})$ with a node $t \notin V$, which will serve as the sink. The node (v_{max_i}, out) will have no incoming edges and serve as the *source* of the flow network. We will use a constant M as an edge capacity which can not be exceeded by the value of the maximum flow of the flow network (see Figure 5)). Since the value of the maximal flow can never exceed the sum of capacities of the edges outgoing of the source, we set $M = \sum_{v_{max_i} \rightarrow v'} x_i(v_{max_i}, v')$.

(a) For $l > max_i$ and $x_i(v_{max_i}, v_l) > 0$: $(v_{max_i}, out) \rightarrow (v_l, in)$ and
    $c((v_{max_i}, out), (v_l, in)) = x_i(v_{max_i}, v_l)$.

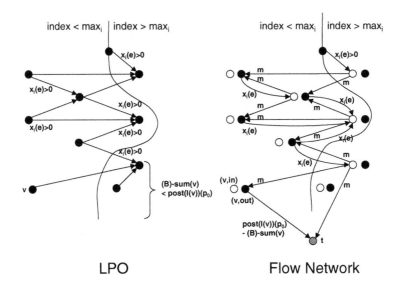

LPO   Flow Network

Fig. 5. The left part shows a part of a labelled partial order where it is indicated on which edges the flow function x_i has positive values. For clearness only the skeleton is regarded. The right part shows the corresponding part of the associated flow network. The "out"-nodes are filled whereas the "in"-nodes are not filled. Each edge carry its capacity

(b) For $j < max_i$ and $x_i(v_j, v_l) > 0$: $(v_j, out) \rightarrow (v_l, in)$ and
$c((v_j, out), (v_l, in)) = x_i(v_j, v_l)$
(c) For $j < max_i$ and $v_j < v_l$: $(v_l, in) \rightarrow (v_j, out)$ and
$c((v_l, in), (v_j, out)) = M$.
(d) For $l > max_i$ and $v_j < v_l$ for some v_j with $j > max_i$: $(v_l, in) \rightarrow t$ and
$c((v_l, in), t) = M$
(e) For $j < max_i$ and $\sum_{v_j < v'} x_i(v_j, v') < W((l(v_j), p))$: $(v_j, out) \rightarrow t$ and
$c((v_j, out), t) = W((l(v_j), p)) - \sum_{v_j < v'} x_i(v_j, v')$.

The following lemma states that for each flow f we can modify x_i into a flow function x_f fulfilling property (A) for all indexes and property (B) for all indexes smaller than max_i with its (B)-sum w.r.t. max_i reduced by $|f|$.

Lemma 4. *Let f be a flow in $G(x_i) = (V(x_i), \rightarrow, c)$. Then the flow function x_f, defined by modifying x_i in the following way, fulfils (A) for all indexes and (B) for all indexes smaller than max_i:*

(a) *For $l > max_i$ and $x_i(v_{max_i}, v_l) > 0$:*
$x_f(v_{max_i}, v_l) = x_i(v_{max_i}, v_l) - f((v_{max_i}, out), (v_l, in))$.
(b) *For $j < max_i$ and $x_i(v_j, v_l) > 0$:*
$x_f(v_j, v_l) = x_i(v_j, v_l) - f((v_j, out), (v_l, in))$.
(c) *For $j < max_i$, $v_j < v_l$ and $x_i(v_j, v_l) = 0$:*
$x_f(v_j, v_l) = x_i(v_j, v_l) + f((v_l, in), (v_j, out))$

(d) For $l > max_i$ such that $v_j < v_l$ for some v_j with $j > max_i$:
  Let J be the maximal index with $v_J < v_l$ and define $x_f(v_J, v_l) = x_i(v_J, v_l) + f((v_l, in), t)$.

Proof. Observe first that by construction all values of x_f are non-negative:
ad (a): $f((v_{max_i}, out), (v_l, in)) \leqslant c((v_{max_i}, out), (v_l, in)) = x_i(v_{max_i}, v_l)$.
ad (b): $f((v_j, out), (v_l, in)) \leqslant c((v_j, out), (v_l, in)) = x_i(v_j, v_l)$.
ad (c): $f((v_l, in), (v_j, out)) > 0$ since $c((v_j, out), (v_l, in)) = 0$.
ad (d): $f((v_l, in), t) > 0$ since $c(t, (v_l, in)) = 0$.

We first show that x_f fulfils part (A) of the flow property. For this we claim that the (A)-sums of x_i and x_f are equal w.r.t. each node v_l. For convenience assume $f(\nu, \mu) = 0$ for $(\nu, \mu) \notin E_\rightarrow$. The argumentation is based on the observation, that by definition for a node v_l with $(t, (v_l, in)) \notin E_\rightarrow$:

$$\sum_{v_j < v_l} f((v_j, out), (v_l, in)) = \sum_{(\nu, (v_l, in)) \in E_\rightarrow} f(\nu, (v_l, in)) = 0.$$

The last equality follows by the definition of flows. In this case:

$$\sum_{v_j < v_l} x_f(v_j, v_l) = \sum_{v_j < v_l} (x_i(v_j, v_l) - f((v_j, out), (v_l, in)))$$

$$= \sum_{v_j < v_l} x_i(v_j, v_l) - \sum_{v_j < v_l} f((v_j, out), (v_l, in))$$

$$= \sum_{v_j < v_l} x_i(v_j, v_l).$$

In case $(t, (v_l, in)) \in E_\rightarrow$ we similarly get the same result.

Finally we show that x_f fulfils part (B) of the flow property for all indexes $j < max_i$. If $(t, (v_j, out)) \notin E_\rightarrow$, we deduce as above that the (B)-sums of x_i and x_f are equal w.r.t. v_j. In the case $(t, (v_j, out)) \in E_\rightarrow$ we get:

$$\sum_{v_j < v_l} f((v_j, out), (v_l, in)) = \left(\sum_{((v_j, out), \mu) \in E_\rightarrow} f((v_j, out), \mu) \right) - f((v_j, out), t)$$

$$= -f((v_j, out), t).$$

As above we deduce

$$\sum_{v_j < v_l} x_f(v_j, v_l) = \sum_{v_j < v_l} x_i(v_j, v_l) - \sum_{v_j < v_l} f((v_j, out), (v_l, in))$$

$$= \sum_{v_j < v_l} x_i(v_j, v_l) + f((v_j, out), t)$$

$$\leqslant \sum_{v_j < v_l} x_i(v_j, v_l) + c((v_j, out), t) = W((l(v_j), p)).$$

$\square$

4.4   The Main Procedure

By Lemma 4, for each flow f in the associated flow network we can decrease the (B)-sum for the index max_i by $|f|$, while (A) for all indexes and (B) for all indexes smaller then max_i remain satisfied. Thus, the main procedure is defined as follows:

- Input: Flow function x_i.
- Compute associated flow network w.r.t. x_i.
- Compute maximal flow f in this flow network.
- Compute x_f.
- Output: Flow function $x_{i+1} = x_f$.

Computing the maximal flow f depends on the applied maximum flow algorithm. As already mentioned, there are maximum flow algorithms taking cubic time and faster. The other steps take at most quadratic time. Altogether the main procedure takes at most cubic time.

Let us mention that the main procedure could be optimized by terminating the maximum flow algorithm as soon as $\sum_{v_{max_i} < v'} x_i(v_{max_i}, v') - |f| \leqslant W((l(v_{max_i}), p))$.

4.5   Termination of the Algorithm

If the algorithm terminates because x_i fulfils (B), then lpo fulfils the flow property for the place p.

It remains to prove that if the algorithm terminates because $max_i = max_{i-1}$, then lpo is not enabled w.r.t. (N, m_0). From proposition 3 this implies that lpo is not executable in (N, m_0).

Theorem 3. *Let f be the maximal flow of the associated flow network w.r.t. x_i. Assume moreover that x_f does not fulfil (B) for the index max_i. Then there is a cut C, such that lpo is not enabled w.r.t. C:*

$$m_0(p) + \sum_{v < C, v \in V} W((l(v), p)) - \sum_{v < C} W((p, l(v))) - \sum_{v \in C} W((p, l(v))) < 0.$$

The proof is based on the following lemma which states that for each flow f and each flow decreasing sequence w.r.t. x_f there is a flow augmenting path w.r.t. f in the flow network associated to x_i.

Lemma 5. *Let f be a flow in the flow network $G(x_i) = (V(x_i), \to, c)$ and let $v^0 = v_{max_i}, w^1, v^1, \ldots, w^k, v^k$ be a flow deceasing sequence w.r.t. x_f. Then*

- *If the flow decreasing sequence is of the first kind,*
  then $(v^0, out) = (v_{max_i}, out), (w^1, in), (v^1, out), \ldots, (w^k, in), t$ is a flow augmenting path w.r.t. f.
- *If the flow decreasing sequence is of the second kind,*
  then $(v^0, out) = (v_{max_i}, out), (w^1, in), (v^1, out), \ldots, (w^k, in), (v^k, out), t$ is a flow augmenting path w.r.t. f.

Proof. Let $v_{max_i}, w^1, v^1, \ldots, w^k, v^k$ be a flow decreasing sequence of the first kind. We have to show that $(v_{max_i}, out), (w^1, in), (v^1, out), \ldots, (w^k, in), t$ is a path in the residual network $(V(x_i), \to_r, c_r)$ of f. We have

- $c_r((w^j, in), (v^j, out)) = c((w^j, in), (v^j, out)) - f((w^j, in), (v^j, out)) = M - f((w^j, in), (v^j, out)) > 0$ since $(w^j, in) \to (v^j, out)$.
- $c_r((v^j, out), (w^{j+1}, in)) = c((v^j, out), (w^{j+1}, in)) - f((v^j, out), (w^{j+1}, in)) = x_i((v^j, out), (w^{j+1}, in)) - f((v^j, out), (w^{j+1}, in)) = x_f(v^j, w^{j+1}) > 0.$
- $c_r((w^k, in), t) = c((w^k, in), t) - f((w^k, in), t) = M - f((w^k, in), t) > 0.$

Let now $v_{max_i}, w^1, v^1, \ldots, w^k, v^k$ be a flow decreasing sequence of the second kind. We show that $(v_{max_i}, out), (w^1, in), (v^1, out), \ldots, (w^k, in), (v^k, out), t$ is a path in the residual network of f.

To show $c_r((w^j, in), (v^j, out)), c_r((v^j, out), (w^{j+1}, in)) > 0$ works as above. Moreover, we get analogously as in the proof of lemma 4:
$c_r((v^k, out), t) = c((v^k, out), t) - f((v^k, out), t) = W((l(v^k), p)) - \sum_{v^k < v'} x_i(v^k, v')$
$-f((v^k, out), t) = W((l(v^k), p)) - \sum_{v^k < v'} x_f(v^k, v') > 0.$ □

From lemma 5 and theorem 2 we get immediately that for the maximal flow f there is no flow decreasing sequence w.r.t. x_f.

Lemma 6. *Let f be the maximal flow of the network $G(x_i) = (V(x_i), \to, c)$ associated to x_i. Assume further x_f does not fulfil property (B) for the index max_i.*

Define W as the set consisting of v_{max_i} and all events $v \in V^0$ such that there exists a sequence $v^0 = v_{max_i}, w^1, v^1, \ldots, w^k, v^k = v$ with

(i) $w^j \neq w^m$ and $v^j \neq v^m$ for $j \neq m$,
(ii) $x_f(v^j, w^{j+1}) > 0$ and $v^j <^0 w^j$.

Define C as the set of all $w \in V$ with $w \notin W$ such that there exists $v \in W$ with $x_f(v, w) > 0$. Then it holds:

(a) If $v_j \in W$ then $j \leqslant max_i$.
(b) $v <^0 C \Leftrightarrow v \in W$.
(c) C is a co-set.
(d) $C' = \{w \in V \mid w \notin W \wedge (v <^0 w \Rightarrow v \in W)\}$ is a cut and $C \subseteq C'$.
(e) For every $v_j \in W$ with $j \neq max_i$ we have $\sum_{v_j <^0 v'} x_f(v_j, v') = W((l(v_j), p))$.

Proof. ad (a): Assume $v_j \in W$ with $j > max_i$. There is a sequence $v^0 = v_{max_i}, w^1, v^1, \ldots, w^k, v^k = v_j$ fulfilling (i) and (ii). Take the smallest index m such that there is $u > max_i$ with $v^m = v_u$. By definition $v^0 = v_{max_i}, w^1, v^1, \ldots, w^m, v^m$ is a flow decreasing sequence of the first kind. This contradicts the assumption (since by Lemma 5 there is no flow decreasing sequence w.r.t. x_f).

ad (b): ($\Longrightarrow$) Let $v <^0 C$. If $v = v_{max_i}$, then $v \in W$ by definition of W. Let $v \neq v_{max_i}$: $v <^0 C$ implies that there is $w \in C$ with $v <^0 w$. By definition of C there exists $v' \in W$ with $x_f(v', w) > 0$. If $v = v'$, we get $v \in W$. Let $v \neq v'$. If $v' = v_{max_i}$, then the sequence v', w, v fulfils (i) and (ii) and therefore $v \in W$. Otherwise by definition of W there is a sequence $v^0, w^1, v^1, \ldots, w^k, v^k = v'$ fulfilling

(i) and (ii). If $v = v^j$ for some j, then $v \in \mathcal{W}$. Let $v \neq v^j$ for all j. If $w^j \neq w$ for all j, then also the sequence $v^0, w^1, v^1, \ldots, w^k, v', w, v$ fulfils (i) and (ii), i.e. $v \in \mathcal{W}$. If $w^j = w$ for some j, let m be the smallest index with $w^m = w$. Then the sequence $v^0, w^1, v^1, \ldots, w^m = w, v$ fulfils (i) and (ii), i.e. $v \in \mathcal{W}$.

$(\Longleftarrow)$ Let $v \in \mathcal{W}$. If $v = v_{max_i}$ then we have: Because x_f does not fulfil (B) for the index max_i, there exists a node $v_j \in V$ with $j > max_i$ such that $x_f(v, v_j) > 0$ and in particular $v <^0 v_j$. According to (a) and the definition of C we have $v_j \in C$ and $v <^0 C$ (in particular C is nonempty). If $v \neq v_{max_i}$ then there is a sequence $v^0, w^1, v^1, \ldots, w^k, v^k = v$ fulfilling (i) and (ii). Since $v^{k-1} \in \mathcal{W}$, we get $x_f(v^{k-1}, w^k) > 0$. If $w^k \notin \mathcal{W}$, this implies $w^k \in C$. From $v = v^k < w^k$ we obtain $v <^0 C$. If $w = w^k \in \mathcal{W}$, there is a sequence $v^0, w^1, v^1, \ldots, w^l, v^l = w$ fulfilling (i) and (ii). From $v <^0 w$ and $w <^0 w^l$ we obtain $v <^0 w^l$. Again if $w^l \notin \mathcal{W}$, we are done, else repeat this procedure. Obviously, this procedure terminates according to the the fact that V is finite and for every $v \in \mathcal{W}$ there is a $w \in V$ satisfying $v <^0 w$, i.e. $\mathcal{W}$ does not contain maximal elements w.r.t. partial order $<^0$.

ad (c): Assume two events $w, w' \in C$ with $w < w'$. From (b) we obtain $w \in \mathcal{W}$, which is a contradiction to $w \in C$.

ad (d): From (b) and (c) we get that $\mathcal{W}$ is a downward closed. From definition of $\mathcal{W}$ we have that it does not contain maximal elements. It is a well known fact that then the set C' is a cut. From (b) we also get $C \subset C'$.

ad (e): Assume $v_j \in \mathcal{W}$ with $j \neq max_i$ and $\sum_{v_j <^0 v'} x_f(v_j, v') \neq W((l(v_j), p))$. From (a) we get $j < max_i$. Since x_f fulfils (B) for all indexes smaller than max_i, this implies $\sum_{v_j <^0 v'} x_f(v_j, v') < W((l(v_j)), p))$. Let $v^0, w^1, v^1, \ldots, w^k, v^k = v_j$ be a sequence fulfilling (i) and (ii). By the above consideration this sequence is a flow decreasing sequence of the second kind. This is a contradiction to the assumption. $\qquad \square$

Lemma 7. *Let f be the maximal flow of the associated flow network $G(x_i)$ w.r.t. x_i and assume x_f does not fulfil (B) for the index max_i. Then C' is a cut satisfying*

$$m_0(p) + \sum_{v < C', v \in V} W((l(v), p)) - \sum_{v < C'} W((p, l(v))) - \sum_{v \in C'} W((p, l(v))) < 0.$$

Proof. We first prove the inequality for the co-set C. Since x_f fulfils (A) we can replace $W((p, l(v)))$ by $\sum_{v' <^0 v} x_f(v', v)$ for each $v <^0 C$ and $v \in C$. Because x_f fulfils statement (b) of the last lemma we can replace $W((l(v), p))$ by $\sum_{v <^0 v'} x_f(v, v')$ for each $v <^0 C$, $v \neq v_{max_i}$. Moreover $m_0(p)$ equals $\sum_{v_0 <^0 v'} x_f(v_0, v')$. Finally we can use $W((l(v_{max_i}), p)) < \sum_{v_{max_i} <^0 v'} x_f(v_{max_i}, v')$.

Altogether it is enough to show that

$$\sum_{v <^0 C} \sum_{v < v'} x_f(v, v') - \sum_{v <^0 C} \sum_{v' <^0 v} x_f(v', v) - \sum_{v \in C} \sum_{v' <^0 v} x_f(v', v) = 0.$$

We claim that in this sum each value $x_f(v, v')$ equals either 0 or is counted once positively and once negatively. The second alternative is obviously fulfiled if $v' < C$

or $v' \in C$. Observe now that for $v < C$ and $x_f(v, v') > 0$ we get by the definition of W and C that $v' <^0 C$ or $v' \in C$. That means if $x_f(v, v')$ does not fulfil the first alternative, it fulfils the second alternative.

Observe $v <^0 C \Leftrightarrow v <^0 C'$. That means, replacing C by C' in the above sum could only change the value of the third sum, namely by values $x_f(v', v)$ with $v \in C' \setminus C$. These values are equal to 0 by the definition of C. $\qquad\square$

5   Strict Executability

In this section we briefly outline a polynomial test for the strict executability of an LPO. The flow property is a necessary and sufficient condition of the executability of a labelled partial order. We extend the flow property to get a necessary and sufficient condition for an LPO to be exactly a run of a marked p/t-net.

Given a partial order $(V, <)$, let $\sqsubset \subseteq <$ be the skeleton of $<$, i.e. the smallest subset of $<$ which fulfils: $\sqsubset^+ = <$.

Definition 16 (Strong flow property). *A labelled partial order* $(V, <, l)$ *fulfils the strong flow property w.r.t.* (N, m_0) *if there exists a family* $X = \{x_p \mid p \in P\}$ *of flows fulfilling the flow property which satisfies: the skeleton* $\sqsubset$ *of* $<$ *is a subset of the relation* $Q_X = \{(v, v') \in V \times V \mid \exists p \in P : x_p(v, v') > 0\}$.

One can easily check that the canonical flow of a run fulfils the strong flow property. Observe that the flow functions of a labelled partial order which fulfil the flow property used for construction of the process in the proof of lemma 3 are the canonical flow functions of the underlying run of the constructed process. Thus, we obtain:

Theorem 4. *A labelled partial order fulfils the strong flow property w.r.t.* (N, m_0) *if and only if it is a run of* (N, m_0).

By Definition 10, the executability of an LPO is a necessary condition for its strict executability. Now, take an LPO $lpo = (V, <, l)$ which is executable in a marked p/t net (N, m_0), i.e. which fulfils the flow property w.r.t. (N, m_0). Denote $X = \{x_p \mid p \in P\}$ the family of flows fulfilling the flow property computed by the algorithm presented in the previous section. The following algorithm decides whether lpo is strictly executable w.r.t. (N, m_0), i.e. whether it is a minimal run of (N, m_0). If the answer is positive, then X is the family of canonical flows of the underlying minimal run.

- Compute skeleton $\sqsubset$ of the partial order $(V, <)$;
- if $\sqsubset \not\subseteq Q_X$ then return "lpo is not strictly executable";
- else for each edge $e \in \sqsubset$: test whether $(V, < \setminus \{e\}, l)$ is executable; if yes, return "lpo is not strictly executable";
- return "lpo is strictly executable".

Due to lack of space, we omit a detailed proof of the correctness of the algorithm and just give some intuition: Observe that after removing a skeleton edge from a partial order $<$ one still gets a partial order. Moreover, any proper subset of a partial order $<$, which is itself a partial order, is a subset of a partial order obtained from $<$ by removing a skeleton edge. Thus, if the algorithm returns "lpo is not strictly executable", then a run was computed which is a proper subset of lpo, i.e. lpo cannot equal a minimal run. If the algorithm returns "lpo is strictly executable", then lpo fulfils the strong flow property and therefore it is a run, and all LPOs which are proper subsets of lpo are not executable, i.e. do not include a run. That means lpo does not properly include any run and therefore it is a minimal run. Obviously the algorithm runs in polynomial time.

6   Conclusion

We have presented a polynomial algorithm for testing whether an LPO is executable in a place/transition net while preserving at least the given amount of concurrency, or in other words, adding no causality (i.e. whether it is a sequentialization of a run). Further, we have formulated a polynomial test deciding whether an LPO is strictly executable, i.e. whether the given amount of concurrency is maximal, or complementary, whether the amount of causality is minimal (i.e. whether the LPO equals a minimal run). It is a question of further research to determine efficiently whether an LPO is executable preserving exactly the given amount of concurrency and causality, i.e. whether it equals a (not necessarily minimal) run. Another interesting question is generalization of so called "legal firing sequence" problem, namely to determine whether an LPO can be executed while preserving at least the given amount of causality (adding no concurrency). Finally, one could combine both approaches into a scenario based specification prescribing minimal/maxiamal level of causality and concurrency.

References

1. J. Desel and W.Reisig. Place/Transition Petri Nets. In *Lectures on Petri nets I: Basic Models*, LNCS 1491, pp. 123–174,1998.
2. L.R. Ford, Jr. and D.R. Fulkerson. Maximal Flow Through a Network. *Canadian Journal of Mathematics* 8, pp. 399–404, 1955.
3. A. Goldberg and S. Rao. Beyond the Flow Decomposition Barrier. *Journal of the ACM* 45/5, pp. 783–797, 1998.
4. U. Goltz and W. Reisig. The Non-Sequential Behaviour of Petri Nets. *Information and Control*, 57(2-3), pp. 125-147, 1983.
5. U. Goltz and W. Reisig: Processes of Place/Transition Nets. - Proc. of ICALP'83, LNCS 154, pp. 264-277, 1983.
6. J. Grabowski. On Partial Languages. *Fundamenta Informaticae* IV.2, pp. 428–498, 1981.
7. A.V. Karzanov. Determining the Maximal Flow in a Network by the Method of Preflows. *Soviet Math. Doc.* 15, pp. 434–437, 1974.
8. A. Kiehn. On the Interrelationship between Synchronized and Non-Synchronized Behavior of Petri Nets. *Journal Inf. Process. Cybern. EIK* 24, pp. 3 – 18, 1988.
9. V. Pratt. Modelling Concurrency with Partial Orders. *Int. Journal of Parallel Programming* 15, pp. 33–71, 1986.
10. W. Vogler. Modular Construction and Partial Order Semantics of Petri Nets. *LNCS* 625, 1992.
11. W. Vogler. Partial words versus processes: a short comparison, Advances in Petri Nets, LNCS 609, Springer 1992, pp. 292-303.

Reference and Value Semantics Are Equivalent for Ordinary Object Petri Nets

Michael Köhler and Heiko Rölke

University of Hamburg, Department of Informatics,
Vogt-Kölln-Str. 30, D-22527 Hamburg
{koehler, roelke}@informatik.uni-hamburg.de

Abstract. The concept of mobile agents imposes a great security risk for information systems. In this paper we propose object nets as a specification formalism for multi-agent systems. Since the general formalism is Turing-powerful not every analysis method that is common for Petri net can be applied. So, we define the subclass of "ordinary" object nets that allows for the application of standard P/T-net techniques, i.e. the computation of boundedness, liveness etc.

1 Introduction

Object Petri nets following the *nets-within-nets* paradigm are a very powerful formalism to describe dynamic multi-levelled systems, e.g. mobile agent systems. It is well known that severe security problems arise in the context of mobile agents (cf. [5]). So, the use of formal methods to overcome security problems is necessary. In [10] the authors showed how the formalism of nets-within-nets can be used to model mobility, especially in the case of mobile agents. Mobile agents are developed using our architecture MULAN [9]. The MULAN-framework offers intuitive modelling even of large agent systems.[1] What was missing is the possibility to profit from analysing tools. This paper undertakes an attempt to build a conceptual background for the transformation of object net systems to P/T-nets. Several requirements have to be met for this transformation to succeed. These requirements are the subject of the following sections.

The need for analysis of object nets is paired with the choice of the firing rule for object nets: There exist two fundamental semantics (i.e. firing rules) for object nets introduced in [20], called reference and value semantics. The difference of reference and value semantics is the concept of "location" for net-tokens which is explicit for value but not for reference semantics, since it is unclear which reference can be considered as the location of a net-token.

As shown in [10] the concept of mobility cannot be expressed adequately by reference semantics due to the possibility of side-effects. Instead value semantics has to be applied. As shown in [12] the concept of locality makes value semantics richer

---

[1] Student projects created agent systems containing more than 200 different reference nets resulting in tens of thousands of net instances at run-time.

G. Ciardo and P. Darondeau (Eds.): ICATPN 2005, LNCS 3536, pp. 309–328, 2005.

than reference semantics – for example the reachablity problem becomes undecidable while boundedness remains decidable. However, the reference semantics can be simulated by a (larger) P/T net, so analysis methods can be applied directly.

Value semantics is adequate from a modelling point of view while reference semantics is adequate from an analytical point of view. In this paper we focus on a restricted class of object nets, the so called *ordinary* object nets. For this class of object nets it can be shown, that value semantics is as expressible as reference semantics. This is shown by providing a direct embedding and simulation of one semantics using the other one.

In this paper we study semantical aspects of the *nets-within-nets* paradigm.[2] The paradigm that allows nets as tokens was introduced by Rüdiger Valk in [19] and extended to the formalism of elementary object net systems in [20, 21] which allows to model a two-levelled system. The formalism has been extended in [11, 12] to an arbitrary nesting structure. A similar approach which allows nested Petri net structures is presented in [15]. For Hypernets [1] net-tokens are restricted to synchronisations of state machines. Reference nets [13] are a specialised nets-within-nets formalism based on reference semantics. Due to the nested structure of object nets the formalism is closely related to mobility calculi like the ambient calculus [4] or to formalism combining mobility calculi and Petri nets like [3].

The paper is structured as follows: Section 2 gives the formal definition of object-net systems. Firing is defined both for value and for reference semantics. Section 3 analyses *located* markings, i.e. markings that describe a unique relation of tokens and their locations: for each net-token there exists exactly one place containing it. Section 4 defines the subclass of *ordinary* object-net systems. It is shown that for this class of object nets all reachable markings are located. Using this result it is shown that reference and value semantics can simulate each other directly. In Section 5 we analyse the processes of ordinary object-net systems. It turns out that the firing relation, the mapping from a process to the original object net and the mappings from reference and value semantics are compatible with each other which results in a three-dimensional cube structure of embeddings. After having cleared the conceptual background we present a case study in Section 6 and present some analysis results. Finally, we give an outlook and conclusion.

2   Object-Net Systems

We define a generalised model of object-net systems, which drops the restriction of [20] to exactly two levels of nesting: Object-Net Systems (Os) are defined to give a precise definition of nets-within-nets using nested multi-set rewriting specifications.

---

[2] As [18] mentioned in the outlook it is a quite natural extension of algebraic Petri nets [16] to allow tokens to be *active* which is impossible for algebraic Petri nets. The canonic way for this extension is to consider nets as active tokens. These tokens are called net-tokens.

2.1   Informal Introduction to Object Nets

There exist two fundamental semantics for object nets introduced in [20], called reference and value semantics. The intuitive meaning of both semantics can be explained using the example Os given in Fig. 1. The example is known as the "α-centauri" example. The name is due to the interpretation that the net token describes a log which is copied and one copy remains on earth while the other one is sent to α-centauri. It then seems somehow counter-intuitive that for reference semantics the state change on α-centauri (the upper branch of the system net) becomes visible immediately on earth (the lower branch).

The arrow from the token on place s_1 expresses that the inner structure of the token is itself a net. The different levels in the object-net system are connected by channels. Transitions inscribed by corresponding channel expressions like on:ch() and :ch() must be fired synchronously. If there is more than one possible partner the choice is non-deterministic. In the Figure each transition pair (t_2, t_{11}) and (t_3, t_{12}) must fire synchronously.

For reference semantics (cf. Fig. 2) the place s_1 initially contains a reference to the object-net: $\mathbf{M} = s_1 + s_{11}$. Firing of t_1 duplicates this reference onto s_2 and s_3 resulting in $\mathbf{M_1} = s_2 + s_3 + s_{11}$. This marking enables the transition pair (t_2, t_{11}) but not (t_3, t_{12}). The resulting marking is $\mathbf{M_2} = s_4 + s_3 + s_{12}$. Since the effect in the object-net is visible in the whole system, the pair (t_3, t_{12}) is now enabled. Firing leads to $\mathbf{M_3} = s_4 + s_5 + s_{13}$.

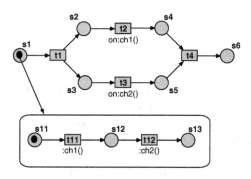

Fig. 1. An Os: The α-centauri example

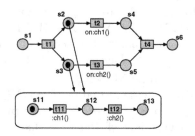

Fig. 2. Firing of transition t_1 w.r.t. reference semantics

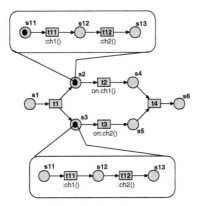

Fig. 3. Firing of transition t_1 w.r.t. value semantics

For value semantics (cf. Fig. 3) we have the *nested* multiset $\mathbf{M} = s_1[s_{11}]$ as the initial marking which corresponds to the initial marking $s_1 + s_{11}$ w.r.t. reference semantics. Firing of t_1 distributes the marking of the net-token. A possible distribution is the marking $\mathbf{M}_1 = s_2[s_{11}] + s_3[\mathbf{0}]$ (where $\mathbf{0}$ denotes the empty multiset) – corresponding to $s_2 + s_3 + s_{11}$ for reference semantics. This marking enables the transition pair (t_2, t_{11}) but not the pair (t_3, t_{12}). The resulting marking is $\mathbf{M}_2 = s_4[s_{12}] + s_3[\mathbf{0}]$. Since the effect in the object net is only local the pair (t_3, t_{12}) is not enabled. So $w = t_1(t_2, t_{11})(t_3, t_{12})$ is a possible firing sequence for reference but not for value semantics.

2.2   Petri Net Notations

A *P/T net structure* is a tuple $N = (P, T, W)$, such that: P is a finite set of places, T is a finite set of transitions, with $P \cap T = \emptyset$, and $W : ((P \times T) \cup (T \times P)) \to \mathbb{N}$ is the arc-weight function. A *marked P/T-net* $N = (P, T, W, \mathbf{M}_0)$ is an P/T net structure (P, T, W) together with an initial marking $\mathbf{M}_0 \in MS(P)$. The term *P/T net* is used both for the unmarked and the marked case. The flow relation is $F := \{(x, y) \mid W(x, y) > 0\}$. Given a net $P(N)$ denotes its places, $T(N)$ its transitions etc. N is called *ordinary* iff $W(x, y) \leq 1$ for all (x, y). For an ordinary P/T net the mapping W coincides with the characteristic function of the flow relation $\chi_F : ((P \times T) \cup (T \times P)) \to \{0, 1\}$. In the case of ordinary nets the relation F is also used to denote the arc weight W and vice versa.

A transition $t \in T$ of a P/T net $N = (P, T, W, \mathbf{M}_0)$ is *enabled* in marking $\mathbf{M}$ iff $\forall p \in P : \mathbf{M}(p) \geq W(p, t)$ holds. The successor marking when firing t is $\mathbf{M}'(p) = \mathbf{M}(p) - W(p, t) + W(t, p)$. The enablement of t in marking $\mathbf{M}$ is denoted by $\mathbf{M} \xrightarrow{t}$. Firing of t is denoted by $\mathbf{M} \xrightarrow{t} \mathbf{M}'$.

A *P/T net* is equivalently characterised as $N = (P, T, \mathbf{pre}, \mathbf{post}, \mathbf{M}_0)$ where the multi-set mappings $\mathbf{pre}, \mathbf{post} : T \to MS(P)$ are $\mathbf{pre}(t)(p) := W(p, t)$ and $\mathbf{post}(t)(p) := W(t, p)$. For notations cf. the appendix.

2.3   Nets as Tokens

An Object-Net System (Os) $OS = (\mathcal{N}, d, \Theta, \mathbf{M}_0)$ consists of a finite set $\mathcal{N}$ of pairwise disjoint P/T nets which includes the black token net $N_\bullet$ and the system-net N_{sn}. The black token net $N_\bullet$ is defined as the object-net with no places and no transitions: $P(N_\bullet) = T(N_\bullet) = \emptyset$. Let $P(OS)$ be the union of all components: $P(OS) := \bigcup_{N \in \mathcal{N}} P(N)$. Analogously for $T(OS)$, $F(OS)$, and $W(OS)$.

Markings are nested multi-sets. Tokens are described as pairs of the marked place p and the marking of $\mathbf{M}$ its net-token which is denoted as $p[\mathbf{M}]$ to emphasize the nesting. The place typing $d : P(OS) \to \mathcal{N}$ is used to define which net-tokens are allowed on a place. A black token is the special net-token $p[\mathbf{0}]$ which can be identified with p. The basic tokens are black tokens, higher-order tokens are generated using net-tokens.

$$\begin{aligned}
\mathcal{P}_0(N) &:= \{p \mid p \in P(N) \wedge d(p) = N_\bullet\} \\
\mathcal{P}_{n+1}(N) &:= \{p[\mathbf{M}] \mid p \in P(N) \wedge \mathbf{M} \in MS(\mathcal{P}_n(d(p)))\}
\end{aligned} \tag{1}$$

Define $\mathcal{P}(N) := \bigcup_{i=0}^{\infty} \mathcal{P}_n(N)$. Each mapping f defined on P can be extended to a mapping $f^\sharp$ on nested markings $\mathcal{P}$ by setting $f^\sharp(p[\mathbf{M}]) = f(p)[f^\sharp(\mathbf{M})]$. In the following $f^\sharp$ is also denoted as f.

A transition $t \in T(OS)$ may be synchronised with transitions of the net-tokens. The resulting synchronisations are nested transitions, i.e. trees:[3]

$$\begin{aligned}
\mathcal{T}_0(N) &:= \{id\} \\
\mathcal{T}_{k+1}(N) &:= \{t[\theta_{N_1}, \ldots, \theta_{N_n}] \mid t \in T(N) \wedge \theta_{N_i} \in \bigcup_{j=0}^{k} \mathcal{T}_j(N_i)\}
\end{aligned} \tag{2}$$

Analogously to markings we identify the minimal synchronisation tree $t[id] := t[id, \ldots, id]$ with the transition t itself. Here id is a "pseudo" transition with $\mathbf{pre}(id) = \mathbf{post}(id) = \mathbf{0}$ (see below). So, every node in the tree has the same degree of branching.

Define $\mathcal{T}(N) := \bigcup_{i=0}^{\infty} \mathcal{T}_n(N)$ and $\mathcal{T}(OS) := \mathcal{T}(\mathcal{N}) := \bigcup_{N \in \mathcal{N}} \mathcal{T}(N)$. A synchronisation structure $\Theta(OS)$ consists of a finite subset of $\mathcal{T}(\mathcal{N})$.

The nesting structure of markings is removed by $\mathrm{fl} : \mathcal{M}_v \to \mathcal{M}_r$ with $\mathcal{M}_r := MS(P(OS))$ where $\mathrm{fl}(p[\mathbf{M}]) := p + \mathrm{fl}(\mathbf{M})$.

The nesting structure of synchronisations is removed by $\mathrm{fl} : \mathcal{T}(OS) \to T(OS)$ where $\mathrm{fl}(t[\theta_1, \ldots, \theta_n]) := t + \mathrm{fl}(\theta_1) + \cdots + \mathrm{fl}(\theta_n)$ and $\mathrm{fl}(id) = \mathbf{0}$.

Let $\Theta \subseteq \mathcal{T}(N)$ be a set of synchronisations. To avoid cycles the set of synchronisations has to contain each transitions eaxactly once: $\mathrm{fl}(\Theta) = T(OS)$ (Note, that minimal synchronisation trees $t[id]$ are allowed).

A marking is a multi-set of system-net tokens: $\mathbf{M} \in \mathcal{M}_v := MS(\mathcal{P}(N_{sn}))$.

Definition 1. *An* Object-Net System *is the tuple $OS = (\mathcal{N}, d, \Theta, \mathbf{M}_0)$, where*

- *$\mathcal{N} = \{N_1, \ldots N_n\}$ is a set of pairwise disjoint P/T-nets $N_i = (P_i, T_i, W_i)$ including the black token net $N_\bullet$ and the system-net N_{sn}.*
- *$d : P \to \mathcal{N}$ is the place typing.*

---

[3] In the graphical representation these trees are formalised by channel inscriptions.

- $\Theta \subseteq T(\mathcal{N})$ *is a finite set of synchronisations with* $\mathrm{fl}(\Theta) = T(OS)$.
- *The initial marking is* $\mathbf{M}_0 \in MS(\mathcal{P}(N_{sn}))$.

We generalise the notion of pre- and post-set for object nets by defining

$$^{(N)}t := {}^{\bullet}t \cap d^{-1}(N) = \{p \in {}^{\bullet}t \mid d(p) = N\}$$
$$t^{(N)} := t^{\bullet} \cap d^{-1}(N) = \{p \in t^{\bullet} \mid d(p) = N\}$$

Definition 2. *For the synchronisation tree* $t[\boldsymbol{\theta}] \in \Theta$ *where* $\boldsymbol{\theta} = (\theta_{N_1}, \ldots, \theta_{N_n})$ *the firing rule is generated inductively from the firing rule of the subtrees* θ_N *(where the multi-set variables* $\mathbf{X}_{p,t,i}$, $\mathbf{X}'_{p,t,i}$ *describe the tokens that are transported and* $\mathbf{Y}_{p,t,i}$, $\mathbf{Y}'_{p,t,i}$ *describe the tokens that are used for synchronisation):*

$$\sum_{p \in {}^{\bullet}t} \sum_{i=1}^{W(p,t)} p[\mathbf{X}_{p,t,i} + \mathbf{Y}_{p,t,i}] \xrightarrow{t[\boldsymbol{\theta}]} \sum_{p \in t^{\bullet}} \sum_{j=1}^{W(t,p)} p[\mathbf{X}'_{p,t,j} + \mathbf{Y}'_{p,t,j}]$$
$$\text{if } \forall N \in \mathcal{N} : \sum_{p \in {}^{(N)}t} \sum_{i=1}^{W(p,t)} \mathbf{X}_{p,t,i} = \sum_{p \in t^{(N)}} \sum_{j=1}^{W(t,p)} \mathbf{X}'_{p,t,j}$$
$$\wedge \sum_{p \in {}^{(N)}t} \sum_{i=1}^{W(p,t)} \mathbf{Y}_{p,t,i} = \mathbf{pre}(\theta_N)$$
$$\wedge \sum_{p \in t^{(N)}} \sum_{j=1}^{W(t,p)} \mathbf{Y}'_{p,t,j} = \mathbf{post}(\theta_N)$$

For the minimal synchronisations $t[\mathbf{id}]$ *this implies* $\mathbf{Y}_{p,t,i} = \mathbf{Y}'_{p,t,i} = 0$.

The marking $\mathbf{M}$ *can be fired by* $\theta \in \Theta$ *to* $\mathbf{M}'$ *iff* $\mathbf{pre}(\theta)$ *is a subterm of* $\mathbf{M}$ *and* $\mathbf{M}'$ *is obtained from* $\mathbf{M}$ *by substituting* $\mathbf{pre}(\theta)$ *with* $\mathbf{post}(\theta)$. *Firing is denoted by* $\mathbf{M} \xrightarrow{\theta} \mathbf{M}'$.

This firing relation formalises value semantics. For reference semantics the "flat" version of OS is needed (for more details cf. [12]).

Definition 3. *The* underlying P/T net of $OS = (\mathcal{N}, d, \Theta, \mathbf{M}_0)$ *is defined as:*

$$\mathrm{fl}(OS) := (P(OS), \Theta, \mathbf{pre}^{\mathrm{fl}}, \mathbf{post}^{\mathrm{fl}}, \mathrm{fl}(\mathbf{M}_0))$$

where $\mathbf{pre}^{\mathrm{fl}}(\theta) := \mathbf{pre}(\mathrm{fl}(\theta))$ *and* $\mathbf{post}^{\mathrm{fl}}(\theta) := \mathbf{post}(\mathrm{fl}(\theta))$.

Reference semantics is obtained by forgetting the nesting structure:

Theorem 1. *For an* Os OS *the mapping* fl *provides a direct embedding of the value semantics, i.e. every firing w.r.t. value semantics is possible w.r.t. reference semantics:*

$$
\begin{array}{ccc}
\mathbf{M}_v & \xrightarrow[OS]{\theta} & \mathbf{M}'_v \\
\mathrm{fl}\downarrow & & \downarrow\mathrm{fl} \\
\mathrm{fl}(\mathbf{M}_v) & \xrightarrow[\mathrm{fl}(OS)]{\theta} & \mathrm{fl}(\mathbf{M}'_v)
\end{array}
$$

Proof. Since $\mathbf{pre}^{\mathrm{fl}}(\theta) := \mathbf{pre}(\mathrm{fl}(\theta))$ and $\mathbf{post}^{\mathrm{fl}}(\theta) := \mathbf{post}(\mathrm{fl}(\theta))$ it is sufficient to show $\mathrm{fl}(\mathbf{pre}(\theta)) = \mathbf{pre}^{\mathrm{fl}}(\theta)$ and $\mathrm{fl}(\mathbf{post}(\theta)) = \mathbf{post}^{\mathrm{fl}}(\theta)$. This is shown by induction over the synchronisation tree.

- For the minimal synchronisation tree we have (since $\sum_{p\in{}^\bullet t} \mathbf{X}_{p,t,i} = \sum_{p\in t^\bullet} (\mathbf{X}'_{p,t,j})$):

$$\begin{aligned}
\mathrm{fl}(\mathbf{pre}(t[\boldsymbol{id}])) &= \mathrm{fl}(\sum_{p\in{}^\bullet t} \sum_{i=1}^{W(p,t)} p[\mathbf{X}_{p,t,i}]) \\
&= \sum_{p\in{}^\bullet t} \sum_{i=1}^{W(p,t)} \mathrm{fl}(p[\mathbf{X}_{p,t,i}]) \\
&= \sum_{p\in{}^\bullet t} \sum_{i=1}^{W(p,t)} p + \mathrm{fl}(\mathbf{X}_{p,t,i}) = \mathbf{pre}(\mathrm{fl}(t[\boldsymbol{id}]))
\end{aligned}$$

$$\begin{aligned}
\mathbf{post}(t[\boldsymbol{id}]) &= \sum_{p\in t^\bullet} \sum_{j=1}^{W(t,p)} p + \mathrm{fl}(\mathbf{X}'_{p,t,j}) \\
&= \sum_{p\in t^\bullet} \sum_{j=1}^{W(t,p)} \mathrm{fl}(p[\mathbf{X}'_{p,t,j}]) \\
&= \mathrm{fl}(\sum_{p\in t^\bullet} \sum_{j=1}^{W(t,p)} p[\mathbf{X}'_{p,t,j}]) = \mathrm{fl}(\mathbf{post}(t))
\end{aligned}$$

- By assumption we have $\mathrm{fl}(\sum_{p\in{}^{(N)}t} \sum_{i=1}^{W(p,t)} \mathbf{Y}_{p,t,i}) = \mathbf{pre}(\mathrm{fl}(\theta_N))$ and also $\mathrm{fl}(\sum_{p\in t^{(N)}} \sum_{j=1}^{W(t,p)} \mathbf{Y}'_{p,t,j}) = \mathbf{post}(\mathrm{fl}(\theta_N))$. Induction on $\theta = t[\boldsymbol{\theta}]$:

$$\begin{aligned}
\mathrm{fl}(\mathbf{pre}(t[\boldsymbol{\theta}])) &= \mathrm{fl}(\sum_{p\in{}^\bullet t} \sum_{i=1}^{W(p,t)} p[\mathbf{X}_{p,t,i} + \mathbf{Y}_{p,t,i}]) \\
&= \sum_{p\in{}^\bullet t} \sum_{i=1}^{W(p,t)} \mathrm{fl}(p[\mathbf{X}_{p,t,i} + \mathbf{Y}_{p,t,i}]) \\
&= \sum_{p\in{}^\bullet t} \sum_{i=1}^{W(p,t)} p + \mathrm{fl}(\mathbf{X}_{p,t,i}) + \mathrm{fl}(\mathbf{Y}_{p,t,i}) \\
&= \sum_{p\in{}^\bullet t} \sum_{i=1}^{W(p,t)} p + \mathrm{fl}(\mathbf{X}_{p,t,i}) + \sum_{N\in\mathcal{N}} \mathbf{pre}(\mathrm{fl}(\theta_N)) \\
&= \mathbf{pre}(\mathrm{fl}(t[\boldsymbol{\theta}]))
\end{aligned}$$

$$\begin{aligned}
\mathrm{fl}(\mathbf{post}(t[\boldsymbol{\theta}])) &= \sum_{p\in t^\bullet} \sum_{j=1}^{W(t,p)} p + \mathrm{fl}(\mathbf{X}'_{p,t,j}) + \sum_{N\in\mathcal{N}} \mathbf{post}(\mathrm{fl}(\theta_N)) \\
&= \sum_{p\in t^\bullet} \sum_{j=1}^{W(t,p)} p + \mathrm{fl}(\mathbf{X}'_{p,t,j}) + \mathrm{fl}(\mathbf{Y}'_{p,t,j}) \\
&= \sum_{p\in t^\bullet} \sum_{j=1}^{W(t,p)} \mathrm{fl}(p[\mathbf{X}'_{p,t,j} + \mathbf{Y}'_{p,t,j}]) \\
&= \mathrm{fl}(\sum_{p\in t^\bullet} \sum_{j=1}^{W(t,p)} p[\mathbf{X}'_{p,t,j} + \mathbf{Y}'_{p,t,j}]) \\
&= \mathrm{fl}(\mathbf{post}(t[\boldsymbol{\theta}]))
\end{aligned}$$

This proves the embedding.   □

The converse (every firing w.r.t. reference semantics is possible w.r.t. value semantics) of Theorem 1, however, does not hold in the general case as seen for the α-centauri example.

3   Located Markings

In the following we consider markings where the localisation of tokens coincide with the type structure induced by the mapping d.

Definition 4. *A marking* $\mathbf{M}_r \in \mathcal{M}_r$ *is located iff for each net* N *there exists exactly one place containing* N *and no place contains the net* N_{sn}:

$$\forall N \in \mathcal{N} \setminus \{N_{sn}, N_\bullet\} : \quad \begin{aligned} \sum_{p\in d^{-1}(N)} |\mathbf{M}_r(p)| &= 1 \wedge \\ \sum_{p\in d^{-1}(N_{sn})} |\mathbf{M}_r(p)| &= 0 \end{aligned}$$

A marking $\mathbf{M}_v \in \mathcal{M}_v$ *is located iff* $\mathrm{fl}(\mathbf{M}_v)$ *is located.*

Therefore for all $N \in \mathcal{N} \setminus \{N_{sn}, N_{\bullet}\}$ the uniquely defined place for which $|\mathbf{M}(p)| = 1$ holds is denoted by $p(N)$.

Note, that if $\mathbf{M}_v$ is located, then there cannot be any recursive nesting, since otherwise there is more than one place containing a net token of type N. Thus, the location of each token is uniquely determined.

Definition 5. *The* localisation $lc(\mathbf{M})$ *is defined recursively starting with the system net:* $lc(\mathbf{M}) := lc_M(\mathbf{M}_{|P(N_{sn})})$ *where*

$$lc_M(p) := \begin{cases} p[lc_M(\mathbf{M}_{|P(N_{d(p)})})], & \text{if } d(p) \neq N_{\bullet} \\ p, & \text{otherwise} \end{cases}$$

For events $t \in T$ we define $lc(t) = t$.

Theorem 2. *For located markings the mapping lc is inverse to fl.*

1. *If $\mathbf{M}_r \in \mathcal{M}_r$ is located, then we have $fl(lc(\mathbf{M}_r)) = \mathbf{M}_r$.*
2. *If $\mathbf{M}_v \in \mathcal{M}_v$ is located, then we have $lc(fl(\mathbf{M}_v)) = \mathbf{M}_v$.*

Proof. 1. Let $\mathbf{M}_r \in \mathcal{M}_r$. Define the relation $R_{\mathbf{M}_r} \subseteq (\mathcal{N} \setminus \{N_{\bullet}\})^2$ by

$$(N_1, N_2) \in R_{\mathbf{M}_r} \iff \exists p_1 \in M : p_1 \in P(N_1) \wedge d(p_1) = N_2$$

If $\mathbf{M}_r$ is located, then the place p_1 such that $d(p_1) = N_2$ with $N_2 \in \mathcal{N} \setminus \{N_{sn}, N_{\bullet}\}$ is uniquely defined, i.e. it is $p_1 = p(N_2)$. So, it follows that $R_{\mathbf{M}_r}$ describes a tree with the system net N_{sn} as the root node (since $\sum_{p \in d^{-1}(N_{sn})} |\mathbf{M}_r(p)| = 0$).

It is easy to see from the definition of lc that the marking is nested along the relation R_M, i.e. all markings of the nets $N' \in (N R_{\mathbf{M}_r}\text{-})$ on paths from the root of the tree are located by $lc_M(\mathbf{M}_{|P(N)})$:

$$fl(lc_M(\mathbf{M}_{|P(N)})) = \sum_{N' \in (N R_{\mathbf{M}_r}\text{-})} \mathbf{M}_{|P(N')}$$

For the whole marking it follows that:

$$fl(lc(\mathbf{M})) = lc_M(\mathbf{M}_{|P(N_{sn})})$$
$$= fl\left(\sum_{N' \in (N_{sn} R_{\mathbf{M}_r}\text{-})} \mathbf{M}_{|P(N')}\right)$$
$$= \mathbf{M}_{|P(N_{sn})} + \mathbf{M}_{|P(N_{\bullet})} + \cdots + \mathbf{M}_{|P(N_n)}$$
$$= M$$

Note, that $\mathbf{M}_{P(N_{\bullet})} = \mathbf{0}$ since $P(N_{\bullet}) = \emptyset$.

2. Let $\mathbf{M}_v \in \mathcal{M}_r$. Define the relation $R_{\mathbf{M}_v} \subseteq (\mathcal{N} \setminus \{N_{\bullet}\})^2$ by

$$(N_1, N_2) \in R_{\mathbf{M}_v} \iff \exists \text{ subterm } (p_1, \mathbf{M}_1) \text{ of } \mathbf{M}_v : p_1 \in P(N_1) \wedge d(p_1) = N_2$$

it is easy to see, that if $\mathbf{M}_v$ is located, then the relations $R_{\mathbf{M}_r}$ and $R_{\mathbf{M}_v}$ are equal, so lc just reconstructs $\mathbf{M}_v$, i.e. $lc(fl(\mathbf{M}_v)) = \mathbf{M}_v$.

$\square$

4   Ordinary Object-Net Systems

As we have seen for the α-centauri example, the converse of Theorem 1 does not hold in general. It will be shown, that for the case of so called *ordinary* ONS the opposite direction can also be proved.

Definition 6. *Let OS be an* Os. *A transitions t is* simple *iff*

$$\forall N \in \mathcal{N} \setminus \{N_\bullet\} : |^{(N)}t| = |t^{(N)}| \leq 1$$

OS is ordinary *iff all its object nets are ordinary, all transitions are simple and the initial marking* $\mathbf{M}_0$ *is located.*

Consider $t[\boldsymbol{\theta}] \in \Theta$. For ordinary Os all arc weights $W(x, y) = 1$ iff $(x, y) \in F$. Additionally, if $|^{(N)}t| > 0$ there is exactly one place $p_N \in {}^\bullet t$ such that $d(p) = N$ and one place $p'_N \in t^\bullet$ such that $d(p') = N$. So, the variable $\mathbf{X}_{p,t}$ can be denoted as $\mathbf{X}_{d(p)}$ (and similar for $\mathbf{X}'_{t,p}$ etc.). Due to this one-to-one correspondence the representation can be simplified further:

- Value semantics: For the synchronisation tree $t[\boldsymbol{\theta}] \in \mathcal{T}_{n+1}(OS)$ the firing rule is generated from the firing rule of the θ_N:

$$\sum_{p \in {}^\bullet t} p[\mathbf{X}_{d(p)} + \mathbf{pre}(\theta_N)] \xrightarrow{t[\boldsymbol{\theta}]} \sum_{p \in t^\bullet} p[\mathbf{X}'_{d(p)} + \mathbf{post}(\theta_N)]$$

For a minimal synchronisation tree this further simplifies to:

$$\sum_{p \in {}^\bullet t} p[\mathbf{X}_{d(p)}] \xrightarrow{t[\boldsymbol{id}]} \sum_{p \in t^\bullet} p[\mathbf{X}'_{d(p)}]$$

- Reference semantics: For the synchronisation tree $t[\boldsymbol{\theta}] \in \mathcal{T}_{n+1}(OS)$:

$$\sum_{p \in {}^\bullet t} p + \mathbf{pre}(\mathrm{fl}(\theta_{d(p)})) \xrightarrow{\mathrm{fl}(t[\boldsymbol{\theta}])} \sum_{p \in t^\bullet} p + \mathbf{post}(\mathrm{fl}(\theta_{d(p)}))$$

For the minimal synchronisation tree this simplifies to:

$$\sum_{p \in {}^\bullet t} p \xrightarrow{\mathrm{fl}(t[\boldsymbol{id}])} \sum_{p \in t^\bullet} p$$

Theorem 3. *If OS is an ordinary* Os, *then all reachable markings are located and all places p with* $d(p) \neq N_\bullet$ *are 1-safe.*

Proof. The initial marking is located by definition. If $\mathbf{M}_r \xrightarrow{t} \mathbf{M}'_r$ then there is exactly one location for N (since $\sum_{p \in d^{-1}(N)} |\mathbf{M}_r(p)| = 1$ for all $N \in \mathcal{N} \setminus \{N_{sn}, N_\bullet\}$), which is either untouched or relocated, since all t are simple.

If $\mathbf{M}_r$ is located, then all places p with $d(p) \neq N_\bullet$ are marked with at most one net-token. Since every reachable marking of a simple Os is located these places are 1-safe. □

Using Theorem 3 we know that all reachable markings are located, we can conclude from Theorem 2:

$$\forall \mathbf{M}_r \in R(\mathrm{fl}(OS)) : \mathrm{fl}(\mathrm{lc}(\mathbf{M}_r)) = \mathbf{M}_r$$
$$\forall \mathbf{M}_v \in R(OS) : \quad \mathrm{lc}(\mathrm{fl}(\mathbf{M}_v)) = \mathbf{M}_v$$

Theorem 4. *For ordinary* Os *OS the mapping* lc *provides a direct embedding of the reference semantics. If* $\mathbf{M}_r$ *is located, then:*

$$
\begin{array}{ccc}
\mathbf{M}_r & \xrightarrow[\mathrm{fl}(OS)]{\theta} & \mathbf{M}'_r \\
\mathrm{lc} \downarrow & & \downarrow \mathrm{lc} \\
\mathrm{lc}(\mathbf{M}_r) & \xrightarrow[OS]{\theta} & \mathrm{lc}(\mathbf{M}'_r)
\end{array}
$$

Proof. Induction over the synchronisation tree $t[\boldsymbol{\theta}]$.

- If $\mathbf{M}_r \xrightarrow{t[id]} \mathbf{M}'_r$ then by monotonicity we can add $\mathrm{fl}(\mathbf{X}_{d(p)})$ with $\mathbf{X}_{d(p)} = \mathrm{lc}(\mathbf{M}_{r|P(d(p))})$.

$$\mathbf{M}_r + \sum_{p\in{}^\bullet t} \mathrm{fl}(\mathbf{X}_{d(p)}) \xrightarrow{t[id]} \mathbf{M}'_r + \sum_{p\in{}^\bullet t} \mathrm{fl}(\mathbf{X}'_{d(p)})$$

Since $\sum_{p\in{}^\bullet t} \mathbf{X}_{d(p)} = \sum_{p\in t^\bullet} \mathbf{X}'_{d(p)} =: X$ the basic tree is:

$$
\begin{aligned}
&= \mathrm{lc}(\mathbf{pre}^{\mathrm{fl}}(t + id_{\mathrm{fl}(\mathbf{X})})) \\
&= \mathrm{lc}(\textstyle\sum_{p\in{}^\bullet t} p + \mathrm{fl}(\mathbf{X}_{d(p)}))) \\
&= \textstyle\sum_{p\in{}^\bullet t} p[\mathbf{X}_{d(p)}]) \\
&\xrightarrow{t[id]} \textstyle\sum_{p\in t^\bullet} p[\mathbf{X}'_{d(p)}]) \\
&= \mathrm{lc}(\textstyle\sum_{p\in t^\bullet} p + \mathrm{fl}(\mathbf{X}'_{d(p)}))) \\
&= \mathrm{lc}(\mathbf{post}^{\mathrm{fl}}(t + id_{\mathrm{fl}(\mathbf{X})}))
\end{aligned}
$$

- Induction: We add $\mathrm{fl}(\mathbf{Y}_{d(p)})$ with $\mathbf{Y}_{d(p)} = \mathbf{pre}(\theta_N)$ and $\mathrm{fl}(\mathbf{X}_{d(p)})$ with $\mathbf{X}_{d(p)} = \mathrm{lc}(\mathbf{M}_{r|P(d(p))}) - \mathbf{Y}_{d(p)}$. Note, that $\mathrm{lc}(\mathbf{M}_{r|P(d(p))}) \geq \mathbf{Y}_{d(p)}$ since $\theta_{d(p)}$ is activated. Let $\mathbf{Y}'_N = \mathbf{post}(\theta_N)$.
  By assumption $\mathbf{pre}(\theta_N) \xrightarrow{\theta_N} \mathbf{post}(\theta_N)$:

$$
\begin{aligned}
&= \mathrm{lc}(\mathbf{pre}^{\mathrm{fl}}(t[\theta_{N_1}, \ldots, \theta_{N_n}] + id_{\mathrm{fl}(\mathbf{X})})) \\
&= \mathrm{lc}(\textstyle\sum_{p\in{}^\bullet t} p + \mathrm{fl}(\mathbf{X}_{d(p)}) + \mathrm{fl}(\mathbf{Y}_{d(p)}))) \\
&= \textstyle\sum_{p\in{}^\bullet t} p[\mathbf{X}_{d(p)} + \mathbf{Y}_{d(p)}]) \\
&\xrightarrow{t[\boldsymbol{\theta}]} \textstyle\sum_{p\in t^\bullet} p[\mathbf{X}'_{d(p)} + \mathbf{Y}'_{d(p)}] \\
&= \mathrm{lc}(\textstyle\sum_{p\in t^\bullet} p + \mathrm{fl}(\mathbf{X}'_{d(p)}) + \mathrm{fl}(\mathbf{Y}'_{d(p)})) \\
&= \mathrm{lc}(\mathbf{post}^{\mathrm{fl}}(t[\theta_{N_1}, \ldots, \theta_{N_n}] + id_{\mathrm{fl}(\mathbf{X})}))
\end{aligned}
$$

This shows the property.   □

$$\mathbf{M}_v \xrightarrow[OS]{\theta} \mathbf{M}'_v \qquad\qquad \mathbf{M}_r \xrightarrow[\text{fl}(OS)]{\theta} \mathbf{M}'_r$$

$$\text{fl}\downarrow \qquad\qquad \downarrow\text{fl} \qquad\qquad\qquad \text{lc}\downarrow \qquad\qquad \downarrow\text{lc}$$

$$\text{fl}(\mathbf{M}_v) \xrightarrow[\text{fl}(OS)]{\theta} \text{fl}(\mathbf{M}'_v) \qquad\qquad \text{lc}(\mathbf{M}_r) \xrightarrow[OS]{\theta} \text{lc}(\mathbf{M}'_r)$$

$$\text{lc}\downarrow \qquad\qquad \downarrow\text{lc} \qquad\qquad\qquad \text{fl}\downarrow \qquad\qquad \downarrow\text{fl}$$

$$\text{lc}(\text{fl}(\mathbf{M}_v)) \xrightarrow[OS]{\theta} \text{lc}(\text{fl}(\mathbf{M}'_v)) = \mathbf{M}'_v \qquad \text{fl}(\text{lc}(\mathbf{M}_r)) \xrightarrow[\text{fl}(OS)]{\theta} \text{fl}(\text{lc}(\mathbf{M}'_r)) = \mathbf{M}'_r$$

Fig. 4. Embeddings extended to an Simulation

Theorem 5. *If OS is a ordinary Os, then reference and value semantics can simulate each other directly.*

Proof. Composition of the two diagrams in Theorem 1 and 4 is shown in Fig. 4. Both diagrams can be further reduced to the following two simulations:

$$\mathbf{M}_v \xrightarrow[OS]{\theta} \text{lc}(\text{fl}(\mathbf{M}'_v)) = \mathbf{M}'_v \qquad\qquad \mathbf{M}_r \xrightarrow[\text{fl}(OS)]{\theta} \text{fl}(\text{lc}(\mathbf{M}'_r)) = \mathbf{M}'_r$$

$$\text{fl}\downarrow \qquad\qquad \uparrow\text{lc} \qquad\qquad\qquad \text{lc}\downarrow \qquad\qquad \uparrow\text{fl}$$

$$\text{fl}(\mathbf{M}_v) \xrightarrow[\text{fl}(OS)]{\theta} \text{fl}(\mathbf{M}'_v) \qquad\qquad \text{lc}(\mathbf{M}_r) \xrightarrow[OS]{\theta} \text{lc}(\mathbf{M}'_r)$$

So, the embeddings in Theorem 1 and 4 also imply a direct simulation.   □

5   Processes of Ordinary Object-Net Systems

In [8] we have given a characterisation of those processes of the reference semantics that can be simulated by the value semantics for the general case. For ordinary object-net systems we know due to Theorem 5 that there is a one-to-one correspondence of reference and value semantics. In the following we will show that this correspondence carries over for processes.

5.1   Basic Definitions

Petri net processes (cf. [6, 2]) describe the behaviour of Petri nets. Processes are themselves Petri nets from the class of *causal nets*, where no branching is allowed for the places. A *run* of a net N is defined as a causal net R with a pair of mappings $\phi = (\phi^P : B \to P, \phi^T : E \to T)$. Extending ϕ^P and ϕ^T to multisets, the run is associated to the net, by requiring the commutativity expressed by: $\phi^P({}^\bullet e) = \mathbf{pre}(\phi^T(e))$ and $\phi^P(e^\bullet) = \mathbf{post}(\phi^T(e))$. That is, ϕ preserves the localities of transitions.[4]

---

[4] Alternatively, a process (R, ϕ) can be constructed from the possible firings, i.e. the enabling of transitions, of the net N by adding transitions according to the enabling condition of the net N. The starting point is given by the initial marking.

Definition 7. *Let $N = (P, T, W, \mathbf{M}_0)$ be a P/T net and $R = (B, E, \lessdot)$ a causal net. Furthermore let $\phi = (\phi^P : B \to P, \phi^T : E \to T)$ be a pair of mappings. Then (R, ϕ) is a* process *of N if the following conditions hold:*

1. *Preservation of the flow relation: $x \lessdot y \Longrightarrow \phi(x) \ F \ \phi(y)$.*
2. *Representation of the initial marking $\mathbf{M}_0$ by $^\circ R$: $\phi^P(^\circ R) = \mathbf{M}_0$.*
3. *Compatibility of ϕ with the arc-weight function:*
   $\phi^P(^\bullet e) = \mathbf{pre}(\phi^T(e))$ *and* $\phi^P(e^\bullet) = \mathbf{post}(\phi^T(e))$.
4. *Representability of R as the limit of finite processes.*

For a run (R, ϕ) of a Petri net N the symmetric and reflexive relations **li** and **co** are defined by **li** := $(\lessdot \cup \lessdot^{-1} \cup id_A)$ and **co** := $(\bar{\mathbf{li}} \cup id_A)$. A ken with respect to **li** is often called a *line*, while a ken with respect to **co** is called a *cut*. If $C \in \text{KEN}(\lessdot)$ and $C \subseteq P$ then C is called a *P-cut* of R.

5.2   Processes of Ordinary Object Nets

The definition of of an object net process is based on the net $\text{fl}(OS)$ defined in Def. 3.

Definition 8. *Let $OS = (\mathcal{N}, d, \Theta, \mathbf{M}_0)$ be an* Os. *The pair (R, ϕ) is a process of OS iff it is a process of $\text{fl}(OS)$.*

Define the set of elements of a *P-cut* C belonging to a net type N:

$$B_N(C) := C \cap \phi^{-1}(d^{-1}(N)) = \{b \in C \mid d(\phi(b)) = N\}$$

Analogously to Theorem 3 we obtain that all reachable *P-cuts* are located:

Lemma 1. *Let (R, ϕ) be a process of an ordinary* Os *OS. For each P-cuts C of R there is exactly one element $b \in C$ carrying a net-token of type N.*

$$\forall N \in \mathcal{N} : |B_N(C)| = 1$$

The uniquely defined element of $B_N(C)$ is denoted by $b(N)$.

Similarly to Def. 5 we define a localisation of *P-cuts* resulting in a nested structure. The restructuring also extends to events, where each e is mapped to an nested event $e^t[\epsilon]$ where ϵ is a nested structure of events which mimics the structure of $\phi(e) = t[\boldsymbol{\theta}]$.

Definition 9. *For a P-cut C of a process R, we define the* localisation $\text{lc}(C)$ *of C as $\text{lc}(C) := \text{lc}_C(B_{N_{sn}}(C))$ where*

$$\text{lc}_C(b) := \begin{cases} b[\text{lc}(B_{d(\phi(b))}(C))], & \text{if } d(\phi(b)) \neq N_\bullet \\ b, & \text{otherwise} \end{cases}$$

For events $e \in E$ we define $\text{lc}(e) := f_e(\phi(e))$ where

$$f_e(t[\boldsymbol{\theta}]) := e^t[f_e(\boldsymbol{\theta})]$$

The process mapping is extended to nested sets by defining $\phi(b[X]) := \phi(b)[\phi(X)]$. Then the localisation commutes with the process map ϕ.

Theorem 6. *Let (R, ϕ) be a process of of an ordinary Os OS. The process map ϕ commutes with lc. For each P-cut C of R we have:*

$$\phi(\mathrm{lc}_C(C)) = \mathrm{lc}_{\phi(C)}(\phi(C))$$

Proof. Induction base for $N_\bullet$:

$$\phi(\mathrm{lc}_C(B_{N_\bullet}(C))) = \phi(B_{N_\bullet}(C)) = \mathrm{lc}_{\phi(C)}(\phi(B_{N_\bullet}(C))$$

Induction step:

$$
\begin{aligned}
\phi(\mathrm{lc}_C(B)) &= \phi\Big(\textstyle\sum_{b\in B}(b, \mathrm{lc}_C(B_{d(\phi(b))}(C))) \Big) \\
&= \textstyle\sum_{b\in B} \phi(b)[\phi(\mathrm{lc}_C(B_{d(\phi(b))}(C))] \\
&= \textstyle\sum_{b\in B} \phi(b)[\mathrm{lc}_{\phi(C)}\big(\phi(B_{d(\phi(b))}(C))\big)] \\
&= \textstyle\sum_{b\in B} \phi(b)[\mathrm{lc}_{\phi(C)}\big(\phi(C \cap \phi^{-1}(d^{-1}(N)))\big)] \\
&= \textstyle\sum_{b\in B} \phi(b)[\mathrm{lc}_{\phi(C)}\big(\phi(C) \cap d^{-1}(N)\big)] \\
&= \textstyle\sum_{b\in B} \phi(b)[\mathrm{lc}_{\phi(C)}\big(\phi(C)_{|P(N)}\big)] \\
&= \mathrm{lc}_{\phi(C)}(\phi(B))
\end{aligned}
$$

This proves the commutativity.   □

Theorem 7. *Let (R, ϕ) a process of an ordinary Os OS. Then we have for all P-cuts C and C' of R:*

$$
\begin{array}{ccc}
C & \xrightarrow[R]{e} & C' \\
\phi\downarrow & & \downarrow\phi \\
\phi(C) & \xrightarrow[\mathrm{fl}(OS)]{\phi(e)} & \phi(C') \\
\mathrm{lc}\downarrow & & \downarrow\mathrm{lc} \\
\mathrm{lc}(\phi(C)) & \xrightarrow[OS]{\phi(e)} & \mathrm{lc}(\phi(C'))
\end{array}
$$

Proof. By definition R is a process of $\mathrm{fl}(OS)$, which shows the first embedding via ϕ. Theorem 4 shows that every step for an ordinary Os can be simulated via lc which is the second embedding.   □

The map lc associates with each process R a nested process which is an object net system: "The semantics of an object net is an object net."

Definition 10. *Let (R, ϕ) with $R = (B, E, <)$ be a process of the Os $OS = (\mathcal{N}, d, \Theta, \mathbf{M}_0)$. Define the* located process $(\mathrm{lc}(R), \phi_R)$ *by the Os*

$$\mathrm{lc}(R) = (\mathcal{N}_R, d_R, \Theta_R, \mathbf{M}_R)$$

where $\mathcal{N}_R = \{R_N \mid N \in \mathcal{N}\}$ with

$$B(R_N) = B \cap \phi^{-1}(P(N))$$
$$E(R_N) = \{e^t \mid t \in \mathrm{fl}(\phi(e)) \wedge t \in T(N)\}$$
$$F(R_N) = \, \lessdot_{\mid(B(R_N)\times E(R_N))\cup(E(R_N)\times B(R_N))}$$

and $d_R(b) = d(\phi(b))$, $\Theta_R(b) = E$, and $\mathbf{M}_R(b) = \mathrm{lc}(^\circ R)$. The process mapping is defined by $\phi_R(b) = \phi(b)$ and $\phi_R(e^t) = \phi(e)$.

Analogously to the previous Theorem we obtain the following embedding when the application order of lc and ϕ is switched.

Theorem 8. *Let* $OS = (\mathcal{N}, d, \Theta, \mathbf{M}_0)$ *be an ordinary Os and* (R, ϕ) *a process of OS. Then we have for all P-cuts C and C' of R:*

$$
\begin{array}{ccc}
C & \xrightarrow[R]{e} & C' \\
{\scriptstyle \mathrm{lc}}\downarrow & & \downarrow{\scriptstyle \mathrm{lc}} \\
\mathrm{lc}(C) & \xrightarrow[\mathrm{lc}(R)]{\mathrm{lc}(e)} & \mathrm{lc}(C') \\
{\scriptstyle \phi_R}\downarrow & & \downarrow{\scriptstyle \phi_R} \\
\phi_R(\mathrm{lc}(C)) & \xrightarrow[OS]{\phi_R(\mathrm{lc}(e))} & \phi_R(\mathrm{lc}(C'))
\end{array}
$$

Proof. Let $C \xrightarrow{e}_{R} C'$ and $\phi(e) = t[\boldsymbol{\theta}]$ then by Def. 9 $\mathrm{lc}(e) = e^t[\boldsymbol{\epsilon}]$. It is easy to see that by the construction in Def. 10 the event $e^t[\boldsymbol{\epsilon}]$ is enabled in $\mathrm{lc}(R)$ for the nested cut $\mathrm{lc}(C)$: $\mathrm{lc}(C) \xrightarrow[\mathrm{lc}(R)]{\mathrm{lc}(e)} \mathrm{lc}(C')$.

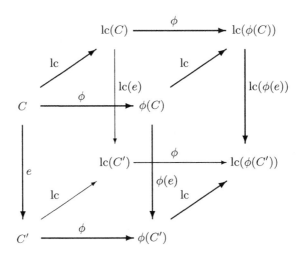

Fig. 5. Process embeddings

Since ϕ_R equals ϕ on places, we have $\phi_R(\mathrm{lc}(C)) = \phi(\mathrm{lc}(C))$. Using Theorem 6 and 7 we know, that $\phi_R(\mathrm{lc}(C)) = \mathrm{lc}(\phi(C))$ can be rewritten by $\phi(e)$.

Since $\phi_R(\mathrm{lc}(e)) = \phi_R(f_e(\phi(e))) = \phi_R(e^t[\epsilon]) = t[\theta] = \phi(e)$ holds, we have $\phi_R(\mathrm{lc}(C)) \xrightarrow[OS]{\phi_R(\mathrm{lc}(e))} \phi_R(\mathrm{lc}(C'))$ for the object-net system OS. $\square$

The cube in Figure 5 summarises all the embeddings of Theorem 7 and 8. The vertical dimension illustrates the firing steps, the dimension from left to right illustrates the mapping from the process to the object system, the dimension from the front to the back illustrates the relation of reference and value semantics.

6   The Household Robot Example, Revisited

In [10] the authors showed how the formalism of nets-within-nets can be used to model mobility, especially in the case of mobile agents. A case study was presented that models a mobile household robot. We adapt this case study for a first approach on how to analyse agent systems. To reach this aim the overall system architecture is simplified while the ideas are still visible. Going along with a better tool support for the analysis we will switch back to the original model.

The household is represented by the system net in Figure 6.[5] The household consists of several rooms (locations): hall, living room, kitchen, next room, and the front yard (dark places). Each room offers special services to the robot: it can fetch coffee in the kitchen, serve it in the living room, fetch mail in the front yard, open and close the door in the hall, and so on (light transitions). The possible movements from one location to another are displayed as dark transitions. Note that moving from room to room is not symmetric in this scenario. For example it is not possible to move directly from the kitchen to the next room. Service transitions are supplemented with additional information (service state/buffer, light places) showing for instance if new mail has arrived, coffee is available and so on. Extraneous actions not accessible for the robot are displayed as thin-lined transitions: arrival of new mail, new assignments for the robot etc.

The door of the house is used to show another possibility of viewing special parts of the system: the state of the door (open/closed) is modelled directly. This system state does not belong to a single service (as for example the state of the mailbox), but is queried by a couple of service transitions including the movements into and out of the house.

This model of the household is filled with life by implementing an appropriate robot agent and defining the desired services for the platforms. The behaviour modelling for this kind of agents has been introduced in [9]. While the Petri net model of the household hides some details – namely the transition inscriptions

---

[5] The use of colour greatly supports the differentiation of different types of places, transitions, or arcs. Unfortunately this is – even in the adapted form of the figures – not so obvious in a black and white representation.

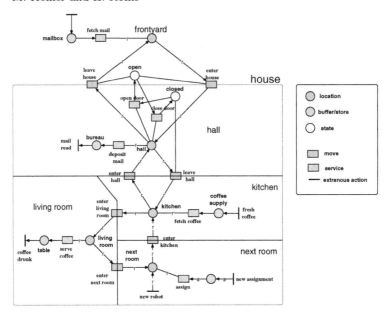

Fig. 6. Household System Net

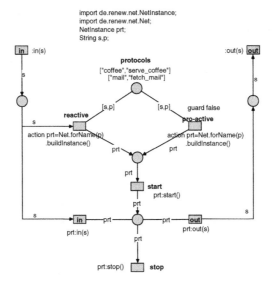

Fig. 7. The robot

e.g. for moving around – the nets for the robot (Fig. 7) and one of its plans
(Fig. 8) are presented in full detail using the syntax of RENEW [14].

Figure 7 shows the interface net of the robot implemented as a MULAN agent.
This kind of agents is explained in [9]. The Figure shows a simplified version still
capable of autonomous, pro- or reactive and reconfigurable behaviour. What is

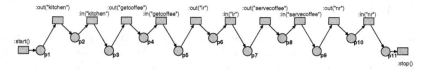

Fig. 8. Plan for serving coffee

omitted in this publication is the platform net, the layer between the overall system (household) and the agent (robot). It is necessary only if one is interested in a dynamically changing environment. Leaving it out we get an architecture of three layers: system - agent - behaviour protocol (plan).

Figure 8 presents a behaviour protocol of the robot. This protocol only consists of sequential actions, therefore we call it a *plan*. Protocols tell the agent what to do and when to do it. Protocol nets can be generated at run-time, which can not be shown here. The plan for serving coffee instruments the robot to move to the kitchen, fetch the coffee, move to the living room, serve the coffee and move back to the next room. Having done this the plan stops.

We have analysed the household system where the house net has been restricted to the kitchen, the next and the living room, since the yard and the hall are not relevant in the restricted scenario. The transitions **new assignment**, **fresh coffee**, and **coffee drunk** have been fused to generate some kind of loop at the system net level. For this example an analysis with INA [17] (integrated in the PEP tool [7]) shows that the simulating P/T net is ordinary, extended simple, bounded, reversible and live. It has no transitions without any pre- or without any post-place. The system is covered by semi-positive P-invariants and thus structurally bounded. It is covered by a semi-positive T-invariant containing each transition once. There are 99 minimal deadlocks. All nonempty traps are initially marked. The system is state-machine decomposable.

Since the system is bounded we know that the design relies on only finitely many resources and since it is live we know e.g. that the robot can offer its service regularly.

7   Conclusion

In this presentation we have introduced the subclass of ordinary object-net systems. This subclass is of special importance because its structure guarantees that each marking is located. We have shown that this implies further, that reference and value semantics can simulate each other directly for this subclass. The structural simulation is also compatible with the concept of a Petri net process – illustrated by the cube in Figure 5. Due to this one-to-one correspondence a formal analysis based on standard tools is possible. Structural and dynamic properties were checked for our household/robot example using the tool INA. Current work is undertaken to investigate extensions of the formalism to allow for high-level concepts as arc inscriptions, bindings etc.

References

1. M. A. Bednarczyk, L. Bernardinello, W. Pawlowski, and L. Pomello. Modelling mobility with Petri hypernets. In J. L. Fiadeiro, P. D. Mosses, and F. Orejas, editors, *Recent Trends in Algebraic Development Techniques (WADT 2004)*, Lecture Notes in Computer Science. Springer-Verlag, 2004.
2. E. Best and C. Fernández. *Nonsequential processes: a Petri net view*. Springer-Verlag, 1988.
3. N. Busi. Mobile nets. *Formal Methods for Open Object-Based Distributed Systems*, pages 51–66, 1999.
4. L. Cardelli, A. D. Gordon, and G. Ghelli. Mobility types for mobile ambients. In *Proceedings of the Conference on Automata, Languages, and Programming (ICALP'99)*, volume 1644 of *Lecture Notes in Computer Science*, pages 230–239. Springer-Verlag, 1999.
5. D. M. Chess. Security issues in mobile code systems. In G. Vigna, editor, *Mobile Agents and Security*, volume 1419 of *Lecture Notes in Computer Science*, pages 1–17. Springer-Verlag, 1998.
6. U. Goltz and W. Reisig. The non-sequential behaviour of Petri nets. *Information and Control*, 57:125–147, 1983.
7. B. Grahlmann and E. Best. PEP — more than a Petri net tool. In T. Margaria and B. Steffen, editors, *Tools and Algorithms for the Construction and Analysis of Systems*, volume 1055 of *Lecture Notes in Computer Science*, pages 397–401. Springer-Verlag, 1996.
8. M. Köhler and B. Farwer. Mobile object-net systems and their processes. *Fundamenta Informaticae*, 59:113–129, 2004.
9. M. Köhler, D. Moldt, and H. Rölke. Modeling the behaviour of Petri net agents. In J. M. Colom and M. Koutny, editors, *International Conference on Application and Theory of Petri Nets*, volume 2075 of *Lecture Notes in Computer Science*, pages 224–241. Springer-Verlag, 2001.
10. M. Köhler, D. Moldt, and H. Rölke. Modelling mobility and mobile agents using nets within nets. In W. v. d. Aalst and E. Best, editors, *International Conference on Application and Theory of Petri Nets 2003*, volume 2679 of *Lecture Notes in Computer Science*, pages 121–140. Springer-Verlag, 2003.
11. M. Köhler and H. Rölke. Concurrency for mobile object-net systems. *Fundamenta Informaticae*, 54(2-3), 2003.
12. M. Köhler and H. Rölke. Properties of Object Petri Nets. In J. Cortadella and W. Reisig, editors, *International Conference on Application and Theory of Petri Nets 2004*, volume 3099 of *Lecture Notes in Computer Science*, pages 278–297. Springer-Verlag, 2004.
13. O. Kummer. *Referenznetze*. Logos Verlag, 2002.
14. O. Kummer, F. Wienberg, M. Duvigneau, J. Schumacher, M. Köhler, D. Moldt, H. Rölke, and R. Valk. An extensible editor and simulation engine for Petri nets: Renew. In J. Cortadella and W. Reisig, editors, *International Conference on Application and Theory of Petri Nets 2004*, volume 3099 of *Lecture Notes in Computer Science*, pages 484 – 493. Springer-Verlag, 2004.
15. I. A. Lomazova. Nested Petri nets – a formalism for specification of multi-agent distributed systems. *Fundamenta Informaticae*, 43(1-4):195–214, 2000.
16. W. Reisig. Petri nets and algebraic specifications. *Theoretical Computer Science*, 80:1–34, 1991.
17. S. Roch and P. Starke. INA – Integrierter Netz-Analysator Version 1.7. Handbuch, Humboldt-University, Institute of Computer Science, Berlin, 1997.

18. M.-O. Stehr, J. Meseguer, and P. C. Ölveczky. Rewriting logic as a unifying framework for Petri nets. In H. Ehrig, G. Juhas, J. Padberg, and G. Rozenberg, editors, *Unifying Petri Nets*, Lecture Notes in Computer Science (Advances in Petri Nets). Springer-Verlag, December 2001.
19. R. Valk. Modelling concurrency by task/flow EN systems. In *3rd Workshop on Concurrency and Compositionality*, number 191 in GMD-Studien, St. Augustin, Bonn, 1991. Gesellschaft für Mathematik und Datenverarbeitung.
20. R. Valk. Petri nets as token objects: An introduction to elementary object nets. In J. Desel and M. Silva, editors, *Application and Theory of Petri Nets*, volume 1420 of *Lecture Notes in Computer Science*, pages 1–25, 1998.
21. R. Valk. Object Petri nets: Using the nets-within-nets paradigm. In J. Desel, W. Reisig, and G. Rozenberg, editors, *Advanced Course on Petri Nets 2003*, volume 3098 of *Lecture Notes in Computer Science*, pages 819–848. Springer-Verlag, 2003.

A   Notations and Basic Definitions

Let $R \subseteq A \times B$ be a relation. A pair $(a, b) \in R$ will also be denoted $a\,R\,b$ in infix notation. The identity relation is defined as $id_A := \{(a, a) \mid a \in A\}$. For $a \in A$ and $b \in B$ define the domain of an element $b \in B$ by $(\_R\,b) := \{a \mid (a, b) \in R\}$ and its co-domain by $(a\,R\_) := \{b \mid (a, b) \in R\}$. We generalise the notion of domain and co-domain to sets $C \subseteq A$ and $D \subseteq B$ by $(C\,R\_) := \{b \mid \exists a \in C : (a, b) \in R\}$ and $(\_R\,D) := \{a \mid \exists b \in D : (a, b) \in R\}$.

Let $R \subseteq A \times A$ be a symmetric and reflexive relation. The set $K \subseteq A$ is a *clique* with respect to R iff all pairs of its elements are in the relation, i.e. for all $x, y \in K$ we have $(x, y) \in R$. A maximal clique is called a *ken* and the set of all kens of R is denoted by $\mathrm{KEN}(R)$.

The definition of Petri nets relies on the notion of multisets. A multiset on the set D is a mapping $\mathbf{A} : D \to \mathbb{N}$. Multisets are generalisations of sets in the sense that every subset of D corresponds to a multiset $\mathbf{A}$ with $\mathbf{A}(x) \leq 1$ for all $x \in D$. The empty multiset $\mathbf{0}$ is defined as $\mathbf{0}(x) = 0$ for all $x \in D$. The cardinality is $|\mathbf{A}| := \sum_{x \in D} \mathbf{A}(x)$. A multiset $\mathbf{A}$ is called *finite* iff $|\mathbf{A}| < \infty$. The multiset sum $\mathbf{A} + \mathbf{B}$ is defined as $(\mathbf{A} + \mathbf{B})(x) := \mathbf{A}(x) + \mathbf{B}(x)$ the difference $\mathbf{A} - \mathbf{B}$ by $(\mathbf{A} - \mathbf{B})(x) := \max(\mathbf{A}(x) - \mathbf{B}(x), 0)$. Equality $\mathbf{A} = \mathbf{B}$ is defined element-wise: $\forall x \in D : \mathbf{A}(x) = \mathbf{B}(x)$. Multisets are partially ordered: $\mathbf{A} \leq \mathbf{B} \iff \forall x \in D : \mathbf{A}(x) \leq \mathbf{B}(x)$ The strict order $\mathbf{A} < \mathbf{B}$ holds iff $\mathbf{A} \leq \mathbf{B}$ and $\mathbf{A} \neq \mathbf{B}$. The notation is overloaded, being used for sets as well as multisets. The meaning will be apparent from its use.

Any mapping $f : D \to D'$ can be generalised to a mapping $f : MS(D) \to MS(D')$ on multisets:

$$f\left(\sum_{i=1}^{n} a_i\right) = \sum_{i=1}^{n} f(a_i)$$

This includes the special case $f(\mathbf{0}) = \mathbf{0}$. These definitions are in accordance with the set-theoretic notation $f(A) = \{f(a) \mid a \in A\}$.

The set of all finite multisets over the set D is denoted $MS(D)$. A multiset $\mathbf{A}$ can be considered as the formal sum $\mathbf{A} = \sum_{x \in D} \mathbf{A}(x) \cdot x$. Finite multisets are the freely generated commutative monoid $(MS(D), +, 0)$. If the set D is finite, then a multiset $\mathbf{A} \in MS(D)$ can be represented equivalently as a vector $\mathbf{A} \in \mathbb{N}^{|D|}$.

$N = (P, T, F)$ is a *Petri net* iff the set of places P and the set of transitions T are disjoint, i.e. $P \cap T = \emptyset$, $F \subseteq (P \times T \cup T \times P)$ is the flow relation. Some commonly used notations for Petri nets are ${}^\bullet y := ({}_-F\, y)$ for the *preset* and $y^\bullet := (y\, F_-)$ for the *postset* of a net element y. The set of minimal elements of a net N is denoted ${}^\circ N := \{x \in P \cup T \mid {}^\bullet x = \emptyset\}$, the set of maximal elements is $N^\circ := \{x \in P \cup T \mid x^\bullet = \emptyset\}$.

A finitely branching Petri net $N = (B, E, <)$ is an *causal net* iff the transitive closure $<^+$ of the flow is acyclic and $|{}^\bullet b| \leq 1$ and $|b^\bullet| \leq 1$ holds for all $b \in B$. For a causal net $N = (B, E, <)$ we define the order $<$ on the net elements $(B \cup E)$ by $< := <^+$.

Particle Petri Nets for Aircraft Procedure Monitoring Under Uncertainty

Charles Lesire[1,2] and Catherine Tessier[1]

[1] Onera-DCSD, Toulouse, France
[2] Supaéro, Toulouse, France
{lesire, tessier}@cert.fr

Abstract. In the framework of the study and analysis of new flight procedures, we propose a new Petri net-based formalism to represent both continuous and discrete evolutions and uncertainties: the *particle Petri net*. This model is based on a particle filtering-like representation of the probabilistic uncertainty on the continuous part of the procedure, and a possibilistic Petri net-inspired approach to deal with the uncertainty on events. After introducing this formalism, we propose an analysis of an approach procedure, and a further application to the on-line tracking of pilots' activities.

1 Introduction

The air traffic is expected to increase by about 40% in the next ten years. This spectacular development makes specialists wonder about how to manage such a traffic, from human, infrastructural, systems and safety points of view.

Some research are dedicated to the improvement of the flight safety and focus on:

1. the design of systems to analyse the pilot's activity [1],
2. the design of systems to help pilots during the flight [2],
3. the analysis of accident reports to help modifying systems and procedures [3],
4. the analysis of conflicts between systems [4],
5. the analysis of flight procedures [5].

Our contribution takes place in the latter domain, i.e. modelling and analysing flight plan and procedures to diagnose some possible conflicts. Though the work of [5] is based on a modular representation of the flight plan using Petri nets and their classical analysis tools, the analysis is mainly based on Petri nets properties, and not really related to procedure safety. Moreover, continuous evolutions are not represented and uncertainties are not dealt with.

After a short state of the art on hybrid system analysis formalisms (next section), we introduce a Petri net-based model of flight procedures (Sect. 3) considering a hybrid evolution of the man–aircraft system. Then we propose to extend this formalism to deal with numerical and discrete uncertainties (particle

G. Ciardo and P. Darondeau (Eds.): ICATPN 2005, LNCS 3536, pp. 329–348, 2005.

Petri nets, see Sect. 5). Finally the analysis of an approach procedure illustrates this formalism (Sect. 6) before concluding with the use of particle Petri nets within a hybrid estimation principle (see Sect. 7).

2   The Pilot-Aircraft System as a Hybrid System

A flight procedure is described as a plan: the aircraft has to fly over some way-points, with given headings, speeds, altitudes... Actually, the plan is split up into segments and the evolution of the aircraft along these segments amounts to a continuous evolution of the aircraft flight parameters.

The pilot has to act on the aircraft systems to put the aircraft in the right configuration according to the segment (e.g. the gear has to be down in a landing segment) and conversely the aircraft sends information to the pilot. This pilot–aircraft interaction is based on events: pressing switches, speaking to the control, information, alarms...

Therefore, the man–aircraft system can be considered as a hybrid system, whose evolution is both continuous (the aircraft motion along the segment) and discrete (the events generated both by the pilot and by the systems).

Several methods are used to deal with hybrid systems monitoring, which are based on automata [6] or on Bayesian nets [7] linked to some numerical evolution models. As far as the extension of Petri nets is concerned to deal with continuous – and consequently hybrid – systems, several approaches have been proposed.

[8] propose to associate a *firing speed* to a *continuous* transition, representing the fact that this transition does not fire immediately: the token passes through the transition at the firing speed. In the same way, [9] associate speeds to arcs to represent the motion of matter from a transition to a place. These two formalisms are dedicated to systems based on matter flows.

In [10], the approach is quite different. The continuous places are linked to differential equations, which are the evolution equations of the system. This formalism, used for instance in [11] to analyse a landing system, can be applied to most hybrid systems.

As far as aircraft procedures modelling is concerned, the continuous evolution of flight parameters is governed by differential equations. Therefore, differential Petri nets seem to be relevant to represent procedures. An example is presented in the next section.

3   The Flight Plan Representation

Usually, the pilot's activity is represented as a three-level control architecture :

1. *navigation*: the several steps of the flight, also called the flight phases,
2. *guidance*: the set of manœuvres dedicated to a phase,
3. *control*: the control of the aircraft motion.

This representation leads to consider the flight plan as a hierarchy: the navigation is the global sequence of phases, each phase is detailed as a guidance procedure, which is structured as a set of controlled segments.

3.1   A Modular Representation

To be consistent with this representation, a modular Petri net representation of the flight plan is relevant. In the *navigation net* (see Fig. 1) each place is associated to a flight phase. Each phase is then detailed in a *guidance net*, composed of a set of segments (see Fig. 2). The navigation net and the guidance nets are linked by *transition fusions* [12].

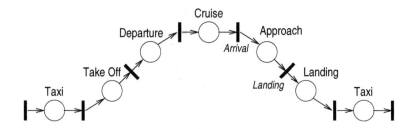

Fig. 1. The *navigation net*

Each guidance net is then a hybrid – differential / discrete – Petri net, where the *control* activity is actually the set of the differential equations representing the aircraft motion within each segment.

Example 1. The Petri net of Fig. 2 models an "approach" procedure from Agen (AGN) to land on runway 14R of the Toulouse-Blagnac airport. This differential Petri net, based on the definition of [10], is composed of differential places (in thin) and discrete places (in bold).

The transitions labelled *Arrival* and *Landing* are merged between Fig. 1 and Fig. 2 Petri nets.

3.2   State Representation

Once the Petri net model of the procedure has been drawn, the tokens representing the state of the pilot–aircraft system have to be defined. Two tokens are considered: a *numerical token*, carrying the state vector of the aircraft, i.e. the continuous parameters that will evolve according to the differential equations of the guidance nets, and a *symbolic token*, carrying the configuration of the aircraft resulting from the pilot's actions, i.e. the discrete parameters.

3.3   Playing a Differential Petri Net

The firing rule of differential Petri nets [10], applied to our procedure model, is illustrated through the following example.

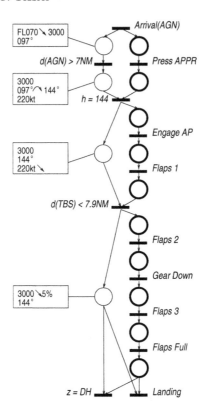

Fig. 2. The "AGN to Toulouse" approach *guidance net*: the differential – thin – places are labelled by a segment description, translated into a differential equation. For instance the last place segment consists in a descent from a 3000-ft altitude at a 5% rate, steering heading 144. The discrete – bold – places represent the *checklist*, i.e. the actions the pilot has to perform during a segment. For instance, during the descent, the pilot has to set flaps2, then gear down and finally flaps3 and flaps full

Example 2. Figure 3 represents the top of the guidance net of Fig. 2 with the initial marking – let us say at time k_0. The value of the numerical vector – carried by the numerical token (in black) – is $\{d = 0, h = 97, z = 7000\}$, where d is the distance of the aircraft to waypoint AGN, h is the heading of the aircraft and z its altitude. The value of the parameter $APPR$ within the symbolic vector – carried by the symbolic token (in white) – is 0, meaning that the $APPR$ mode is not engaged.

Let K be the time of the *Press APPR* event occurrence. At k_0, no transition is enabled: the guard is not satisfied $(d < 7)$ and $k_0 \neq K$.

The numerical token evolves according to the differential equation associated to the first segment, which represents a descent from 7000 to 3000ft steering heading 97, and at k_1 $(k_1 < K)$ the numerical token carries vector $\{d = 7.1, h = 97, z = 3000\}$. $d > 7NM$, therefore the guard is satisfied and the transition fires.

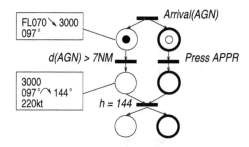

Fig. 3. Initial marking of the approach guidance net

Then the continuous evolution of the numerical token goes on according to the second segment equation, i.e. a turn from heading 97 to heading 144 at speed 220kt. At time K, the *Press APPR* event occurs, and the corresponding transition fires.

Finally, at time k_2, the numerical token matches condition $h = 144$, and the corresponding transition fires.

The successive markings of the sequence are represented on Fig. 4.

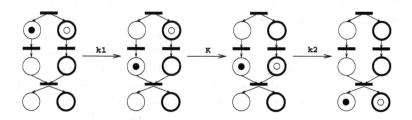

Fig. 4. Markings reached with the example of Fig. 3

Although this formalism is proper to model and simulate flight procedures, the next section discusses the necessity to consider numerical and symbolic uncertainties within the evolution process.

4   Uncertainty in the Procedures

4.1   Numerical Uncertainty

As far as numerical parameters are concerned, uncertainty exists because: (1) the evolution models do not reflect perfectly what actually happens and (2) the numerical values are imprecise. These two kinds of uncertainties are usually dealt with through an estimator, whose aim is to estimate the parameter values according to a model noise and a measurement noise.

The particle filter [13] allows the state x_k at time k of a dynamic system subject to deterministic and random inputs to be estimated from observations

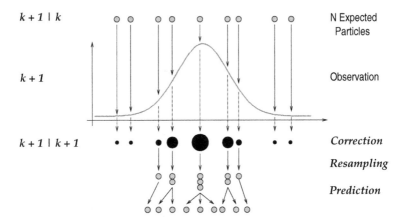

Fig. 5. Particle filtering [14–slide 8]

z_k spoilt with stochastic errors. It is based on a discretization of the uncertainty on the state value: the probability distribution function of the estimate $\hat{x}_{k|k}$ – meaning the state estimated at time k knowing the measurement at time k – is represented by a set of N particles $x_{k|k}^{(1)}, \ldots, x_{k|k}^{(N)}$ (see Fig. 5). The estimation is achieved through a two-step process.

The first step – called *prediction* – consists in estimating the next particles $\hat{x}_{k+1|k}^{(1)}, \ldots, \hat{x}_{k+1|k}^{(N)}$ according to the noisy evolution model. The second step – called *correction* – is based on a comparison of the expected particle values with the observation, represented as a Gaussian value on the figure: the closer the expected particle values are to the most probable value of the observation, the bigger weight they are assigned. Then N new particles $\hat{x}_{k+1|k+1}^{(1)}, \ldots, \hat{x}_{k+1|k+1}^{(N)}$ are generated from a resampling of the weighted corrected particles.

The particle filter is well suited to our problem as Petri nets are mainly discrete and particle filters can easily model non linear evolutions. This will lead to the numerical part of particle Petri nets, that is presented in Sect. 5.

4.2   Symbolic Uncertainty

As far as symbolic parameters are concerned, uncertainty can be discussed through the example of Fig. 4: we can notice that the marking of Fig. 6 has not been reached; this is due to the choice of K. What the checklist actually says is that the pilot has to perform the *Press APPR* action during a (some) segment(s). Consequently, when simulating the pilot's behaviour, we have to consider all the possible times they will possibly press *APPR*. A way of dealing with this uncertainty could be by starting a new simulation for each time the event might happen – within $[k_0, k_2]$ – and repeat this for each event, which is far too complex.

Symbolic uncertainty could be dealt with considering an imprecise firing time of the symbolic transitions, based on timed Petri nets [15] or stochastic Petri nets

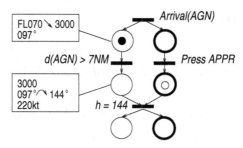

Fig. 6. A "reachable" marking that is not reached during the simulation

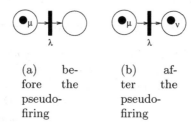

(a) be-
fore   the
pseudo-
firing

(b) af-
ter   the
pseudo-
firing

Fig. 7. Pseudo-firing within a possibilistic Petri net

[16]. Nevertheless, as there is no knowledge about this time, it is quite difficult to quantify this uncertainty [17].

The formalism described in [18] allows uncertainty on the marking of a place to be considered: possibilistic Petri nets allow a possibility measure to be associated to the fact that a token is in a place. A possibility is also associated to transition firing.

Let us consider the possibilistic Petri net of Fig. 7. In Fig. 7(a), the possibility value associated to the token is μ, meaning that the possibility that the token is really in this place is μ. The possibility of firing the transition is λ. The *pseudo-firing* of the transition results in the marking of Fig. 7(b): the token is marking the output place with a possibility ν, with $\nu = \lambda.\mu$.

Now let us consider the marking of Fig. 7(b). The transition can be pseudo-fired again, which changes the possibilities of the tokens: let ν' be the possibility of the second token after the second pseudo-firing of the transition. ν' is given by the following equation

$$\nu' = \max(\lambda.\mu, \nu) \ . \tag{1}$$

Remark 1. The fact that nothing is known about the transition firing time can be represented by the fact that all the transitions are completely enabled, i.e. $\lambda = 1$. Moreover, if the value of μ has not changed, one can notice that (1) implies $\nu' = \nu$.

The pseudo-firing within possibilistic Petri nets will inspire the symbolic part of the particle Petri nets.

5   Particle Petri Nets

5.1   Definition

Definition 1. *A particle Petri net is defined as a tuple* $\mathcal{P} = <P, T, A, D, C, E,$ $\mathcal{M}_0^0 >$ *where*

1. *P is the set of places, partitioned in* numerical *and* symbolic *places,*

$$P = P_N \cup P_S \text{ and } P_N \cap P_S = \emptyset$$

   with P_N (resp. P_S) the set of numerical (resp. symbolic) places,
2. *T is the set of transitions, partitioned in* numerical *and* symbolic *transitions,*

$$T = T_N \cup T_S \text{ and } T_N \cap T_S = \emptyset$$

   with T_N (resp. T_S) the set of numerical (resp. symbolic) transitions,
3. *A is the set of arcs, $A \subset P \times T \cup T \times P$,*
4. *D is the set of differential equations associated to the places of P_N: for each numerical place $p \in P_N$, the differential equation $D(p)$ represents the continuous evolution of a numerical token when p is marked, i.e. the evolution of continuous parameters,*
5. *C is the set of conditions associated to numerical transitions: condition $C(t)$ is a boolean function of the numerical token values,*
6. *E is the set of effects associated to symbolic transitions: the effect $E(t)$ is an assignment of some parameters carried by a symbolic token,*
7. *$\mathcal{M}_0^0$ is the initial marking of the Petri net.*

The marking $\mathcal{M}(p)$ of a place p consists in a set of tokens, defined as follows:

1. A *particle* $\pi_k^{(i)}$ is a numerical token at time k. It is one of the possible numerical vectors representing the continuous state of the aircraft at time k;
2. A *configuration* $\delta_k^{(i)}$ is a symbolic token at time k. It is one of the possible symbolic vectors representing the discrete state of the aircraft at time k.

Numerical places can only be marked by particles while symbolic places can only be marked by one configuration. Numerical uncertainty is represented by a set of possible particles within a (or several) numerical place(s) whereas the symbolic uncertainty is represented by a set of symbolic places marked by one possible configuration.

Example 3. The particle Petri net of the Toulouse-Blagnac approach has the same structure as the hybrid Petri net of Fig. 2, with:

1. numerical places, represented by thin circles, whose differential equations are described through a label (for instance descending from flight level 70 to 3000 ft steering heading 97),
2. symbolic places, represented by bold circles,
3. numerical transitions, such as the transition whose condition is *d(AGN)>7NM*,

4. symbolic transitions, such as the transition whose label is *Press APPR* and effect is to set parameter *APPR* to 1 within the configuration.

The main differences between particle Petri nets and hybrid Petri nets are on the marking and on the firing rules, which are defined in the following section.

5.2   Firing Rules

Firing a Symbolic Transition. According to the possibilistic Petri net formalism, the fact that a configuration is active – a symbolic token is marking the corresponding place – can be considered as a 1-valued possibility, whereas the fact that a configuration is inactive – the corresponding place is not marked – is represented as a 0-valued possibility.

Then we can apply a pseudo-firing that represents the possibility that an event may occur and change the configuration accordingly.

Definition 2. *A symbolic transition t is enabled if and only if each of its input places is marked.*

$$t \in T_S \text{ is enabled } \Leftrightarrow \begin{cases} \forall p \in \bullet t \cap P_N, \exists \pi \in \mathcal{M}(p) \\ \forall p \in \bullet t \cap P_S, \exists \delta \in \mathcal{M}(p) \end{cases} . \tag{2}$$

Firing a symbolic transition amounts to update the possibility of the tokens, as in [18]. In particle Petri nets, the configurations can only have possibility values of 0 or 1, which induces the firing rules:

$$\forall p \in \bullet t \cap P_N, \forall \pi \in \mathcal{M}(p), \forall p' \in t \bullet \cap P_N, \quad \mathcal{M}(p') := \mathcal{M}(p') \cup \{\pi\} . \tag{3}$$
$$\forall p \in \bullet t \cap P_S, \forall \delta \in \mathcal{M}(p), \forall p' \in t \bullet \cap P_S, \quad \mathcal{M}(p') := \mathcal{M}(p') \cup \{E(t)(\delta)\} . \tag{4}$$

where $E(t)(\delta)$ is configuration δ modified by effect $E(t)$.

Remark 2. When computing the discrete evolution at time k, we can notice that a fired symbolic transition remains enabled, as the tokens are not removed from the input places. We can also notice that firing this transition twice (or more) has no effect: the input tokens are still in their places and the output tokens are already present within the next places (with a possibility of 1 that cannot evolve according to (1)). This problem is solved associating a mark to the transition: when a transition fires for the first time, it is marked. A transition is enabled only if (2) is satisfied and if not marked. The mark is removed when the markings of the input places are modified.

Example 4. Let us consider the marking shown in Fig. 8(a). The configuration is $\delta^{(1)}$ such as $APPR = 0$, meaning that the approach mode is not selected. The symbolic transition is enabled, as its input place is marked by $\delta^{(1)}$. Therefore the transition fires, and the resulting marking is shown in Fig. 8(b): the configuration $\delta^{(2)}$, such as $APPR = 1$ is added to the next place, $\delta^{(1)}$ remains in the input place, and the transition is marked (which is represented by a white transition).

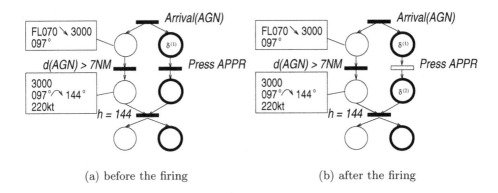

(a) before the firing   (b) after the firing

Fig. 8. Firing of a symbolic transition

Firing a Numerical Transition.

Definition 3. *A numerical transition is enabled if and only if each of its input places is marked by a token satisfying the associated condition.*

$$t \in T_N \text{ is enabled} \Leftrightarrow \begin{cases} \forall p \in \bullet t \cap P_N, \exists \pi \in \mathcal{M}(p) \ /C(t)(\pi) \\ \forall p \in \bullet t \cap P_S, \exists \delta \in \mathcal{M}(p) \end{cases} \tag{5}$$

where $C(t)(\pi)$ is true if particle π satisfies condition $C(t)$.

Firing a numerical transition t consists in moving all the particles satisfying condition $C(t)$ from the numerical input places to the numerical output places and copying the input configurations to the symbolic output places:

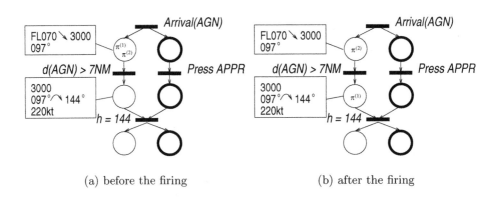

(a) before the firing   (b) after the firing

Fig. 9. Firing of a numerical transition

$$\forall p \in \bullet t \cap P_N, \forall \pi \in \mathcal{M}(p) \,/\, C(t)(\pi), \quad \mathcal{M}(p) := \mathcal{M}(p)\backslash\{\pi\} \;. \tag{6}$$

$$\forall p' \in t \bullet \cap P_N, \quad \mathcal{M}(p') := \mathcal{M}(p') \cup \{\pi\} \;. \tag{7}$$

$$\forall p \in \bullet t \cap P_S, \forall \delta \in \mathcal{M}(p), \forall p' \in t \bullet \cap P_S, \quad \mathcal{M}(p') := \mathcal{M}(p') \cup \{E(t)(\delta)\} \;. \tag{8}$$

Example 5. Let us consider the Petri net of Fig. 9(a) with initial marking $\pi^{(1)} = \{d = 7.1, h = 97, z = 3200\}$ and $\pi^{(2)} = \{d = 7, h = 96, z = 3200\}$. The numerical transition is enabled as particle $\pi^{(1)}$ satisfies condition $d > 7$. Firing this transition generates the marking of Fig. 9(b).

5.3   Evolution of the Marking

According to the previously defined firing rules, the evolution of the marking through the whole particle Petri net can be simulated. At time k, the marking is $\mathcal{M}_k^0$, meaning the initial estimated marking at time k.

The evolution of the marking consists in a two-step process, alternating a continuous evolution of the particle values and a discrete evolution of the Petri net markings:

Isoparticle Evolution of the Marking. This evolution is processed at time k when at least one transition is enabled. It consists in applying the firing rules previously defined to compute the evolution of the marking. Let $\mathcal{M}_k^f$ be the final state of the marking at time k.

Isomarking Evolution of the Particles. This evolution is processed at time k when no more transition is enabled. It consists in computing the continuous evolution of the particle values, according to the differential equations associated to the places they are marking. The differential equations are discretized so that the isomarking evolution computes the particles at time $k + 1$ according to the discretized differential equations. Let $\mathcal{M}_{k+1}^0$ be the result of the isomarking evolution of the particles (Fig. 10).

To illustrate the particle Petri nets presented in this section and their evolution processes, an application to an approach procedure analysis is presented in the next section.

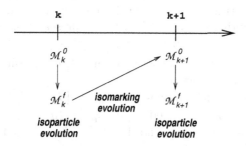

Fig. 10. The evolution process: isoparticle evolution of the marking and isomarking evolution of the particles

6   Example: Simulation of the Toulouse Approach Procedure

This section first illustrates the firing process of particle Petri nets through the evolution of an initial marking within the guidance Petri net of Fig. 2, then a way to use this evolution to analyse the procedure operation is proposed.

6.1   Simulation of the Activities

Let us consider the initial marking $\mathcal{M}_0^0$ at time $k = 0$ shown in Fig. 11(a). The particles are $\pi_0^{(1)} = \{d = 0.1, h = 97, z = 7000\}$, $\pi_0^{(2)} = \{d = 0, h = 96, z = 7000\}$ and $\pi_0^{(3)} = \{d = 0, h = 97, z = 6900\}$, representing the initial uncertainties on three parameters within the numerical vector. The initial configuration is $\delta_0^{(1)}$ such as $APPR = 0$, meaning that the approach mode is not selected.

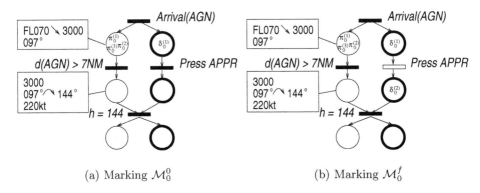

(a) Marking $\mathcal{M}_0^0$ (b) Marking $\mathcal{M}_0^f$

Fig. 11. First step of the evolution: isoparticle evolution of the marking at $k = 0$

As explained in Example 4, the symbolic transition whose label is *Press APPR* fires, and the resulting marking is shown in Fig. 11(b): $\delta_0^{(2)}$ is such as $APPR = 1$. As no transition is enabled anymore, an isomarking evolution is processed. The resulting marking $\mathcal{M}_1^0$, whose representation is the same as Fig. 11(b), is $\pi_1^{(1)} = \{d = 0.6, h = 97, z = 6900\}$, $\pi_1^{(2)} = \{d = 0.5, h = 96, z = 6900\}$ and $\pi_1^{(3)} = \{d = 0.5, h = 97, z = 6800\}$. At time 1, no transition is enabled, and the isomarking evolution goes on.

At time 50, the particles values are $\pi_{50}^{(1)} = \{d = 7.1, h = 97, z = 3200\}$, $\pi_{50}^{(2)} = \{d = 7, h = 96, z = 3200\}$ and $\pi_{50}^{(3)} = \{d = 7, h = 97, z = 3100\}$. The numerical transition whose condition is $d > 7$ is enabled, and the resulting marking is shown in Fig. 12(a) (see Example 5).

Then no more transition is enabled, and the isomarking evolution goes on. At time $k = 51$, particles $\pi^{(2)}$ and $\pi^{(3)}$ have evolved according to the "descent" equation, while particle $\pi^{(1)}$ has evolved according to the "turn" equation. The

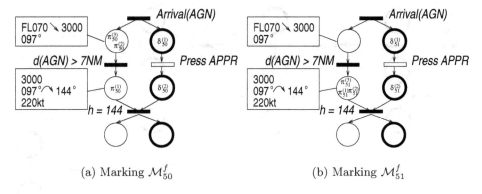

(a) Marking $\mathcal{M}_{50}^f$ (b) Marking $\mathcal{M}_{51}^f$

Fig. 12. Evolution of the markings between $k = 50$ and $k = 51$

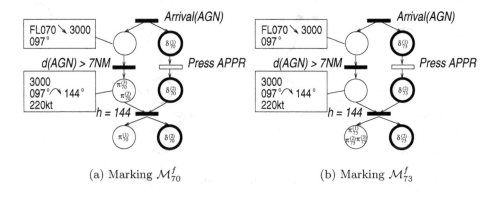

(a) Marking $\mathcal{M}_{70}^f$ (b) Marking $\mathcal{M}_{73}^f$

Fig. 13. Evolution of the markings between $k = 70$ and $k = 73$

computed particles are $\pi_{51}^{(1)} = \{d = 7.3, h = 99, z = 3200\}$, $\pi_{51}^{(2)} = \{d = 7.4, h = 96, z = 3000\}$ and $\pi_{51}^{(3)} = \{d = 7.4, h = 97, z = 2800\}$. The numerical transition $d > 7$ becomes enabled again, and the firing results in marking $\mathcal{M}_{51}^f$ represented on Fig. 12(b).

The numerical transition whose condition is $h=144$ is enabled at time $k = 70$, when the particles values are $\pi_{70}^{(1)} = \{d = 7.7, h = 144, z = 3200\}$, $\pi_{70}^{(2)} = \{d = 7.8, h = 136, z = 3000\}$ and $\pi_{70}^{(3)} = \{d = 7.8, h = 140, z = 2800\}$. The transition fires, moving particle $\pi^{(1)}$ to the next numerical place and copying configuration $\delta^{(2)}$ to the next symbolic place (see Fig. 13(a)).

At time $k = 71$, particle $\pi^{(3)}$ can pass through the transition, and so can particle $\pi^{(2)}$ at time $k = 73$. As configuration $\delta^{(2)}$ has already been copied to the output symbolic place, the markings of the symbolic places are not modified. The resulting marking of this example (Fig. 13(b)) is then $\mathcal{M}_{73}^f$ where tokens values are $\pi_{73}^{(1)} = \{d = 7.7, h = 144, z = 3200\}$, $\pi_{73}^{(2)} = \{d = 8.2, h = 144, z = 3000\}$,

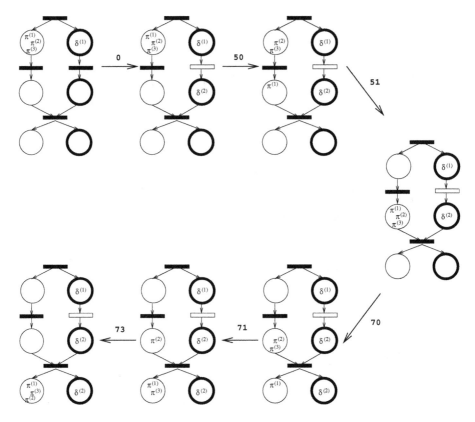

Fig. 14. The reachable markings from $\mathcal{M}_0^0$

$\pi_{73}^{(3)} = \{d = 8, h = 144, z = 2800\}$, $\delta_{73}^{(1)}$ such as $APPR = 0$ and $\delta_{73}^{(2)}$ such as $APPR = 1$.

To summarize, the successive markings reached during the simulation are represented on Fig. 14.

The next section suggests some ways to deduce properties from the reachable markings.

6.2   Analysis of the Procedure

Once the flight plan has been defined by a set of modular particle Petri nets, these nets have to be studied to determine the correctness of the procedure, in regard to its achievement and to flight safety.

Transition Liveness. A useful property to analyse flight procedures is the liveness of transitions. This property guarantees that a transition can fire from any reachable marking of the Petri nets, ensuring that the aircraft will proceed correctly with the flight plan. For instance, the liveness of *Landing* transition of Fig. 2 ensures that the aircraft will actually enter the landing phase.

To be consistent with the particle representation of numerical uncertainty, the notion of π-liveness is defined as follows:

Definition 4. *A transition t is π-live from the marking $\mathcal{M}$ if t is live from the marking $\mathcal{M}/\pi$, i.e. the marking containing the particle π and the configurations of $\mathcal{M}$.*

This means that particle π passes through transition t during the evolution of marking $\mathcal{M}$.

Thus, the analysis of the π-liveness of a transition allows to distinguish the particles that are consistent with the procedure (the particles for which the *crucial* transitions – transitions whose firing matters – are π-live) from the *conflictual* particles whose associated aircraft states do not match the nominal procedure (and this non-matching matters for safety [19]).

Example 6. Let us consider the marked Petri net of Fig. 13(a) with the particle values $\pi_{70}^{(2)} = \{d = 7.8, h = 136, z = 3000\}$ and $\pi_{70}^{(3)} = \{d = 7.8, h = 140, z = 2800\}$. Let us suppose that the isomarking evolution computes the next values $\pi_{70}^{(2)} = \{d = 8, h = 144, z = 3000\}$ and $\pi_{70}^{(3)} = \{d = 8, h = 148, z = 2800\}$. Then the numerical transition whose condition is $h = 144$ is enabled by $\pi^{(2)}$: this transition is $\pi^{(2)}$-live. On the contrary, $\pi^{(3)}$ does not enable the transition, and the aircraft goes on turning. As the transition will never be enabled by $\pi^{(3)}$, $\pi^{(3)}$ is considered as a conflictual particle.

Safety Zones. In flight safety and procedures design, flight zones are associated to each segment, to be sure that the aircraft is safe within this zone, in regard to the ground and to the traffic. This zone is well defined according to the description of each segment.

The analysis of safety zones is dedicated to check whether the position represented by a particle value is within the safety zone associated to the segment it follows.

Thus, the particles that do not respect segment safety zones can be determined, and finally, as for liveness analysis, the possible aircraft states that do not match the nominal procedure can be determined.

Example 7. Let us consider the second segment (a turn from heading 97 to 144 at 3000ft and 220kt), and its associated "safety zone" defined by the *constraints* $z > 2500$ and $180 < s < 250$, meaning that the altitude *must be* greater than 2500ft and the speed between 180kt and 250kt. In the example presented before, all the particles satisfy these constraints, which ensures that the aircraft remains on a safe trajectory.

These two analysis principles allow to extract the conflictual particles from the set of particles. This information may help the procedure designer either to confirm or to update the procedure.

7   Towards Particle Petri Nets for Pilot's Activity Tracking

As far as flight safety is concerned, systems to help the pilot during the flight are currently studied. In this paper, a formalism has been proposed to model and analyse aircraft procedures. In this section, some ideas are given about how to use this formalism within an estimation process to track the pilot's activities during the flight. Tracking – and consequently analysing – their activities can be used as an input for other systems (such as [2]) to help the pilot with conflictual situations.

7.1   The Estimation Principle

Several approaches have been proposed to track human activities, based on automata [20], on Bayesian nets [21] or on Petri nets [22].

The latter paper proposes a symbolic estimator for activity tracking. The estimation principle consists in a two-step process: a prediction step, which computes the expected activities, and a correction step, which updates prediction according to a new observation.

This principle is close to the particle filter principle, whose process has been explained through Fig. 5: a prediction of the expected particles and an update according to the observation.

Thus the estimation process described in this section is both based on the particle filter and on [22]. Its principle, previously introduced in [23], is detailed in Fig. 15: from the current expected situation $Sit(k|k)$ – meaning the expected situation at time k knowing the measurement at time k –, the prediction computes the further expected situations $Sit(k+i|k)$. The correction step computes

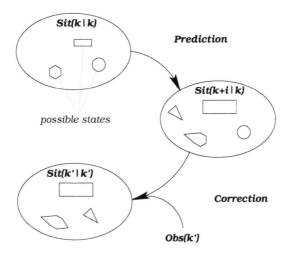

Fig. 15. The estimation principle

the corrected situation $Sit(k'|k')$ according to the observation at time k'. The following section presents the particle Petri net-based prediction of the pilot's activities, then the correction step is introduced.

7.2   Particle Petri Nets to Predict Pilot's Activity

Particle Petri nets allow to deal with uncertainty within the prediction process, i.e the uncertainty on the numerical models, the uncertainty on the observed data and the uncertainty on the pilot's actions.

The current situation $Sit(k|k)$ is a marking of the modular particle Petri net, composed of some probable values of the aircraft numerical parameters – the particles – and some possible aircraft configurations.

From this marking, the set of expected situations $Sit(k+i|k)$ is computed, applying the previously defined firing rules and evolution processes of the particle Petri nets.

As this computation can be quite complex, a temporal horizon H_P is defined. Its aim is to limit the computation of the reachable markings. Its value is determined according to the structure of the guidance nets, the information needed during prediction and the discretization time step. Then the computed expected situations are the markings $\mathcal{M}_i^f$ for i within $[k, k + H_P]$.

Example 8. Let the current situation $Sit(0|0)$ be the marking of the guidance net of Fig. 11(a): the pilot is starting the approach procedure with three probable particles. Let $H_P = 70$ be the prediction horizon of the estimation process. Then the expected situations computed by the prediction step are $\mathcal{M}_0^f$ (Fig. 11(b)), $\mathcal{M}_{50}^f$ (Fig. 12(a)), $\mathcal{M}_{51}^f$ (Fig. 12(b)) and $\mathcal{M}_{70}^f$ (Fig. 13(a)).

7.3   Correction of the Markings

The correction is dedicated to the update of the expected situations according to a new observation. In the pilot's activity tracking domain, the observation consists in a set of aircraft observed parameters $Z(k')$ – meaning the observation vector at time k'. $Z(k')$ can be split into a numerical vector $Z_N(k')$ containing the continuous parameters of the aircraft and a symbolic vector $Z_S(k')$ containing the discrete parameters of the aircraft.

The correction step consists in the comparison of:

1. the expected particles with the numerical observation, which is called *numerical correction*, and
2. the expected configurations with the symbolic observations, which is called *symbolic correction*.

The numerical correction is inspired from the particle filter update: the expected particles within situation $Sit(k'|k)$ – the expected situation at time k' knowing the observation at time k – are weighted according to the numerical observation $Z_N(k')$, which is spoilt with a measurement noise.

The symbolic correction consists in determining the configurations that match the symbolic observation. These configurations are ranked from the "best"

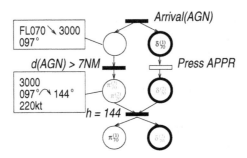

Fig. 16. Inconsistent predicted marking (see Fig. 13(a))

matching to the "worst", according to *rules* [24]. This step is currently under study.

Finally, the numerical correction and the symbolic correction are compared to select the *consistent* markings in regard to the structure of the Petri nets. Once this update is done, the prediction can go on with the updated situation $Sit(k'|k')$.

As far as the *inconsistent* markings resulting from correction are concerned, they may reveal some crucial points to focus on: they may be due to an observation error – a system giving a parameter value is out of order – or to a conflictual situation between the aircraft state and what the pilot has done. For instance, if the estimated situation is the marking of Fig. 7.3, and the corrected marking is $(\pi^{(1)}, \delta^{(1)})$, some conflictual situation can be suspected, as this marking is inconsistent: the pilot may have forgotten to press the APPR button.

In the case of inconsistent markings, some information may be sent to the pilot to help them with the conflictual situation [2].

8   Conclusion

In the framework of modelling and analysing aircraft procedures, a new formalism has been defined: the particle Petri net. Its structure is based on a hybrid modelling of the marking, which includes both particles – representing the possible continuous states of the aircraft – and configurations – representing the possible discrete states of the aircraft.

The evolution of this marking is based on a two-step process: an isomarking evolution of the particles that consists in computing the particles values at the next time step, and an isoparticle evolution of the marking that consists in applying the firing rules to the tokens.

An application to modelling an approach procedure has illustrated the formalism. Then some analysis properties have been proposed to determine the states that are conflictual with the procedure.

Finally, an application to an online estimation of the pilot's activity has been introduced, based on a prediction of the pilot's activities using particle Petri net reachable markings and a correction of the prediction from the incoming observation. In this scope, uncertainty management within particle Petri net allows to detect and predict conflictual situations – thanks to inconsistency analysis – in order to help the pilot to prevent or to cure them.

References

1. Callantine, T.: Activity tracking for pilot error detection from flight data. In: EAM'02, Glasgow, UK (2002)
2. Dehais, F., Tessier, C., Chaudron, L.: GHOST: experimenting conflicts counter-measures in the pilot's activity. In: IJCAI'03, Acapulco, Mexico (2003)
3. Wiegmann, D., Shappel, S.: Applying the human factors analysis and classification system (HFACS) to the analysis of commercial aviation accident data. In: Aviation Psychology, Colombus, OH (2001)
4. Song, L., Kuchar, K.: Describing, predicting, and mitigating dissonance between alerting systems. In: Human Error, Safety, and System Development, Linköping, Sweden (2001)
5. Ruckdeschel, W., Onken, R.: Modelling of pilot behaviour using Petri nets. In: ATPN'94, Zaragoza, Spain (1994)
6. Grastien, A., Cordier, M.O., Largouët, C.: Extending decentralized discrete-event modelling to diagnose reconfigurable systems. In: DX'04, Carcassonne, France (2004)
7. Lerner, U., Moses, B., Scott, M., McIlraith, S., Koller, D.: Monitoring a complex physical system using a hybrid dynamic Bayes net. In: UAI'02, Edmonton, AB (2002)
8. Alla, H., David, R.: A modeling and analysis tool for discrete event systems – continuous Petri net. Performance Evaluation 33 (1998)
9. Horton, G., Kulkarni, V., Nicol, D., Trivedi, K.: Fluid stochastic Petri nets: theory, applications and solution. Operational Research 105 (1998)
10. Champagnat, R., Pingaud, H., Valette, R.: An extension of high-level Petri nets for modelling batch systems. In: CSCC'99, Athens, Greece (1999)
11. Villani, E., Pascal, J.C., Miyagi, P., Valette, R.: Differential predicate transition Petri nets and objects, an aid for proving properties in hybrid systems. In: ADHS'03, Saint-Malo, France (2003)
12. Christensen, S., Petrucci, L.: Towards a modular analysis of coloured Petri nets. In: ATPN'92, Sheffield, UK (1992)
13. Carpenter, J., Clifford, P., Fearnhead, P.: An improved particle filter for non-linear problems. Technical report, University of Oxford (1997)
14. Lehmann, E.: Particle filter, Ph.D. Coursework. Technical report, Australian National University (2003)
15. Wang, J.: Timed Petri nets, theory and application. Kluwer Academic Publishers (1998)
16. Bause, F., Kritzinger, P.: Stochastic Petri nets – an introduction to the theory. Viewer Verlag (2002)
17. Chachoua, M., Pacholczyk, D.: A symbolic approach to uncertainty management. Applied Intelligence 13 (2000)
18. Cardoso, J., Valette, R., Dubois, D.: Possibilistic Petri nets. IEEE Trans. on Systems, Man and Cybernetics 29 (1999)

19. Tessier, C., Chaudron, L., Fiorino, H., Müller, H.J.: Agents' conflicts: new issues. In Tessier, C., Chaudron, L., Müller, H.J., eds.: Conflicting agents - Conflict management in multi-agent systems. Multiagent systems, artificial societies and simulated organizations 1. Kluwer Academic Publishers (2000) 1–30

20. Vu, V., Brémond, F., Thonnat, M.: Automatic video interpretation: a novel algorithm for temporal scenario recognition. In: IJCAI'03, Acapulco, Mexico (2003)

21. Intille, S., Bobick, A.: Visual recognition of multi-agent action using binary temporal relations. In: CVPR'99, Fort Collins, CO (1999)

22. Tessier, C.: Towards a commonsense estimator for activity tracking. In: AAAI Spring Symposium, Palo Alto, CA (2003)

23. Lesire, C.: A numerical/symbolic estimator for activity tracking. A preliminary report. In: KR'04 Doctoral Consortium, Whistler, BC (2004)

24. El-Sayed, M., Pacholczyk, D.: A qualitative reasoning with nuanced information. In: JELIA'02, Consenza, Italy (2002)

On the Expressive Power of Petri Net Schemata

W. Reisig

Department of Computer Science, Humboldt-Universität zu Berlin

Abstract. High-level Petri nets are frequently represented as *Petri net schemas*, with places, transitions and arcs inscribed by *terms*. A concrete system is then gained by *interpreting* the symbols in those terms. The behavior of a concrete system is a *transition system*. The composition of all those transition systems represents the behavior of the Petri net schema.

This paper characterizes the expressive power of (a basic class of) Petri net schemas. It turns out that quite simple as well as quite general requirements at a transition system suffice to generate it by such a Petri net schema.

1   Introduction

Among the various versions of high level Petri nets there is the particular version of *basic Petri net schemata*: The places, transitions and arcs of a net are inscribed by terms, constructed from constant- and function symbols of a signature, Σ. To turn such a schema into a full-fledged Petri net, the symbols in the inscriptions must be interpreted by concrete constants and functions, respectively.

Fig. 1 shows an example of a schema: a, b, c, f_1 and f_4 are constant symbols, l and r are unary function symbols. Fig. 2 shows an interpretation. For the sake of simplicity, each constant symbol is interpreted by itself. There are two additional constant symbols (and, hence, constants) f_2 and f_3, not shining up in Fig. 1. The function symbols l and r are interpreted by functions from P to F, as described in Fig. 2.

For the sake of simplicity, these functions are also denoted as l and r, respectively. The interpretation of Fig. 2 turns the schema of Fig. 1 into the well-known system of three dining philosophers; with an additional, never used fork, f_4.

Fig. 3 shows a different interpretation: Intuitively formulated, the philosophers a and c no longer share a fork, f_3. Instead, philosopher a employs f_4 as his right fork. This yields a high-level Petri net quite different form the net gained form the interpretation as in Fig. 2. Summing up, different interpretations of the symbols in a net schema's inscriptions yield different concrete system nets.

In this paper we are interested in the expressive power of basic Petri net schemata. This is useful e.g. to estimate the relevance of various extensions of such schemata. Fig. 1 shows a schema of the most simple form. One may additionally include symbols for sets of constants, set valued functions and variables in transition inscriptions, allowing for different occurrence modes of a transition.

G. Ciardo and P. Darondeau (Eds.): ICATPN 2005, LNCS 3536, pp. 349–364, 2005.

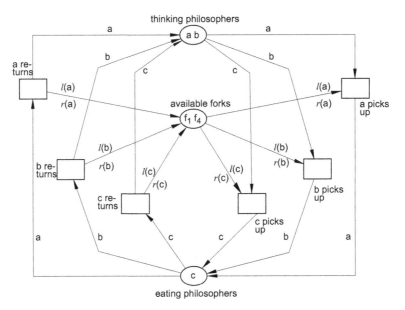

Fig. 1. A Petri net schema

$P = \{a, b, c\}$ $l(a) = r(b) = f_1$
$F = \{f_1, f_2, f_3, f_4\}$ $l(b) = r(c) = f_2$
$l, r : P \to F$ $l(c) = r(a) = f_3$

$P = \{a, b, c\}$ $l(a) = r(b) = f_1$
$F = \{f_1, f_2, f_3, f_4\}$ $l(b) = r(c) = f_2$
$l, r : P \to F$ $l(c) = f_3$
    $r(a) = f_4$

Fig. 2. Interpretation of Fig. 1   **Fig. 3.** Another interpretation of Fig. 1

In fact, all these extensions have turned out useful and have frequently been employed, e.g. for the representation of Distributed Algorithms as in [Rei97].

The conventional way to characterize the expressive power of a formalism is its classification in the Chomsky Hierarchy, or in a complexity class of computable functions. However, a Petri net schema does not characterize a formal language, nor does it compute a function. So, the expressive power of Petri net schemata can not be expressed in conventional terms.

Nielsen, Rozenberg and Thiagarajan in [NRT92] address the expressive power of elementary Petri nets, i.e. Petri nets where each place never carries more than one "black dot" token. They describe the semantics of an elementary Petri net as a transition system – as is commonly done – and characterize a class of transition systems in quite general and abstract terms. Then they show that those transition systems exactly describe the semantics of elementary Petri nets.

We similarly proceed for a basic class of Petri net schemata, constructing a transitions system for a Petri net schema in two steps: A high level Petri net N, i.e. a Petri net schema together with an interpretation of its inscriptions, yields a transition system: Its nodes are the reachable states of N, and its arcs are

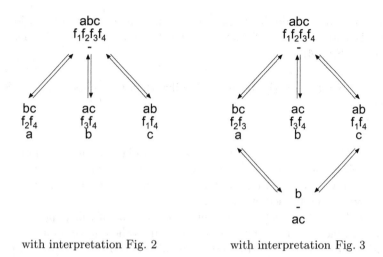

with interpretation Fig. 2   with interpretation Fig. 3

Fig. 4. The transition system to Fig. 1

the steps as defined by the occurrence rule for Petri nets. Hence, this transition system describes the *behavior* of N. As an example, Fig. 4 shows two transition systems to the schema of of Fig. 1, interpreted as in Fig. 2 and in Fig. 3. A node in Fig. 4 consists of three lines, representing (from up to down) the tokens on the places *thinking philosophers*, *available forks*, and *eating philosophers*, respectively.

As two transition systems together form again a transition system, each Petri net schema characterizes a transition system – in fact quite a large one, composed of the behavior of all transition systems of the schema's interpretations.

The problem of this paper now reads:

Which transition systems are characterized by basic Petri net schemata? (1)

We will answer (1) for the case of Petri net schemata without symbols for sets of constants, set valued functions or variables in arc inscriptions.

This characterization turns out intuitively quite simple, as well as quite general. A transition system A is characterized by a basic Petri net schema if it meets six requirements:

1. All nodes of A are structures of the same signature.
2. If R is a node of A and if S is isomorphic to R, then S is a node of A, too.
3. If $S \longrightarrow S'$ is a step of A then the universes of S and S' are equal, and the function symbols are interpreted alike in S and S'.
4. There exists a bound, $k \in \mathbb{N}$ such that to each node S of A there exist at most k successor nodes S', i.e. at most k arcs shaped $S \longrightarrow S'$.
5. All initial states of A can be represented by the same finite set of terms.

Those requirements are merely technicalities. The decisive 6th requirement is the existence of a finite set T of terms which must suffice to "describe" the transition system. We don't require any details of how this description would look like. But any such description must meet the following requirement for all nodes R and S of A:

6. If R and S interpret the terms in T alike, then to each step $S \longrightarrow S'$ there exists a step $R \longrightarrow R'$ such that both steps yield the "same amount of change".

This result immediately rises the quest for similar characterizations of other versions of Petri net schemata, in particular schemata with symbols for sets of constants, set valued functions and variables in transition inscriptions. These questions will be addressed elsewhere.

The problem (1), as well as the central proof ideas of this paper, are not entirely new. In fact, Yuri Gurevich in [Gur85] raises the question for a "computation model that is more powerful and more universal than standard computation models". He compiles a list or requirements and motivates why every reasonable notion of "algorithm" would meet these requirements. Gurevich then in [Gur00] proved that any set of states and steps, fulfilling the requirements, can in fact be generated by a *sequential abstract state machine*. [Rei03] critically examines Gurevich's requirements, as well as Gurevich's proof of the above mentioned theorem.

The above list of 6 requirements for transition systems to be representable by basic Petri net schemata resembles Gurevich's requirements. Likewise, the forthcoming proof outline in Section 4 picks up decisive arguments of Gurevich's proof, in the version of [Rei03].

The rest of this paper is organized as follows: Section 2 introduces basic Petri net schemata and their interpretation. The presentation slightly deviates from what is common in Petri nets, but it is more convenient for the sequel, and, I assume, less technical than usual presentations. Section 3 introduces basic schematic transition systems: The above mentioned list of 6 requirements is made precise, together with some fundamental Lemmata for such transition systems. Section 4 finally outlines the proof that each basic schematic transition system can be represented as a basic Petri net schema.

2   Basic Petri Net Schemata

We start out with the algebraic notions as required throughout the paper. 2.2 provides the syntax of basic Petri net schemata; 2.3 gives the semantics.

2.1   Structures and Signatures

This subsection provides elementary notions and notions on structures, signatures and terms, as they are common in General Algebra. In this paper we will require structures including both function as well as predicates.

Definition 1 (structure). *Let U be a set.*

i. *A function shaped $f : U^n \longrightarrow U$ is a function over U. n is the* arity *of f. A function f is a* constant *in U if its arity is 0.*

ii. *A subset $p \subseteq U$ is a* predicate *over U.*

iii. *Let F be a finite set of functions and let P be a finite set of predicates over U. Then $S = (U, F, P)$ is a* structure. *U is the* universe *of S.*

We define a version of *signatures* that fits the above structures:

Definition 2 (signature). *Let Φ and Π be finite, disjoint sets of symbols. For each $f \in \Phi$ assume a natural number, its* arity. *Then $\Sigma = (\Phi, \Pi)$ is a* signature.

A symbol $f \in \Phi$ is a constant symbol *if its arity is 0, and a* function symbol *if its arity exceeds 0. A symbol $p \in \Pi$ is a* predicate symbol.

Each structure can be assigned a signature:

Definition 3 (Σ-structure). *Let $S = (U, F, P)$ be a structure and let $\Sigma = (\Phi, \Pi)$ be a signature such that the n-ary symbols $f \in \Phi$ are bijectively assigned the n-ary functions $f_S \in F$, and the predicate symbols $p \in \Pi$ are bijectively assigned the predicates $p_S \in P$. Then S is a Σ-structure.*

By $str(\Sigma)$ we denote the set of all Σ-Structures.

In the rest of this paper we assume a signature $\Sigma = (\Phi, \Pi)$.

In this paper we stick to *ground terms*, i.e. variable free terms:

Definition 4 (terms). *The set T_Φ of terms over Φ is the smallest set of symbol sequences such that*

i. *Each constant symbol $t \in \Phi$ is a term over Σ,*

ii. *For each n-ary $f \in \Phi$ and all $t_1, \ldots, t_n \in T_\Phi$, $f(t_1, \ldots, t_n) \in T_\Phi$.*

Definition 5 (interpretation). *Let $S = (U, F, P)$ be a Σ-structure.*

i. *Each term $t \in T_\Phi$ denotes an element $t_S \in U$, inductively defined by*
   1. *t_S if t is a constant symbol*
   2. *$f_S(t_{1S}, \ldots, t_{nS})$ if $t = (t_1, \ldots, t_n)$. t_S is the* interpretation *of t in S.*

ii. *For a set $T \subseteq T_\Phi$, let $T_S = \{t_S | t \in T\}$*

We need Boolean terms in the sequel, defined as follows:

Definition 6 (bool terms). *Assume "$=$", "$\wedge$" and "$\neg$" not in $\Phi \cup \Pi$. Then the set of* boolean terms *of Φ is the smallest set of symbol sequences such that*

i. *For all $t, t' \in T_\Phi$ $t = t'$ is a boolean term,*

ii. *If β and β' are boolean terms, so are $\neg\beta$ and $\beta \wedge \beta'$.*

The interpretation β_S of a boolean term β in a Σ-structure S is obvious; we refrain from a formal definition.

2.2   Syntax of Petri Net Schemata

As a technical convenience, we compose a basic Petri net schema (such as Fig. 1) from *transitions* together with their environments. Each transition, in turn, is a collection of *arc pairs*, one for each place. Each place is a predicate symbol $p \in \Pi$ and each arc is a set of $V \subseteq T_\Phi$ of terms. Fig. 5 and Fig. 6 outlines these constructs, depicting an arc pair and a transition of the schema of Fig. 1. Here the technicalities:

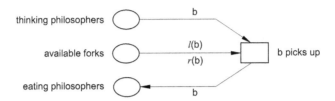

Fig. 5. Graphical outline of a transition (∅-inscribed arcs and set braces are skipped, as usual)

Definition 7 (Σ-transition).

   i. Let $p \in \Pi$ and let $V, W \subseteq T_\Phi$ be finite sets of terms. Then $a = (p, V, W)$ is an arc pair for p.

  ii. Let $\beta \in T_\Phi$ be a boolean term. For each $p \in \Pi$ let a_p be an arc pair for p. Then $(\beta, \{a_p | p \in \Pi\})$ is a Σ-transition. β is its guard.

A *basic Petri net schema* is now a finite set N of Σ-transitions, together with a symbolic representation I of an initial state:

Definition 8 (basic Petri net schema). *For each $p \in \Pi$ let $I_p \subseteq T_\Phi$ be a finite set of terms, and let $I =_{def} \{I_p | p \in \Pi\}$. Let N be a finite set of Σ-transitions. Then $M = (N, I)$ is a basic Petri net schema based on Σ.*

The graphical representation of basic Petri net schemata is obvious: Each $p \in \Pi$ is depicted as a circle, and each Σ-transition t as a box, inscribed by the guard. An arc pair $a_p = (p, V, W)$ of t is depicted by a V-inscribed arrow from p to t and a W-inscribed arrow form t to p. Fig. 6 shows an example. The guard true is usually skipped.

Fig. 6. Graphical outline of an arc pair

As a matter of convenience, ∅-inscribed arrows as well as set braces in arc inscriptions are skipped, as in the representation of a transition in Fig. 5.

The graphical representation of the basic Petri net schema in Fig. 1 is now obvious.

2.3   Semantics of Basic Petri Net Schemata

For a transition based on Σ, a Σ-structure S interprets the inscription of an arc as a subset of element of the universe U of S, to be removed from or augmented to the adjacent place, respectively. In addition, S likewise interprets the inscription of each place as a subset of U, i.e. as a marking.

We first define the steps (S, S') defined by a single transition. The steps of a basic Petri net schema are then just the steps of all its transitions.

Definition 9 (step of transition). *Let $b = (\beta, K)$ be a Σ-transition and let S be a Σ-structure.*

i. *b is enabled at S if $\beta_S = true$ and for all $(p, V, W) \in K$ holds: $V_S \subseteq p_S$, and $W_S \cap p_S = \emptyset$.*

ii. *Let b be enabled at S and let S' be a Σ-structure, defined for each $p \in \Pi$ by $p_{S'} = (p_S \setminus V_S) \cup W_S$, with $(p, V, W) \in K$. For $f \in \Phi$ let $f_{S'} = f_S$. Then (S, S') is a step of b.*

iii. *Let $sem(b)$ denote the set of all steps of b.*

Definition 10 (step of a basic Petri net schema). *Let $M = (N, I)$ be a basic Petri net schema. Then the set sem_M of all steps of all transitions in N is the semantics of M.*

3   Basic Schematic Transition Systems

Here we give the general notion of transition systems, and introduce the six requirements mentioned in the introduction, in greater detail.

3.1   Signature Based Transition Systems

The starting point of this section is the usual notion of nondeterministic transition systems, together with a set of initial states.

Definition 11 (transition system). *Let* states *be a set, let* init $\subseteq$ states, *and let $\tau \subseteq$* states $\times$ states. *Then $A = ($* states, init $, \tau)$ *is a transition system.*

As usual, one may define a behavior of a transition system as a finite or infinite sequence $S_0 S_1 S_2 \ldots$ of states S_i, starting with an initial state $S_0 \in init$. We don't dwell onto this notion, hence don't introduce it formally.

We will stick to transition systems where each state is a Σ-structure, for a signature $\Sigma = (\Phi, \Pi)$ as assumed in Section 2.1:

Definition 12 (transition system based on Σ). *Let $A = ($* states, init $, \tau)$ *be a transition system with* states $\subseteq str(\Sigma)$. *Then A is based on Σ.*

3.2   Isomorphism Closed Transition Systems

From General Algebra it is well known that the language of Σ-terms can not distinguish isomorphic Σ-structures. So we will consider states of transition systems up to isomorphism:

Definition 13 (isomorphic structures). *Let* $S = (U, F, P)$, $S' = (U', F', P')$ *be two Σ-structures. S and S' are isomorphic, written $S \simeq S'$, if there exists a bijective mapping* $h : U \to U'$ *such that*

- *For all* $p \in \Pi$ *and all* $d \in U$, $d \in p_S$ *iff* $h(d) \in p_{S'}$.
- *For all* $t = f(t_1, \ldots, t_n) \in T_\Phi$, $h(f_S(t_{1S}, \ldots, t_{nS})) = f_{S'}(h(t_{1S}), \ldots, h(t_{nS}))$.

Definition 14 (isomorpic closed transition system). *Let* $A = ($states, init, $\tau)$ *be a transition system based on Σ. A is* isomorphism closed *iff for all Σ-structures R and S holds:*

i. If $R \in$ states *and* $R \simeq S$ *then* $S \in$ states.

ii. If $R \in$ init *and* $R \simeq S$ *then* $S \in$ init.

3.3   Structure Preserving Transition Systems

We stick to transition systems that preserve universes and operations.

Definition 15. *Let* $A = ($states, init, $\tau)$ *be a transition system based on Σ.*

i. A preserves universes *if for all steps* $(S, S') \in \tau$, *the universes of S and of S' are identical.*

ii. A preserves operations *if for all steps* $(S, S') \in \tau$ *and all* $f \in \Phi$, $f_S = f_{S'}$.

3.4   Bounded Branching

In contrast to [Gur00] and [Rei03], we consider here a nondeterministic version of transition systems: Each state S may have not just one, but a set of successor states. We assume nondeterminism not only be finite (each state has finitely many successors), but even bounded (each state has at most k successors).

Definition 16 (bounded branching). *Let* $A = ($states, init, $\tau)$ *be a transition system. A is* boundedly branching *if there exist an integer $k \in \mathbb{N}$ such that to each state $S \in$ states there exists at most k states S' with $(S, S') \in \tau$.*

3.5   Initial Normalization

We require all initial states S to be representable by the same set of terms.

Definition 17 (Initial normalization). *Let* $A = ($states, init, $\tau)$ *be a transition system based on Σ. To each $p \in \Pi$, let $T_p \subseteq T_\Sigma$ be a finite set of terms. Then A is* initially normalized *with $\{T_p \mid p \in \Pi\}$ iff for each $S \in$ init and each $p \in \Pi$, $p_S = \{t_S \mid t \in T_p\}$.*

3.6   Bounded Exploration

Here we formally define the decisive 6th requirement: The existence of a finite set T of terms, capable to describe all steps. We start with the *update* $\Delta(S, S')$ of a step (S, S'): The update describes the *change*, i.e. the difference, between S and S':

Definition 18 (update of a τ-step). *Let $A = (\text{states}, \text{init}, \tau)$ be a transition system based on Σ, preserving operations and universes. Let $S, S' \in$ states.*

i. *For each $p \in \Pi$, $\Delta_p(S, S') =_{def} (p, p_S \backslash p_{S'}, p_{S'} \backslash p_S)$ is the p-update of (S, S').*

ii. *The set $\Delta(S, S') =_{def} \{\Delta_p(S, S') \mid p \in \Pi\}$ is the p-update of (S, S').*

Definition 19 (bounded exploration). *Let $A = (\text{states}, \text{init}, \tau)$ be a transition system based on Σ, preserving operations and universes.*

i. *Let $T \subseteq T_\Phi$ such that for all $R, S \in$ states holds: If for all $t \in T$ $t_R = t_S$ then to each step $(R, R') \in \tau$ there exists a step $(S, S') \in \tau$ with $\Delta(R, R') = \Delta(S, S')$. In this case, T is* characteristic *for A.*

ii. *A is* bounded exploration *if there exists a finite characteristic set $T \subseteq T_\Phi$.*

3.7   Basic Schematic Transition Systems

A transition system A fulfilling all above described properties will be denoted as basic schematic, because it will turn at that its semantics, τ, can be described as the semantics sem_M of a basic Petri net schema M.

Definition 20 (basic schematic transition system). *Let $A = (\text{states}, \text{init}, \tau)$ be an isomorphism closed, finitely branching, initially normalized and bounded exploration transition system based on Σ, preserving operations and universes. Then A is* basic schematic.

4   Expressiveness Theorem

We are now ready to state the central theorem of this paper: Every basic schematic transition system can be mimicked by a basic Petri net schema.

4.1   To Be Shown

In addition to a signature $\Sigma = (\Phi, \Pi)$ as in Section 2.1 we assume in the entire rest of this section

- a basic schematic transition system $\mathcal{A} = (states, init, \tau)$ based on Σ, and
- a finite set $T \subseteq T_\Phi$ of terms, characteristic for $\mathcal{A}$.

We have to show the following

Theorem. *There exists a basic Petri net schema M based on Σ with initial marking $\{I_p \mid p \in \Pi\}$ such that*

i. $\tau = sem_M$

ii. For each state $S \in states$ holds for each $p \in \Pi$: $S \in init$ iff $p_S = I_{p_S}$.

The basic Petri net schema to be constructed will be *characteristic*, i.e. will be constructed only from terms in the set T of characteristic terms:

Definition 21 (characteristic basic PN schema). *A basic Petri net schema M based on Σ is characteristic iff for each arc pair (p, V, W) of each transition of M holds: $V, W \subseteq T$.*

Lemma 1 (Isomorphism). *Let R, S be two Σ-algebras.*

i. Let $h : R \rightarrow S$ be a homomorphism. Then for all $t \in T_\Sigma$ holds $h(t_R) = t_S$

ii. Let $h : R \rightarrow S$ be an isomorphism. Then for all $t, v, \in T_\Sigma$ holds: $t_R = v_R$ iff $t_S = v_S$.

Proof.

i. By induction on the structure of T_Σ: If t is a constant symbol, the property follows from Def. 5. Otherwise, let $t = f(t_1, \ldots, t_n)$ and assume the property for $t_1, \ldots, t_n$. Then

$$h(t_R) = h(f(t_1, \ldots, t_n)_R) = h(f_R(t_{1R}, \ldots, t_{nR}))$$
$$= h(f_R(t_{1S}, \ldots, t_{nS})) = f_S(t_{1S}, \ldots, t_{nS})$$
$$= f(t_1, \ldots, t_n)_S = t_S.$$

ii. $t_R = v_R$ iff $h(t_R) = h(v_R)$ (as h is bijective) iff $t_S = v_S$ (by $i.$). ∎

4.2   Updates of Basic Petri Net Schemata

In this subsection we consider elementary properties of basic Petri net schemata: Each step yields a set of *updates*, i.e. the change of tokens in the net's places. Isomorphic steps yield isomorphic updates. For a characteristic Petri net schema holds even more: States that can not be distinguished by the characteristic terms, yield equal sets of updates.

Definition 22 (update of transitions). *Let $b = (\beta, K)$ be a Σ-transition, let $p \in \Pi$, $(p, V, W) \in K$, and let (S, S') be a step of b. Then (p, V_S, W_S) is the p-update of b for (S, S').*

p-updates can be described by the help of set differences:

Lemma 2 (p-update). *Let $b = (\beta, K)$ be a Σ-transition and let $(S, S') \in sem(b)$. Then for each $p \in \Pi$, $(p, p_S \backslash p_{S'}, p_{S'} \backslash p_S)$ is the p-update of b for (S, S').*

Proof. There exists $V, W \subseteq T_\Sigma$ with $(p, V, W) \in K$. Then $p_{S'} = (p_S \backslash V_S) \cup W_S$, by Definition 9.*ii*. Then $V_S = p_S \backslash p_{S'}$ and $W_S = p_{S'} \backslash p_S$, by Definition 9.*i*. Hence the proposition. ∎

Each basic Petri net schema M yields a unique set $delta(M, S, S')$ of updates for each step (S, S):

Definition 23 (update of a PN step). *Let $M = (N, I)$ be a basic Petri net schema based on Σ, and let (S, S') be a step of M. Then let $delta(M, S, S') =_{def} \{(p, A, B) \mid ex.\ b \in N$ and $p \in \Pi$ such that (p, A, B) is the p-update of b for $(S, S')\}$, called the* update of (S, S') by M.

The forthcoming delta-isomorphism-Lemma shows that isomorphic steps of isomorphic states of a basic Petri net schema yield isomorphic updates:

Lemma 3 (delta-isomorphism). *Let M be a basic Petri net schema over Σ, let (R, R') and $(S, S') \in sem_M$ and let $h : R \to S$ be an isomorphism. Then $(p, A, B) \in delta(M, R, R')$ iff $(p, h(A), h(B)) \in delta(M, S, S')$.*

Proof. Let $M = (N, I)$.
$(p, A, B) \in delta(M, R, R')$
iff ex. $b \in N$ and (p, A, B) is the p-update of b for (R, R')
iff ex. $b \in N$ with $(p, V, W) \in b$, $A = V_R$ and $B = W_R$
iff ex. $b \in N$ with $(p, V, W) \in b$, $h(A) = h(V_R)$ and $h(B) = h(W_R)$
iff ex. $b \in N$ with $(p, V, W) \in b$, $h(A) = V_S$ and $h(B) = W_S$
iff ex. $b \in N$ and $(p, h(A), h(B))$ is the p-update of b for (S, S')
iff $(p, h(A), h(B)) \in delta(M, S, S')$.

1st equivalence:	Definition 23
2nd equivalence:	Definition 22
3rd equivalence:	as h is an isomorphism
4th equivalence:	Lemma 1
5th equivalence:	Definition 22
6th equivalence:	Definition 23

∎

The delta-Coincidence-Lemma relates updates of a characteristic Petri net schema to the characteristic terms: Two states generate the same updates if they interpret the characteristic terms equally:

Lemma 4 (delta-coincidence). *Let M be a characteristic basic Petri net schema over Σ and let (R, R') and $(S, S') \in sem(M)$ with $t_R = t_S$ for all $t \in T$. Then $delta(M, R, R') = delta(M, S, S')$.*

Proof. in analogy to the proof of Lemma 3 we argue as follows:
    Let $M = (N, I)$.
$(p, A, B) \in delta(M, R, R')$
iff ex. $b \in N$ and (p, A, B) is the p-update of b for (R, R')
iff ex. $b \in N$ with $(p, V, W) \in b$, $A = V_R$ and $B = W_R$
iff ex. $b \in N$ with $(p, V, W) \in b$, $A = V_S$ and $B = W_S$
iff ex. $b \in N$ and (p, A, B) is the p-update of b for (S, S')
iff $(p, A, B) \in delta(M, S, S')$.

1st equivalence: Definition 23
2nd equivalence: Definition 22
3rd equivalence: the Lemma's assumption on T
4th equivalence: Definition 22
5th equivalence: Definition 23   ∎

4.3   The Transition of a Step

In this subsection we consider elementary properties of basic schematic transition systems.

Each step (S, S') of the transition system $\mathcal{A}$ yields an update (p, A, B) on each predicate $p \in \Pi$. The next Lemma shows that the subsets A and B of the universe of S can be represented by characteristic terms in T. This is not at all trivial. The proof exploits and combines arguments on isomorphism closedness, bounded exploration and universe preservation of $\mathcal{A}$.

As a first and fundamental property of an basic schematic transition system we show that isomorphic states yield isomorphic updates:

Lemma 5 (Δ-Isomorphism). *Let (R, R') and (S, S') be steps of $\mathcal{A}$ and assume a mapping h that constitutes isomorphisms $h : R \to S$ and $h : R' \to S'$. Then $\Delta_p(\tau, R, R') = (p, A, B)$ iff $\Delta_p(\tau, S, S') = (p, h(A), h(B))$.*

Proof. $(p, A, B) \in \Delta_p(R, R')$ iff $p_R \backslash p_{R'} = A$ and $p_{R'} \backslash p_R = B$ iff $h(p_R) \backslash h(p_{R'}) = h(A)$ and $h(p_{R'}) \backslash h(p_R) = h(B)$ iff $p_S \backslash p_{S'} = h(A)$ and $p_{S'} \backslash p_S = h(B)$ iff $(p, h(A), h(B)) \in \Delta_p(S, S')$.

1st equivalence: Definition 18.i
2nd equivalence: h is an isomorphism
3rd equivalence: Lemma 1
4th equivalence: Definition 18.i ∎

Lemma 6 (characteristic subsets). *Let $(S, S') \in \tau$, let $p \in \Pi$ and let $\sigma = (p, A, B)$ be the p-update of (S, S'). Then there exist sets $T^A, T^B \subseteq T$ of characteristic terms, with $T^A{}_S = A$ and $T^B{}_S = B$.*

Proof. By contradiction, assume an element $e \in (A \cup B) \backslash T_S$. Let c be any item, not in the universe U of S, and replace in S each occurrence of e by c. This yields a Σ-structure, R, with universe $V = U \backslash \{e\} \cup \{c\}$. Obviously $R \simeq S$, hence $R \in states$ (as $\mathcal{A}$ is isomorphism closed).

We construct two contradicting arguments:

Firstly, $e \notin T_S$, hence $t_R = t_S$ for all $t \in T$. This implies a step $(R, R') \in \tau$ with $\Delta(R, R') = \Delta(S, S')$, because $\mathcal{A}$ is bounded exploration. Hence, in particular σ is the p-update of (R, R').

Secondly, $e \notin V$, hence e is not in the universe of R and of R', because $\mathcal{A}$ is universe preserving. Hence $e \notin (A \cup B)$. Hence σ is not the p-update of (R, R').
                                                        ∎

For a step (S, S'), the existence of sets T^A and T^B of characteristic terms, as stated in the above Lemma, lies the ground for the construction of a Σ-transition, that later will turn out to mimic the step (S, S'), as well as all steps equivalent to this step.

To each step (S, S') of $\mathcal{A}$ we now construct an unconditional transition $(\overline{S, S'})$, describing the updates (S, S') by help of characteristic terms:

Definition 24 (transition of (S, S')). *Let $S, S' \in$ states. Let $\sigma = (p, A, B)$ be the p-update of (S, S') and let $T^A, T^B \subseteq T$ with $T^A{}_S = A$ and $T^B{}_S = B$. Then $b_p = (p, T^A, T^B)$ is the p-arcpair for (S, S'), and $(\overline{S, S'}) =_{def} \{b_p \mid p \in \Pi\}$ is the unconditional transition for (S, S').*

4.4   Updates of Characteristic PN Schemata

Lemma 7 ($delta((\overline{S, S'}), S, S')$). *Let $(S, S') \in \tau$. Then $delta((\overline{S, S'}), S, S') = \Delta(\tau, S, S')$.*

Proof. $(p, A, B) \in delta((\overline{S, S'}), S, S')$
iff (p, A, B) is the p-update of $(\overline{S, S'})$ for (S, S'), with $A = p_S \backslash p_{S'}$ and $B = p_{S'} \backslash p_S$
iff $(p, A, B) = \Delta_p(\tau, S, S')$ with $A = p_S \backslash p_{S'}$ and $B = p_{S'} \backslash p_S$
iff $(p, A, B) = \Delta(\tau, S, S')$

1st equivalence: Definition 23
2nd equivalence: Lemma 2
3rd equivalence: Definition 18   ∎

Lemma 8 (delta-Δ-coincidence). *Let $R, R', S, S' \in$ states, $(R, R') \in \tau$. For each $t \in T$ let $t_R = t_S$. Then*
$delta((\overline{R, R'}), S, S') = \Delta(S, S')$.

Proof. $delta((\overline{R, R'}), S, S') = delta((\overline{R, R'}), R, R') = \Delta(R, R') = \Delta(S, S')$.

1st equation: Lemma 4
2nd equation: Lemma 7
3rd equation: Definition 19   ∎

Lemma 9 (QRS). *Let $Q, Q', R, R', S, S' \in$ states, let $(R, R') \in \tau$, let $delta((\overline{R, R'}), Q, Q') = \Delta(Q, Q')$ and assume a mapping h such that $h : Q \to S$ and $h : Q' \to S'$ are isomorphisms. Then $delta((\overline{R, R'}), S, S') = \Delta(S, S')$.*

Proof. Let $h : S \to Q$ be an isomorphism. Then

$(p, A, B) \in delta((\overline{R, R'}), S, S')$
iff $(p, h(A), h(B)) \in delta((\overline{R, R'}), Q, Q')$
iff $(p, h(A), h(B)) \in \Delta(\tau, Q, Q')$
iff $(p, h^{-1}(h(A)), h^{-1}(h(B))) \in \Delta(\tau, S, S')$
iff $(p, A, B) \in \Delta(\tau, S, S')$

1st equivalence: Lemma 3
2nd equivalence: the Lemma's assumption
3rd equivalence: Lemma 5
4th equivalence: h is an isomorphism   ∎

Lemma 10 (Δ-τ). *Let M be a Petri net schema based on Σ, and let $S, S' \in$ states such that $delta(M, S, S') = \Delta(\tau, S, S')$. Then $(S, S') \in sem_M$ iff $(S, S') \in \tau$.*

Proof. $(S, S') \in sem_M$
iff ex. transition $b = (\beta, K) \in M$ such that $(S, S') \in sem(b)$
iff for all $p \in \Pi$ $(p, p_S \backslash p_{S'}, p_{S'} \backslash p_S)$ is the p-update of some transition b of M for (S, S')
iff for all $p \in \Pi$ $(p, p_S \backslash p_{S'}, p_{S'} \backslash p_S) \in delta(M, S, S')$
iff for all $p \in \Pi$ $(p, p_S \backslash p_{S'}, p_{S'} \backslash p_S) \in \Delta(S, S')$
iff $(S, S') \in \tau$

1st equivalence: Definition 10
2nd equivalence: Lemma 2
3rd equivalence: Definition 23
4th equivalence: the Lemma's assumption
5th equivalence: Definition 18   ∎

The existence of the finite set T of characteristic terms implies an equivalence on terms and on states with finitely many equivalence classes, such that all states in one class behave essentially alike: One transition will suffice to describe a step of all states in one equivalence class.

Definition 25 (equivalence $\approx$).

  i. *Each $S \in$ states defines an equivalence $\sim_S$ on T, defined for $t, t' \in T$ by $t \sim t'$ iff $t_S = t'_S$.*

  ii. *Let $\approx$ be an equivalence on states, defined for all $R, S \in$ states by $R \approx S$ iff $\sim_R = \sim_S$.*

As an immediate consequence of this definition, equivalent states can't distinguish terms in T. Furthermore, isomorphic states are equivalent.

Lemma 11 (equivalence). *Let $R, S \in$ states.*

  i. *If $R \approx S$ then for all $t, t' \in T$: $t_R = t'_R$ iff $t_S = t'_S$.*

  ii. *If $R \simeq S$ then $R \approx S$.*

Proof. *i.* follows from the first part of Definition 25. For *ii.*, let $h : R \to S$ be an isomorphism. Then for all $t \in T$: $t_R = t'_R$ iff $h(t_R) = h(t'_R)$ iff $t_S = t'_S$.

1st equation: h is an isomorphism.
2nd equation: Lemma 1   ∎

The *semantics lemma* characterizes steps in τ by the help of transitions:

Lemma 12 (semantics). *Let* $R, R', S, S' \in states$, $(R, R') \in \tau$, $R \approx S$ *and* $R' \approx S'$. *Then* $(S, S') \in sem_{\overline{(R,R')}}$ *iff* $(S, S') \in \tau$.

Proof. 1st case: The universes U_R and U_S of R and S are disjoint. Then for each characteristic term $t \in T$, replace in U_S the value of t_S by t_R. This is well defined as to proposition *i.* of the above equivalence Lemma 11. Let Q denote the new state. Obviously, $t_Q = t_R$ for each $t \in T$. Then Lemma 8 implies $delta((\overline{R, R'}), Q, Q') = \Delta(\tau, Q, Q')$. Furthermore, by construction of Q there exists a mapping h such that $h : Q \to S$ and $h : Q' \to S'$ are isomorphisms. Then Lemma 9 implies $delta((\overline{R, R'}), S, S') = \Delta(\tau, S, S')$. Then Lemma 10 implies the proposition.

2nd case: The universes U_R and U_S are not disjoint. Then replace each element in $U_R \cap U_S$ by some new element in U_S. The resulting state, Q, is isomorphic to S and its universe is disjoint to U_R. The second proposition of Lemma 11 implies $Q \approx S$. This reduces the 2nd case to the 1st case. ∎

4.5   Guards and Final Proof

For each state S we define a boolean term, the *guard of* S, that is true at S and at all states equivalent to S:

Definition 26 (S-guard). *Let* $S \in states$.

i. *For two terms* $t, t' \in T$, *the boolean term* $t = t'$ *is an* S-*guard in case* $t_S = t'_S$. *The term* $\neg(t = t')$ *is an* S-*guard in case* $t_S \neq t'_S$.

ii. *Let* β^S *denote the conjunction of all* S-*guards constructed from terms in* S.

The S-guard in fact characterizes the equivalence class of S:

Lemma 13 (guard). *Let* $R, S \in states$. *Then* $\beta^S{}_R = true$ *iff* $R \approx S$.

Proof. follows immediately from Lemma 11.*i.* ∎

We are now ready to prove the Theorem as follows:

Proof of the Theorem. $\approx$ has finitely many equivalence classes. Let $R_1, \ldots, R_n$ be representants of these classes. Obviously, to each step $(S, S') \in \tau$ there exist states $R, R' \in \{R_1, \ldots, R_n\}$ with $S \approx R$ and $S' \approx R'$. Let L be the set of all steps (R, R') with $R, R' \in \{R_1, \ldots, R_n\}$ and some $(S, S') \in \tau$ with $S \approx R$ and $S' \approx R'$. $(R, R') \in \tau$ by bounded exploration. Let N be the set of all transitions $(\beta^R, (\overline{R, R'}))$ with $(R, R') \in L$. Assume $\mathcal{A}$ is initially normalized with I. Then let $M = [N, I]$.

Then $(S, S') \in sem_M$
iff for some $R \approx S$ and $R' \approx S'$ with $(R, R') \in L$ holds: $(S, S') \in sem_{(\beta^R, (\overline{R,R'}))}$
iff for some $R \approx S$ and $R' \approx S'$ with $(R, R') \in L$ holds: $(S, S') \in sem_{\overline{(R,R')}}$
iff $(S, S') \in \tau$.

1st equivalence: construction of L and Definition 9
2nd equivalence: Lemma 13
3rd equivalence: Lemma 12 and construction of L ∎

5   Conclusion

As mentioned in the Introduction already, the Theorem and its proof as given above, are a variant of a theory of Abstract State Machines [Gur00]. This theory seems to expand to a wide range of system models. In particular, in a forthcoming paper we will show how other variants of Petri net schemata can be captured by similar means.

Acknowledgement

I appreciate the referees' numerous valuable comments.

References

[Gur85]   Y. Gurevich. A new thesis. *American Mathematical Society Abstracts*, 6(4):317, August 1985.

[Gur00]   Y. Gurevich. Sequential Abstract-State Machines Capture Sequential Algorithms. *ACMTCL*, 1(1):77–111, July 2000.

[NRT92]   M. Nielsen, G. Rozenberg, and P. S. Thiagarajan. Elementary Transition Systems. *Theoretical Computer Science*, 96:pp 3–333, 1992.

[Rei97]   W. Reisig. *Distributed Algorithms*. Springer-Verlag, 1997.

[Rei03]   W. Reisig. On Gurevich's theorem on sequential algorithms. *Acta Informatica*, 39(5):pp 273–305, 2003.

Determinate STG Decomposition
of Marked Graphs*

Mark Schäfer[1], Walter Vogler[1], and Petr Jančar[2,**]

[1] Institut für Informatik, Universität Augsburg
{mark.schaefer, walter.vogler}@informatik.uni-augsburg.de
[2] Centre for Applied Cybernetics, Technical University of Ostrava
petr.jancar@vsb.cz

Abstract. STGs give a formalism for the description of asynchronous circuits based on Petri nets. To overcome the state explosion problem one may encounter during circuit synthesis, a nondeterministic algorithm for decomposing STGs was suggested by Chu and improved by one of the present authors. To find the best possible result the algorithm might produce, it would be important to know to what extent nondeterminism influences the result, i.e. to what extent the algorithm is determinate.

The result of the algorithm clearly depends on the partition of output signals that has to be chosen initially. In general, it also depends on the order of computation steps. We prove that for live marked graphs — a subclass of Petri nets of definite practical importance in the area of circuit design — the decomposition result depends only on the signal partition. In the proof, we also characterise redundant places in these marked graphs as shortcut places; this easy-to-apply graph-theoretic characterisation is of independent interest.

1   Introduction

Signal Transition Graphs (STG) are a formalism for the description of asynchronous circuits. An STG is a labelled Petri net where the labels denote signal changes between logical high and logical low. Signals are subdivided into input signals, which are produced by the environment, and output signals, which the circuit should produce as specified by the STG. The synthesis of circuits from STGs is supported by several tools, e.g. PETRIFY [CKK+97], and it often involves the generation of the reachability graph, which may have a size exponential in the size of the STG (state explosion). To cope with this problem, Chu suggested a nondeterministic method for decomposing an STG into several smaller ones [Chu87]. While there are strong restrictions on the structure and labelling of

---

* This work was partially supported by the DFG-project 'STG-Dekomposition' Vo615/7-1.
** This author is supported by the Czech Ministry of Education, Grant No. 1M6840770004.

G. Ciardo and P. Darondeau (Eds.): ICATPN 2005, LNCS 3536, pp. 365–384, 2005.

STGs in [Chu87], the improved decomposition algorithm given in [VW02] works under – comparatively moderate – restrictions on the labelling only.

Roughly, the decomposition algorithm works as follows; see [VW02] for details. Initially, a partition of the output signals has to be chosen, and for each set in this partition a component producing the respective output signals is constructed. The result clearly depends on this partition, so we will only consider the case that it has been fixed, and we will concentrate on the construction of one component. To construct a component, one finds a set of signals that (at least initially) can be regarded as irrelevant for the output signals under consideration; then, one takes a copy of the original STG and turns each transition corresponding to an irrelevant signal into an internal (λ-labelled) transition; finally, one tries to remove all internal transitions by so-called secure transition contractions and deletions of (structurally) redundant places, resulting in the final component.

The aim is to find components with small reachability graphs. In principle, this requires to consider all possible sequences of contractions and deletions; but if the algorithm is determinate, i.e. nondeterminism does not influences the result, it is sufficient to consider only one sequence, which greatly increases efficiency. Our main contribution is a determinacy result for a subclass of STGs, where a part of the result applies to STGs in general.

In general, one might find during the processing of a component that additional signals are relevant; then, one has to start anew from a suitably modified copy of the original STG – which eventually gives a correct component as proven in [VW02]. Even in simple cases, the order of operations may influence for which signals this backtracking is performed, resulting in different components as shown in [VW02–Fig. 7]. Since this does not give much hope for a general determinacy-result, we will not consider backtracking in this paper; we will mostly concentrate on the subclass of live marked graphs, for which backtracking is never needed as already noted in [VW02–p. 178].

Although marked graphs are a rather restricted subclass of Petri nets, our results for this subclass are non-trivial. Marked graphs are definitely of practical importance for asynchronous circuit and particularly prominent in benchmark examples studied in the respective community.

As a result of the above considerations, we can abstract from all signals or signal changes, and study the problem under which circumstances the following algorithm is determinate: given an unlabelled Petri net where some transitions are marked as internal, apply secure transition contractions and redundant place deletions as long as possible.

We will show that for live marked graphs the algorithm is determinate, i.e. it produces a unique component (up to isomorphism). Part of this result applies to general Petri nets, for which we show that secure transition contractions satisfy a weak diamond property. We give an easy-to-apply graph-theoretic characterisation of redundant places in marked graphs as so-called shortcut places; our result is a small generalisation of a result in [CCJS94] and our contribution is a much simpler proof. This result is an important ingredient to prove our main result.

The paper is organised as follows. In the next section, Petri nets and their basic notions are introduced, as well as redundant places and secure transition contractions. In Section 3, we characterise redundant places in marked graphs as shortcut places. The other contribution is proven in Section 4. We conclude with Section 5.

2   Basic Definitions

Definition 1. A *Petri net* is a 4-tuple $N = (P, T, W, M_N)$ with

- P the finite set of *places*, T the finite set of *transitions* with $P \cap T = \emptyset$,
- $W : P \times T \cup T \times P \to \mathbb{N}_0$ the *weight function*,
- M_N the *initial marking*, where a *marking* is a function $P \to \mathbb{N}_0$

A Petri net can be considered as a bipartite graph with weighted and directed edges between its nodes. A marking is a function which assigns a number of *tokens* to each place; for a (sub)set Q of places we define $M(Q) = \sum_{p \in Q} M(p)$ (where the sum is zero if $Q = \emptyset$). A *node* is a place or a transition.   □

Definition 2. Let N be a Petri net. The *preset* of a node x is denoted as $^\bullet x$ and defined by $^\bullet x = \{y \in P \cup T \mid W(y, x) > 0\}$, the *postset* of a node x is denoted as $x^\bullet$ and defined by $x^\bullet = \{y \in P \cup T \mid W(x, y) > 0\}$. We say that there is an *arc* from each $y \in {}^\bullet x$ to x. We write $^\bullet x^\bullet$ as shorthand for $^\bullet x \cup x^\bullet$. All these notions are extended to sets as usual.   □

Whenever a Petri net N, N', N_1, etc. is introduced, the corresponding tuples $(P, T, W, M_N), (P', T', W', M_{N'})$, (P_1, T_1, W_1, M_{N_1}) etc. are introduced implicitly. In a graphical representation of a Petri net, places are drawn as circles, transitions as rectangles, the weight function as directed arcs xy (labelled with $W(x, y)$ if $W(x, y) > 1$) and a marking of a place as a number or as a set of small dots drawn in the interior of the corresponding circle. We will regard isomorphic Petri nets as equal.

Definition 3. Let N be a Petri net. A *path* w is a sequence $x_0 x_1 \dots x_n$, $n \geq 0$ of different nodes such that $W(x_i, x_{i+1}) > 0 \ \forall i = 0, \dots, n - 1$. A *cycle* c is a sequence $x_0 x_1 \dots x_n x_0$, $n \geq 1$ with $x_0 \dots x_n$ is a path and $W(x_n, x_0) > 0$. Frequently, we will treat paths and cycles like sets consisting of the respective nodes. By the marking of a path (cycle resp.) we mean the marking (i.e. the sum of the tokens) of the set of its places.

Definition 4. Let N be a Petri net. A transition t is *enabled under a marking* M if $M(p) \geq W(p, t) \ \forall p \in {}^\bullet t$, which is denoted by $M[t\rangle$. An enabled transition can *fire* or *occur* yielding a new marking M', which is written as $M[t\rangle M'$ if $M[t\rangle$ and $M'(p) = M(p) - W(p, t) + W(t, p) \ \forall p \in P$.

A transition sequence $v = t_0 t_1 \dots t_n$ is *enabled under a marking* M if $M[t_0\rangle M_0[t_1\rangle M_1 \dots M_{n-1}[t_n\rangle M_n$, and we write $M[v\rangle$, $M[v\rangle M_n$ resp., v is called *firing sequence* if $M_N[v\rangle$. The empty transition sequence is written as λ and enabled under every marking.

M' is called *reachable from M* if a transition sequence v with $M[v\rangle M'$ exists. The set of all markings reachable from M is denoted by $[M\rangle$. For $[M_N\rangle$ we just write *reachable markings* (of N).

A transition t is called *live under a marking M* if for every $M' \in [M\rangle$ there exists an $M'' \in [M'\rangle$ with $M''[t\rangle$, t is *live* if it is live under M_N and N is *live* if every $t \in T$ is live. A transition t is *dead under a marking M* if there is no $M' \in [M\rangle$ with $M'[t\rangle$. $\qquad\square$

Definition 5. A place p of a Petri net N is *bounded* if for some $k \in \mathbb{N}, M(p) \leq k$ holds for every reachable marking M. N is *bounded* if every place is bounded.

A marking M is a *home marking* of N if it is reachable from every reachable marking. N is called *reversible* if M_N is a home marking. $\qquad\square$

Definition 6. A Petri net N is a *marked graph* (MG) (or *T-system*) if:

1. $\forall p \in P.\ |{}^\bullet p| = 1 = |p^\bullet|$
2. $\forall x, y \in P \cup T.W(x, y) \leq 1$ $\qquad\square$

Due to this, we often identify ${}^\bullet p$ and t if ${}^\bullet p = \{t\}$, and analogously for $p^\bullet$.

Definition 7. Let N be a Petri net and $p \in P$. Place p is called *implicit* if it can be removed from N without changing the set of firing sequences.

Place p is *(structurally) redundant* [Ber87] if there is a set of places Q – called *reference set* – with $p \notin Q$, a valuation $V : Q \cup \{p\} \to \mathbb{N}$ and some $d \in \mathbb{N}_0$ which satisfy the following properties for all transitions t:

1. $V(p)M_N(p) - \sum_{q \in Q} V(q)M_N(q) = d$
2. $V(p)(W(t, p) - W(p, t)) - \sum_{q \in Q} V(q)(W(t, q) - W(q, t)) \geq 0$
3. $V(p)W(p, t) - \sum_{q \in Q} V(q)W(q, t) \leq d$

We call V *balanced* if, for all transitions $t \in T$, $V(p)(W(t, p) - W(p, t)) - \sum_{q \in Q} V(q)(W(t, q) - W(q, t)) = 0$. $\qquad\square$

When constructing a component in our STG decomposition, we would like to remove implicit places. Implicitness is hard to decide, and therefore we actually consider only structural redundancy, since checking this does not require to generate the reachability graph.

Remark: It is well-known that the reachability problem (RP) for Petri nets is EXPSPACE-hard. This even holds for SPZ-RP (Single-Place-Zero RP) where we ask if a given place p can be emptied; we can also assume arc-weights to be 1. Given an instance of SPZ-RP, we can add a fresh t and the arcs $(p, t), (t, p)$, and observe that p is implicit iff no marking with zero tokens in p is reachable. This shows EXPSPACE-hardness of the implicitness problem. On the other hand, redundancy can be solved by linear programming.

The proof that a redundant place p is indeed implicit argues that initially the valuated token number of p is at least d greater than the valuated token sum on Q by the first item, and that this difference can only get greater when firing transitions by the second item; the third item says that each transition needs at most d 'valuated tokens' more from p than from the places in Q. This shows that for the enabling of a transition the presence or absence of p does not matter.

Since deletion of p preserves the firing sequences it also preserves liveness. In general implicitness does not imply redundancy, but we will show that these notions coincide for live marked graphs.[1]

Throughout this paper, if a place p $(p', p_1, \ldots)$ is considered to be redundant, a corresponding reference set Q $(Q', Q_1, \ldots)$ and valuation function V $(V', V_1, \ldots)$ are implicitly given. If only some valuation function V is given, the reference set is implicitly determined as its support by $Q = \{p \in P \mid V(p) > 0\}$.

Furthermore, it is useful to distinguish between different types of redundant places as introduced in the following definition.

Definition 8. Let p be a place of a Petri net N.

- p is an *extended duplicate* of place $p' \in P$ if $\forall t \in T.\ W(p,t) = W(p',t) \wedge W(t,p) = W(t,p')$ and $M_N(p) \geq M_N(p')$.
- p is a *loop-only place* place if $\forall t \in T.\ M_N(p) \geq W(p,t) \leq W(t,p)$.
- If N is a marked graph, p is a *shortcut place* if a path $w = {}^\bullet p \ldots p^\bullet$ exists containing at least one place but not p and satisfying $p \notin w$ and $M_N(p) \geq M_N(w \cap P)$. □

Definition 9. Let N be a Petri net and $t \in T$. If t is not incident to an arc with weight greater than 1 and ${}^\bullet t \cap t^\bullet = \emptyset$, we define the *t-contraction of N*, denoted by $\overline{N}^t$ or just $\overline{N}$, as follows:

$$\overline{T} = T - \{t\} \qquad \overline{P} = \{(p,\star) \mid p \notin {}^\bullet t^\bullet\} \cup \{(p_1, p_2) \mid p_1 \in {}^\bullet t, p_2 \in t^\bullet\}$$
$$\overline{W}((p_1,p_2),t') = W(p_1,t') + W(p_2,t')$$
$$\overline{W}(t',(p_1,p_2)) = W(t',p_1) + W(t',p_2)$$
$$\overline{M}((p_1,p_2)) = M(p_1) + M(p_2)$$

In this definition $\star \notin P \cup T$ is a dummy element used to make all places of $\overline{N}$ to be pairs; we assume $M(\star)$, $W(\star, t')$ and $W(t', \star)$ to be 0.

If more than one contraction is applied to a net N, e.g. $\overline{\overline{N}^{t_1}}^{t_2}$, this is denoted by $\overline{N}^{t_1, t_2}$.

A t-contraction is called *secure* iff $({}^\bullet t)^\bullet \subseteq \{t\}$ or ${}^\bullet(t^\bullet) = \{t\}$. □

The rationale for secure transition contractions is explained in [VW02]. In this paper, arbitrary contractions in general Petri nets are considered in Theorem 20; otherwise, we consider marked graphs where all contractions are secure.

---

[1] [CCJS94] shows that the second redundancy item characterizes that p is *structurally implicit*, i.e. each marking of the other places can be extended to p such that p is implicit.

3   Redundant Places in Marked Graphs

This section deals with redundant and implicit places in live marked graphs. The main result will be that redundant and implicit places coincide in live marked graphs and furthermore they are either loop-only places or shortcut places.

We start with two propositions about redundant places in general.

Proposition 10.

1. *Extended duplicates, loop-only places and shortcut places are redundant.*
2. *If p is a redundant place of a Petri net N, it is a loop-only place iff some reference set Q is empty.*

Proof. (1) For an extended duplicate p of place p' set $Q = \{p'\}$, $V(p) = V(p') = 1$. For a loop-only place p set $Q = \emptyset$, $V(p) = 1$. For a shortcut place p with corresponding path w, set $Q = w \cap P$, $V(p) = 1$ and $V(q) = 1$ for $q \in Q$.

(2) The first direction follows from the proof of part (1). Therefore assume the reference set Q to be empty. Since p is redundant we get immediately $\forall t \in T$:

$$V(p)M_N(p) = d$$
$$V(p)(W(t,p) - W(p,t)) \geq 0$$
$$V(p)W(p,t) \leq d$$

Dividing by $V(p)$ and combining the first and the last (in)equation yields: $\forall t \in T. M_N(p) \geq W(p,t)$, $W(t,p) \geq W(p,t)$, which is equivalent to the definition of a loop-only place. $\square$

The first part of the following proposition was used in an alternative proof of Theorem 13, and we think that it is of independent interest. The second part will be applied below.

Proposition 11.   1. *Let p be a redundant place of a live Petri net N with at least one home marking. Then V is balanced.*
2. *If, in an arbitrary net N, p is redundant under a marking $M \in [M_N\rangle$ with a balanced valuation, it is also redundant under M_N with the same valuation. In particular, if p is a shortcut place under M, it is also one under M_N.*

Proof. 1) Let M_H be a home marking of N. Using part 2 of Definition 7, it can be shown that $\forall t \in T. M_1[t\rangle M_2 \Rightarrow V(p)M_1(p) - \sum_{q \in Q} V(q)M_1(q) \leq V(p)M_2(p) - \sum_{q \in Q} V(q)M_2(q)$ (*).

Let $M_H[v_1\rangle M[v_2\rangle M_H$, such that v_1 contains every transition $t \in T$ at least once. Such a sequence v_1 exists because N is live, v_2 exists because M_H is a home marking. Together with (*) we get:

$$V(p)M_H(p) - \sum_{q \in Q} V(q)M_H(q)$$

$$\leq V(p)M(p) - \sum_{q \in Q} V(q)M(q)$$

$$\leq V(p)M_H(p) - \sum_{q \in Q} V(q)M_H(q)$$

Since N is live, there exists a marking $M_1 \in [M_H\rangle$ for each transition t with $M_1[t\rangle M_2$ and

$$V(p)M_1(p) - \sum_{q \in Q} V(q)M_1(q) = V(p)M_2(p) - \sum_{q \in Q} V(q)M_2(q)$$

Together with $M_2(s) = M_1(s) - W(s,t) + W(t,s) \ \forall s \in P$ this leads to:

$$V(p)M_1(p) - \sum_{q \in Q} V(q)M_1(q)$$

$$= V(p)(M_1(p) - W(p,t) + W(t,p)) - \sum_{q \in Q} V(q)(M_1(q) - W(q,t) + W(t,q))$$

$$= V(p)M_1(p) - \Big(\sum_{q \in Q} V(q)M_1(q) \Big) + V(p)(W(t,p) - W(p,t))$$

$$- \sum_{q \in Q} V(q)(W(t,q) - W(q,t))$$

$$\Rightarrow V(p)(W(t,p) - W(p,t)) - \sum_{q \in Q} V(q)(W(t,q) - W(q,t)) = 0$$

This implies directly that V is balanced.

2) Items 2 and 3 of Definition 7 do not depend on the marking and item 1 follows directly from the valuation being balanced. If p is a shortcut place then the respective path induces a balanced valuation V (as observed in the proof of 10) and, since item 1 can be transferred from M to M_N, the marking of this path is at most the marking of p also under M_N.   □

Before we prove the main theorem of this section, we note an easy lemma about liveness in marked graphs.

Lemma 12. *Let c be a cycle of a marked graph N. For every reachable marking M, $M(c) = M_N(c)$. If N is live, c is initially marked.*

Proof. Let $M[t\rangle M'$. We show that $M(c) = M'(c)$. For $t \in c$ this is trivially true, since all edge weights are 1 in marked graphs. Otherwise, t is not adjacent to any place of c, since N is a marked graph.

The second statement now follows easily; if c is not marked under M_N it is not marked under any reachable marking and therefore no transition of c can ever fire, a contradiction.   □

Actually, marked graphs are live if and only if every cycle is initially marked, see e.g. [DE95]. This is a deeper result, which we do not need here. In fact, our proof of the next theorem has the advantage that it does not require profound knowledge about marked graphs, and we only proved the above lemma to demonstrate that our proof of Theorem 13 is indeed elementary.

Theorem 13. *Let N be a live marked graph and $p \in P$. The following properties are equivalent:*

1. *p is a redundant place*
2. *p is an implicit place*
3. *p is a loop-only place or a shortcut place*

Proof. "1→2" even holds for arbitrary Petri nets – as we observed already –, and "3→1" follows from Proposition 10.

"2→3": Let p be an implicit place but not a loop-only one. We define $\{t_i\} = {}^{\bullet}p$ and $p^{\bullet} = \{t_o\}$, obviously $t_i \neq t_o$, otherwise $M_N(p) = 0$, which contradicts with liveness. Let N' be the net obtained from N by deleting t_i and all incident arcs. Observe that p is also implicit in N', since the set of firing sequence of N' coincides with the set of those firing sequences of N which do not contain t_i.

In N', starting from the initial marking we fire transitions until a maximal set D of transitions is dead.[2] From this marking fire every transition not in D at least once; we denote the marking reached by M. Observe that $(*)$ M can be reached in N by the same firing sequence.

Since t_o can fire at most $M_N(p)$ times in N', we must have $t_o \in D$. Furthermore, there exists a $p_1 \in {}^{\bullet}t_o$, $p_1 \neq p$ with $M(p_1) = 0$. If not, p would be the only place in ${}^{\bullet}t_o$ preventing the firing of t_o, hence would not be implicit in N'.

This implies ${}^{\bullet}p_1 \in D$; otherwise p_1 would have been marked when every transition not in D fired once. Now there is an unmarked place p_2 in ${}^{\bullet}({}^{\bullet}p_1)$ and so on. This leads either to a cycle not containinig any tokens, which is by $(*)$ a contradiction to N being live (cf. Lemma 12); or ends up in a place p' with an empty preset in N', hence $p' \in t_i{}^{\bullet}$ and so we have constructed an unmarked path from t_i to t_o not containing p. Therefore p is a shortcut place under M in N, cf. $(*)$, and we are done by Proposition 11.2.   □

Remark: Javier Esparza pointed out to us that a weaker version of this theorem could be proved as follows. Assume p is a redundant place of a live and bounded marked graph N (or more generally: free-choice net N); then the removal of p results again in a live and bounded marked graph N', which is (roughly speaking) strongly connected by [Bes87]; in particular the transitions ${}^{\bullet}p$ and $p^{\bullet}$ are connected by a path in N'. This result is close to the above theorem, but it is in fact not useful for the purpose of the present paper, since it does not make any statements about the marking of such a path; the pure existence of a path is not sufficient for a place to be redundant.

A result very close to Theorem 13 can be found in [CCJS94]. The difference is that strong connectedness is assumed there – an assumption that we do

---

[2] D does not neccessarily contain all transitions, since we do not assume boundedness or connectedness.

not need. Furthermore, the proof in [CCJS94] makes heavy use of deep results about marked graphs, while our direct proof only needs elementary knowledge. [CCJS94] also considers some form of decomposition of marked graphs; we will discuss the relationship to our approach at the end of the next section.

To determine whether a place is structurally redundant, one can set up an instance of linear programming [STC98]. Our theorem leads to a more efficient algorithm for live marked graphs as already noted in [CCJS94]: to check whether place p is structurally redundant, regard each place p_1 as an edge from ${}^\bullet p_1$ to $p_1^\bullet$, weighted according to the initial marking. Remove the edge corresponding to p and determine the shortest path from ${}^\bullet p$ to $p^\bullet$; if its length (i.e. its cumulated weight) is at most $M_N(p)$, p is redundant. With the basic version of Dijkstra's algorithm, this takes time $O(n^2)$, where n is the number of transitions.

Actually, in [CCJS94] the addition of implicit places is considered; for deciding whether a given place is redundant we note the following improvement.

Dijkstra's algorithm determines all distances from ${}^\bullet p$ in increasing order; hence, the algorithm can already be finished with a negative answer, if all transitions with a distance of no more than $M_N(p)$ have been found and if $p^\bullet$ is not among them. If $M_N(p) = 0$, one can delete all edges corresponding to initially marked places, and simply check for a path from ${}^\bullet p$ to $p^\bullet$ in the remainder e.g. with depth first search in time linear in the number of transitions and places.

4   Determinacy of Petri Net Operations

In this section the determinacy of the decomposition method — with its operations of secure transition contraction and redundant place deletion — is studied. For this we view these Petri net operations as a terminating reduction system, such that determinacy is related to confluence and local confluence.

The notion 'reduction system' comes from the field of term rewriting. The following definition and lemma are taken from [BN98], where a detailed introduction can be found.

Definition 14. Let A be a nonempty set with $a, a', \ldots \in A$.

1. A reduction system is a pair $(A, \to)$ with $\to \subseteq A \times A$. The relation $\to$ is called *reduction* or *reduction rule*; $\to^*$ denotes the reflexive and transitive closure of $\to$, and $\to^=$ the reflexive closure.
2. A reduction $\to$
    (a) is *terminating* if there exists no infinite chain $a_0 \to a_1 \to a_2 \ldots$
    (b) is *confluent* if $a \to^* a_1, a \to^* a_2$ implies $a_1 \to^* a', a_2 \to^* a'$ for some a'
    (c) is *locally confluent* if $a \to a_1, a \to a_2$ implies $a_1 \to^* a', a_2 \to^* a'$ for some a'
    (d) has the *diamond property* if $a \to a_1, a \to a_2$ implies $a_1 \to a', a_2 \to a'$
3. An element a is
    (a) in *normal form* if $\neg\exists a'. a \to a'$
    (b) a *normal form of* a' if $a' \to^* a$ and a is in normal form.   □

Lemma 15.

1. *A terminating relation is confluent iff it is locally confluent.*
2. *If → is terminating and confluent, every element has a unique normal form.*

Next, we model the behaviour of the decomposition algorithm as a reduction system. As explained in the introduction, we can restrict ourselves to the processing of one net, where repeatedly structurally redundant places are removed and transitions from a distinguished set are securely contracted. Also, we concentrate on live marked graphs, although the reduction rules below are actually defined for general nets; Theorem 20 gives a result for general Petri nets.

Definition 16. Let $MGR := \{(N, \Lambda) | N$ is a live marked graph, $\Lambda \subseteq T\}$, where Λ denotes the set of internal transitions to be contracted. We define the following reduction rules on MGR.

1. $(N, \Lambda) \to_{stc} (\overline{N}^t, \Lambda - \{t\})$, where secure contraction of $t \in \Lambda$ is applied.
2. $(N, \Lambda) \to_{rpd} (N', \Lambda)$ if N' is obtained from N by deleting a redundant place.
3. $\to_{red} = \to_{stc} \cup \to_{rpd}$ □

These reductions are well-defined according to the following proposition.

Proposition 17. *Applying* $\to_{red}$ *preserves the marked graph properties (Definition 6) as well as liveness.*

Proof. Deleting a redundant place does not change the firing sequences of the net and therefore liveness is preserved. Since the other places are not affected, the marked graph properties remain valid.

Let $p' = (p_1, p_2)$ be a place resulting from a secure transition contraction. Since p_1 has exactly one transition in its preset, so has p', and analogously for the postset. Since the contraction of a transition t shortens each cycle c containing t but leaves $M_N(c)$ unchanged, the cycles of $\overline{N}^t$ still contain at least one token each, and thus $\overline{N}^t$ is live.   □

Furthermore, $\to_{red}$ is a terminating reduction, as noted in [VW02] for general Petri nets: only finite nets are considered, $\to_{stc}$ reduces the number of transitions, this stays the same under $\to_{rpd}$, and $\to_{rpd}$ reduces the number of places.

Each normal form of $(N, \Lambda) \in MGR$ is a possible result of the decomposition algorithm; thus, by Lemma 15, it suffices to show that $\to_{red}$ is locally confluent in order to prove decomposition to be determinate, because in this case every element of MGR has a unique normal form; recall that we regard isomorphic nets as equal.

To show the local confluence of $\to_{red}$, we need to show the local confluence for every of the three combinations of $\to_{stc}$ and $\to_{rpd}$.

Local Confluence of $\to_{stc}$

We will show now the local confluence for secure transition contractions in live marked graphs. Before that, a result for arbitrary transition contractions in arbitrary Petri nets similar to local confluence is given, namely Theorem 20, which is something like a weak diamond property.

Table 1. Structures of possible places after two transition contractions. This table is obtained from all syntactically possible places by omitting cases which contains a leading $\star$, e.g. $(\star, (p, \star))$. Here, p is only a placeholder for an arbitrary place; in Table 2 all possible allocations are considered

Group	Structure
1	$((p, \star), \star)$
2	$((p, p), \star)$
3	$((p, \star), (p, \star))$
4	$((p, \star), (p, p))$
5	$((p, p), (p, \star))$
6	$((p, p), (p, p))$

Definition 18. Let N be a Petri net and N' a Petri net obtained from N by arbitrary transition contractions. Each $p' \in P'$ is a structured tuple with components from $P \cup \{\star\}$. $\mathfrak{M}_N^{N'}(p')$ is defined as the multi-set of those places $p \in P$ occurring in p'. $\qquad\square$

As an example: Let N be a Petri net with $P = \{p_1, p_2, \ldots, p_n\}$, then
$\mathfrak{M}_N^{N'}(((((p_1, \star), (p_2, \star)), \star), (((p_1, \star), (p_3, p_4)), \star))) = \{2 \cdot p_1, p_2, p_3, p_4\}$.

Lemma 19. *Let N be a Petri net, $N' = \overline{N}^{t_1, t_2}$ and $p_1', p_2' \in P'$. If $\overline{N}^{t_2, t_1}$ is defined as well, $\mathfrak{M}_N^{N'}(p_1') = \mathfrak{M}_N^{N'}(p_2')$ implies $p_1' = p_2'$.*

Proof. This proof works with the Tables 1 and 2. In the first one, all possibilities for the structure of a place after two transition contractions are listed. In the latter one these 6 cases are instantiated resulting in 30 combinations of places from the original net.

As indicated in Table 2 many of the combinations are actually not posssible for simple reasons. For example, if (p_1, p_1) is part of the place then $p_1 \in {}^\bullet t_1$ and $p_1 \in t_1{}^\bullet$, a contradiction since a contraction of a transition with a loop is not defined. As another example, case 23 drops out, because p_1 belongs to the preset of t_1 due the occurrence of (p_1, p_2), and on the other hand p_1 is element of its postset, due to the occurrence of (p_2, p_1). Therefore p_1 forms a loop with the first contracted transition. With the same argumentation cases 24 and 28 are impossible.

The remaining impossible cases 25, 27, and 30 are considered in more detail.

Case 25 leads either to a loop after contracting t_1 or to an arc with weight 2 after contracting t_2, see Figure 1. Case 27 is very similar to the previous one, only the pre- and postsets of t_1 are exchanged.

At last case 30 remains which is more complicated but nevertheless turns out to be impossible, see Figure 2.

In summary, it sufficies to consider the cases 1, 3, 5, 10 and 15 (also shown in Table 3, middle column, the last column is used later). We distinguish three cases for $\mathfrak{M}_N^{N'}(p_1')$.

1. $\mathfrak{M}_N^{N'}(p_1') = \{p_1\} = \mathfrak{M}_N^{N'}(p_2')$. This is only possible if both p_1' and p_2' are in the form of case 1 which implies $p_1' = p_2'$.

Table 2. All combinatory possible places (up to isomorphism) after contraction of t_1 and then t_2. This table is obtained from Table 1 by instantiating p. The places p_i are pairwise different. The places which have an 'type error'-entry are not possible, since a place is treated as being and at the same time as not being adjacent to a contracted transition; 'initial loop' means that there is a loop at one of the transitions initially. Rows with a leading ▲ are considered in greater detail in the text

No.	Group	# Places	Example	Possible	If not, why?
▲ 1	1	1	$((p_1, \star), \star)$	●	
2	2	1	$((p_1, p_1), \star)$	-	initial loop $p_1 - t_1$
▲ 3	2	2	$((p_1, p_2), \star)$	●	
4	3	1	$((p_1, \star), (p_1, \star))$	-	initial loop $p_1 - t_2$
▲ 5	3	2	$((p_1, \star), (p_2, \star))$	●	
6	4	1	$((p_1, \star), (p_1, p_1))$	-	type error
7	4	2	$((p_1, \star), (p_1, p_2))$	-	type error
8	4	2	$((p_1, \star), (p_2, p_1))$	-	type error
9	4	2	$((p_2, \star), (p_1, p_1))$	-	initial loop $p_1 - t_1$
▲ 10	4	3	$((p_1, \star), (p_2, p_3))$	●	
11	5	1	$((p_1, p_1), (p_1, \star))$	-	type error
12	5	2	$((p_1, p_1), (p_2, \star))$	-	initial loop $p_1 - t_1$
13	5	2	$((p_1, p_2), (p_1, \star))$	-	type error
14	5	2	$((p_2, p_1), (p_1, \star))$	-	type error
▲ 15	5	3	$((p_1, p_2), (p_3, \star))$	●	
16	6	1	$((p_1, p_1), (p_1, p_1))$	-	initial loop $p_1 - t_1$
17	6	2	$((p_1, p_1), (p_1, p_2))$	-	initial loop $p_1 - t_1$
18	6	2	$((p_1, p_1), (p_2, p_1))$	-	initial loop $p_1 - t_1$
19	6	2	$((p_1, p_2), (p_1, p_1))$	-	initial loop $p_1 - t_1$
20	6	2	$((p_2, p_1), (p_1, p_1))$	-	initial loop $p_1 - t_1$
21	6	2	$((p_1, p_1), (p_2, p_2))$	-	initial loop $p_1 - t_1$ and $p_2 - t_1$
22	6	2	$((p_1, p_2), (p_1, p_2))$	-	loop after contracting t_1
23	6	2	$((p_2, p_1), (p_1, p_2))$	-	initial loop $p_1 - t_1$
24	6	3	$((p_1, p_2), (p_3, p_1))$	-	initial loop $p_1 - t_1$
▲ 25	6	3	$((p_1, p_2), (p_1, p_3))$	-	loop after contracting t_1 or weight 2 after contracting t_2 first
26	6	3	$((p_1, p_1), (p_2, p_3))$	-	initial loop $p_1 - t_1$
▲ 27	6	3	$((p_2, p_1), (p_3, p_1))$	-	loop after contracting t_1 or weight 2 after contracting t_2 first
28	6	3	$((p_2, p_1), (p_1, p_3))$	-	initial loop $p_1 - t_1$
29	6	3	$((p_2, p_3), (p_1, p_1))$	-	initial loop $p_1 - t_1$
▲ 30	6	4	$((p_1, p_2), (p_3, p_4))$	-	loop or weight 2 after contracting t_2 first

2. $\mathfrak{M}_N^{N'}(p_1') = \{p_1, p_2\} = \mathfrak{M}_N^{N'}(p_2')$. Hence, $p_1', p_2' \in \{((p_1, p_2), \star), ((p_2, p_1), \star), ((p_1, \star), (p_2, \star)), ((p_2, \star), (p_1, \star))\}$. If a fixed p_1' from this set occurs in the net $\overline{N}^{t_1, t_2}$ it is not possible that a different element from this set occurs, too;

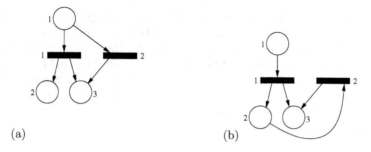

(a)   (b)

Fig. 1. Case 25 - $p' = ((p_1, p_2), (p_1, p_3))$. p_1 has to be an element of $^\bullet t_1$ p_2 and p_2, p_3 have to be elements of $t_1^\bullet$. Then there are 4 cases: 1) $p_1 \in {}^\bullet t_2, p_3 \in t_2^\bullet$: loop after contracting t_1, see (a)   2) $p_1 \in {}^\bullet t_2, p_1 \in t_2^\bullet$: initial loop $p_1 - t_1$ 3) $p_2 \in {}^\bullet t_2, p_1 \in t_2^\bullet$: loop after contracting t_1 4) $p_2 \in {}^\bullet t_2, p_3 \in t_2^\bullet$: weight 2 after contracting t_2, see (b)

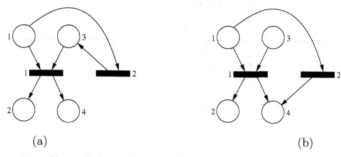

(a)   (b)

Fig. 2. Case 30 - $((p_1, p_2), (p_3, p_4))$. p_1 and p_3 have to be in the preset of the first transition to be contracted (t_1), p_2 and p_4 in the postset. For the connection to t_2 there are several possibilities; all of them satisfy that p_1 or p_2 (or both) are in the preset and p_3 or p_4 (or both) are in the postset, which leads to 9 sub-cases. Exemplarily two of them are considered. (a) leads to an arc with weight 2 when t_2 is contracted first and (b) leads to a loop. The other cases are similar to these ones or contain them

for example: if $p_1' = ((p_1, p_2), \star)$ there is no place $p_1'' = ((p_2, p_1), \star)$, since the existence of p_1' implies that p_1 is an element of $^\bullet t_1$ but the existence of p_1'' implies p_1 is an element of $t_1^\bullet$; a contradiction, since the contraction was possible. With similar argumentations one can exclude the other combinations. Hence, $\mathfrak{M}_N^{N'}(p_1') = \mathfrak{M}_N^{N'}(p_2')$ implies $p_1' = p_2'$ for this case.

3. $\mathfrak{M}_N^{N'}(p_1') = \{p_1, p_2, p_3\} = \mathfrak{M}_N^{N'}(p_2')$. Analogous to the second case we obtain twelve possible structures for p_1', p_2' resp. which all exclude each other as places of P', see the following table.

1	$((p_1, p_2), (p_3, \star))$	7	$((p_1, \star), (p_2, p_3))$
2	$((p_1, p_3), (p_2, \star))$	8	$((p_1, \star), (p_3, p_2))$
3	$((p_2, p_1), (p_3, \star))$	9	$((p_2, \star), (p_1, p_3))$
4	$((p_2, p_3), (p_1, \star))$	10	$((p_2, \star), (p_3, p_1))$
5	$((p_3, p_1), (p_2, \star))$	11	$((p_3, \star), (p_1, p_2))$
6	$((p_3, p_2), (p_1, \star))$	12	$((p_3, \star), (p_2, p_1))$

Without loss of generality, assume $p_1' = ((p_1, p_2), (p_3, \star))$ (case 1) or $p_1' = ((p_3, \star), (p_1, p_2))$ (case 11). $(p_3, \star)$ implies that p_3 is not adjacent to t_1, and therefore the existence of such a place excludes the existence of places 2,4-10. The remaining cases 3 and 12 can be excluded, since (p_2, p_1) implies $p_1 \in t_1{}^\bullet$ whereas p_1' implies $p_1 \in {}^\bullet t_1$; in this case p_1 would be a loop place which is a contradiction. Case 1 cannot coexist with case 11, since the latter implies $(p_1, p_2) \in t_2{}^\bullet$ whereas the former case implies $(p_1, p_2) \in {}^\bullet t_2$ after contracting t_1, also a contradiction. $\qquad \square$

Table 3. Possible places after two transition contractions. In the middle column one can find the places from Table 2 which turned out to be possible according to Definition 9. In each case there exists a place in $\overline{N}^{t_2, t_1}$ which uses the same places from N as the one in the middle column. This place is shown in the last column; for line 4 and 5 there are two possibilities, but only one of them exists

No.	$\overline{N}^{t_1, t_2}$	$\overline{N}^{t_2, t_1}$
1	$((p_1, \star), \star)$	$((p_1, \star), \star)$
2	$((p_1, p_2), \star)$	$((p_1, \star), (p_2, \star))$
3	$((p_1, \star), (p_2, \star))$	$((p_1, p_2), \star)$
4	$((p_1, \star), (p_2, p_3))$	$((p_1, p_2), (p_3, \star))$ / $((p_2, \star), (p_1, p_3))$
5	$((p_1, p_2), (p_3, \star))$	$((p_1, \star), (p_2, p_3))$ / $((p_1, p_3), (p_2, \star))$

Theorem 20. *Let N be a Petri net and $t_1, t_2 \in T$. If both $\overline{N}^{t_1, t_2}$ and $\overline{N}^{t_2, t_1}$ are defined then they are isomorphic (even if the contractions are not secure).*

Proof. For this proof Table 3 is used again; the last column shows the place of $N_2 = \overline{N}^{t_2, t_1}$, which uses the same places from N as the place from $N_1 = \overline{N}^{t_1, t_2}$ in the middle column. If there are two possibilities, only one of them exists. For lines 1-3, it is quite clear that these places exist in N_2, for line 4 see Figure 4: since the place $((p_1, \star), (p_2, p_3))$ exists in $\overline{N}^{t_1, t_2}$, N must contain the net fragment (a); observe that exactly one of the dotted arcs exists but not both (in this case contracting t_1 would generate an arc with weight 2). Depending on which arc exist in $\overline{N}^{t_2, t_1}$, exactly one of the places in the last column exists. Line 5 is analogous.

We define a relation $f \subseteq P_1 \times P_2 \cup T_1 \times T_2$ by $f|_{T_1 \times T_2} = Id$ and $(p_1', p_2') \in f \Leftrightarrow \mathfrak{M}_N^{N_1}(p_1') = \mathfrak{M}_N^{N_2}(p_2')$. We will show that f is an isomorphism.

a) f is a partial function: $(p_1', p_2'), (p_1', p_2'') \in f \Rightarrow \mathfrak{M}_N^{N_2}(p_2') = \mathfrak{M}_N^{N_2}(p_2'')$. Lemma 19 implies $p_2' = p_2''$.

b) f is total (surjective): After two contractions each place $p_1' \in P_1$ has a structure shown in Table 3, middle column, and $\mathfrak{M}_N^{N_1}(p_1') = \mathfrak{M}_N^{N_2}(p_2')$ holds for the corresponding place p_2' in the last column. Analogous for surjective.

c) f is injective: $f(p_1') = f(p_1'') \Rightarrow \mathfrak{M}_N^{N_1}(p_1') = \mathfrak{M}_N^{N_1}(p_1'')$. From Lemma 19 follows $p_1' = p_1''$.

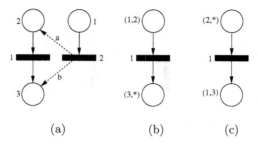

$$(a) \qquad\qquad (b) \qquad (c)$$

Fig. 3. For line 4 from Table 3. Since the place $((p_1, \star), (p_2, p_3))$ exists in $\overline{N}^{t_1, t_2}$, N must contain the net fragment (a); observe that exactly one of the dotted arcs exists but not both (in this case contracting t_1 would generate an arc with weight 2). If arc a, b resp. exists, contracting t_2 first results in (b), (c) respectively; the next contraction results in $((p_1, p_2), (p_3, \star))$, $((p_2, \star), (p_1, p_3))$ resp. as it is written in the last column

d) f preserves the structure, i.e. $W_1(p_1', t) = W_2(f(p_1'), f(t))$, $W_1(t, p_1') = W_2(f(t), f(p_1'))$ $\forall p_1' \in P_1, t \in T_1$. This follows from the definition of transition contraction. Since the weight of an arc incident to a composite place is the sum of the related weights of the component places, we derive that $W_1(p_1', t_1) = \sum_{p \in \mathfrak{M}_N^{N_1}(p_1')} W(p, t_1) = \sum_{p \in \mathfrak{M}_N^{N_2}(f(p_1'))} W(p, t_1) = W_2(f(p_1'), f(t_1))$. Observe that for every place p_1' of N_1 shown in Table 3, $\mathfrak{M}_N^{N_1}(p_1')$ is a *set*. Analogous for the second case. □

The proof for the following lemma uses Theorem 20; if this is not applicable, we show that – since $N \in MGR$ – in N_1 and N_2 loop-only places can be deleted such that the contraction of t_2 and t_1 resp. is applicable afterwards. After the contraction, extended duplicates can be deleted such that the results are isomorphic.

Lemma 21. *For* $(N, \Lambda) \in MGR$, *let* $(N, \Lambda) \to_{stc} (N_1, \Lambda_1)$ *and* $(N, \Lambda) \to_{stc} (N_2, \Lambda_2)$. *Then, there exists* $(N', \Lambda') \in MGR$ *with* $(N_1, \Lambda_1) \to_{red}^* (N', \Lambda')$ *and* $(N_2, \Lambda_2) \to_{red}^* (N', \Lambda')$.

Proof. Let the contractions concern transition t_1 and t_2. If both $\overline{N}^{t_1, t_2}$ and $\overline{N}^{t_2, t_1}$ are defined, Theorem 20 implies that the results are isomorphic. In this case even the diamond property is fulfilled.

Therefore assume that w.l.o.g. $\overline{N}^{t_1, t_2}$ is not defined. Since $N_1 = \overline{N}^{t_1}$ is defined by hypothesis, the contraction of t_2 is not possible in N_1, although it is possible in N. Since N_1 is a marked graph — in particular no arc weight becomes greater than 1 —, the contraction of t_1 in N must have generated a loop place adjacent to t_2, because t_1 and t_2 form a cycle with two places in N. Since N is a live marked graph, this cycle contains at least one token making the loop place redundant.

This situation is schematically shown in Figure 4(a): each place represents a set of places connected to t_1 and t_2 in the same way, e.g. places of type 1 are in the preset of t_1 and not adjacent to t_2. Figure 4(b) and (c) depict the results of

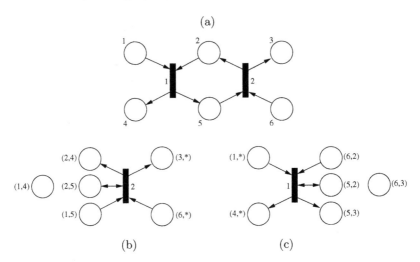

Fig. 4. (a) Scheme of a net fragment where contraction generates a loop (b) After t_1-contraction (c) After t_2-contraction

contracting t_1 and t_2 resp. in the same way, e.g. places of type $(2,4)$ are pairs (p, p') with p of type 2 and p' of type 4.

Places of type $(2,5)$ and $(5,2)$ are loop-only places, which can be removed as noted above; afterwards, the other transition contraction becomes possible. These contractions give places of types $((1,4), *), ((1,5), (3, \star)), ((1,5), (2,4)),$ $((6, *), (2,4)), ((6, *), (3, *))$ in the first case and $((1, *), (4, *)), ((1, *), (5,3)),$ $((6,2), (5,3)), ((6,2), (4, *)), ((6,3), *)$ in the second. We will argue that the resulting nets are isomorphic after removal of some redundant places.

As noted in the proof of Theorem 20, the connections of these places to the remaining transitions are determined by their at most four components, and analogously for the initial marking. In particular, places of type $((1,5), (2,4))$ are connected in the same way as places of type $((1,4), *)$ in the first case – since t_1 and t_2 are not present anymore – and they carry even more tokens, since at least one of a type-2 and a type-5 place is marked in N. Therefore, places of type $((1,5), (2,4))$ are extended duplicates, and so are places of type $(6,2), (5,3))$; we remove them in the two nets.

For the other types, we find a matching between $((1,4), *)$ and $((1, *), (4, *)),$ $((1,5), (3, *))$ and $((1, *), (5,3))$ etc., which matches each place of type $((1,4), *)$ to the place of type $((1, *), (4, *))$ with the same component-places etc. By the above, this gives an isomorphism between the remaining nets when the above extended duplicates are removed. □

Local Confluence of $\to_{rpd}$

We will now proceed to the next part of the local confluence proof. Although the local confluence of redundant place deletion might seem rather obvious, in fact some effort is already needed to prove it at least for marked graphs.

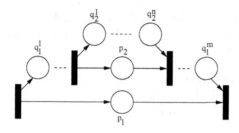

Fig. 5. Two redundant places p_1, p_2 with $p_1 \notin Q_2, p_2 \in Q_1$

Let p_1, p_2 be redundant places of $N \in MGR$ with $p_1 \neq p_2$. If one of them, lets say p_1, is a loop-only place, then $p_2 \notin Q_1 = \emptyset$ and $p_1 \notin Q_2$, because p_1 is only adjacent to one transition. This case obviously fulfils the diamond property, since the deletion of one of the redundant places does neither affect the other one nor its reference set.

Due to Theorem 13 we can now assume that p_1 and p_2 are shortcut places and the reference sets consist of the places of the corresponding paths.

We will distinguish three cases: 1) $p_1 \notin Q_2, p_2 \notin Q_1$, 2) $p_1 \notin Q_2, p_2 \in Q_1$ (w.l.o.g.) and 3) $p_1 \in Q_2, p_2 \in Q_1$.

The first case is treated as above. For the second case take a look at Figure 5. Since p_1 is not a loop-only place, p_2 lies on a Q_1-path $w_1 = {}^\bullet p_1 q_1^1 \ldots q_1^m p_1{}^\bullet$. Since p_2 is not a loop-only place either, a Q_2-path $w_2 = {}^\bullet p_2 q_2^1 \ldots q_2^n p_2{}^\bullet$ exists. This implies that there is a path w connecting ${}^\bullet p_1$ and $p_1{}^\bullet$ and using only places from $q_1^1 \ldots q_1^m$ excluding p_2 and from $q_2^1 \ldots q_2^n$. $M_N(p_1) \geq \sum_{i=1}^{m} M_N(q_1^i)$ and $M_N(p_2) \geq \sum_{i=1}^{n} M_N(q_2^i)$ (Definition 7(1)) directly imply that $M_N(p_1) \geq \sum_{i=1}^{m} M_N(q_1^i) - M_N(p_2) + \sum_{i=1}^{n} M_N(q_2^i)$; hence, w also shows that p_1 is redundant; the corresponding reference set does not contain p_2 and we are done by case (1).

The last case $p_1 \in Q_2, p_2 \in Q_1$ is impossible, because it implies

$$M_N(p_1) \geq \sum_{q \in Q_1 \backslash \{p_2\}} M_N(q) + M_N(p_2) \qquad M_N(p_2) \geq \sum_{q \in Q_2 \backslash \{p_1\}} M_N(q) + M_N(p_1)$$

From this we get immediately:

$$M_N(p_1) = M_N(p_2) \quad \text{and} \quad \sum_{q \in Q_1 \backslash \{p_2\}} M_N(q) = \sum_{q \in Q_2 \backslash \{p_1\}} M_N(q) = 0 \quad (*)$$

Since $p_1 \in Q_2$, there are Q_2-paths ${}^\bullet p_2 \ldots {}^\bullet p_1$ and $p_1{}^\bullet \ldots p_2{}^\bullet$ not using p_1, and analogously there are Q_1-paths ${}^\bullet p_1 \ldots {}^\bullet p_2$ and $p_2{}^\bullet \ldots p_1{}^\bullet$ not using p_2. Therefore, either a cycle c using only places from $(Q_1 \cup Q_2) \backslash \{p_1, p_2\}$ exists which contradicts N being live by Lemma 12, since $(*)$ implies $M_N(c) = 0$; or $(Q_1 \cup Q_2) \backslash \{p_1, p_2\} = \emptyset$. In the latter case, p_1 and p_2 are extended duplicates of each other with the same initial marking; thus, removing either of them gives the same net up to isomorphism.

Altogether the following lemma holds.

Lemma 22. *Let* $(N, \Lambda) \rightarrow_{rpd} (N_1, \Lambda_1)$ *and* $(N, \Lambda) \rightarrow_{rpd} (N_2, \Lambda_2)$ *for some* $(N, \Lambda) \in MGR$. *Then an* $(N', \Lambda') \in MGR$ *exists with* $(N_1, \Lambda_1) \rightarrow_{rpd}^{=} (N', \Lambda')$ *and* $(N_2, \Lambda_2) \rightarrow_{rpd}^{=} (N', \Lambda')$.

Observe that two steps of $\rightarrow_{rpd}$ fulfil the diamond property or lead to iso-morphic results; in particular we have not used $\rightarrow_{stc}$.

Local Confluence of $\rightarrow_{stc}$ and $\rightarrow_{rpd}$

Lemma 23. *Let* $(N, \Lambda) \rightarrow_{rpd} (N_1, \Lambda_1)$ *and* $(N, \Lambda) \rightarrow_{stc} (N_2, \Lambda_2)$ *for some* $(N, \Lambda) \in MGR$. *Then, there exists an* $(N', \Lambda') \in MGR$ *with* $(N_1, \Lambda_1) \rightarrow_{red}^{*}$ (N', Λ') *and* $(N_2, \Lambda_2) \rightarrow_{red}^{*} (N', \Lambda')$.

Proof. Let p be the redundant place and t the transition to be contracted. In live marked graphs p is either a loop-only place or a shortcut place.

In the first case t and p are not adjacent because the contraction of t is possible for (N, Λ), i.e. p forms a loop with another transition and the operations can be performed independently.

If p is a shortcut place, there are the following possibilities: 1) t is neither adjacent to p nor part of the path making p redundant; then both operations are independent of each other again. 2) t is part of the path but not adjacent to p. The contraction of t shortens the path but does not interrupt it, and also the sum of the markings remains unchanged; hence, the two operations are independent. 3) t is adjacent to the path *and* p – leading to two sub-cases, one of them shown in Figure 6(a). In the other one, analogously the path starts from t and $p \in t^{\bullet}$.

We will only consider the case depicted in (a), with the results of contraction and deletion shown in (b) and (c) resp. Each place (p_s, p_{xi}) in (b) is a shortcut

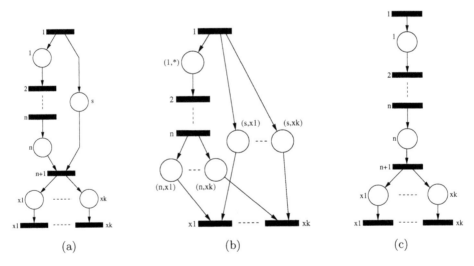

(a)   (b)   (c)

Fig. 6. Confluence of shortcut place deletion and transition contraction. (a) $p \equiv p_s$ is a shortcut place of $\{p_1, \ldots, p_n\}$ and $t \equiv t_{n+1}$ is the transition to be contracted. The net in (b) is obtained by contracting t_{n+1}, (c) by deleting p_s

place of $\{(p_1, *), \ldots, (p_{n-1}, *), (p_n, p_{xi})\}$ because they give a path and the initially marking of this path as well as $M_N(p_s)$ are increased by the same value $M_N(xi)$. Therefore, these shortcut places can be deleted yielding a Petri net which also results from (c) when contracting t. $\qquad \square$

Altogether, our results can be collected in the central theorem of this section.

Theorem 24. *The reduction rule* $\rightarrow_{red}$ *is confluent and terminating for live marked graphs.*

Corollary 25. *The STG-decomposition algorithm of [VW02] is determinate for live marked graphs.*

In [CCJS94] a decomposition of strongly connected live marked graphs into two components is considered. In this approach the nets are unlabelled, while our STG decomposition is directed by the labelling with signal transitions; therefore the decomposition of [CCJS94] is not applicable in our setting.

What is interesting is that in the decomposition of [CCJS94] a whole subnet is removed and this could be used in our setting to remove several internal transitions together. A result of [CCJS94] implies that this removal preserves the language, but this does not immediately imply that subnet removal can be used to determine correct STG decompositions in the sense of [VW02]. In fact, the correctness criterion of [VW02] is of bisimulation type, but does not imply language equivalence. Furthermore, redundant place deletion and secure transition contractions always lead to a correct decomposition while subnet removal presupposes liveness and strong connectedness. Liveness is a precondition for determinacy of STG decomposition but not for its correctness.

Nevertheless subnet removal might be closely related to redundant place deletion and secure transition contractions. If one could show some sort of coincidence this might lead to an alternative proof of our determinacy result. Such a result would not imply that subnet removal is more efficient; the latter involves solving an all-pairs shortest paths problem, which takes time of $O(n^3)$ where n is the number of removed internal transitions *plus* the number of "neighbouring" non-internal transitions.

5   Conclusion

We have shown that the STG decomposition algorithm presented in [VW02] is determinate if applied to live marked graphs, a subclass of considerable interest in the area of circuit design. The proof of this result is based on several statements, and only one of them could be shown for general Petri nets. It would be clearly interesting to generalise some other partial results to other net classes. We currently look at nets where the marked-graph requirements are only violated 'in a few places'; such nets also turn up often in circuit design. A problematic point is that our proofs rely several times on the liveness characterisation of marked graphs via the markings of cycles.

Related to the determinacy result, but also of independent interest, is the conceptionally and algorithmically easy characterisation of redundant places in live marked graphs, for which we provided a new easy proof. Again, we would like to generalise this result; it is clear that in S-Systems [DE95] — which coincide with finite automata — no place can be redundant if every place has at least one transition in its postset, and we currently consider a generalisation to free-choice nets, which is not obvious at all.

Acknowledgement: We thank the anonymous referees for their comments which helped to improve the paper.

References

[Ber87]   G. Berthelot. Transformations and decompositions of nets. In W. Brauer et al., editors, *Petri Nets: Central Models and Their Properties*, Lect. Notes Comp. Sci. 254, 359–376. Springer, 1987.

[Bes87]   E. Best. Structure theory of Petri nets: The free choice hiatus. In W. Brauer et al., editors, *Petri Nets: Central Models and Their Properties*, Lect. Notes Comp. Sci. 254, 168–205. Springer, 1987.

[BN98]   F. Baader and T. Nipkow. *Term Rewriting and All That.* Cambridge University Press, Cambridge, 1998.

[CCJS94]   J. Campos, J. M. Colom, H. Jungnitz, and M. Silva. Approximate throughput computation of stochastic marked graphs. In *IEEE Transactions on Software Engineering 20*, pages 526–535, 1994.

[Chu87]   T.-A. Chu. *Synthesis of Self-Timed VLSI Circuits from Graph-Theoretic Specifications.* PhD thesis, MIT, 1987.

[CKK+97]   J. Cortadella, M. Kishinevsky, A. Kondratyev, L. Lavagno, and A. Yakovlev. Petrify: a tool for manipulating concurrent specifications and synthesis of asynchronous controllers. *IEICE Trans. Information and Systems*, E80-D, 3:315–325, 1997.

[DE95]   J. Desel and J. Esparza. *Free Choice Petri Nets.* Cambridge University Press, Cambridge, 1995.

[STC98]   M. Silva, E. Teruel, and J.M. Colom. Linear algebraic and linear programming techniques for the analysis of place/transition net systems. In *Lectures on Petri Nets I; Basic Models*, LNCS 1491, 309–373. Springer, 1998.

[VW02]   W. Vogler and R. Wollowski. Decomposition in asynchronous circuit design. In J. Cortadella et al., editors, *Concurrency and Hardware Design*, Lect. Notes Comp. Sci. 2549, 152 – 190. Springer, 2002.

Timed-Arc Petri Nets vs.
Networks of Timed Automata

Jiří Srba*

BRICS**
Department of Computer Science,
Aalborg University, Fredrik Bajersvej 7B,
9220 Aalborg East Denmark
srba@brics.dk

Abstract. We establish mutual translations between the classes of 1-safe timed-arc Petri nets (and its extension with testing arcs) and networks of timed automata (and its subclass where every clock used in the guard has to be reset). The presented translations are very tight (up to isomorphism of labelled transition systems with time). This provides a convenient characterization from the theoretical point of view but is not always satisfactory from the practical point of view because of the possible non-polynomial blow up in the size (in the direction from automata to nets). Hence we relax the isomorphism requirement and provide efficient (polynomial time) reductions between networks of timed automata and 1-safe timed-arc Petri nets preserving the answer to the reachability question. This makes our techniques suitable for automatic translation into a format required by tools like UPPAAL and KRONOS. A direct corollary of the presented reductions is a new PSPACE-completeness result for reachability in 1-safe timed-arc Petri nets, reusing the region/zone techniques already developed for timed automata.

1   Introduction

One of the major challenges in theoretical computer science is to establish a precise relationship among a wide range of modelling formalisms available nowadays and to compare their descriptive power and verification capabilities. Recently, various models to describe concurrent systems with real-time features attracted lots of attention. The most wide-spread models include timed automata of Alur and Dill [3] and several timed extensions of Petri nets (see [10, 29] for overviews). Verification tools for timed automata have been developed (including tools like UPPAAL [21], KRONOS [12] and CMC [19]) as well as tools for timed Petri nets (including tools like ROMEO [24] and TINA [6]).

---

* The author is supported in part by grants ITI-1M0021620808, 1ET-408050503, and GACR 201/03/1161.

** Basic Research in Computer Science,
    Centre of the Danish National Research Foundation.

G. Ciardo and P. Darondeau (Eds.): ICATPN 2005, LNCS 3536, pp. 385–402, 2005.

In this paper we focus on timed-arc Petri nets [7, 17], a model where time (age) is associated with tokens and transitions are labelled by time intervals which restrict the age of tokens that can be used to fire them. We consider the weak (non-urgent) semantics. The reachability problem for general timed-arc Petri nets is undecidable [26], even in the case where tokens in different places are not required to age at the same rate [23]. On the other hand, coverability and boundedness are decidable [25, 1]. We compare the timed-arc Petri net model with timed automata. Unlike the other models of timed Petri nets, timed-arc Petri nets still suffer from a lack of analysis and verification tools and hence automatic translations to already existing tools (e.g. to UPPAAL timed automata) are of a relatively high interest.

Our Contribution. We provide a comparison of the (sub/super)-classes of networks of timed automata (also called concurrent timed automata) and 1-safe timed-arc Petri nets up to isomorphism of labelled transitions systems with time (and hence also up to timed bisimilarity). The usefulness of the presented translations is documented by a PSPACE-completeness result for reachability of 1-safe timed-arc Petri nets where we directly reuse the recent results for reachability in networks of timed automata by Aceto and Laroussinie [2]. At the practical level, we describe polynomial time (and size) translations between the mentioned models up to reachability and we show a technique to reduce the general type of synchronization function in networks of timed automata into the synchronization policy accepted by UPPAAL (handshake synchronization). Last but not least our results provide a natural motivation for the study of the extension of timed-arc Petri net with testing arcs and of the subclass of timed automata where clocks used in guards are mandatorily reset by the same transition.

Related Work. Techniques to translate Petri nets extended with time features into equivalent timed automata have not been often studied in the past. Only recently, there has been some focus on translating different variants of timed-transitions Petri nets into timed automata [22, 16, 8]. A work most related to ours is by Cassez and Roux [14]. In their paper they propose (independently of our work) a structural translation of bounded timed-transitions Petri nets into communicating timed automata. Each transition in their model is translated into a single timed automaton and transition firing is supervised by an additional timed automaton. Our model is, however, different from timed-transitions Petri nets (time features are associated to tokens, not to transitions) and we provide a translation that does not rely on additional UPPAAL features like arrays of integers that are used in [14]. As concluded in [11], timed-transitions Petri nets express timed behaviour and timed-arc Petri nets express time behaviour *and* time constraints.

Considering the class we study (timed-arc Petri nets), the only related work we are aware of is by Sifakis and Yovine [27]. They provide a different (non-structural) translation of timed-arc Petri nets (with urgent behaviour) into timed automata (with invariants), hence achieving essentially the result of our Corollary 3, except for the fact that their translation causes an exponential blow up in the size of the timed automaton whereas our translation is structural and can be implemented in polynomial time (and space).

2   Basic Definitions

2.1   Labelled Transition Systems with Time

A *rooted labelled transition system with (real) time* (or simply LTS) is a tuple $T = (S, Act, \longrightarrow, s^0)$ where S is a set of *states*, Act is a set of *actions* such that $Act \cap \mathbb{R}^+ = \emptyset$, $\longrightarrow \subseteq S \times (Act \cup \mathbb{R}^+) \times S$ is a *transition relation*, written $s \xrightarrow{a} s'$ for $(s, a, s') \in \longrightarrow$ where $a \in Act$, and $s \xrightarrow{\epsilon(r)} s'$ for $(s, r, s') \in \longrightarrow$ where $r \in \mathbb{R}^+$, and $s^0 \in S$ is a distinguished *initial state*.

Hence LTS have two kinds of transitions: the standard ones under the visible actions from Act and time-elapsing ones under the actions $\epsilon(r)$, where for all $r \in \mathbb{R}^+$ the symbol $\epsilon(r)$ represents a special action name called a *time delay*.

We write $s \longrightarrow s'$ whenever $s \xrightarrow{a} s'$ for some $a \in Act$ or $s \xrightarrow{\epsilon(r)} s'$ for some $r \in \mathbb{R}^+$. Let $\longrightarrow^*$ denote the reflexive and transitive closure of $\longrightarrow$. By $Reach(s^0) \stackrel{\text{def}}{=} \{s \in S \mid s^0 \longrightarrow^* s\}$ we denote the set of all *reachable states* in T.

Let $T_1 = (S_1, Act_1, \longrightarrow_1, s_1^0)$ and $T_2 = (S_2, Act_2, \longrightarrow_2, s_2^0)$ be two LTS. We say that T_1 and T_2 are isomorphic (and write $T_1 \cong T_2$) whenever there is a bijection $f : Reach(s_1^0) \to Reach(s_2^0)$ such that for all $a \in Act$, $r \in \mathbb{R}^+$ and $s, s' \in Reach(s_1^0)$ it is the case that $s \xrightarrow{a}_1 s'$ iff $f(s) \xrightarrow{a}_2 f(s')$, and $s \xrightarrow{\epsilon(r)}_1 s'$ iff $f(s) \xrightarrow{\epsilon(r)}_2 f(s')$. Hence the reachable parts of the transition systems are identical up to renaming of states.

2.2   Time Domains and Time Intervals

In this paper we consider a continuous time domain, i.e., time values are from the set of nonnegative real numbers $\mathbb{R}^+$. The set of natural numbers (including 0) is denoted by $\mathbb{N}$ and is used in guards.

Remark 1. We can use also discrete time in the time domain and/or rational numbers in guards and this does not influence the results presented in this paper.

The set $\mathcal{I}$ of *time intervals* is defined by the following abstract syntax where a and b range over $\mathbb{N}$ such that $a < b$.

$$ I ::= [a, b] \mid [a, a] \mid (a, b] \mid [a, b) \mid (a, b) \mid [a, \infty) \mid (a, \infty) $$

Let $I \in \mathcal{I}$. Given a time point $r \in \mathbb{R}^+$, the validity of the expression $r \in I$ is defined in the usual way, e.g., $r \in [a, b)$ iff $a \leq r < b$ and $r \in (a, \infty)$ iff $a < r$.

Remark 2. It is easy to see that any intersection of finitely many time intervals $I_1, \ldots, I_n \in \mathcal{I}$ (denoted by $\cap_{i=1}^n I_i$) is either empty ($\emptyset$) or it belongs to $\mathcal{I}$.

2.3   Timed-Arc Petri Nets

A *(labelled) timed-arc Petri net* (TAPN) is a tuple $N = (P, T, F, c, Act, \lambda)$, where P is a finite set of *places*, T is a finite set of *transitions* such that $T \cap P = \emptyset$, $F \subseteq (P \times T) \cup (T \times P)$ is a *flow relation*, $c : F|_{P \times T} \rightharpoonup \mathcal{I}$ is a *time constraint* assigning a time interval to every arc from a place to a transition, $\mathcal{A}ct$ is a set of *labels* (*actions*), and $\lambda : T \to \mathcal{A}ct$ a *labelling function*.

We also define $^\bullet t \overset{\text{def}}{=} \{p \mid (p,t) \in F\}$ and $t^\bullet \overset{\text{def}}{=} \{p \mid (t,p) \in F\}$. Let $N = (P, T, F, c, \mathcal{A}ct, \lambda)$ be a TAPN. A *marking* M on the net N is a function $M : P \to \mathcal{B}(\mathbb{R}^+)$ where $\mathcal{B}(\mathbb{R}^+)$ denotes the set of finite multisets on $\mathbb{R}^+$. Each place is thus assigned a certain number of tokens, and each token is annotated with a real number (*age*). Let $B \in \mathcal{B}(\mathbb{R}^+)$ and $a \in \mathbb{R}^+$. We define $B + a$ in such a way that we add the value a to every element of B, i.e., $B + a \overset{\text{def}}{=} \{b + a \mid b \in B\}$. As *initial markings* we allow only markings with all tokens of age 0. A *marked TAPN* is a pair (N, M_0) where N is a timed-arc Petri net and M_0 is an initial marking.

Let us now define the dynamics of TAPN. We introduce two types of transition rules: *firing* of a transition and *time-elapsing*.

Let $N = (P, T, F, c, \mathcal{A}ct, \lambda)$ be a TAPN, M a marking and $t \in T$.

- We say that t is *enabled* by M iff $\forall p \in {}^\bullet t.\ \exists x \in M(p).\ x \in c(p,t)$.
- If t is enabled by M then it can *fire*, producing a marking M' such that:

$$\forall p \in P.\ M'(p) = \Big(M(p) \smallsetminus C^-(p,t) \Big) \cup C^+(t,p)$$

where C^- and C^+ are chosen to satisfy the following equations (note that there may be more possibilities and that all the operations are on multisets):

$$C^-(p,t) = \begin{cases} \{x\} & \text{if } p \in {}^\bullet t \text{ s.t. } x \in M(p) \wedge x \in c(p,t) \\ \emptyset & \text{otherwise} \end{cases}$$

$$C^+(t,p) = \begin{cases} \{0\} & \text{if } p \in t^\bullet \\ \emptyset & \text{otherwise.} \end{cases}$$

Then we write $M[t\rangle M'$. Note that the new tokens added to the places in $t^\bullet$ are of the initial age 0.

- We define a *time-elapsing* transition $\epsilon(r)$, for $r \in \mathbb{R}^+$, as follows:

$$M[\epsilon(r)\rangle M' \quad \text{iff} \quad \forall p \in P.\ M'(p) = M(p) + r.$$

A marked TAPN (N, M_0) where $N = (P, T, F, c, \mathcal{A}ct, \lambda)$ generates a LTS

$$T(N, M_0) \overset{\text{def}}{=} (P \to \mathcal{B}(\mathbb{R}^+), \mathcal{A}ct, \longrightarrow, M_0)$$

where states are markings on N, and the transition relation $\longrightarrow$ is defined as follows:

$$M \xrightarrow{a} M' \text{ whenever } M[t\rangle M' \text{ for some } t \in T \text{ such that } \lambda(t) = a$$
$$M \xrightarrow{\epsilon(r)} M' \text{ whenever } M[\epsilon(r)\rangle M'.$$

In standard P/T Petri nets there is a simple construction to ensure that a transition can be fired only if a token is present in a certain place, without removing the token. This is done by adding two arcs: one from the place to the transition and one in the opposite direction. A similar construction, however, does not work in TAPN, as consuming a (timed) token resets its age. In order to recover this possibility we shall explicitly add so called *testing arcs*. Testing arcs (called *read arcs* in e.g. [28, 13]) were also investigated in connection with partial order semantics for (untimed) P/T nets.

A *timed-arc Petri net with testing arcs* is a tuple $N = (P, T, F, c, Act, \lambda, F*, c*)$ such that $(P, T, F, c, Act, \lambda)$ is a timed-arc Petri net, $F* \subseteq P \times T$ is a set of *testing arcs*, and $c* : F* \to \mathcal{I}$ is a function which assigns a time interval to every testing arc from $F*$. We define $^*t \stackrel{\text{def}}{=} \{p \mid (p, t) \in F*\}$.

The dynamics of TAPN with testing arcs is defined as in the case of TAPN with the only difference that for a transition t to be enabled, it has to satisfy an extra condition, namely

$$\forall p \in {}^*t. \; \exists x \in M(p). \; x \in c*(p, t).$$

In other words, a necessary condition for a transition to fire is that all places which are connected via testing arcs to the transition contain a token satisfying the constraint on the testing arc. The transition can then fire according to the rules defined above, which means that the testing arcs do not consume any tokens and hence do not influence their age.

A *1-safe marking* is a marking M having at most one token in every place, i.e., $|M(p)| \leq 1$ for all $p \in P$. A *1-safe marked TAPN* is a marked TAPN (N, M_0) where all markings from $Reach(M_0)$ are 1-safe markings. In case of 1-safe nets we will require that $F \cap F* = \emptyset$. This is without loss of generality as the conditions on testing arcs can be inserted (by Remark 2) onto standard arcs whenever we know that every place contains at most one token.

We shall use the following abbreviations: TAPN for timed-arc Petri nets; TAPN* for timed-arc Petri nets with testing arcs; 1-TAPN for 1-safe timed-arc Petri nets; and 1-TAPN* for 1-safe timed-arc Petri nets with testing arcs. The size of (N, M_0) where $N = (P, T, F, c, Act, \lambda, F*, c*)$ is the size of N (the size of the description of N) plus the size of M_0, formally $size(N, M_0) \stackrel{\text{def}}{=} |P| + |T| + |F| + |Act| + |F*| + \sum_{p \in P} |M_0(p)|$.

2.4 Concurrent Timed Automata

Let C be a finite set of *clocks*. A *(time) valuation* of clocks from C is a function $v : C \to \mathbb{R}^+$. Let v be a valuation and $r \in \mathbb{R}^+$. We define a valuation $v + r : C \to \mathbb{R}^+$ by $(v + r)(x) \stackrel{\text{def}}{=} v(x) + r$ for every $x \in C$. For every set $R \subseteq C$ we define a valuation $v[R := 0] : C \to \mathbb{R}^+$ by $v[R := 0](x) \stackrel{\text{def}}{=} v(x)$ for $x \in C \setminus R$ and $v[R := 0](x) \stackrel{\text{def}}{=} 0$ for $x \in R$. A *clock guard* is a partial function $g : C \hookrightarrow \mathcal{I}$ assigning a time interval to selected clocks. We say that a valuation v satisfies a guard g (written $v \models g$) iff $v(x) \in g(x)$ for all $x \in dom(g)$.

Remark 3. The definition of a clock guard given above enables to encode also boolean combinations of clock constraints in such a way that conjunction of several guards for a certain clock is replaced by intersection using Remark 2 and disjunction is replaced by multiple edges (see e.g. [5]). We also do not consider difference constraints since from the expressive point of view there are techniques for their replacement with simple constraints [5] and because of the well accepted observation that difference constraints are not crucial for modelling real-life systems [9].

A *timed automaton* (TA) is a tuple $A = (S, \mathcal{A}ct, C, \longrightarrow, s^0)$ where S is a finite set of *control states*, $\mathcal{A}ct$ is a finite set *actions*, C is a finite set of *clocks*, $\longrightarrow \subseteq S \times (C \hookrightarrow \mathcal{I}) \times \mathcal{A}ct \times 2^C \times S$ is a finite *transition relation* written $s \xrightarrow{g,a,R} s'$ for $(s, g, a, R, s') \in \longrightarrow$, and $s^0 \in S$ is an *initial* control state.

A *configuration* of a timed automaton A is a pair (s, v) where s is a control state $(s \in S)$ and v is a time valuation on C $(v : C \rightarrow \mathbb{R}^+)$. An *initial configuration* of A is (s^0, v^0) such that $v^0(x) \overset{\text{def}}{=} 0$ for all $x \in C$. A given timed automaton $A = (S, \mathcal{A}ct, C, \longrightarrow, s^0)$ determines a LTS

$$T(A) \overset{\text{def}}{=} (S \times (C \rightarrow \mathbb{R}^+), \mathcal{A}ct, \longrightarrow, (s^0, v^0))$$

where states are configuration of A and the transition relation $\longrightarrow$ is defined by

$(s, v) \xrightarrow{a} (s', v[R := 0])$ whenever there is a transition $s \xrightarrow{g,a,R} s'$ in A s.t. $v \models g$

$(s, v) \xrightarrow{\epsilon(r)} (s, v + r)$ for all $r \in \mathbb{R}^+$.

In modelling real-time systems using timed automata, we often design several subcomponents and then define the composed behaviour of the whole system. This is usually done by parallel composition with a certain communication scheme (see e.g. the tools UPPAAL [21] and KRONOS [12]). Assume that the system consists of n concurrent components. As suggested e.g. in [2] a general synchronization scheme in the style of Arnold-Nivat [4] can be described by a so called *synchronization function* $\phi : (\mathcal{A}ct \cup \{\bullet\})^n \hookrightarrow \mathcal{A}ct$ which is a partial function where $\bullet$ denotes a distinguished symbol of *inactivity* such that $\bullet \notin \mathcal{A}ct$. The components different from $\bullet$ are called *active* components and we require that $(\bullet, \bullet, \dots, \bullet) \notin dom(\phi)$, i.e., at least one component in every synchronization tuple is active. By $|\phi|$ we understand the cardinality of the set $dom(\phi)$ and by $width(\phi)$ we denote the maximum number of active components over all tuples from $dom(\phi)$.

Let $A_1, \dots, A_n$ be timed automata where (for all i, $1 \leq i \leq n$) $A_i = (S_i, \mathcal{A}ct, C, \longrightarrow_i, s_i^0)$ and where $\mathcal{A}ct$ and C are fixed sets of actions and clocks, respectively. Let ϕ be a synchronization function. A *concurrent timed automaton* (CTA) with a synchronization function ϕ is a parallel composition of $A_1, \dots, A_n$ denoted by $A = (A_1 \| \cdots \| A_n)_\phi$.

A CTA A determines an LTS

$$T(A) \overset{\text{def}}{=} (S_1 \times \cdots \times S_n \times (C \rightarrow \mathbb{R}^+), \mathcal{A}ct, \longrightarrow, (s_1^0, \dots, s_n^0, v^0))$$

where states are Cartesian products of the control states of the individual automata together with a clock valuation, and the transition relation $\longrightarrow$ is defined by

- $(s_1, \ldots, s_n, v) \xrightarrow{a} (s'_1, \ldots, s'_n, v')$ whenever there is a tuple $(a_1, \ldots, a_n) \in dom(\phi)$ such that:
  - for all i, $1 \leq i \leq n$, with $a_i = \bullet$ we have $s'_i = s_i$, and we set $R_i \overset{\text{def}}{=} \emptyset$
  - for all i, $1 \leq i \leq n$, with $a_i \in \mathcal{A}ct$ we have (in A_i) a transition rule $s_i \xrightarrow{g_i, a_i, R_i} s'_i$ s.t. $v \models g_i$
  - $v' = v[R := 0]$ where R is defined by $R \overset{\text{def}}{=} \cup_{i=1}^{n} R_i$,
  - $a = \phi(a_1, \ldots, a_n)$
- $(s_1, \ldots, s_n, v) \xrightarrow{\epsilon(r)} (s_1, \ldots, s_n, v + r)$ for all $r \in \mathbb{R}^+$.

This means that the composition of timed automata can perform a synchronization step whenever the individual automata can perform transitions labelled according to the function ϕ. Moreover, if some automaton resets a clock, the clock is also reset in the concurrent timed automaton. The size of a concurrent automaton A is the sum of the sizes of all its components A_i plus the size of the synchronization function ϕ, formally $size(A) \overset{\text{def}}{=} |\phi| + \sum_{i=1}^{n} |A_i|$ where $|A_i|$ for $A_i = (S_i, \mathcal{A}ct, C, \longrightarrow_i, s_i^0)$ is equal to $|S_i| + |\mathcal{A}ct| + |C| + | \longrightarrow_i |$.

We now define subclasses of TA and CTA called *mandatory reset* TA (mrTA) and *mandatory reset* CTA (mrCTA). The intuition is that every clock which is tested in a guard while performing a transition must be mandatorily reset on that transition. Formally, a mandatory reset TA is a TA such that every transition $s \xrightarrow{g, a, R} s'$ in the automaton satisfies $dom(g) \subseteq R$. A CTA is called a mandatory reset CTA iff each of its components is a mrTA.

Remark 4. The idea of mandatory reset in the case of one clock only was considered by Laroussinie, Markey and Schnoebelen in [20]. Their definition forms a natural subclass of timed automata and we will provide further justification of this notion in what follows.

3   From Automata to Nets

We shall now demonstrate that the class of LTS generated by CTA is contained (up to isomorphism) in the class of LTS generated by 1-TAPN*, i.e., we present a construction that for a given CTA A algorithmically defines a marked 1-TAPN* (N, M_0) such that $T(A) \cong T(N, M_0)$.

Let $\{A_i = (S_i, \mathcal{A}ct, C, \longrightarrow_i, s_i^0)\}_{i=1}^{n}$ be a collection of TA and let ϕ be an arbitrary synchronization function. Without loss of generality we assume that the sets of control states S_i are pairwise disjoint for all i, $1 \leq i \leq n$, and that the function ϕ satisfies that whenever $(a_1, \ldots, a_n) \in dom(\phi)$ then for each a_i different from $\bullet$ there exists at least one transition rule in A_i labelled with the action a_i.

Remark 5. Without loss of generality we also assume that every TA A_i contains a distinguished clock $delay_i \in C$ which is never used in any guard of any timed automaton but is always reset in all transition rules of the automaton A_i. This means that the clock $delay_i$ measures how much time has elapsed (in the i'th component) from the last occurrence of a transition labelled by an action from Act. These additional clocks do not influence the behaviour of the concurrent automaton A and are considered only for technical reasons in order to establish isomorphism of the respective labelled transition systems. They are not necessary if we were interested in weaker equivalence notions (e.g. in timed bisimilarity).

We now consider a CTA $A = (A_1 \| \cdots \| A_n)_\phi$ with the initial configuration $(s_1^0, \ldots, s_n^0, v^0)$ where $v^0(x) \overset{\text{def}}{=} 0$ for all $x \in C$, and with the special clocks $delay_i$ according to Remark 5. We shall construct a 1-TAPN* $N \overset{\text{def}}{=} (P, T, F, c, Act, \lambda, F*, c*)$ with an initial marking M_0 such that it generates an LTS isomorphic to $T(A)$. The intuition is that every tuple $(a_1, \ldots, a_n)$ from $dom(\phi)$ defines a set of transitions in the net, which simulate the effect of the synchronization function. The set of transitions enumerates all possible combinations of transition rules available in the components of the CTA and labelled by the corresponding actions. Places in the net represent control states and there are also special places designed for storing tokens representing clocks. An extra place called e (which will never become marked) is added for a technical convenience.

$$P \overset{\text{def}}{=} C \cup \{s \mid s \in \cup_{i=1}^n S_i\} \cup \{e\}$$
$$T \overset{\text{def}}{=} \{t_{(a_1,\ldots,a_n),(w_1,\ldots,w_n)} \mid (a_1,\ldots,a_n) \in dom(\phi) \text{ and for all } i, 1 \leq i \leq n,$$
$$w_i = \bullet \text{ if } a_i = \bullet; \text{ otherwise } w_i = (s_i, g_i, a_i, R_i, s_i') \text{ whenever there is}$$
$$\text{a transition rule } s_i \xrightarrow{g_i, a_i, R_i}_i s_i' \text{ in } A_i\}$$

For every transition $t_{(a_1,\ldots,a_n),(w_1,\ldots,w_n)} \in T$ we define $\lambda(t_{(a_1,\ldots,a_n),(w_1,\ldots,w_n)}) \overset{\text{def}}{=} \phi(a_1, \ldots, a_n)$. The set of arcs associated with every transition from T is given as follows. Let $t = t_{(a_1,\ldots,a_n),(w_1,\ldots,w_n)} \in T$ where every w_i is either $\bullet$ or of the form $(s_i, g_i, a_i, R_i, s_i')$ for all i, $1 \leq i \leq n$. We define a set $J \subseteq \{1, \ldots, n\}$ of active components of t by $i \in J$ iff $w_i \neq \bullet$, and the set of all reset clocks by $R \overset{\text{def}}{=} \cup_{i \in J} R_i$. For each clock $x \in C$ such that there is at least one g_i where $x \in dom(g_i)$ we also define the combined guard

$$I_x \overset{\text{def}}{=} \bigcap_{i \in J \, \wedge \, x \in dom(g_i)} g_i(x)$$

which expresses a combined requirement (time interval) of all active components of A on the clock x. In the case that these requirements are not consistent, we have $I_x = \emptyset$ according to Remark 2. Let us now construct the set of arcs of N incident with t.

– For every w_i, $i \in J$, we add the following arcs: $(s_i, t) \in F$ with $c(s_i, t) = [0, \infty)$ and $(t, s_i') \in F$.
   (The intuition is that a token in a place s_i means that A_i is in the control state s_i and the defined arcs represent the change of control states of the active automata; the age of the tokens is irrelevant here.)

- For every clock $x \in C$ with at least one g_i, $i \in J$, such that $x \in dom(g_i)$ we add the following arcs (the age of the token in place x represents the corresponding clock value):
  - $(x, t) \in F$ with $c(x, t) = I_x$ and $(t, x) \in F$ whenever $I_x \neq \emptyset$ and $x \in R$ (The clock x is used in a guard and reset afterwards.)
  - $(x, t) \in F*$ with $c*(x, t) = I_x$ whenever $I_x \neq \emptyset$ and $x \notin R$ (The clock x is used in a guard but not reset.)
  - $(e, t) \in F$ with $c(e, t) = [0, \infty)$ whenever $I_x = \emptyset$ (If the guards for x are inconsistent, t is disabled: the place e will never become marked.)
- For every clock $x \in R$ such that there is no g_i, $i \in J$, where $x \in dom(g_i)$ we add the arcs: $(x, t) \in F$ with $c(x, t) = [0, \infty)$ and $(t, x) \in F$. (The clock x is not used in any guard but it is reset.)

The construction is schematically depicted in Figure 1 using the standard Petri net notation (testing arcs are drawn by dashed arrows).

For every configuration $(s_1, \ldots, s_n, v)$ of A reachable from its initial state we define a corresponding marking $M_{(s_1, \ldots, s_n, v)}$ in N by

$$M_{(s_1, \ldots, s_n, v)}(p) \stackrel{\text{def}}{=} \begin{cases} \{v(delay_i)\} & \text{if } p = s_i \text{ for some } i, 1 \leq i \leq n \\ \{v(p)\} & \text{if } p \in C \\ \emptyset & \text{otherwise.} \end{cases}$$

The initial marking M_0 of N is then given by $M_0 \stackrel{\text{def}}{=} M_{(s_1^0, \ldots, s_n^0, v^0)}$. It is easy to see that (N, M_0) is a 1-safe net.

Theorem 1. *Every CTA can be transformed into 1-TAPN* preserving isomorphism of LTS.*

Proof. It can be verified that for any reachable configuration $(s_1, \ldots, s_n, v)$ of A whenever we have $(s_1, \ldots, s_n, v) \stackrel{a}{\longrightarrow} (s_1', \ldots, s_n', v')$ then also $M_{(s_1, \ldots, s_n, v)} \stackrel{a}{\longrightarrow} M_{(s_1', \ldots, s_n', v')}$ and vice versa (any transition from $M_{(s_1, \ldots, s_n, v)}$ corresponds to a transition from $(s_1, \ldots, s_n, v)$). The same holds also for time-elapsing transitions under the actions $\epsilon(r)$ where $r \in \mathbb{R}^+$. $\square$

Corollary 1. *Every mrCTA can be transformed into 1-TAPN preserving isomorphism of LTS.*

Proof. By inspecting the translation presented above we notice that no testing arcs are used, provided that the CTA A is with mandatory reset. $\square$

Remark 6. In general the translation from CTA to 1-TAPN* (and from mrCTA to 1-TAPN) produces more than an exponential blow up in the size of the 1-TAPN* (caused by the number of transitions; in general this number can be $k^{\Omega(width(\phi))}$ where $k \stackrel{\text{def}}{=} \max_{1 \leq i \leq n} | \longrightarrow_i |$, which is $k^{\Omega(n)}$ if ϕ is of the maximal

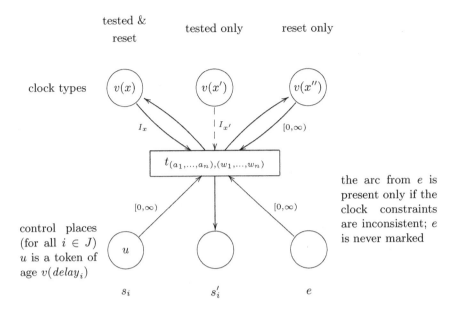

Fig. 1. Schematic construction of the net N

width n). However, considering synchronization functions of constant width (and this is often a common practice — e.g. UPPAAL uses only synchronization of width 2), the translation defines a 1-TAPN* of polynomial size with regard to the size of the CTA.

Corollary 2. *Every CTA resp. mrCTA where the width of the synchronization function is constant can be transformed in polynomial time into a 1-TAPN* resp. 1-TAPN (of polynomial size) preserving isomorphism of LTS.*

4   From Nets to Automata

In this section we present a structural translation from 1-TAPN* to CTA preserving isomorphism of LTS. Let $N = (P, T, F, c, Act, \lambda, F*, c*)$ be 1-TAPN* and M_0 its initial marking.

Remark 7. Without loss of generality assume that $P = \{p_1, \ldots, p_n, fired\}$ where $p_1, \ldots, p_n$ are the standard places of the net and *fired* is a newly added place together with the following arcs (for all $t \in T$): $(fired, t) \in F$ and $(t, fired) \in F$ such that $c(fired, t) = [0, \infty)$ and $M_0(fired) = \{0\}$. The intuition is that the place *fired* will always contain exactly one token (representing the time elapsed from the most recent transition firing). This token has no influence on the behaviour of the net.

We shall construct a CTA A (having n parallel components) such that $T(A)$ is isomorphic to $T(N, M_0)$. Let us define a set *Interv* of all time intervals[1] that appear as a time constraint in N, i.e., $Interv \stackrel{\text{def}}{=} range(c) \cup range(c*)$. For all i, $1 \leq i \leq n$, we define a timed automaton A_i representing a place of the net N and storing the information whether a token is present in the place p_i (control state in_i) or not (control state out_i), and what is its age (the age will be stored in a clock x_i). In this reduction every transition of the net N will correspond to a certain tuple in the synchronization function ϕ. Let $Act \stackrel{\text{def}}{=} \{(remove, I) \mid I \in Interv\} \cup \{(reset, I) \mid I \in Interv\} \cup \{(test, I) \mid I \in Interv\} \cup \{(add), (null)\}$ and let $C \stackrel{\text{def}}{=} \{x_1, \ldots, x_n\}$. We define $A_i \stackrel{\text{def}}{=} (S_i, Act, C, \longrightarrow_i, s_i^0)$ such that

- $S_i \stackrel{\text{def}}{=} \{in_i, out_i\}$,
- for all $I \in Interv$ we have the following transitions:

$$in_i \xrightarrow{g,(remove,I),\{x_i\}} out_i \qquad in_i \xrightarrow{g,(reset,I),\{x_i\}} in_i \qquad in_i \xrightarrow{g,(test,I),\emptyset} in_i$$

$$out_i \xrightarrow{g',(add),\{x_i\}} in_i \qquad in_i \xrightarrow{g',(null),\emptyset} in_i \qquad out_i \xrightarrow{g',(null),\{x_i\}} out_i$$

such that $g(x_i) \stackrel{\text{def}}{=} I$ and $g(x_j) \stackrel{\text{def}}{=} undef$ for all j, $1 \leq j \neq i \leq n$; and $g'(x_j) \stackrel{\text{def}}{=} undef$ for all j, $1 \leq j \leq n$,

- $s_i^0 \stackrel{\text{def}}{=} out_i$ if $M_0(p_i) = \emptyset$, and $s_i^0 \stackrel{\text{def}}{=} in_i$ otherwise.

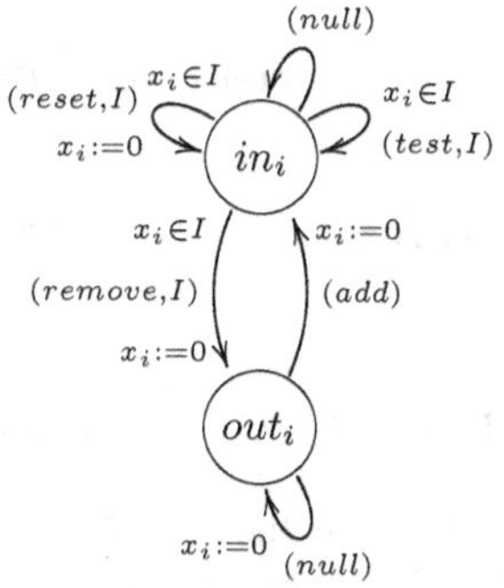

$$a_i \stackrel{\text{def}}{=} \begin{cases} (remove, c(p_i, t)) & \text{if } p_i \in {}^\bullet t \smallsetminus t^\bullet \\ (reset, c(p_i, t)) & \text{if } p_i \in {}^\bullet t \cap t^\bullet \\ (test, c*(p_i, t)) & \text{if } p_i \in {}^* t \\ (add) & \text{if } p_i \in t^\bullet \smallsetminus {}^\bullet t \\ (null) & \text{otherwise.} \end{cases}$$

Fig. 2. Timed automaton A_i and definition of the synchronization tuples

A picture of the timed automaton A_i is depicted in Figure 2. Let us now consider a CTA $A \stackrel{\text{def}}{=} (A_1 \| \cdots \| A_n)_\phi$ where for every transition $t \in T$ we define $\phi(a_1, \ldots, a_n) \stackrel{\text{def}}{=} \lambda(t)$ such that the tuple $(a_1, \ldots, a_n)$ is given in Figure 2. For the remaining tuples not defined in Figure 2, the partial function ϕ is undefined. Note that because N is a 1-safe net, we can freely assume that ${}^* t \cap t^\bullet = \emptyset$.

---

[1] For technical convenience, only one *Interv* set is defined. The construction can be further optimized by considering a separate $Interv_p$ (for every place p of the net) containing only the relevant intervals.

Remark 8. In case that a certain place is not involved in firing the transition t, we use the action $(null)$ in the synchronization tuple. The reason is again that we aim at proving isomorphism of the labelled transition systems: if a token is present in such a place, performing the action $(null)$ does not have any influence; if there is no token in the place then the action $(null)$ resets the corresponding clock x_i and hence we know that all the empty places have the clock x_i set to the age of the token in the Petri net place *fired* according to Remark 7. If we aimed at relating the transition systems up to e.g. timed bisimilarity, the $(null)$ actions can be replaced with •.

Theorem 2. *Every 1-TAPN* can be transformed in polynomial time into a CTA (of polynomial size) preserving isomorphism of LTS.*

Proof. Every marking M reachable in (N, M_0) defines a unique configuration $(s_1, \ldots, s_n, v)$ of the CTA A such that

- $s_i \stackrel{\text{def}}{=} in_i$ and $v(x_i) \stackrel{\text{def}}{=} r$ if $M(p_i) = \{r\}$
- $s_i \stackrel{\text{def}}{=} out_i$ and $v(x_i) \stackrel{\text{def}}{=} r$ if $M(p_i) = \emptyset$ and $M(\textit{fired}) = \{r\}$.

It is now easy to verify that every transition (including time elapsing) from M in the net can be matched by a transition under the same action from $(s_1, \ldots, s_n, v)$ such that the mapping defined above is preserved and vice versa. $\square$

Corollary 3. *Every 1-TAPN can be transformed in polynomial time into a mrCTA (of polynomial size) preserving isomorphism of LTS.*

Proof. Observe that the construction translates 1-TAPN into mandatory reset CTA. $\square$

Theorem 3. *Reachability for 1-TAPN* (and 1-TAPN) is PSPACE-complete.*

Proof. The hardness of the problem follows from PSPACE-hardness of the reachability problem for (untimed) 1-safe Petri nets [15]. The containment follows from Theorem 2 and from the fact that reachability is decidable in PSPACE for CTA [2]. $\square$

5   Reducing the Width of Synchronization Function

So far we have characterized the correspondence between the classes of automata and nets up to isomorphism. This has established a precise relationship w.r.t. to expressiveness, however, when transforming automata to nets, the reduction does not work in polynomial time. Moreover, when transforming nets to automata (in polynomial time), we use a synchronization function of width n, hence the reduction does not provide a direct way to verify problems for 1-TAPN* (and

1-TAPN) by means of UPPAAL, which allows for handshake synchronization (width 2) only. Even the possibility of broadcast communication does not seem to help in this case.

Most of the practical approaches to verification of timed systems focus on reachability. In this section we will hence describe a polynomial time reachability preserving reduction from CTA with arbitrary synchronization function into CTA with synchronization function of width 2.

Let $A = (A_1 \| \cdots \| A_n)_\phi$ be a CTA with an arbitrary synchronization function ϕ such that $A_i = (S_i, \mathcal{A}ct, C, \longrightarrow_i, s_i^0)$.

Remark 9. Without loss of generality we can assume that the CTA A contains a distinguished clock $delay \in C$ which is never used in any guard of any A_i but is always reset in all transition rules of every single automaton. This means that the clock $delay$ measures how much time has elapsed from the last occurrence of some transition labelled by an action from $\mathcal{A}ct$, but does not influence the behaviour of the concurrent automaton. We also assume that there are at least two active components in every synchronization tuple from $dom(\phi)$ (if not we can add an additional "dummy" component).

We will construct a CTA A' with a synchronization function ϕ' of width 2 such that reachability in A is reducible in polynomial time into reachability in A'.

First, we define timed automata A_i' (small modifications of A_i where new intermediate states are inserted for every transition in A_i). Let the new sets of actions and clocks be defined by $\mathcal{A}ct' \stackrel{\text{def}}{=} \mathcal{A}ct \cup \{!a@i, ?a@i \mid a \in \mathcal{A}ct, 1 \le i \le n\} \cup \{\tau\}$ and $C' \stackrel{\text{def}}{=} C \cup \{z\}$ where τ is a fresh action and z is a fresh clock. The intuition is that actions of the form $!a@i$ and $?a@i$ (or $!a_i@i$ and $?a_i@i$) where $a, a_i \in \mathcal{A}ct$ are designed for synchronization with the i'th component. For all i, $1 \le i \le n$, the automata A_i' are defined by

$$A_i' \stackrel{\text{def}}{=} (S_i \cup \{v(s_i, g_i, a_i, R_i, s_i') \mid (s_i, g_i, a_i, R_i, s_i') \in \longrightarrow_i\}, \mathcal{A}ct', C', \Longrightarrow_i, s_i^0)$$

such that $v(s_i, g_i, a_i, R_i, s_i')$ are newly added states and the transition relation $\Longrightarrow_i$ is given by:

- $s_i \stackrel{g_i, ?a_i@i, \emptyset}{\Longrightarrow}_i v(s_i, g_i, a_i, R_i, s_i')$ and
- $v(s_i, g_i, a_i, R_i, s_i') \stackrel{g', ?a_i@i, R_i}{\Longrightarrow}_i s_i'$ where $g'(x) \stackrel{\text{def}}{=} undef$ for all $x \in C'$

for all $(s_i, g_i, a_i, R_i, s_i') \in \longrightarrow_i$. The following picture illustrates the transformation.

For a given tuple $(a_1, \ldots, a_n) \in (\mathcal{A}ct \cup \{\bullet\})^n$ let $J(a_1, \ldots, a_n) \stackrel{\text{def}}{=} \{i \in \{1, \ldots, n\} \mid a_i \neq \bullet\}$ be the set of active components. We define a new parallel component of A', a timed automaton A'_{n+1} given by

$$A'_{n+1} \stackrel{\text{def}}{=} (S'_{n+1}, \mathcal{A}ct', C', \Longrightarrow_{n+1}, init)$$

where $S'_{n+1} \stackrel{\text{def}}{=} \{ test(a_1, \ldots, a_n, i), \ reset(a_1, \ldots, a_n, i) \mid (a_1, \ldots, a_n) \in dom(\phi),$ $i \in J(a_1, \ldots, a_n)\} \cup \{init\}$ with the following transitions for all $(a_1, \ldots, a_n) \in dom(\phi)$ (assume that $J(a_1, \ldots, a_n) = \{i_1, \ldots, i_m\}$ such that $i_1 < i_2 < \cdots < i_m$):

- $init \stackrel{g', a, \{z\}}{\Longrightarrow}_{n+1} test(a_1, \ldots, a_n, i_1)$ where $a = \phi(a_1, \ldots, a_n)$
- $test(a_1, \ldots, a_n, i_\ell) \stackrel{g_z, !a_{i_\ell}@i_\ell, \emptyset}{\Longrightarrow}_{n+1} test(a_1, \ldots, a_n, i_{\ell+1})$ for all ℓ, $1 \leq \ell < m$
- $test(a_1, \ldots, a_n, i_m) \stackrel{g_z, !a_{i_m}@i_m, \emptyset}{\Longrightarrow}_{n+1} reset(a_1, \ldots, a_n, i_1)$
- $reset(a_1, \ldots, a_n, i_\ell) \stackrel{g_z, !a_{i_\ell}@i_\ell, \emptyset}{\Longrightarrow}_{n+1} reset(a_1, \ldots, a_n, i_{\ell+1})$ for all ℓ, $1 \leq \ell < m$
- $reset(a_1, \ldots, a_n, i_m) \stackrel{g_z, !a_{i_m}@i_m, \emptyset}{\Longrightarrow}_{n+1} init$

where g' is the empty guard ($g'(x) \stackrel{\text{def}}{=} undef$ for all $x \in C'$), and $g_z(z) \stackrel{\text{def}}{=} [0, 0]$ and $g_z(x) \stackrel{\text{def}}{=} undef$ for all $x \in C' \smallsetminus \{z\}$. It first decides which tuple from the domain of ϕ is going to be synchronized upon and then performs twice the sequence of actions $!a_{i_1}@i_1, \ldots, !a_{i_m}@i_m$ and becomes $init$ again. The intuition is that the action $!a_{i_\ell}@i_\ell$ can synchronize only with the corresponding action $?a_{i_\ell}@i_\ell$ in the component i_ℓ. Because of the guard g_z no time elapsing steps are possible during the pairwise synchronization, otherwise the whole system gets stuck (and hence cannot reach the requested configuration). During the first sequence of actions $!a_{i_1}@i_1, \ldots, !a_{i_m}@i_m$ the guards of the active components are consecutively verified and in the second round the requested clocks are reset. (Note that it is not possible to do both guard verification and resetting of clocks in one run as clocks reset in some earlier synchronized components can still be used in guards later on.) A graphical representation of the automaton A_{n+1} is depicted in Figure 3 (only one loop for a particular $(a_1, \ldots, a_n) \in dom(\phi)$ is included). The synchronization function ϕ' is defined exactly for the following tuples:

- $\phi'(\bullet, \ldots, \bullet, a) \stackrel{\text{def}}{=} a$ for all $a \in \mathcal{A}ct$
  (only the last component can make moves without synchronizing with other components; in the initial state $init$ it selects an element $(a_1, \ldots, a_n)$ from $dom(\phi)$ to be used)
- $\phi'(\bullet, \ldots, \bullet, ?a@i, \bullet, \ldots, \bullet, !a@i) \stackrel{\text{def}}{=} \tau$ for all $a \in \mathcal{A}ct$ and $1 \leq i \leq n$ such that $?a@i$ is at the i'th coordinate
  (the last component can handshake with the i'th component).

Obviously, ϕ' is of width 2 as required. Let us define $A' \stackrel{\text{def}}{=} (A'_1 \| \cdots \| A'_n \| A'_{n+1})_{\phi'}$.

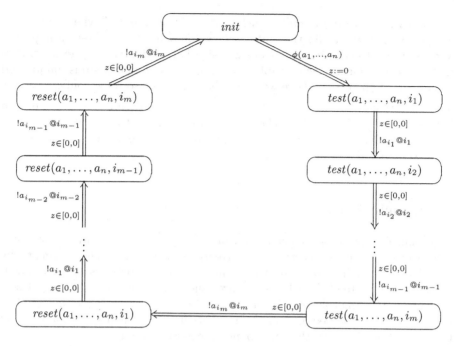

Fig. 3. Automaton A_{n+1}

Theorem 4. *Reachability for CTA with an arbitrary synchronization function is reducible in polynomial time (and space) into reachability for CTA with synchronization function of width 2. The number of parallel components and of clocks is increased by one.*

Proof. Let $c = (s_1, \ldots, s_n, v)$ be a reachable configuration in A. A corresponding configuration $f(c)$ in A' is defined by $f(c) \stackrel{\text{def}}{=} (s_1, \ldots, s_n, init, v')$ where $v'(x) \stackrel{\text{def}}{=} v(x)$ for all $x \in C$ and $v'(z) \stackrel{\text{def}}{=} v(delay)$. Let $c^0 = (s_1^0, \ldots, s_n^0, v^0)$ be the initial configuration of A. We claim that a given configuration c is reachable (in A) from c^0 if and only if $f(c)$ is reachable (in A') from $f(c^0)$. The claim follows from the following observation.

Let $c = (s_1, \ldots, s_n, v)$ be an arbitrary configuration of A. It is easy to see that whenever $c \xrightarrow{a} c'$ for some $c' = (s_1', \ldots, s_n', v')$ by using a tuple $(a_1, \ldots, a_n)$ from $dom(\phi)$ then also $f(c) \xrightarrow{a} \circ (\xrightarrow{\tau})^{2m} f(c')$ where m is the number of active components in $(a_1, \ldots, a_n)$. In A' we can first perform the transition $init \stackrel{g', a, \{z\}}{\Longrightarrow} test(a_1, \ldots, a_n, i_1)$ where i_1 is the first active component. This transition is followed by a unique continuation according to the path in Figure 3, all guards (of the corresponding component automata) during the first part of the path are satisfied by our assumption that $c \xrightarrow{a} c'$ in A. The corresponding clocks are reset during the second part of the path, until $f(c')$ is finally reached. On the other hand, if $f(c)$ performs a sequence of moves starting with selecting a tuple $(a_1, \ldots, a_n)$ from $dom(\phi)$ such that we finally reach c'' where in the last component appears again the state $init$, we know that no time-elapsing steps were performed because of the clock z which is tested to zero by every transition. Moreover, for a selected tuple $(a_1, \ldots, a_n)$ this computation is unique and by similar arguments as above we know that there is also some configuration c' in A such that $c \overset{\phi(a_1, \ldots, a_n)}{\longrightarrow} c'$ and $f(c') = c''$. Similarly, time elapsing steps on both sides can be directly matched in the other automaton. As it can be seen from the construction, only one extra parallel component and one additional clock were introduced.   □

6   Conclusion

The main results of the paper are outlined in Figure 4. We have identified a naturally motivated subclass of concurrent (networks of) timed automata by introducing the mandatory reset feature and a natural superclass of 1-safe timed-arc Petri nets extended with testing arcs such that the corresponding classes coincide up to isomorphism of labelled transition systems. We have then studied more practically oriented questions of reachability and demonstrated polynomial time reductions between the corresponding classes in Figure 4.

Our study justifies that it is interesting to investigate the extension of timed-arc Petri nets with testing arcs. This feature is present in the untimed Petri net model (using a standard trick) but it is missing when time features are added. We claim that all the classes CTA, mrCTA, 1-TAPN and 1-TAPN* are polynomially equivalent w.r.t. reachability. The answer to the question of polynomial equivalence w.r.t. reachability for (unbounded) TAPN and TAPN* seems to be also positive, even though more involved techniques need to be used to establish the reduction. In the future work we plan to consider TAPN extended with urgency and see how they compare to networks of timed automata with invariants. We will also investigate which TCTL properties are preserved by the polynomial time reductions presented in this paper and what is the relationship between TAPN and Merlin's time Petri nets.

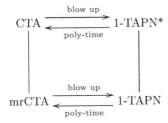

arbitrary ϕ: up to isomorphism

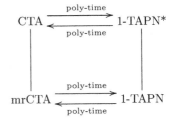

constant width ϕ: up to isomorphism
arbitrary ϕ: up to reachability

Fig. 4. Summary of results

Finally, let us draw our attention to an interesting observation. Already for concurrent finite automata (untimed), the reachability problem is PSPACE-complete [18] and adding time features (CTA) does not increase the theoretical complexity of the problem [2]. The results of this paper enable to draw a similar conclusion also for the case of 1-safe Petri nets: the reachability problem for untimed 1-safe Petri nets (a formalism which can model concurrency) is PSPACE-complete [15] and adding time features does not increase its complexity.

Acknowledgments. I would like to thank Luca Aceto and Ole H. Jensen for reading a preliminary draft of this paper, and the anonymous referees for their comments and suggestions.

References

[1] P.A. Abdulla and A. Nyln. Timed Petri nets and BQOs. In *Proc. of the 22nd International Conference on Application and Theory of Petri Nets (ICATPN'01)*, volume 2075 of *LNCS*, pages 53–70. Springer-Verlag, 2001.

[2] L. Aceto and F. Laroussinie. Is your model checker on time? On the complexity of model checking for timed modal logics. *Journal of Logic and Algebraic Programming*, 52–53:7–51, 2002.

[3] R. Alur and D. Dill. A theory of timed automata. *Theoretical Computer Science*, 126(2):183–235, 1994.

[4] A. Arnold. *Finite Transition Systems*. Prentice-Hall, 1994.

[5] B. Berard, A. Petit, V. Diekert, and P. Gastin. Characterization of the expressive power of silent transitions in timed automata. *Fundamenta Informaticae*, 36(2-3):145–182, 1998.

[6] B. Berthomieu, P-O. Ribet, and F. Vernadat. The tool tina - construction of abstract state spaces for petri nets and time petri nets. *International Journal of Production Research*, 2004. To appear.

[7] T. Bolognesi, F. Lucidi, and S. Trigila. From timed Petri nets to timed LOTOS. In *Proc. of the IFIP WG 6.1 Tenth International Symposium on Protocol Specification, Testing and Verification (Ottawa 1990)*, pages 1–14. 1990.

[8] S. Bornot, J. Sifakis, and S. Tripakis. Modeling urgency in timed systems. In *International Symposium: Compositionality - The Significant Difference, Malente (Holstein, Germany)*, volume 1536 of *LNCS*, 1998.

[9] P. Bouyer. Untameable timed automata! In *Proc. of the 20th Annual Symposium on Theoretical Aspects of Computer Science (STACS'03)*, volume 2607 of *LNCS*, pages 620–631. Springer-Verlag, 2003.

[10] Fred D.J. Bowden. Modelling time in Petri nets. In *Proc. of the Second Australia-Japan Workshop on Stochastic Models*, 1996. http://www.itr.unisa.edu.au/~fbowden/pprs/stomod96/.

[11] M. Boyer and M. Diaz. Non equivalence between time Petri nets and time stream Petri nets. In *Proc. of the 8th International Workshop on Petri Nets and Performance Models (PNPM'99)*, pages 198–207. IEEE Computer Society, 1999.

[12] M. Bozga, C. Daws, O. Maler, A. Olivero, S. Tripakis, and S. Yovine. Kronos: A model-checking tool for real-time systems. In *Proc. of the 10th International Conference on Computer-Aided Verification (CAV'98)*, volume 1427 of *LNCS*, pages 546–550. Springer–Verlag, 1998.

[13] N. Busi and G.M. Pinna. Process semantics for place/transition nets with inhibitor and read arcs. *Fundamenta Informaticae*, 40(2–3):165–197, 1999.

[14] F. Cassez and O.H. Roux. Structural translation of time petri nets into timed automata. In *Workshop on Automated Verification of Critical Systems (AVoCS'04)*, ENTCS. Elsevier, 2004.

[15] A. Cheng, J. Esparza, and J. Palsberg. Complexity results for 1-safe nets. *Theoretical Computer Science*, 147(1-2):117–136, 1995.

[16] S. Haar, F. Simonot-Lion, L. Kaiser, and J. Toussaint. Equivalence of timed state machines and safe time Petri nets. In *Proc. of the 6th International Workshop on Discrete Event Systems (WODES'02)*, pages 119–126, 2002.

[17] H.M. Hanisch. Analysis of place/transition nets with timed-arcs and its application to batch process control. In *Proc. of the 14th International Conference on Application and Theory of Petri Nets (ICATPN'93)*, volume 691 of *LNCS*, pages 282–299, 1993.

[18] D. Kozen. Lower bounds for natural proof systems. In *Proc. of the 18th Annual Symposium on Foundations of Computer Science*, pages 254–266. IEEE, 1977.

[19] F. Laroussinie and K.G. Larsen. CMC: A tool for compositional model-checking of real-time systems. In *Proc. of the FIP TC6 WG6.1 Joint International Conference on Formal Description Techniques for Distributed Systems and Communication Protocols (FORTE XI) and Protocol Specification, Testing and Verification (PSTV XVIII)*, pages 439–456. Kluwer, B.V., 1998.

[20] F. Laroussinie, N. Markey, and Ph. Schnoebelen. On model checking durational Kripke structures. In *Proc. of the 5th International Conference on Foundations of Software Science and Computation Structures (FOSSACS'02)*, volume 2303 of *LNCS*, pages 264–279. Springer-Verlag, 2002.

[21] K.G. Larsen, P. Pettersson, and W. Yi. UPPAAL in a Nutshell. *International Journal on Software Tools for Technology Transfer*, 1(1–2):134–152, 1997.

[22] D. Lime and O.H. Roux. State class timed automaton of a time Petri net. In *Proc. of the 10th International Workshop on Petri Net and Performance Models (PNPM'03)*, pages 124–133, 2003.

[23] M. Nielsen, V. Sassone, and J. Srba. Properties of distributed timed-arc Petri nets. In *Proc. of the 21st International Conference on Foundations of Software Technology and Theoretical Computer Science (FSTTCS'01)*, volume 2245 of *LNCS*, pages 280–291. Springer-Verlag, 2001.

[24] O. Roux, D. Lime, and G. Gardey. software studio for time Petri net analysis. http://www.irccyn.ec-nantes.fr/irccyn/d/en/equipes/TempsReel/logs/software-2-romeo.

[25] V. Valero Ruiz, D. de Frutos Escrig, and O. Marroquin Alonso. Decidability of properties of timed-arc Petri nets. In *Proc. of the 21st International Conference on Application and Theory of Petri Nets (ICATPN'00)*, volume 1825 of *LNCS*, pages 187–206. Springer-Verlag, 2000.

[26] V. Valero Ruiz, F. Cuartero Gomez, and D. de Frutos Escrig. On non-decidability of reachability for timed-arc Petri nets. In *Proc. of the 8th International Workshop on Petri Net and Performance Models (PNPM'99)*, pages 188–196, 1999.

[27] J. Sifakis and S. Yovine. Compositional specification of timed systems. In *Proc. of the 13th Annual Symposim on Theoretical Aspects of Computer Science (STACS'96)*, volume 1046 of *LNCS*, pages 347–359. Springer-Verlag, 1996.

[28] W. Vogler. Partial order semantics and read arcs. *Theoretical Computer Science*, 286(1):33–63, 2002.

[29] J. Wang. *Timed Petri Nets, Theory and Application*. Kluwer Academic Publishers, 1998.

Specifying and Analyzing Software Safety Requirements of a Frequency Converter Using Coloured Petri Nets

Lisa Wells[1] and Thomas Maier[2]

[1] Department of Computer Science, University of Aarhus,
IT-Parken, Aabogade 34, DK-8200 Aarhus N, Denmark
wells@daimi.au.dk
[2] Danfoss Drives A/S, Ulsnaes 1, DK-6300 Graasten, Denmark
tm@danfoss.com

Abstract. Safety-critical systems are systems that can cause undesired loss or damage to life, property, or the environment. Standards for developing safety-critical software often recommend that semi-formal or formal methods should be used to specify, analyze, and verify the behavior of safety-critical software. This paper presents results from a project in which Coloured Petri Nets were used to specify and analyze software safety requirements of a frequency converter being developed by Danfoss Drives. Frequency converters are used to control the speed of motors. The analysis of the model revealed behavior which could lead to hazardous situations or unnecessary failures. Prototype tool support was developed for validating the behavior of an Java-based executable software architecture prototype against the CP-net that specified the desired behavior of the software.

1   Introduction

Safety-critical systems are systems that can cause undesired loss or damage to life, property, or the environment, and safety-critical software is any software that can contribute to such loss or damage [1]. Since such systems have the potential to cause extensive damage, there are many standards and guidelines describing processes, techniques, and methods for developing safety-critical systems. The IEC 61508 [2] is one such standard for achieving functional safety of programmable electronic safety-related systems. The standard contains recommendations regarding which techniques and measures should be used throughout the entire life-cycle of a product (including both hardware and software) – from initial safety requirement specification, to design, realization, test, installation, commissioning, operation, maintenance and decommissioning.

Danfoss Drives produces frequency converters which are used to control the speed of motors, e.g. for elevators, cranes, and conveyor belts. Frequency converters are safety-critical in that their behavior can lead to hazardous situations, e.g., if an elevator continues to run even though the emergency stop button has

G. Ciardo and P. Darondeau (Eds.): ICATPN 2005, LNCS 3536, pp. 403–422, 2005.

been pushed, or if the speed of a conveyor belt is not kept below a given threshold during a routine maintenance check. In order to meet the needs of their customers, Danfoss Drives is developing a frequency converter that is to be evaluated and certified by international safety certification authorities. As part of the certification process, Danfoss will have to conform to the development process described in IEC 61508.

IEC 61508 recommends that semi-formal methods should be used in order to avoid mistakes during the specification of requirements, and to express and structure the requirements such that they are "clear, precise, unambiguous, verifiable, testable, maintainable and feasible". In order to comply with this recommendation, Danfoss used one of the recommended semi-formal methods, namely finite state machines in the form of a Statechart [3] model, to make a somewhat informal specification of system-level safety requirements in the initial project proposal that has been approved by the safety certification authorities. The Statechart was informal in that it was drawn in a generic drawing tool rather than in a CASE tool supporting Statecharts. By creating the Statechart in this way, Danfoss could not take advantage of any of the benefits that a CASE tool offers, such as automatic syntax checking, or static and dynamic verification that can identify safety-related problems such as deadlock or unreachable states.

Danfoss was interested in investigating the advantages of using semi-formal methods and associated CASE tools for the specification of safety requirements. This is one of the problems that was addressed in a collaborative research project between Danfoss Drives, ISIS Katrinebjerg, the Department of Computer Science at the University of Aarhus, and Systematic Software Engineering. As part of this project, software safety requirements, which are more detailed than system-level safety requirements, were specified in two different formalisms, namely Statecharts and Coloured Petri Nets (CPNs or CP-nets) [4, 5]. The goal of this work was to create a software safety requirement specification that fulfilled the requirements of IEC 61508 and, more importantly, that was thorough, detailed, understandable, and suitable as a basis for further software development steps.

This paper presents the CPN model that specifies the software safety requirements for a frequency converter. Analysis of the model identified potentially hazardous behavior in the frequency converter, and this hazardous behavior has been flagged for additional analysis during later development phases at Danfoss. The paper also presents a technique and prototype tool support for validating the behavior of a Java program against a CPN-based requirement specification.

The paper is structured as follows. Section 2 describes the hardware and software of frequency converters. It also presents the original system safety requirements that were to be refined into software safety requirements. Section 3 presents the CPN model of the software safety requirements. Section 4 discusses the results of the analysis of the CPN model and the technique for validating a Java program against the CPN-based executable requirement specification. Section 5 discusses related work.

2   Frequency Converter

Danfoss Drives is currently developing a frequency converter with integrated safety functions, where the safety functions are directly used to control and stop the attached motor. Such frequency converters are increasingly applied in the process industry and production, where they replace classical electromechanical safety devices such as power relays. An advantage with safety-related frequency converters is that a higher up-time of processes and installations can be achieved. Section 2.1 describes the hardware in a frequency converter. Section 2.2 describes the safety-related software of a frequency converter. Section 2.3 describes the system safety requirements that were defined before the project started. Section 2.4 describes how the system safety requirements were to be refined into software safety requirements.

2.1   Hardware

The hardware structure of the frequency converter is shown in Fig. 1. The two blocks *PWM Generator* and *Power Electronics* make up the normal, "non safety-related" part of a frequency converter. These two blocks convert the main power supply to the desired output frequency, and thereby control the speed of the attached motor.

The safety functionality is achieved by an additional subsystem on the *Safe Board* composed of *Channels 1* and *2*, each consisting of a microprocessor (*uP*), a *Switch off* path, a number of *Digital Inputs*, and one *Speed Information* input. The two microprocessors can, independently from each other, activate its own switch-off path to stop the rotational torque in the motor. The two *Channels* cross-monitor each other through *Feedbacks 1* and *2* and through the *Cross Com-*

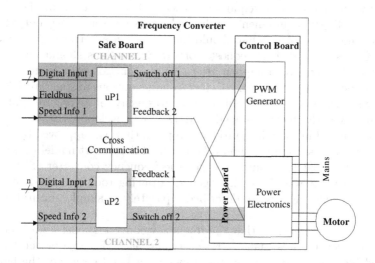

Fig. 1. Hardware structure of a frequency converter with safety functions

munication connection. Channel 1 is also connected to a PROFIbus® fieldbus with fail-safe properties (PROFIsafe®). The two *Channels* are physically diverse to provide a maximal protection against common cause failures by factors such as electromagnetic fields.

2.2   Software

A number of so-called *designated safety functions* (DSF, or safety function) can be realized on the basis of the physical capabilities of the frequency converter. The logic of these safety functions is implemented in software that runs on the two microprocessors on the *Safe Board*. A specific safety function is activated upon reception of signals either at digital inputs at each of the *Channels*, or via the fieldbus connection at *Channel 1*.

The simplest safety function is a so-called 'uncontrolled stop' which immediately stops torque generation in the motor. Another safety function is a 'controlled stop', where the stop of torque generation is delayed, allowing the non-safety-related part of the frequency converter to ramp the motor down in a controlled way. A more complex example is the 'safe speed' where an uncontrolled stop is made if the motor speed exceeds a set limit.

All diagnostic functionality with respect to cross monitoring and self monitoring of the *Channels* is implemented in software. On detection of a dangerous failure, an appropriate fault reaction is initiated, and the motor is stopped.

2.3   System Safety Requirements

The software that runs on the two microprocessors on the *Safe Board* is safety-critical since it can contribute to loss or damage to the environment of the frequency converter through its affect on the speed and control of the attached motor. System-level safety requirements were already defined at the outset of the project. These requirements addressed issues such as, when output to the motor should be enabled, what should happen when an error occurs (either in hardware or software), how requests for safety functions should be made and handled, and what should happen after a safety function completes.

As mentioned previously, one of the recommendations of standard IEC 61508 is that semi-formal methods should be used to specify safety requirements. In order to comply with this recommendation, Danfoss developed a Statechart model that was included in the initial product proposal that was approved by the certification authorities. Figure 2 shows the informal Statechart model that specifies the system-level safety requirements for the frequency converter. The model is informal in that it was drawn in a generic drawing tool, and the states, transitions, and event triggers are described separately in simple, natural-language texts.

The Statechart specifies that the frequency converter must always be in one of three top-level states, namely No dangerous failure, Fail-safe or the Final state (denoted by a dot in a circle in the upper right-hand corner of the figure). If any kind of error occurs, then the frequency converter must enter Fail-safe state,

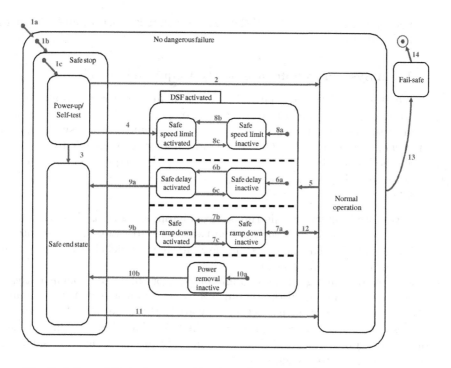

Fig. 2. Informal Statechart specification of system-level safety requirements

and the output to the motor must be disabled, i.e. the motor must be stopped. Examples of errors are short circuits in the hardware, exceeding the delay period when the 'controlled stop' DSF is active, or if the external temperature is too high. The only way to leave Fail-safe state is to turn the frequency converter off (Transition 14), and thereby enter Final state.

If no errors have occurred, then the frequency converter must be in No dangerous failure state, and more specifically, in one of its three composed states: Normal operation, DSF activated or Safe stop. In Safe stop state, output to the motor is always disabled. After turning the power on (Transitions 1a, 1b, 1,c), the frequency converter must be in Power-up/Self test state in which the frequency converter is initialized. If initialization is successful, and if there are no requests for safety functions, then the frequency converter will enter Normal operation state via Transition 2. If a request for a DSF exists at start up, then Safe end state may be entered via Transition 3. Normal operation can only be reached from Safe end (via Transition 11) after all requests for DSFs have been removed, and after a user has confirmed that all DSF requests have been removed. If a safety function is requested during Normal operation, then an appropriate substate of DSF activated is entered, i.e. Safe speed limit activated, Safe delay activated, Safe ramp ramp down activated, or Power removal inactive, depending on which DSF is requested. It is possible to request several different DSFs simultaneously, in which case more than one of these states may be active at the same time. When one of the DSFs completes (Transition 9a, 9b or 10b), then Safe end state is reached, and output to the motor is disabled.

2.4   From System to Software Safety Requirements

Using the system safety requirements already defined at the beginning of the project, one of the goals of the project was to specify software safety requirements. The software safety requirements were a refinement of the system safety requirements. Again, the IEC 61508 standard highly recommended that semi-formal methods should be used to define software safety requirements. Examples of the software safety requirements that needed to be clarified and formalized are as follows:

- Safety-related software would run on two microprocessors, as shown in Fig. 1.
- The software running on each of the two microprocessors must essentially fulfill the requirements specified in the Statechart model shown in Fig. 2, in addition to other requirements.
- The software running on the two microprocessors must run concurrently.
- The software running on the two microprocessors must regularly communicate and synchronize to ensure that they are in the same state. For example, requests for a DSF must be present at the redundant digital inputs at each of the microprocessors, and the redundant software must enable/disable the output to the motor at more or less the same time.
- If the internal states of the software on the two microprocessors is significantly different for a significant period of time, then an error has occurred.
- Diagnostics must be run regularly to ensure that hardware failures, such as short circuits, have not occurred. The two microprocessors must coordinate to run diagnostics, and the results of the diagnostics must be compared.

Timed Petri nets are listed as one of the recommended semi-formal methods for specifying software safety requirements. Given that the software safety requirements needed to specify behavior that included concurrency, synchronization, states, and discrete state changes, (Coloured) Petri Nets was an obvious choice as one of the specification languages.

3   Specifying Software Safety Requirements with CPN

This section describes the hierarchical CPN model that specifies software safety requirements for the safe frequency converter. All of the requirements that are specified in the Statechart model from Fig. 2 are included in the CPN model. Those requirements have been specified more formally, and the specification is much more detailed. In addition, the CPN model specifies requirements that are not addressed in the Statechart model, such as diagnostics and synchronization of the state of the software on the two microprocessors. Section 3.1 provides an overview of the model. Section 3.2 describes the specification of the software requirements for the safe board in the frequency converter.

3.1   Model Overview

Figure 3 provides an overview of the model. The model was created in CPN Tools [6], and the model consists of 28 modules (also called *pages* in CPN terminology). Each node in the figure corresponds to one page in the model. An arc from one

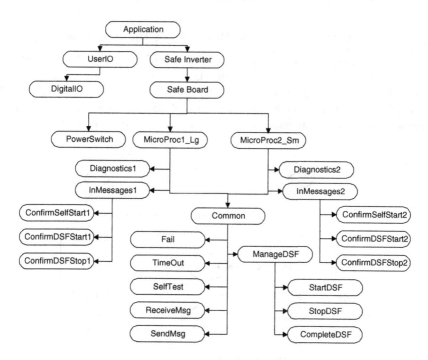

Fig. 3. Module hierarchy of the CPN model

node to another indicates that the page represented by the source node contains a so-called *substitution transition* whose detailed behavior is described on the page of the destination node. The page represented by the destination node is called a *subpage* of the page with the corresponding substitution transition. The model contains 75 places and 31 regular transitions, i.e. transitions that are not substitution transitions. Of the 75 places, 14 places are unique, and the 61 remaining places are so-called *port places* which conceptually glue places on different pages together.

The Application page (at the top of Fig. 3) is the most abstract representation of the frequency converter and its environment. This page has two subpages, namely User IO and Safe Inverter, modeling the means for user input/output, i.e. the digital inputs (page Digital IO) described in Fig. 1, and the frequency converter itself, respectively. Due to time constraints, neither the redundant speed information from the motor nor the ProfiSAFE bus from Fig. 1 were modeled. The use of the ProfiSAFE bus and redundant speed information is, however, optional for users, which made the absence of these components from the model acceptable within the context of this project. Obviously these components would have to be included in a complete requirement specification for the frequency converter. The Safe Inverter consists only of the Safe Board – the Power and Control boards from Fig.1 are also not modeled. The Safe Board and its two microprocessors MicroProc1_Lg and MicroProc2_Sm are described in detail below.

3.2   Specifying Safety-Critical Behavior in the Safe Board Module

The most abstract representation of the safe board from Fig. 1 is modeled on the Safe Board page which is shown in Fig. 4. This page has three subpages, namely the pages PowerSwitch, MicroProc1_Lg and MicroProc2_Sm as indicated by the small tags at the lower left-hand corner of the corresponding substitution transitions. The place InternalIO represents the internal communication channels

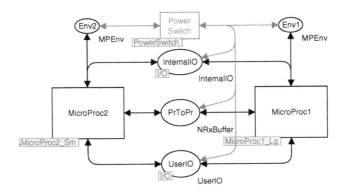

Fig. 4. The SafeBoard page

between the safe board and the power and control boards, i.e. the switch off and feedback channels in Fig.1. The User IO place represents the values of the digital inputs. The PrToPr place represents the processor-to-processor communication channel, i.e. the cross-communication channel in Fig.1. In the following these three places will be collectively referred to as the *communication* places. Tokens on the communication places represent either messages that are being passed between two entities, or the voltage (24v (high), 0v (low), or error) of the corresponding digital circuits.

The internal state and environment of the software on the two microprocessors is modeled by tokens on the places Env1 and Env2 in Fig. 4. Figure 5 shows an example of the tokens that can be found on an Environment place. Each microprocessor has an ID (1 or 2). A microprocessor has a State, such as *Ready, Sending, Waiting* or *Failed*. A number of parameters for the frequency converter can be set by users, such as whether the frequency converter can automatically start after successful initialization (AutoRelease(true)), whether it is necessary to confirm the removal of a request for a safety function (ConfirmRemoval(false)), and

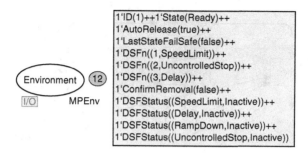

Fig. 5. Examples of tokens representing the software state on microprocessor 1

Fig. 6. The Fail page

which safety function is mapped to each of the n=3 digital inputs, for example, DSFn(2,UncontrolledStop) indicates that the uncontrolled stop safety function is mapped to digital input number 2. Furthermore, there are tokens indicating the status of each of four safety functions, e.g. DSFStatus(SpeedLimit,Inactive) indicates that the speed limiting safety function is currently *inactive*. Whenever an event takes place in a microprocessor one or more tokens are removed and added to the corresponding Environment place.

The use of a single place to model the state of the software on a microprocessor is necessitated by Statechart Transition 13 from No dangerous failure to Fail safe in Fig.2. Transition 13 indicates that a dangerous failure can occur at any time and in any state, and it must always be possible to reach the Fail safe state. If the internal state of a microprocessor was represented by tokens on a number of places, then it would have been difficult to create an understandable CPN that modeled this functionality since extra transitions would have to be connected to each of these places in order to remove or update the appropriate tokens when a dangerous failure is detected. Using one place to represent the internal state of a microprocessor does not reduce the concurrency in the model. It does, however, make it very easy to model the fact that something similar to Fail-safe state can be reached from virtually any state in the CP-net. Figure 6 shows the Fail page that models the detection of dangerous failures. Two kinds of failures can occur: dangerous failures that are detected and handled properly (transition Handle Failure), and sudden failures that occur spontaneously and which cannot be handled properly (transition Sudden Failure).

The MicroProc1_Lg page, shown in Fig. 7, is the most abstract representation of the software to be run on microprocessor 1. A similar page models microprocessor

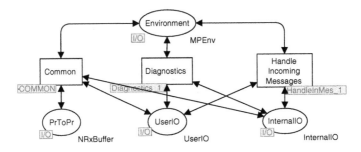

Fig. 7. The MicroProc1_Lg page

2. In this page, the Environment port place is conceptually glued to the Env1 place on the SafeBoard page shown in Fig. 4. The two microprocessors have the same functionality for some tasks, as modeled by the COMMON page and its subpages, but the functionality of the two microprocessors differs for Diagnostics and Handle Incoming Messages. When it comes to diagnostics, microprocessor 1 will act as a master and initiate all diagnostic routines, and microprocessor 2 will act as a slave and will only run diagnostics after a request has been made by microprocessor 1. The two microprocessors will also send different kinds of messages in different situations, such as during initialization and diagnostics, and will therefore have to handle different kinds of incoming messages.

The functionality that is common for the two microprocessors is shown in Fig. 8 which shows the COMMON page. Each microprocessor can TimeOut if a response to a message is not received in time. Output to the motor can be enabled or disabled (Enable PWM) when appropriate. Safety functions can be started when they are requested, they can be stopped when a request is removed, and they can complete (ManageDSF and its subpages). A SelfTest is run once immediately after the frequency converter is turned on. After most of the aforementioned events occur, a message must be sent (SendMsg) to the other microprocessor to indicate that the internal state has changed. Messages that are received (RcvMsg) are either acknowledgements or messages indicating that the state has changed in the other microprocessor. When the state has changed in the other microprocessor an acknowledgement must be sent if the current microprocessor is (or will soon be) in the same state, otherwise an error has occurred, an appropriate message must be sent back, and the frequency converter must enter a fail-safe state in which output to the motor is disabled (transition Handle Failure in Fig. 6).

In this model, regular transitions can only be found on pages that do not have subpages, i.e. on the pages represented by nodes that do not have outgoing arcs in Fig. 3. Furthermore, each page with regular transitions contains at most 3 transitions, and for most pages it is rare that there is more than one enabled transition on the page at any one time. Almost every transition in the model updates tokens on an Environment place for one of the microprocessors, and many transitions also add or remove tokens on the communication places, as can be seen by the arcs connected to these places in Figs. 7 and 8.

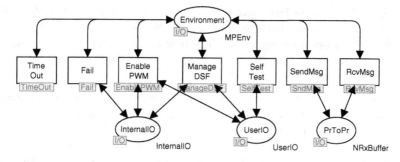

Fig. 8. The COMMON page for common functionality for the two microprocessors

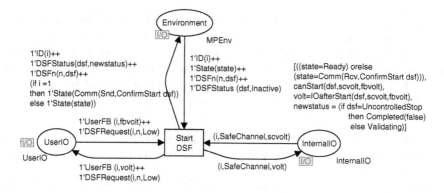

Fig. 9. The StartDSF page for starting a safety function

A fairly typical example of a page with a regular transition is the StartDSF page as shown in Fig. 9. The arc and guard expressions are quite detailed in order to make explicit what conditions must be fulfilled in order for the corresponding event to occur. For example the voltage on digital input number n (n=1,2, or 3) for microprocessor i must be low (as indicated by DSFRequest(i,n,Low) on the arc from place User IO to transition Start DSF), digital input number n must be mapped to the safety function *dsf*, and the safety function *dsf* must be inactive (as indicated by DSFn(n,dsf) and DSFStatus(dsf,Inactive) on the arc from place Environment to transition Start DSF). A number of functions are used in the guard to ensure that additional conditions are fulfilled. Given the complexity of the inscriptions and the fact that each transition is connected to an Environment place and one or more communication places, the model would have been quite illegible if pages contained more than one or two transitions.

4   Analysis and System Validation

This section discusses the analysis of the CPN model. Section 4.1 discusses the results of the simulations that were run and the experiences we had discussing the CPN model with project team members who were unfamiliar with CP-nets. Section 4.2 comments on state space analysis. Section 4.3 presents the tool for validating a Java program against a CPN-based requirement specification.

4.1 Simulation

Simulations of the model were run for three main purposes: for debugging the model, for analyzing the behavior of the model, and for discussing the software requirement specification with the project team. The model was developed iteratively whereby a CPN expert added increasing detail to the model, and problems and ambiguities regarding expected behavior were discussed with a domain expert from Danfoss. For these discussions it was rarely necessary to examine the CP-net directly.

Simulations were also run in order to investigate the specified behavior of the frequency converter. The goal of running the simulations was to identify potentially hazardous behavior, such as deadlocks or situations in which failures occurred but were undetected. Even though an exhaustive investigation of the behavior of the model was not performed, a number of important problems were identified through the construction and simulation of the model. Examples of these problems are:

- A simple diagnostic algorithm that was proposed for use with peer (rather than master-slave) microprocessors could lead to deadlock or unnecessary errors.
- Outdated messages in message queues between the two microprocessors could lead to hazards, such as enabling output to the motor after an error occurred in one microprocessor.
- If requests for safety functions are not removed prior to turning off the frequency converter, then there are ambiguities regarding the subsequent power-up phase.
- Identification of new sequences of events that could lead to dangerous failures.
- The system safety requirements, including the Statechart in Fig. 2, did not specify when or how the frequency converter could be turned off if a dangerous failure did *not* occur.

Figure 10 shows a simplified scenario in which an outdated message can lead to a potentially hazardous situation. In this scenario, a user requests a DSF, both microprocessors start the DSF, and microprocessor 2 adds an acknowledgement to a DSF requested message to the message queue for microprocessor 1. Before microprocessor 1 handles the acknowledgement, the user removes the request for the DSF, and then immediately requests the DSF again, after which microprocessor 2 fails. When microprocessor 1 retrieves the outdated acknowledgement from its message queue, a potentially hazardous situation arises because the failure of microprocessor 2 has not yet been discovered. Presumably the failure would be discovered reasonably quickly, e.g. through diagnostics, however, Danfoss had not considered the problem of outdated messages before it was illustrated through simulation.

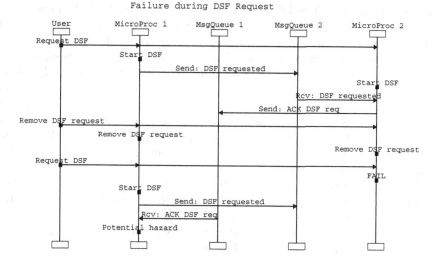

Fig. 10. Scenario in which an outdated message leads to a potential hazard

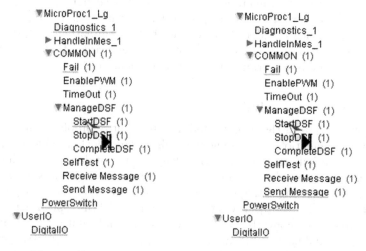

Fig. 11. Using page names to run step-by-step simulations

The problems that were identified had to be discussed with the project team. As several members of the project team were not familiar with CP-nets, we faced the usual challenges of discussing a (Coloured) Petri net with people who are unfamiliar with the formalism. In this case, the way in which the net was constructed proved to have unexpected benefits when discussing the net with the project team. Each page name provides a good indication of the action(s) that was modeled on the page. In CPN Tools, enabled transitions and the pages that they are on are highlighted (in green). In Fig. 11, each text is the name of a page in the net, and a line under a page name indicates that the page contains an enabled transition. The left-hand side of Fig. 11 shows that the PowerSwitch may be pressed (in this case, to turn the power off), DigitalIO actions may occur (to request a safety function or to remove a request), and the following actions can occur in microprocessor 1: it may Fail, Diagnostics_1 can be started, and the StartDSF action will start a recently requested safety function. In the left-hand side of Fig. 11, the *Single Step* simulation tool can be clicked on the StartDSF page name to fire the enabled transition on the page. The right side of Fig. 11 shows how the simulation feedback is updated after the transition on the StartDSF page is fired – it is no longer possible to start diagnostics, and the transition on the SendMessage page is enabled. Since pages rarely contain more than one enabled transition at a time, the highlighted page names provide a good indication of exactly which actions can occur in a given state. By running simulations in this manner, non-CPN experts had no difficulty in discussing and examining the functionality of the CPN model together with CPN experts.

The problems and ambiguities that were identified in the specified behavior were simulated, as described above, for the project team, and decisions subsequently had to be made about how to deal with the problems. The deadlock problems in the diagnostic routine were solved by using a master-slave relationship between the two microprocessors (an earlier version of the CPN model modeled the two microprocessors as peers). Due to time constraints, most of the other problems that were identified were not fixed by updating the CPN model. Rather they have been flagged as potential problem areas that will probably require further analysis during software development at Danfoss.

4.2   State Space Analysis

State space analysis of the model could help to identify problems that were not found during simulation, such as the possibility of deadlock, the identification of additional hazardous situations, or the identification of dangerous failures that are not handled properly. Attempts were made to generate the state space for the model presented in Sect. 3, but the state space was too large to be generated. One of the goals of the project was to construct a model that specifies the behavior of the frequency converter as accurately as possible, since the requirement specification would have to be evaluated and approved by safety certification authorities. Unfortunately, this resulted in a model that was too complex for state space analysis. Many kinds of events can happen in almost all states: safety functions can be requested or requests for safety functions can be removed, diagnostics can be started, and the frequency converter can fail or be turned off. Allowing so many concurrently enabled events contributes significantly to the state explosion problem for this particular model. The model was modified such that limits were put on the number of times that these kinds of events could occur, but it was still not possible to generate a full state space. This is common problem when using (Coloured) Petri nets within industrial projects. It is difficult create a model that provides a sufficiently detailed representation of the system while at the same time being abstract enough to allow for meaningful state space analysis.

4.3   Validation of an Executable Architecture Prototype

One of the challenges in any software development project is ensuring that the developed software conforms to the requirement specification. One of the advantages of using a semi-formal method for specifying software safety requirements is that there may be methods both for validating the behavior of the model itself and for ensuring that the developed software conforms to the specified behavior. For example, in the visualSTATE tool, that Danfoss uses for creating Statecharts, there are both static and dynamic analysis methods that can be used to identify live- and deadlock and states and transitions that are never activated. Furthermore, the tool can be used to generate code that implements the functionality of a Statechart, thereby ensuring that the software fulfills the requirements specification.

Danfoss does not intend to generate code from visualSTATE, so another one of the goals of this project was to investigate and develop techniques for ensuring that safety-critical software fulfills the corresponding software safety requirements. In other words, we were interested in closing the gap between a semi-formal requirement specification and a software implementation. We focused on techniques for specifying and validating a software architecture (rather than the final software) for the frequency converter. A software architecture was developed using a technique similar to Krutchen's 4+1 technique [7] in which an architecture is described from different viewpoints. These viewpoints describe different aspects of the software architecture such as the structuring of software components, the distribution of software components on hardware components, and use scenarios.

The architecture was defined largely by UML diagrams, including class, package, deployment, and sequence diagrams. Once the architecture had been designed, an executable architecture prototype [8] was implemented as skeleton classes in Java. A number of important use scenarios, such as initialization during power-up and requesting safety functions were implemented as simple Java programs that exercised the architecture prototype by emulating external events of the frequency converter, e.g. pressing the power button or requesting a safety function by activating a digital input, by calling appropriate methods in the executable architecture prototype. Given this architecture prototype, Danfoss was interested in developing techniques for ensuring that the architecture fulfilled the software safety requirements, including those specified by the CPN model.

A technique and prototype facilities for validating a Java program against a CPN-based requirement specification expressed in CPN Tools has been developed. With this technique, a mapping between methods in the Java program and transitions in the CPN model must be made. A method may correspond to a single transition, a method may not correspond to any transitions, a method may correspond to a sequence of transition occurrences, and a sequence of method calls may correspond to a single transition. Currently this mapping must be made manually. An execution of the architecture prototype corresponds to a sequence of calls to methods in the prototype. Such an execution is legal if the

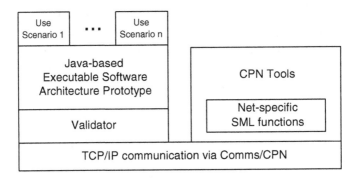

Fig. 12. Validating prototype against CPN-based requirement specification

sequence of method calls corresponds to a legal occurrence sequence that starts in the initial marking of the CPN model.

The structure of the architecture validation facilities is shown in Fig. 12. On the left is the architecture prototype. The architecture is tested by running the executable use scenarios. An additional Java *Validator* class sends method names to CPN Tools, and waits for responses from CPN Tools. The simulator of CPN Tools was slightly modified so that sequences of transitions could be specified and fired if they were enabled. The net-specific Standard ML functions in CPN Tools specify the mapping between method names and transitions, and fire sequences of transitions.

The facilities work as follows. Each of the relevant methods in the architecture prototype must be instrumented so that they send their own method name to the *Validator* object which then sends the method name to CPN Tools. The architecture prototype is blocked until a response is received from the *Validator*. If the method corresponds to a sequence of enabled transitions, then the transitions are fired, and an appropriate response is returned to the *Validator*. If the method corresponds to a transition that is not enabled, then the response from CPN Tools indicates that the architecture prototype has attempted to perform a sequence of actions that do not fulfill the requirement specification. Similarly, if an unknown method name is received in the simulator, then it returns an appropriate response.

Textual feedback provides information about the success or failure of an attempt to validate a use scenario. Figure 13 shows output generated by the architecture prototype and the *Validator*. All methods that are called in the architecture prototype are logged by printing the method name in the console window. For example, the first method that is called is the `powerUp` method in the `Controller` object. The third line in Fig. 13 shows that the `Controller.powerUp` method was successfully validated in the CPN simulator. Not all methods are instrumented to be validated, as indicated by a missing `validate...` line after the method name. The upper half of Fig. 13 shows the successful validation of a use scenario in which a safety function is requested via digital input and thereafter activated. The bottom half of the figure shows an unsuccessful validation

```
$ java safeinverter/scenarios/DSFExecution_2_1_1DigitalIO
Controller.powerUp
    validate (Controller.powerUp) with CPN...OK
Controller.selfCheck
    validate (Controller.selfCheck) with CPN...OK
DigitalIO.selfCheck
DigitalIO.requestDSF
    validate (DigitalIO.requestDSF) with CPN...OK
SafeDelay.activate
    validate (SafeDelay.activate) with CPN...OK
Timer.run

$ java safeinverter/scenarios/DSFExecution_2_1_1ProfiSafe
Controller.powerUp
    validate (Controller.powerUp) with CPN...OK
Controller.selfCheck
    validate (Controller.selfCheck) with CPN...OK
ProfiSafe.selfCheck
ProfiSafe.requestDSF
    validate (ProfiSafe.requestDSF) with CPN...Unknown event
```

Fig. 13. Output when validating the architecture prototype against the CPN model

of a use scenario as the `ProfiSafe.requestDSF` method was instrumented to be validated, but it was not mapped to any transitions (because the PROFIsafe fieldbus is not modeled in the CP-net). Standard simulation reports are generated in CPN Tools to provide a trace of the transitions that occur when validating a particular use scenario. Such trace files are particularly useful when investigating illegal behavior in the architecture.

Due to time constraints, a full architecture prototype could not be implemented and validated. However, the proof-of-concept validation tools were useful, and the technique certainly has the potential for being useful for validating software against a CPN-based requirement specification.

5   Related Work

IEC 61508 recommends the use of the following semi-formal methods for requirement specification: finite state machines/state transition diagrams, timed Petri nets, logic/function block diagrams, sequence diagrams, data flow diagrams, and decision/truth tables. Of these methods only finite state machines and timed Petri nets are executable. In the following, we will restrict our discussion of related work to the use of these methods for specifying and analyzing the behavior of safety-critical systems.

Petri nets and related formalisms have been used to model and analyze different aspects of safety-critical systems. Stochastic Well-formed nets and Stochastic Activity networks have been used to analyze the reliability of safety-critical systems [9, 10]. Leveson and Stolzy developed an algorithm for starting with hazardous states in a PN model and working backward to see how to change the model to make the hazardous states unreachable [1]. Petri nets have also been combined with other kinds of formal methods for the purpose of analyzing different kinds of behavior. One technique is based on converting CP-nets to prototype verification system (PVS) specifications [11] in order to use the theorem-proving techniques of PVS as a means to avoid the state explosion problem that is often experienced when generating state spaces for CP-nets. In another technique, Z specifications are associated with transitions in a Petri net in order to analyze the concurrent behavior of a system while at the same time ensuring that the PN model of the system fulfilled the Z specification of the data-handling aspects of the system [12]. An observation that can be found in most of these works, is that Petri nets are not necessarily a practical modeling language for real systems either because it can be difficult to create understandable, detailed PN models of real systems or because the state explosion problem is often met when generating state spaces for complex Petri nets. While these observations are certainly true, this and other case studies [13, 14] indicate that the process of building a (possibly incomplete) executable model of a system and running simulations can help to identify problems and resolve inconsistencies even though an exhaustive analysis of system behavior cannot be performed.

CP-nets have also been used to evaluate software architectures. In [15], software architectures are modeled as CP-nets and non-functional qualities such as reliability and efficiency of the proposed architecture are estimated using simulation-based performance analysis. To the best of our knowledge no other work provides a technique for validating an executable software architecture against a CPN-based requirement specification.

Several variations of Statecharts have also been used to specify and analyze behavior of safety-critical systems. Safecharts [16] are a variety of Statecharts for modeling behavior of safety-critical systems. Tools and techniques for analyzing the completeness of safety-related requirements were originally developed for a Statechart-like specification language [17], and methods and tools for these techniques have been developed for traditional Statecharts [18]. As with Petri nets, Statecharts have been combined with other formal methods, such as Z [19] and extended time graphs [20] in an attempt to find formalisms that complement each other and that can be used to analyze different kinds of behavior. Statecharts have the advantages that they are more widely used in industry than Petri nets, and there is support for generating code from models thus ensuring consistency between the model and the code. On the other hand, Statecharts do not support modeling multiple instances of the same functionality (as is useful for modeling the common functionality of the two microprocessors in this project), and they have lacked a well-defined and widely accepted semantics.

6   Conclusion

This paper has presented a case study in which Coloured Petri Nets were used to specify and analyze software safety requirements for a frequency converter. Analysis of the model revealed several ways in which the specified behavior could lead to hazardous situations or unnecessary failures. Each of the problem areas have been flagged for additional analysis as development of the frequency converter progresses. Relying on the simulation feedback and flexibility of the simulation tools in CPN Tools, it was possible to discuss and examine the behavior of the model with non-CPN experts while hiding most of the detailed net structure. Prototype tools for validating Java programs against corresponding CPN-based requirement specifications have also been developed.

While the results of the project have been very satisfactory, there are many interesting topics for future work. Requirement completeness analysis is based on state machine representations of system behavior [17], and this type of analysis could be performed using state spaces for CP-nets. Future work could be done to reduce the size of the state space, either through state reduction techniques or by limiting the behavior of the model, in order to perform at least partial requirement completeness analysis. The tools for validating a Java process against a CPN-based requirement specification could also be improved significantly: using Aspect-oriented programming or the Java debugger would remove the need to instrument Java classes with code for activating the validator, it should be possible to specify and fire transitions with specific variable bindings (rather than just transitions), and it would be advantageous if the simulation feedback in the GUI for CPN Tools could be updated when a simulation is controlled by an external process.

References

1. Leveson, N.: Safeware: System Safety and Computers. Addison-Wesley (1995)
2. International Electrotechnical Commission: Functional Safety of Electrical/ Electronic/ Programmable Electronic Safety-Related Systems. 1st edn. (1998-2000) International Standard IEC 61508, Parts 1-7.
3. Harel, D.: Statecharts: A visual formalism for complex systems. Science of Computer Programming **8** (1987) 231–274
4. Jensen, K.: Coloured Petri Nets: Basic Concepts, Analysis Methods and Practical Use. Vol. 1, Basic Concepts. Monographs in Theoretical Computer Science. Springer-Verlag (1997) 2nd corrected printing.
5. Kristensen, L.M., Christensen, S., Jensen, K.: The practitioner's guide to coloured Petri nets. International Journal on Software Tools for Technology Transfer **2** (1998) 98–132
6. CPN Tools, http://www.daimi.au.dk/CPNTools/.
7. Kruchten, P.: The 4+1 view model of architecture. IEEE Software **12** (1995)
8. Bardram, J., Christensen, H.B., Hansen, K.M.: Architectural prototyping: An approach for grounding architectural design and learning. In: Proceedings of the 4th Working IEEE/IFIP Conference on Software Architecture. (2004) 15–24
9. Bobbio, A., Ciancamerla, E., Franceschinis, G., Gaeta, R., Minichino, M., Portinale, L.: Methods of increasing modelling power for safety analysis, applied to a turbine digital control system. In Stuart Anderson, Sandro Bologna, M.F., ed.: Computer Safety, Reliability and Security: 21st International Conference, SAFECOMP 2002. Volume 2434 of LNCS., Springer (2002)
10. Campelo, J., Yuste, P., Rodriguez, F., Gil, P., Serrano, J.: Hierarchical reliability and safety models of fault tolerant distributed industrial control systems. In Massimo Felici, Karama Kanoun, A.P., ed.: Computer Safety, Reliability and Security: 18th International Conference, SAFECOMP 1999. Volume 1698 of LNCS., Springer (1999)

11. Son, H.S., Seong, P.H.: Development of a safety critical software requirements verification method with combined CPN and PVS: a nuclear power plant protection system application. Reliability Engineering and System Safety **80** (2003) 19–32

12. Heiner, M., Heisel, M.: Modeling safety-critical systems with Z and Petri nets. In Massimo Felici, Karama Kanoun, A.P., ed.: Computer Safety, Reliability and Security: 18th International Conference, SAFECOMP 1999. Volume 1698 of LNCS., Springer (1999)

13. Lorentsen, L., Tuovinen, A.P., Xu, J.: Modelling of features and feature interactions in nokia mobile phones using coloured petri nets. In Esparza, J., Lakos, C., eds.: Application and Theory of Petri Nets: 23rd ICATPN. Volume 2360 of LNCS., Springer Verlag (2002) 299–313

14. Jrgensen, J.B.: Coloured Petri nets in development of a pervasive health care system. In van der Aalst, W.M., Best, E., eds.: Application and Theory of Petri Nets: 24th International Conference on Applications and Theory of Petri Nets (ICATPN 2003). Volume 2679 of LNCS., Springer (2003) 256–275

15. Fukuzawa, K., Saeki, M.: Evaluating software architectures by coloured Petri nets. In: Proceedings of the 14th international conference on Software Engineering and Knowledge Engineering, ACM Press (2002) 263–270

16. Dammag, H., Nissanke, N.: Safecharts for specifying and designing safety critical systems. In: Proceedings of the 18th IEEE Symposium on Reliable Distributed Systems, IEEE (1999)

17. Heimdahl, M., Leveson, N.: Completeness and consistency checking of software requirements. IEEE Transactions on Software Engineering **22** (1996)

18. Pap, Z., Majzik, I., Pataricza, A.: Checking general safety criteria on UML Statecharts. In Voges, U., ed.: Computer Safety, Reliability and Security: 20th International Conference, SAFECOMP 2001. Volume 2187 of LNCS., Springer (2001)

19. Winter, K., Santen, T., Heisel, M.: An agenda for specifying software components with complex data models. In Ehrenberger, W., ed.: Computer Safety, Reliability and Security: 19th International Conference, SAFECOMP 2000. Volume 1516 of LNCS., Springer (2000)

20. van Katwijk, J., et al.: Specification and verification of a safety shell with statecharts and extended timed graphs. In: Computer Safety, Reliability and Security: 19th International Conference, SAFECOMP 2000. Volume 1943., Springer (2000)

Achieving a General, Formal and Decidable Approach to the OR-Join in Workflow Using Reset Nets

Moe Thandar Wynn[1], David Edmond[1], W.M.P. van der Aalst[1,2], and A.H.M. ter Hofstede[1]

[1] Center for IT Innovation, Queensland University of Technology,
P.O. Box 2434, Brisbane Qld 4001, Australia
{m.wynn,d.edmond, a.terhofstede}@qut.edu.au
[2] Department of Technology Management, Eindhoven University of Technology,
P.O. Box 513, NIL-5600 MB, Eindhoven, The Netherlands
w.m.p.v.d.aalst@tm.tue.nl

Abstract. Workflow languages offer constructs for coordinating tasks. Among these constructs are various types of splits and joins. One type of join, which shows up in various incarnations, is the OR-join. Different approaches assign a different (often only intuitive) semantics to this type of join, though they do share the common theme that synchronisation is only to be performed for active threads. Depending on context assumptions this behaviour may be relatively easy to deal with, though in general its semantics is complicated, both from a definition point of view (in terms of formally capturing a desired intuitive semantics) and from a computational point of view (how does one determine whether an OR-join is enabled?). In this paper the concept of OR-join is examined in detail in the context of the workflow language YAWL, a powerful workflow language designed to support a collection of workflow patterns and inspired by Petri nets. The OR-join's definition is adapted from an earlier proposal and an algorithmic approach towards determining OR-join enablement is examined. This approach exploits a link that is proposed between YAWL and Reset nets, a variant of Petri nets with a special type of arc that can remove all tokens from a place.

Keywords: OR-join, YAWL, Workflow patterns, synchronizing merge, Petri nets, Reset nets.

1 Introduction

Workflow specifications should capture various aspects of business models such as the flow of control, the flow of data, the structure of the organisation, and the use of resources (see e.g.[13]). The control flow perspective captures the execution interdependencies between the tasks of a business process. In-depth analysis and comparison of a number of commercially available workflow management systems has been performed [4]. The findings demonstrate that the interpretation of even the basic control flow constructs is not uniform and it is often unclear how the more complex requirements could be supported. The authors propose 20 workflow patterns to address control flow requirements in a language independent style. YAWL (Yet Another Workflow

G. Ciardo and P. Darondeau (Eds.): ICATPN 2005, LNCS 3536, pp. 423–443, 2005.
© Springer-Verlag Berlin Heidelberg 2005

Language) is a result of this analysis, it provides direct support for most patterns [3]. YAWL has a formal semantics specified as a transition system. Although YAWL exploits concepts from Petri nets, it also provides direct support for those patterns hard to realise in Petri nets. One of these patterns corresponds to the synchronising merge or the OR-join, the focus of this paper. In practice, there is a need for a construct like the OR-join as is evident from e.g. the fact that some commercial systems support OR-join like constructs. However, experience with these systems shows that it is difficult to select a suitable semantics and implement it efficiently. Workflow management systems like InConcert, eProcess, and WebSphere MQ Workflow have solved problems related to the OR-join using syntactical restrictions. IBM WebSphere MQ Workflow [17] (formerly known as MQSeries Workflow and FlowMark and also used as a basis for the new BPEL standard) offers full support for the OR-join but in order to do this it requires the workflow to be acyclic, i.e., the only way to introduce loops is by executing the entire (sub)process [2]. Other systems like Eastman and Domino Workflow seem to use a non-local semantics similar to the one used in YAWL. Such a non-local semantics may lead to unexpected results. Moreover, a non-local semantics may result in poor performance as is stated in the manual of Eastman: "Parallel instances can accumulate at a Join workstep if the instances are routed to the workstep by preprocessing rules. These instances will eventually be joined by a RouteEngine subprocess (thread) that examines Join worksteps for such instances. This Join scavenger thread reduces system efficiency, so routing to Join worksteps using preprocessing rules should be avoided" [9]. These examples illustrate the practical relevance of the OR-join and serve as a motivation for the work reported in this paper. For a more complete discussion on workflow systems' support for OR-join semantics, we refer to [2, 4, 14, 15].

The OR-join is a control flow construct that sometimes behaves like an AND join and sometimes like an XOR join based on the current context. Variants and interpretations of the OR-join have been proposed in the literature. In [18], several possible interpretations of OR-join semantics in the context of Event-driven Process Chains (EPCs) are discussed. If there is a matching OR-split, the OR-join semantics is taken to be "wait for the completion of all paths activated by the matching split". If there is no matching split, there could be at least three interpretations of an OR join: wait-for-all, first-come and every-time [18]. In [2], the authors highlight the technical, conceptual and practical problems with the formal semantics of the OR-join in Event driven Process Chains (EPCs). The authors suggest that there is no sound formal semantics for EPCs that is fully compliant with the informal semantics and that any formal semantics for EPCs will impose some restrictions or will deviate from the informal semantics to some extent. The authors demonstrate the problems using vicious circles, which are formed when two or more OR-joins are in a feedback loop and each OR-join waits for the other OR-join to complete first. On the other hand, in [15] a semantic framework for formally defining the non-local semantics of EPCs including the OR-join is proposed. The author states that "a single transition relation cannot precisely capture the informal semantics of EPCs". It is proposed that the non-local semantics be defined as a pair of transition relations and a semantic definition using techniques from fixed point theory is presented [15]. The current OR-join approach in YAWL [3] is intended to be a generalised approach and the formal semantics of the OR-join is defined by ignoring all other OR-joins. This approach is described as "ad hoc in some way" [15].

The contributions of this paper are threefold. Firstly, we re-examine the OR-join semantics as proposed in [3], because its behaviour is non-intuitive in the context of OR-joins depending on other OR-joins and composite tasks (they cannot be treated like black boxes). Secondly, for the purposes of the OR-join definition and analysis, we propose an abstract view on YAWL, one which is formalised in terms of *Reset nets* [5, 6, 7, 8, 10, 11, 12]. Reset nets are considered the most suitable formalism as reset arcs provide direct support for the cancellation feature in YAWL (another concept introduced to YAWL as a result of the workflow patterns and the difficulty of realising this feature in Petri nets). Thirdly, the mapping of YAWL nets to Reset nets is exploited to find an algorithmic solution to the non-trivial problem of OR-join enablement. Note that the contribution of this paper is not limited to YAWL. Many systems and languages struggle with the semantics and implementation of the OR-join. This paper provides suitable semantics and gives a concrete algorithm to support an efficient implementation.

This rest of the paper is organised as follows. In Section 2, we introduce the current OR-join semantics in YAWL, discuss the problems with this semantics and propose alternative treatments for OR-joins depending on other OR-joins in a YAWL net. In Section 3, the definitions of EWF-nets (Extended Workflow Nets) and Reset nets are presented together with the proposed abstractions to enable EWF-net to Reset net mappings. In Section 4, we propose a new semantics for the OR-join in YAWL. In Section 5, we propose an algorithm for OR-join analysis based on well-known backwards search techniques. Section 6 concludes the paper.

2   Current Semantics of the OR-Join in YAWL

In this section, we first outline the challenges associated with the non-local semantics of the OR-join. In particular, we show how ignoring other OR-joins during the analysis can lead to counter-intuitive results. We then propose some alternative treatments for OR-joins on the path to other OR-joins.

2.1   The OR-Join in YAWL

A YAWL model is made up of tasks, conditions and a flow relation between tasks and conditions. In YAWL, tasks may be directly connected graphically. The splits, joins, conditions and cancellation symbols for YAWL are shown in Figure 1. YAWL uses the terms tasks and conditions to avoid confusion with Petri net terminology (transitions and places).

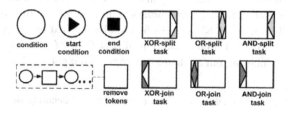

Fig. 1. Splits, joins, conditions and cancellation in YAWL

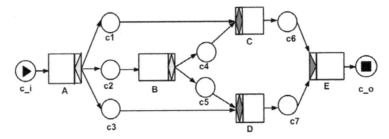

Fig. 2. A YAWL net with an OR-split task B and two OR-join tasks C and D

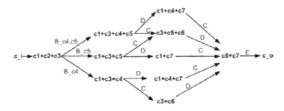

Fig. 3. Reachability graph of the YAWL net in Figure 2 (assuming some OR-join behaviour)

If there is a cancellation set associated with a task, the execution of the task removes all the tokens from the conditions and tasks in the cancellation set. Cancelling a task is achieved by removing tokens from internal conditions of the task. An OR-join task is enabled at a marking iff at least one of its input conditions is marked and it is not possible to reach a marking that marks all currently marked input conditions (possibly with fewer tokens) and at least one that is currently unmarked. If it is possible to place tokens in the unmarked input conditions of an OR-join in the reachable markings from the current marking, then the OR-join task should not be enabled and wait until either more input conditions are marked or until it is no longer possible to mark more input conditions.

The example in Figure 2 demonstrates an unstructured YAWL net with AND-split task A, AND-join task E, OR-split task B and OR-join tasks C and D. This example demonstrates the different behaviours of OR-joins in the context of two different markings. First consider a marking $M = c1+c2+c3$ where there is a token in input condition $c1$ of OR-join task C and in input condition $c3$ of OR-join task D. To determine whether tasks C and/or D should be enabled at M, we need to find out whether tokens could be put into $c4$ or $c5$ in the reachable markings from M. The reachability graph of Figure 3 shows the reachable markings from the initial marking $M_0 = c_i$ to the end marking $M = c_o$.[1] We can see that by executing task B, we can reach markings $c1 + c3 + c5$ or $c1 + c3 + c4 + c5$ that mark $c5$, an unmarked input condition of task D in M. Also, markings $c1 + c3 + c4$, $c1 + c3 + c4 + c5$ could be reached by executing task B and they mark $c4$, an unmarked input condition of task C in M. As we can reach a new marking from M which can put a token in an unmarked input condition of the OR-join

---

[1] Note the overloading of notation, i.e., here c_o is a multiset denoting the marking with one token in condition c_o.

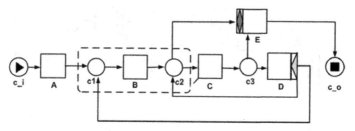

Fig. 4. Cancellation task C with an infinite loop

tasks C and D, neither task C or D should be enabled at M. If we consider a marking $M' = c1 + c3 + c4$, where all the input conditions of C (i.e., $c1$ and $c4$) are marked, then C would be enabled at M'. We will also enable task D at M' as it is not possible for another token to arrive at input condition $c5$. Note that in the scenario where we move from M to M', task D was not enabled in M and, although no tokens were added to the input conditions of this task, it got enabled in M'.

Now, let us consider OR-joins in the light of cancellation. In Figure 4, we describe a YAWL net with (i) task C removing tokens from the conditions $c1$, $c2$ and task B when firing, and (ii) an OR-join task E. At a marking $M = c2$, we marked one of the input conditions of E and we need to perform an analysis to decide whether both $c2$ and $c3$ could be marked in a reachable marking from M. We can observe the following sequence of reachable markings from M: $c2 \rightarrow^C c3 \rightarrow^D c1 + c2 \rightarrow^B 2c2 \rightarrow^C c3$. This is due to the cancellation feature of C, removing tokens from $c2$ when firing. We can conclude that it is not possible to reach a bigger marking $c2 + c3$ from M and therefore, E should be enabled at M. Let us consider a different situation where task C does not have a cancellation set associated with it. From marking $M = c2$, we can observe the following sequence of reachable markings: $c2 \rightarrow^C c3 \rightarrow^D c1 + c2 \rightarrow^B 2c2 \rightarrow^C c2 + c3$. As we can reach $c2 + c3$ which marks more input places of the OR-join task E, the analysis will conclude that task E should not be enabled at M. This example demonstrates the possible effect that the cancellation feature of a task can have on the OR-join enablement analysis.

From the above examples, it is obvious that the OR-join semantics requires careful analysis and the decision to enable an OR-join cannot be made locally. Any OR-join algorithm must evaluate all the reachable markings from a current marking to determine whether there is a possibility of a token arriving at an input condition of an OR-join which is not currently marked (while all input conditions which were already marked remain marked though possibly with fewer tokens). This algorithm potentially needs to be applied every time the marking changes and the OR-join analysis could place a significant load on any workflow engine required to execute it, cf. the quote from the manual of Eastman [9] in the introduction.

2.2   Problems with Current OR-Join Semantics in YAWL

Two problems may be identified with the current OR-join semantics of YAWL which are related to the treatment of OR-joins and composite tasks preceding an OR-join under consideration.

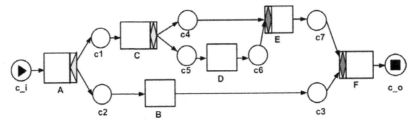

Fig. 5. A YAWL net with two OR-join tasks E and F

The current OR-join semantics ignores other OR-joins when analysing whether a particular OR-join should be enabled at a given marking [3]. In Figure 5, there are two OR-join tasks, E and F in the YAWL net. Consider a marking $M = c1 + c3$ where the analysis for the OR-join task, F is performed. After executing task C, it is possible to reach either $c3 + c4$, $c3 + c5$ or $c3 + c4 + c5$. One possible occurrence sequence is $c1 + c3 \rightarrow^C c3 + c4 + c5 \rightarrow^D c3 + c4 + c6 \rightarrow^E c3 + c7$. Hence, $M' = c3 + c7$ is a reachable marking from M. However, the current OR-join semantics ignores other OR-joins on the path to F, so task E and the associated conditions will not be taken into account, and M' is therefore not considered as a reachable marking during the OR-join analysis of F. As a result, the analysis will conclude that there is no possibility of another token arriving in $c7$ and F would be enabled at M. This behaviour is probably not what one would expect from this specification. It could also result in multiple executions of task F and more than one token could be produced for c_o e.g. ($c1 + c_o$). A YAWL model which can produce a token for the output condition c_o while still having tokens in the other conditions is considered as not having proper completion and is therefore not sound [1]. We have seen that as the analysis of a given OR-join does not consider the possibility of a token arriving from a path which has an OR-join, this could result in premature enabling and multiple execution of OR-join tasks when they are nested.

The other problem is that the OR-join semantics in [3] does not treat composite tasks as "black boxes", i.e., the semantics is based on the "unfolding" of the YAWL model. This semantics implies that a YAWL net at a lower level cannot be considered as a black box, thus impacting the OR-join analysis at a higher level net. Consider a specification where task B in Figure 5 is a composite task with an OR-join. When evaluating whether an OR-join should be enabled at a given marking, the analysis will be performed at lower level nets that make up a YAWL specification. This also applies for composite tasks which can deadlock. Consider marking $c2 + c7$. If task B contains a subprocess that will deadlock, then F is enabled. If B has proper completion, then F is not enabled. This also demonstrates that in the current semantics, composite tasks cannot be treated as black boxes.

2.3 Optimistic and Pessimistic Approaches

Instead of ignoring other OR-join tasks altogether during the analysis, we propose two alternative treatments for those OR-joins: treat them either as XOR-joins (*optimistic*) or as AND-joins (*pessimistic*). Both optimistic and pessimistic approaches achieve the desired behaviour for an OR-join analysis by delaying enablement when there is a possibility of more tokens arriving to unmarked input conditions of the OR-join. We believe that these two alternatives result in an analysis which is more closely related to the informal semantics of OR-joins and still allow for sound semantics (i.e., avoid the fixpoint problems discussed in [2]).

The treatment of an OR-join on the path to another OR-join as an XOR-join is an *optimistic* approach. Consider a marking $M = c1 + c3$ in Figure 5 where an OR-join analysis for task F would be performed. Instead of ignoring the other OR-join task E during the analysis, task E will be treated as an XOR-join task. This will mean that the occurrence sequence $c1 + c3 \rightarrow^C c3 + c4 \rightarrow^E c3 + c7$ would be considered. As a result, task F is not enabled at M. This interpretation of OR-join task E as an XOR-join, prevents F from being enabled prematurely and it matches more closely with the informal semantics of OR-joins.

The treatment of an OR-join on the path to another OR-join as an AND-join is a *pessimistic* approach, as this now requires tokens in all input conditions of the AND-join before enabling. Consider again $M = c1 + c3$ in Figure 5 where an OR-join analysis for task F would be performed. This time, instead of ignoring task E, it will be treated as an AND-join task. Due to the OR-split behaviour of task C, tokens can be present in $c4$ or $c5$ or both after firing C. This occurrence sequence $c1 + c3 \rightarrow^C c3 + c4 + c5 \rightarrow^D c3 + c4 + c6 \rightarrow^E c3 + c7$ is possible. As a token can be put in $c7$ while $c3$ remains marked, F is not enabled at M. This preserves the same informal semantics as an optimistic approach, and both approaches result in delaying the enablement of the OR-join task F.

In some cases, we observe that treating other OR-joins on the path as XOR-joins using an optimistic approach is more appropriate for the analysis. Consider a scenario where task C in Figure 5 is an XOR-split task rather than the OR-split task. Let us consider a marking $c1 + c3$ and that we treat task E as an AND-join task. As it is not possible for task E to fire due to the XOR-split and AND-join combination, the OR-join analysis will conclude that F should be enabled. As a result, task F could be executed more than once and the YAWL net does not have proper completion. The analysis will reach the same conclusion as the current semantics in YAWL where the semantics ignores the OR-join dependencies.

We have also found that when OR-joins are in conflict, there might not be a satisfactory treatment for OR-joins. Let N be a YAWL net and $o_1 \, {}^\backprime o_2$ be two OR-join tasks. We define o_1 and o_2 to be *in conflict* iff o_1 is on a directed path to o_2 and o_2 is on a directed path to o_1. We have in Figure 6 an unusual situation described as a vicious circle in [15] where the OR-joins are in conflict and it is unclear what the exact informal semantics of the model should be. In Figure 6, there are two OR-join tasks B and C which are in conflict with each other. Condition $c3$ is an output condition of C and an input condition of B and $c4$ is an output condition of B and an input condition of C. Figure 6 is inspired by [15]. Consider a marking $c1 + c2$ where an OR-join analysis is carried out for task B and C. Using the *optimistic* approach, we treat task C as an XOR-join task during the analysis for B. As a result, we can find a reachable marking $c1 + c3 + c6$, which marks both input conditions of B. Therefore, B should not be enabled at $c1 + c2$. Similarly, we will treat B as an XOR-join task for the analysis of task C and there is a reachable marking $c2 + c4 + c5$. Therefore, task C should not be enabled at $c1 + c2$. As a result of

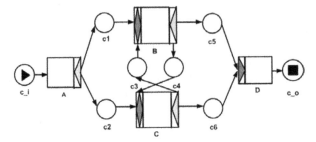

Fig. 6. OR-join tasks B and C in conflict

this *optimistic* approach, the YAWL net will *deadlock* because of the OR-join semantics using the *optimistic* approach. Using the *pessimistic* approach, we treat task C as an AND-join task during the analysis for B. At the marking $c1 + c2$, it is not possible to enable C due to the AND-join semantics, and therefore, task B will be enabled and can be fired. This will enable task C and after firing C, tokens will be placed in $c3$ and $c6$. Therefore, tasks B and C could potentially keep firing alternatingly thus resulting in a potentially infinite number of firings of task D. The same is true for the analysis of task C. We can see that the *pessimistic* approach would also result in improper completion. The original semantics that ignores other OR-joins would also result in a similar behaviour to the *pessimistic* approach. In this case, all three approaches deviate from the informal semantics of the OR-join and it is not possible to define the formal semantics accurately.

From the above discussions, it can be seen that there is no ideal treatment for non-local OR-join semantics in YAWL. Any formal semantics will impose some restrictions or deviate from the informal semantics to some extent. In our opinion, the XOR-join treatment of other OR-joins matches more closely the informal semantics of the OR-join. Consider the YAWL example in Figure 5 with a marking $M = c3 + c4$. If we treat E as an XOR-join during the analysis for task F, the outcome would be that F is not enabled at M because it is possible to reach a marking $M' = c3 + c7$ by executing E first. On the other hand, AND-join treatment of E will result in F being enabled at M and could result in F being executed twice. Hence, we chose to use the optimistic approach (XOR-join treatment) for our formal semantics.

3   Establishing a Formal Foundation

The formal semantics of YAWL is expressed in terms of a transition system [3] and while inspired by Petri nets, YAWL should not be seen as an extension of these. New concepts were introduced in YAWL to suitably deal with the workflow patterns [4]. YAWL constructs such as OR-join, cancellation and multiple instances are not directly supported by Petri nets. To perform an OR-join analysis, a multiple instances task does not effect the analysis but cancellation plays an important role (as shown in Figure 4). This cancellation feature of YAWL is theoretically closely related to Reset nets, which are Petri nets with reset arcs. For an OR-join analysis, we propose to map a YAWL model represented as an EWF-net (Extended Workflow Net) to a Reset net. In this section, we first present the definitions of EWF-nets and then discuss the proposed abstractions to the EWF-nets. We then present the definition and firing rules for Reset nets.

3.1   EWF-Nets

A YAWL model is formally defined as a nested collection of EWF-nets [3]. As we will show later, it suffices to consider only one EWF-net in isolation when evaluating an OR-join.

Definition 1 (EWF-net [3]). *An extended workflow net (EWF-net) N is a tuple $(C^c$ i^c o^c T^c F^c split^c join^c rem^c nofi) such that[2]*

- C *is a set of conditions and T is a set of tasks,*
- i $\in C$ *is the unique input condition and* o $\in C$ *is the unique output condition,*
- $F \subseteq (C \setminus \{o\} \times T) \cup (T \times C \setminus \{i\}) \cup (T \times T)$ *is the flow relation,*
- *every node in the graph $(C \cup T^c F)$ is on a directed path from i to o,*
- *split: $T \to \{AND^c XOR^c OR\}$ specifies the split behaviour of each task and join: $T \to \{AND^c XOR^c OR\}$ specifies the join behaviour of each task,*
- *rem: $T \nrightarrow \mathbb{P}(T \cup C \setminus \{i^c o\})$ specifies the additional tokens to be removed by emptying a part of the workflow;*
- *nofi: $T \nrightarrow \mathbb{N} \times \mathbb{N}^{inf} \times \mathbb{N}^{inf} \times \{dynamic, static\}$ specifies the multiplicity of each task (minimum, maximum, threshold for continuation, and dynamic/static creation of instances).*

In an EWF-net, it is possible for two tasks to have a direct connection. We will add an implicit condition $c_{(t_1, t_2)}$ between two tasks $t_1^c t_2$ if there is a direct connection from t_1 to t_2. We denote as C^{ext} the set of conditions extended to include implicit conditions, and denote the extended flow relation as F^{ext}. We now define an explicit extended workflow net (E2WF-net) using C^{ext} and F^{ext} as follows:

Definition 2 (E2WF-net). *Let $N = (C^c$ i^c o^c T^c F^c split^c join^c rem^c nofi) be an EWF-net, the corresponding explicit EWF-net (E2WF-net) is defined as $(C^{ext c}$ i^c o^c T^c F$^{ext c}$ split^c join^c rem^c nofi) where*

$C^{ext} = C \cup \{c_{(t_1, t_2)} \mid (t_1^c t_2) \in F \cap (T \times T)\}$ *and*
$F^{ext} = (F \setminus (T \times T))$
$\qquad \cup \{(t_1^c c_{(t_1, t_2)}) \mid (t_1^c t_2) \in F \cap (T \times T)\}$
$\qquad \cup \{(c_{(t_1, t_2)}^c t_2) \mid (t_1^c t_2) \in F \cap (T \times T)\}.$

Let N be an E2WF-net and $x \in C^{ext} \cup T$, we use •x and x• to denote the set of inputs and outputs of a node i.e. •$x = \{y \mid (y^c x) \in F^{ext}\}$ and $x• = \{y \mid (x^c y) \in F^{ext}\}$. A marking is denoted by M and, just as with ordinary Petri nets, it can be interpreted as a vector, function, and multiset. M is an m-vector, where m is the total number of conditions. This vector can also be seen as a function $M : C^{ext} \to \mathbb{N}$, where $M(c)$ returns the number of tokens in a condition c of a marking M. Functions mapping some domain (in this case C) onto $\mathbb{N}$ can also be seen as multisets, i.e., M is a multiset over

---

[2] Note that we are using basic mathematical notations such as $\nrightarrow$ for a partial function, $\mathbb{P}$ for powerset, $\mathbb{N}$ for natural numbers, and $\mathbb{N}^{inf}$ for $\mathbb{N} \cup \{inf\}$.

C. Since a marking is a multiset, we can use notations such as $M \leq M'$, $M + M'$, and $M - M'$. $M \leq M'$ iff $\forall_{c \in dom(M)} M(c) \leq M'(c)$. $M + M'$ and $M - M'$ are a multisets such that for any $c \in dom(M)$: $(M + M')(c) = M(c) + M'(c)$ and $(M - M')(c) = M(c) - M'(c)$.

Tasks are the active components of an E2WF-net and when a task t fires at a marking M, it changes the state and reaches a new marking M', denoted as $M \rightarrow^t M'$. A YAWL specification supports hierarchy and a composite task is mapped onto an EWF-net. As we will abstract from composition, we refer the reader to [3] for a formal definition of a YAWL specification.

3.2   Abstractions

We propose to abstract the constructs in YAWL that do not affect an OR-join analysis. They include multiple instances, composite tasks and internal conditions of a task. We can assume that if a multiple instances task is enabled and executed, it will complete and put tokens into the appropriate output conditions of the task. Similarly, with the state transitions and internal conditions within a task, we can abstract from these transitions and only consider the input and output conditions of a task. In the mappings to Reset nets, we will introduce one place for each task which indicates whether a task is currently executing and as a result, abstract from the internal conditions of a task. We also propose to treat EWF-nets as *flat* nets, and ignore the hierarchical structure for the purpose of an OR-join analysis. In other words, when deciding whether an OR-join should be enabled at a given marking, we will not be considering the effect of deadlock within a composite task. We assume that a YAWL subnet which is used as a composite task at a given level is sound. Therefore, if a composite task can be enabled and executed, it will terminate at some time, and tokens will be placed in the appropriate output condition(s) of the composite task. As a result, even if there is an OR-join task in the composite task, it will not influence the decision to enable an OR-join task at a higher level. We recognise that due to the semantics of only considering tasks at the same level, the OR-join task could wait and result in a deadlock if a composite task is not sound and could deadlock. Because of these proposed abstractions from an EWF-net, we are now able to map to a Petri net like formalism. During an OR-join analysis, we are only required to consider the split and join behaviours of tasks and the cancellation set that is associated with a task. To support the cancellation feature of an EWF-net, we propose to map an EWF-net onto a Reset net.

3.3   Reset Nets

A Reset net is a Petri net with special reset arcs, that can clear the tokens in selected places. Reset arcs do not change the requirements of enabling a transition but when a transition fires, they will remove tokens from the specified places. The reset arcs are used to underpin the *rem* function that models the cancellation feature of EWF-nets, cf. Definition 1. This approach allows us to leverage existing literature and techniques in the area of Petri nets and Reset nets in particular [5, 6, 7, 8, 10, 11, 12].

Definition 3 (Reset net). *A Petri net is a tuple (P, T, F) where P is a set of places, T is a set of transitions, $P \cap T = \emptyset$ and $F \subseteq (P \times T) \cup (T \times P)$. A Reset net is a tuple*

$(P^\chi\, T^c\, F^c\, R)$ where $(P^\chi\, T^c\, F)$ is a Petri net and $R \in T \nrightarrow \mathbb{P}(P)$ is the set of reset arcs associated with every transition $t \in T$.

In the remainder of the paper, when we use the expression $F(x^c\, y)$, it denotes 1 if $(x^c\, y) \in F$ and 0 if $(x^c\, y) \notin F$. A reachable marking M' is defined by first removing tokens needed for enabling t from its input places ($\bullet t$), then removing all tokens from reset places and then finally adding tokens to the output places of t ($t\bullet$). The notation $M[P]$ denotes function restriction and restricts M to a set of places P, i.e., a projection.

Definition 4 (Enabling and firing Reset nets). *Let $(P^\chi\, T^c\, F^c\, R^c\, M)$ be a marked Reset net. A transition $t \in T$ is enabled iff $\bullet t \le M$. Firing t at marking M reaches marking M', denoted by $M \to^t M'$, iff $\bullet t \le M$ and $M' = (M - \bullet t)[P \setminus R(t)] + t\bullet$.*

Definition 5 (Occurrence sequence). *Let $((P^\chi\, T^c\, F^c\, R)^c\, M_0)$ be a marked Reset net. Let $M_1^c ...^c M_n$ be markings of the reset net and let $t_0^c\, t_1^c ...^c\, t_{n-1}$ be transitions in T. Sequence $s = M_0 t_0 M_1 ... t_{n-1} M_n$ is an occurrence sequence iff $M_i \to^{t_i} M_{i+1}$ for all $i^c\, 0 \le i \le n - 1$. A marking M'' is reachable from a marking M, written $M \to^* M''$, iff there is an occurrence sequence with initial marking M and final/last marking M''.*

To conclude this section, we define the notion of backward firing. This notion will be used to analyze coverability and is required for the OR-join analysis as is described in the remainder of this paper.

Definition 6 (Backward firing). *Let $(P^\chi\, T^c\, F^c\, R)$ be a Reset net and let M and M' be a markings of this net. $M' \dashrightarrow^t M$ if and only if it possible to fire a transition t backwards starting from M and resulting in M'.* [3]

$$M' \dashrightarrow^t M \Leftrightarrow \forall p \in R(t) : M(p) \le F(t^c\, p) \,\wedge$$
$$M'(p) = \begin{cases} (M(p) \overset{\cdot}{-} F(t^c\, p)) + F(p^c\, t) & \text{if } p \in P \setminus R(t) \\ F(p^c\, t) & \text{if } p \in R(t). \end{cases}$$

For any reset place p, $M(p) \le F(t^c\, p)$ because it is emptied when firing and then $F(t^c\, p)$ tokens are added. We do not require $M(p) = F(t^c\, p)$ because the aim is coverability and not reachability. M', i.e., the marking before (forward) firing t, should at least contain the *minimal* number of tokens required for enabling and resulting in a marking of at least M. Therefore, only $F(p^c\, t)$ tokens are assumed to be present in a reset place p.

4   Linking YAWL to Reset Nets

In this section, we describe how an EWF-net could be transformed into a Reset net. After the abstractions from multiple instances, composite tasks and internal places in a YAWL net, we can consider a YAWL net as having tasks with various split and join behaviours and possible cancellation sets and explicit and implicit conditions. For an

---

[3] For any natural numbers a, b: $a \overset{\cdot}{-} b$ is defined as $\max(a - b, 0)$.

EWF-net without OR-join tasks, there is then a straight-forward mapping into a Reset net. For an EWF-net with OR-join tasks, we propose to use the *optimistic* treatment whereby other OR-joins on the path are replaced with XOR-joins, and perform the necessary transformations.

4.1   Semantics of an EWF-Net Without OR-Joins

For every task t in an E2WF-net, we split t into t_{start} and t_{end} to support the various split and join constructs in YAWL. The number of t_{start} transitions depends on the join behaviour of a task and the number of t_{end} transitions depends on the split behaviour. Figure 7 illustrates the approach taken in the transformation.

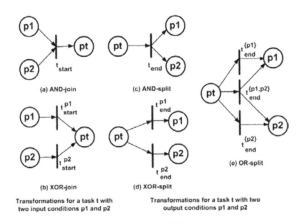

Fig. 7. Reset net transformations for YAWL split and join behaviours

Definition 7 (E2WF-Reset net). *Let* $N = (C^{ext}, i, o, T, F^{ext}, split, join, rem, nofi)$ *be an E2WF-net without OR-joins. A corresponding E2WF-Reset net is a tuple* (P, T', F', R) *such that*

$P = C^{ext} \cup \{p_t | t \in T\}$ *is a set of places,*
$T' = T_{start} \cup T_{end}$ *such that*
$T_{start} = \{t_{start} | t \in T \land join(t) = AND\}$
$\qquad \cup \{t^p_{start} | t \in T \land join(t) = XOR \land p \in \bullet t\},$
$T_{end} = \{t_{end} | t \in T \land split(t) = AND\}$
$\qquad \cup \{t^p_{end} | t \in T \land split(t) = XOR \land p \in t\bullet\}$
$\qquad \cup \{t^x_{end} | t \in T \land split(t) = OR \land x \subseteq t \bullet \land x \neq \emptyset\},$
$F' = \{(p, t_{start}) | t \in T \land join(t) = AND \land p \in \bullet t\}$
$\qquad \cup \{(t_{start}, p_t) | t \in T \land join(t) = AND\}$
$\qquad \cup \{(p_t, t_{end}) | t \in T \land split(t) = AND\}$
$\qquad \cup \{(t_{end}, p) | t \in T \land split(t) = AND \land p \in t\bullet\}$
$\qquad \cup \{(p, t^p_{start}) | t \in T \land join(t) = XOR \land p \in \bullet t\}$
$\qquad \cup \{(t^p_{start}, p_t) | t \in T \land join(t) = XOR \land p \in \bullet t\}$
$\qquad \cup \{(p_t, t^p_{end}) | t \in T \land split(t) = XOR \land p \in t\bullet\}$

$$\cup \{(t^p_{end}, p)|t \in T \wedge split(t) = XOR \wedge p \in t\bullet\}$$
$$\cup \{(p_t, t^x_{end})|t \in T \wedge split(t) = OR \wedge x \subseteq t\bullet \wedge x \neq \emptyset\}$$
$$\cup \{(t^x_{end}, p)|t \in T \wedge split(t) = OR \wedge p \in x \wedge x \subseteq t\bullet \wedge x \neq \emptyset\},$$
$R \in T' \nrightarrow \mathbb{P}(P)$ and $dom(R) \subseteq T_{end}$ such that
$$t \in T \wedge split(t) = AND$$
$$\Rightarrow R(t_{end}) = \{p_{t'}|t' \in rem(t) \cap T\} \cup (rem(t) \cap C^{ext}),$$
$$t \in T \wedge split(t) = XOR \wedge p \in t\bullet$$
$$\Rightarrow R(t^p_{end}) = \{p_{t'}|t' \in rem(t) \cap T\} \cup (rem(t) \cap C^{ext}),$$
$$t \in T \wedge x \subseteq t\bullet \wedge x \neq \emptyset \wedge split(t) = OR$$
$$\Rightarrow R(t^x_{end}) = \{p_{t'}|t' \in rem(t) \cap T\} \cup (rem(t) \cap C^{ext}).$$

The set of reset places for a given transition t_{end} has been defined in R to support the cancellation feature in YAWL. A place p_t is also introduced to represent an internal place between t_{start} and t_{end}. The flow relation F' is also modified so that the newly introduced places in P and transitions T' are properly connected.

The function *marked* returns the set of marked conditions in an EWF-net for a given marking M.

Definition 8 (Marked). *For a marking M of an E2WF-Reset net:*
$marked(M)=\{c \in dom(M) \mid M(c) > 0\}$.

The $\sqsubseteq$ relation indicates that M marks fewer or the same places as M'. This is a looser notion of smaller markings than $\leq$, because only the marking of places is considered and the number of tokens in a place is ignored. The notation $\sqsubset$ is used to indicate that M marks strictly less places than M'. The notation $M[C]$ restricts M to a set of conditions C, i.e., a projection. For instance, $M[t\bullet] \sqsubset M'[t\bullet]$, represents a comparison between M and M' that is restricted to the output places of t.

Definition 9 ($\sqsubseteq$). *Let M, M' be two markings of an E2WF-Reset net and C a set of conditions: $M \sqsubseteq M'$ iff $marked(M) \subseteq marked(M')$, $M \sqsubset M'$ iff $M \sqsubseteq M'$ and $\neg(M' \sqsubseteq M)$.*

We now define how a given marking M in an E2WF-net can be linked to a marking M^* in the corresponding E2WF-Reset net. For all the conditions that exist in an E2WF-net, they will be marked exactly the same in M^* and zero tokens for the newly introduced places in the E2WF-Reset net i.e. $M = M^*$.

Definition 10 (M^*). *Let (N, M) be a marked E2WF-net and N^* be the corresponding marked E2WF-Reset net of N, then M corresponds in a natural way to a marking of N^*. This marking marks all the places in N^* which correspond to conditions in N with the same number of tokens. We will refer to this as the corresponding marking and denote it as M^*.*

We define the enabling and firing rules for tasks in an E2WF-net using the transition firing rules as defined for Reset nets. Executing a task of an E2WF-net corresponds to executing the corresponding start and end transitions t_{start} and t_{end} of the E2WF-Reset net.

Definition 11 (Enabling and firing E2WF-net). *Let* $(N \cdot M)$ *be a marked E2WF-net and* $(N^* \cdot M^*)$ *be the corresponding marked E2WF-Reset net. A task* t *is enabled at* $(N \cdot M)$ *iff* $\bullet t \leq M^*$. *Firing* t *at* M *reaches* M', *denoted by* $M \rightarrow^t M'$ *iff for the corresponding start transition* t_{start} *and end transition* t_{end}, *we have* $M^* \rightarrow^{t_{start}} M'' \rightarrow^{t_{end}} M^{*'}$.

Note that this definition allows us to transfer typical Petri-net concepts such as reachability to E2WF-nets.

We are seeking a predicate *superM* to determine whether we can reach a marking that marks more places than M for a certain set of places. From a given marking M and a given set of places P', we can determine whether it is possible to reach a marking from M which marks more places in P'. If we define $P' = \bullet o\text{-}j$, a set of input conditions of an OR-join, then we can determine whether a bigger marking (restricted to places in P') exists for a given marking M (in which case the OR-join is not enabled).

Definition 12 (superM). *Let* $N = (P \cdot T \cdot F \cdot R \cdot M)$ *be a marked E2WF-net and* $P' \subseteq P$ *be a set of places for consideration, superM*$(N \cdot M \cdot P')$ *holds iff there is a marking* M' *such that* $M \rightarrow^* M'$ *and* $M[P'] \sqsubset M'[P']$.

4.2   Semantics of an EWF-Net with OR-Joins

The transformation from an EWF-net with OR-join tasks into an E2WF-OJ is identical to E2WF-Reset net transformation for all tasks that are not OR-join tasks. The additional steps to incorporate OR-join tasks are include creating a set OJ for the t_{start} transition of each OR-join task in the E2WF-net and adding t_{start} transitions in OJ into T_{start}.

Definition 13 (E2WF-OJ). *Let* N *be an EWF-net with OR-joins and* N^{ext} *be the E2WF-net of* N, *the corresponding E2WF-OJ is a tuple* $(P \cdot T'' \cdot F'' \cdot R \cdot OJ)$ *such that* P, T', T_{start}, T_{end}, F', *and* R *are as defined in Definition 7 and* T'', F'', OJ *are defined as follows:*

$$T'' = T'_{start} \cup T_{end},$$
$$T'_{start} = T_{start} \cup \{t_{start} | t \in T \land join(t) = OR)\},$$
$$F'' = F' \cup \{(p \cdot t_{start}) | t \in T \land join(t) = OR \land p \in \bullet t\} \cup$$
$$\{(t_{start} \cdot p_t) | t \in T \land join(t) = OR\}, and$$
$$OJ = \{t_{start} | t \in T \land join(t) = OR\}.$$

The function OJ-Remove is used to transform E2WF-OJ by replacing the join behaviour of all the OR-join tasks in an E2WF-net to XOR-join and removing the OR-join task in question. This effectively converts an E2WF-OJ into an E2WF-Reset net so that we can use the transition firing rules and superM predicate defined for Reset nets.

Definition 14 (OJ-Remove function). *Let* $N' = (P \cdot T \cdot F \cdot R \cdot OJ)$ *be an E2WF-OJ for an EWF-net* N *and* $j \in OJ$ *be an OR-join task under consideration. The function OJ-Remove*$(N' \cdot j)$ *returns* $(P' \cdot T' \cdot F' \cdot R')$ *such that*

$$P' = P,$$
$$T' = (T \setminus OJ) \cup \{t^p_{start} | t \in OJ \setminus \{j\} \land p \in \bullet^N t\},$$

$$F' = F \cap ((P' \times T') \cup (T' \times P'))$$
$$\cup \{(p^c t^p_{start}) | p \in \bullet^N t \ \wedge \ t \in OJ \setminus \{j\}\}$$
$$\cup \{(t^p_{start}{}^c p_t) | p \in \bullet^N t \ \wedge \ t \in OJ \setminus \{j\}\}$$
$$R' = R.$$

The firing rules for an E2WF-OJ are defined as follows. The firing rule for a transition t which is not an OR-join is the same as for Reset nets. For transitions $o\text{-}j$ that are OR-joins in E2WF-net, (i.e. $o\text{-}j \in OJ$), the firing rule is defined in two steps. We first use the OJ-Remove function to transform other OR-joins (except $o\text{-}j$) into XOR-joins and produce an equivalent E2WF-Reset net. We then check whether $superM$ holds. If $superM$ holds then the OR-join, $o\text{-}j$, should not be enabled at M. Otherwise, $o\text{-}j$ is enabled at M.

Definition 15 (Enabling rule). *Let* $(P^c T^c F^c R^c OJ^c M)$ *be a marked E2WF-OJ. A transition* $t \in T \setminus OJ$ *is enabled at* M *iff* $\bullet t \le M$. *A transition* $o\text{-}j \in OJ$ *is enabled at marking* M *iff at least one of its input places is marked and* $superM(OJ\text{-}Remove(P^c T^c F^c R^c OJ^c o\text{-}j)^c M^c \bullet o\text{-}j)$ *does not hold.*

Definition 16 (Forward firing). *When a transition* t *of an E2WF-OJ is enabled at a marking* M', *it can fire and a new marking* M *is reached.*

$$M' \rightarrow^t M \Leftrightarrow \forall p \in P : M'(p) \ge F(p^c t) \ \wedge$$
$$M(p) = \begin{cases} M'(p) - F(p^c t) + F(t^c p) & \text{if} \quad p \in P \setminus R(t) \\ F(t^c p) & \text{if} \quad p \in R(t). \end{cases}$$

Definition 17 (Reachable markings). *We denote* $M \rightarrow M'$ *iff there is a* $t \in T$ *such that* $M \rightarrow^t M'$. *We denote* $M \rightarrow^* M''$ *if there is an occurrence sequence from* M *to* M''.

We will now describe how the transformations will be performed for an EWF-net with two OR-join tasks C and D as shown in Figure 8. The shaded place indicates the explicit condition c_{BD} which has been added for the implicit condition between tasks B and D. Figure 9 shows an equivalent Reset-net for the E2WF-net in Figure 8 for OR-join analysis of task D. The OR-join task C is on the path to task D and the OJ-Remove function is applied to treat task C as an XOR-join task. Also note that OR-join task D has been removed from the net by the OJ-Remove function.

Consider a marking $M = c1 + c_{BD}$ of N where OR-join analysis for task D would be performed. The input places of task D are $c4$ and c_{BD}. We need to investigate whether

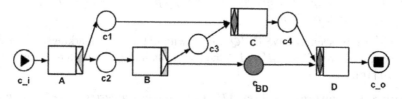

Fig. 8. An E2WF-net N with OR-join tasks C and D

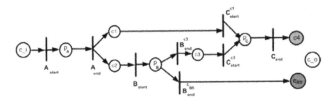

Fig. 9. An E2WF-Reset net for OR-join analysis of task D in Figure 8

it is possible to reach a marking that marks both $c4$ and c_{BD}. We can observe the sequence $c1 + c_{BD} \to^{C^{c1}_{start}} p_C + c_{BD} \to^{C_{end}} c4 + c_{BD}$ exists and that we can reach $M' = c4 + c_{BD}$ from M. Therefore, *superM* predicate holds as $M \to^* M'$ and $M[\{c4^c c_{BD}\}] \sqsubseteq M'[\{c4^c c_{BD}\}]$. The OR-join analysis for task D will conclude that D should not be enabled at marking M as it is possible to reach a marking from M that marks more input places of the OR-join than M does.

5   OR-Join Algorithm Proposal

The main objective of the OR-join algorithm is to determine, for a given OR-join, whether or not a marking M' is reachable from a given marking M that marks more input places of that OR-join exists. We perform this analysis by first transforming an EWF-net (with OR-joins) into an E2WF-Reset net for a given OR-join task and then by calling the OR-join algorithm. Our algorithm is based on backward search techniques for Well-Structured Transition Systems (WSTSs) [5, 7, 10, 11, 12]. The algorithm works backwards by computing the predecessor markings for a given marking, as opposed to the forward approach used in coverability tree algorithms. A Reset net can be represented as a WSTS and the backwards algorithm has been successfully applied to solve the coverability problems for Reset nets [7, 16].

5.1   Backward Algorithm for OR-Join Analysis

WSTSs are "a general class of infinite state systems for which decidability results rely on the existence of a well-quasi-ordering between states that is compatible with the transitions." [12]. The existence of a well-quasi-ordering over an infinite set of states ensures the decidability of termination and coverability properties [7, 12].

Definition 18 (Well-Structured Transition System [7]). *A well-structured transition system (WSTS) is a structure $S = \langle Q^c \to^c \leq \rangle$ such that $Q = \{m^c ...\}$ is a set of states, $\to \subseteq Q \times Q$ is a set of transitions, $\leq \subseteq Q \times Q$ is a well-quasi-ordering (wqo) on the set of states, satisfying the simple monotonicity property, $m \to m'$ and $m_1 \geq m$ imply $m_1 \to m'_1$ for some $m'_1 \geq m'$.*

Reset nets can be seen as a WSTS $\langle Q^c \to^c \leq \rangle$ with Q the set of markings, $M \to M'$ if for some t, we have $M \to^t M'$ and $\leq$ the corresponding $\leq$ order on markings (which is a wqo) [16].

Definition 19 (Upward-closed set [12]). *Given a quasi-ordering $\leq$ on X, an upward-closed set is any set $I \subseteq X$ such that $y \geq x$ and $x \in I$ entail $y \in I$. To any $x \in X$*

we associate $\uparrow x =^{def} \{y | y \geq x\}$. *A basis of an upward-closed* I *is a set* I^b *such that* $I = \bigcup_{x \in I^b} \uparrow x$.

Given a WSTS $\langle Q, \rightarrow, \leq \rangle$ and a set of states $I \subseteq Q$, $Pred(I)$, $pb(I)$ and $Pred^*(I)$ can be defined [16]. The immediate predecessors of I: $Pred(I) = \{x | x \rightarrow y \wedge y \in I\}$, all predecessor states of I, $Pred^*(I) = \{x | x \rightarrow^* y \wedge y \in I\}$ and $pb(I) = \bigcup_{y \in I} pb(y)$ where $pb(y)$ yields a finite basis of $\uparrow Pred(\uparrow\{y\})$ (i.e., $pb(y)$ yields a finite set such that $\uparrow pb(y) = \uparrow Pred(\uparrow\{y\})$) [16]. The coverability problem for a Reset net is as follows: given two markings x and y can we reach $y' \geq y$ starting from x [16]. Provided that $\leq$ is decidable and $pb(y)$ exists and can be effectively computed [12], the backwards reachability analysis can be performed to decide the coverability [7, 10, 16]. $\{y\}$ is a basis of upward closed set $\uparrow\{y\}$ and we can determine that y is coverable from x if there exists a $x' \in Pred^*(\uparrow\{y\})$ such that $x' \leq x$ (because $\leq$ is a wqo). As $\uparrow\{y\}$ is upward-closed, $Pred^*(\uparrow\{y\})$ is upward-closed [12]. We can compute a finite basis of $Pred^*(\uparrow\{y\})$ as the limit of the sequence $I_0 \subseteq I_1 \subseteq ...$ where $I_0 =^{def} \{y\}$ and $I_{n+1} =^{def} I_n \cup pb(I_n)$ [16]. The sequence eventually stabilises at some I_n when $\uparrow I_{n+1} = \uparrow I_n$ and we have reached a stabilisation point that has the property $\uparrow I_n = Pred^*(\uparrow\{y\})$ [16]. The coverability question now becomes: is there an $x' \in \uparrow I_n$ such that $x' \leq x$. I_n is a finite basis for $Pred^*(\uparrow\{y\})$ and the coverability question can now be answered by testing whether there exists a $x' \in I_n$ such that $x' \leq x$.

We now present the procedures that operationalise the coverability question for Reset nets. The procedure **Coverable** returns a Boolean value to indicate whether a marking t is coverable from a marking s of a Reset net [16].

PROCEDURE Coverable (Marking x, y): Boolean
Marking x';
BEGIN
   for $x' \leq x$ **do**
      if $x' \in$ **FiniteBasisPred**$^*(\{y\})$ **then return** TRUE; **end if**;
   end for;
   return FALSE;
END

The procedure **FiniteBasisPred*** returns a set of markings which represents a finite basis of all predecessors and is based on the method described in [16].

PROCEDURE FiniteBasisPred* (SET Marking I): SET Marking
SET Marking K, K_{next};
BEGIN
   $K := I$; $K_{next} := K \cup \mathbf{pb}(K)$;
   while not IsUpwardEqual(K, K_{next}) **do**
      $K := K_{next}$; $K_{next} := K \cup \mathbf{pb}(K)$;
   end while;
   return K;
END

The procedure call **IsUpwardEqual**($K^c K_{next}$) is used to detect whether the stabilisation point has been reached i.e. $\uparrow K_{next} = \uparrow K$, cf. [11]. The procedure **pb**(I) returns $pb(I)$ such that $pb(I) = \bigcup_{x \in I} pb(x)$ [16].

PROCEDURE pb (SET Marking I): SET Marking
Set Marking $Z = \emptyset$; Marking M;
BEGIN
   for $M \in I$ **do** $Z := Z \cup$ **pb**(M); **end for**;
   return Z;
END

pb(M) is effectively computed for Reset nets by "executing the transitions backwards and setting a place to the minimum number of tokens required to fire the transition if it caused a reset on this place" [16].[4] Note that, in our case, this minimum is one as we do not have weighted arcs. We will make use of backward firing rule as defined in Definition 6. For each transition $t \in T$, we determine whether an M' exists such that $M' \dashrightarrow^t M$. Hence, $pb(M) = \{M' | \exists_{t \in T} M' \dashrightarrow^t M\}$.

PROCEDURE pb (Marking M): SET Marking
SET Marking $Z = \emptyset$;
BEGIN
   for $t \in T$ **do**
      if $M[R(t)] \leq t \bullet [R(t)]$ **then**
         $Z := Z \cup \{(M \dotminus t \bullet + \bullet t)[P \setminus R(t)] + (M + \bullet t)[R(t)]\}$;
      end if;
   end for;
   return Z;
END

We can then apply the coverability findings of a Reset net to the OR-join analysis. Let ($N^c M$) be a marked E2WF-net, o-j be the OR-join task under consideration, X be $\bullet$ o-j, N' be the corresponding E2WF-Reset net and Y be a set of markings such that each marking in Y has only one token in each of the marked input places of o-j in M and one token in exactly one of the unmarked input places of the o-j in M. To determine whether o-j should be enabled at M, we need to determine whether there exists a $M' \in Pred^*(M_w)$ such that $M' \leq M$ for each of the markings $M_w \in Y$ (coverability question). Each marking M_w in Y satisfies the condition $M[X] \sqsubset M_w[X]$, i.e. M_w has tokens in more input places of the OR-join o-j and if M_w can be reached from M, the OR-join is not enabled. The procedure **OrJoinEnabled** is called with parameters M and X and it returns a Boolean value to indicate whether the OR-join should be enabled at M.

---

[4] Note that the algorithm described in [16] is incorrect. On Page 105 in [16], $pb(M)$ is defined in a rather naive way. Applying $pb(M)$ to the empty marking yields a counter example, since it is not a finite basis for $\uparrow Pred^*(\uparrow\{M\})$.

PROCEDURE OrJoinEnabled (Marking M, SET Place X): Boolean
SET Marking Y; Marking M_w;
BEGIN
$\quad Y = \{q + \sum_{p \in X : M(p) > 0} p \mid q \in X \ \wedge \ M(q) = 0\};$
$\quad$**for** $M_w \in Y$ **do**
$\quad\quad$**if** Coverable$(M \triangleleft M_w)$ **then return** FALSE; **end if**;
$\quad$**end for**;
$\quad$**return** TRUE;
END

5.2   A Worked Example

Throughout the paper we have shown several examples where it is a non-trivial task to decide if an OR-join is enabled or not. Clearly, the algorithm can be applied successfully to these situations. To illustrate its inner working in some detail we use one last example.

Consider a marking $M = c1 + c7$ in Figure 10 where the OR-join analysis for task G is carried out. It is possible to have an occurrence sequence, $c1 + c7 \rightarrow^B c_{BB} + c3 + c7 \rightarrow^E c_{BB} + c5 + c7 \rightarrow^B c_{BB} + c3 + c5 + c7 \rightarrow^D c4 + c5 + c7 \rightarrow^F c6 + c7$. As a result, $c6 + c7$ is a reachable marking from $c1 + c7$ and the OR-join should not be enabled at marking M. The evaluation will start with a call to the procedure **OrJoinEnabled**$(c1 + c7 \triangleleft \{c6 \triangleleft c7\})$. $Y := \{c6 + c7\}$ and for $M_w = c6 + c7$, we will obtain a finite basis of all the predecessors of $c6 + c7$. Figure 12 illustrates the backwards reachability analysis [11], with the basis of the predecessor markings for $c6 + c7$. It can be seen that $c1 + c7$ is a predecessor of $c6 + c7$. $M' \leq M$ includes the following markings $\{c1 \triangleleft c7 \triangleleft c1 + c7\}$. As $M' = c1 + c7$ is in the predecessors for

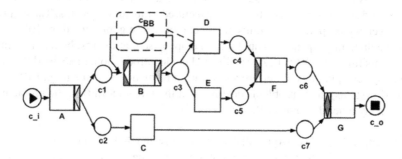

Fig. 10. A YAWL net with an OR-join task G and cancellation

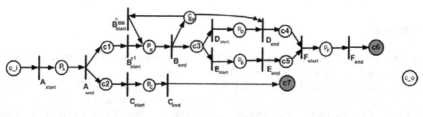

Fig. 11. A corresponding Reset net for Figure 10 (note the double-headed arrow denoting the reset arc from C_{BB} to D_{end})

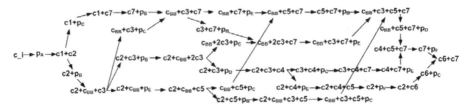

Fig. 12. Illustration of backwards reachablility analysis

$c6 + c7$, the procedure will return FALSE, concluding that the OR-join should not be enabled at M.

6   Conclusion

This paper focuses on the OR-join construct in YAWL and proposes a new semantics. The decision to enable an OR-join task cannot be made locally: an OR-join task should only be enabled when there is at least one token in one of the input conditions and there is no possibility of a token arriving at one of the yet unmarked input conditions of the OR-join. Otherwise, the OR-join task should wait for synchronisation. Instead of ignoring other OR-joins on the path, we propose two alternative approaches (optimistic or pessimistic) for OR-joins which are on the path of other OR-joins. Reset nets are used as formal basis for OR-join analysis to support cancellation features. This is made possible by the fact that we can abstract from the concepts of YAWL such as multiple instances, composite task and internal state transitions of a task. We present transformation rules from a YAWL model with OR-joins to a Reset net for a specific OR-join analysis. We then propose an OR-join evaluation algorithm which is based on the backward search techniques for Well-Structured Transition Systems. The algorithm does not yet exploit potential optimisation techniques as e.g. presented in [10].

To conclude the paper, we would like to emphasise that the results reported in this paper are not limited to YAWL. As is indicated in the introduction, many workflow management systems, but also other process-aware information systems (e.g., ERP, CRM, and PDM systems), have problems dealing with the OR-join. In fact, the problem surfaces in many other domains [19].

Acknowledgements. We would like to especially thank Philippe Schnoebelen and Jerome Leroux for their valuable input on the issue of decidability of OR-join algorithm and for many useful references provided in the area of Reset nets.

References

1. W.M.P. van der Aalst. The Application of Petri Nets to Workflow Management. *The Journal of Circuits, Systems and Computers*, 8(1):21–66, 1998.
2. W.M.P van der Aalst, J. Desel, and E. Kindler. On the Semantics of EPCs: A Vicious Circle. In M. Rump and F.J. Nüttgens, editors, *Proceedings of the EPK 2002: Business Process Management using EPCs*, pages 71–80, Trier, Germany, 2002. Gesellschaft für Informatik, Bonn.

3. W.M.P. van der Aalst and A.H.M ter Hofstede. YAWL: Yet Another Workflow Language. *Information Systems*, 30(4):245–275, June 2005.

4. W.M.P van der Aalst, A.H.M ter Hofstede, B.Kiepuszewski, and A.P.Barros. Workflow Patterns. *Distributed and Parallel Databases*, 14:5–51, 2003.

5. P.A. Abdulla, K. Cerans, B. Jonsson, and Y.-K. Tsay. General decidability theorems for infinite-state systems. In *Proceedings of the 11th Annual IEEE Symposium on Logic in Computer Science (27 - 30 July)*, pages 313–321, New Brunswick, NJ, July 1996. IEEE Computer Society.

6. P. Darondeau. Unbounded Petri net Synthesis. In J. Desel, W. Reisig, and G. Rozenberg, editors, *Lectures on Concurrency and Petri Nets, Advances in Petri Nets*, volume 3098 of *Lecture Notes in Computer Science*, pages 413–428, Eichstätt,Germany, 2003. Springer-Verlag.

7. C. Dufourd, A. Finkel, and Ph. Schnoebelen. Reset Nets Between Decidability and Undecidability. In K. Larsen, S. Skyum, and G. Winskel, editors, *Proceedings of the 25th International Colloquium on Automata, Languages and Programming*, volume 1443 of *Lecture Notes in Computer Science*, pages 103–115, Aalborg, Denmark, July 1998. Springer-Verlag.

8. C. Dufourd, P. Jančar, and Ph. Schnoebelen. Boundedness of Reset P/T Nets. In J. Wiedermann, P. van Emde Boas, and M. Nielsen, editors, *Lectures on Concurrency and Petri Nets*, volume 1644 of *Lecture Notes in Computer Science*, pages 301–310, Prague, Czech Republic, July 1999. Springer-Verlag.

9. Eastman Software. *RouteBuilder Tool User's Guide*. Eastman Software, Inc, Billerica, MA, USA, 1998.

10. A. Finkel, J.-F. Raskin, M. Samuelides, and L. van Begin. Monotonic Extensions of Petri Nets: Forward and Backward Search Revisited. *Electronic Notes in Theoretical Computer Science*, 68(6):1–22, 2002.

11. A. Finkel and Ph. Schnoebelen. Fundamental Structures in Well-Structured Infinite Transition Systems. In C.L. Lucchesi and A.V. Moura, editors, *Theoretical Informatics: Third Latin American Symposium, Campinas, LATIN'98 (20 - 24 April)*, volume 1380 of *Lecture Notes in Computer Science*, pages 102–118, Campinas, Brazil, 1998. Springer-Verlag.

12. A. Finkel and Ph. Schnoebelen. Well-structured Transition Systems everywhere! *Theoretical Computer Science*, 256(1–2):63–92, April 2001.

13. S. Jablonski and C. Bussler. *Workflow Management: Modeling Concepts, Architecture, and Implementation*. International Thomson Computer Press, London, UK, 1996.

14. B. Kiepuszewski. *Expressiveness and Suitability of Languages for Control Flow Modelling in Workflows*. Phd thesis, Queensland University of Technology, Brisbane, Australia, 2003.

15. E. Kindler. On the Semantics of EPCs: A Framework for Resolving the Vicious Circle. In J. Desel, B. Pernici, and M. Weske, editors, *Proceedings of 2nd International Conference on Business Process Management*, volume 3080 of *Lecture Notes in Computer Science*, pages 82–97, Potsdam, Germany, 2004. Springer-Verlag.

16. M. Leuschel and H. Lehmann. Coverability of Reset Petri Nets and other Well-Structured Transition Systems by Partial Deduction. In J. Lloyd et al., editors, *Proceedings of Computational Logic 2000*, volume 1861 of *Lecture Notes in Artificial Intelligence*, pages 101–115, London, UK, 2000. Springer-Verlag.

17. F. Leymann and D. Roller. *Production Workflow: Concepts and Techniques*. Prentice-Hall PTR, Upper Saddle River, New Jersey, USA, 1999.

18. P. Rittgen. Modified EPCs and their Formal Semantics. Technical Report 99/19, Institute of Information Systems, University Koblenz-Landau, Koblenz, Germany, 1999.

19. A. Yakovlev, M. Kishinevsky, A. Kondratyev, L. Lavagno, and M. Pietkiewicz-Koutny. On the Models for Asynchronous Circuit Behaviour with OR Causality. *Formal Methods in System Design*, 9(3):189–233, 1996.

The ProM Framework:
A New Era in Process Mining Tool Support

B.F. van Dongen, A.K.A. de Medeiros, H.M.W. Verbeek, A.J.M.M. Weijters,
and W.M.P. van der Aalst

Department of Technology Management, Eindhoven University of Technology,
P.O. Box 513, NL-5600 MB, Eindhoven,
The Netherlands
b.f.v.dongen@tue.nl

Abstract. Under the umbrella of buzzwords such as "Business Activity Monitoring" (BAM) and "Business Process Intelligence" (BPI) both academic (e.g., EMiT, Little Thumb, InWoLvE, Process Miner, and MinSoN) and commercial tools (e.g., ARIS PPM, HP BPI, and ILOG JViews) have been developed. The goal of these tools is to extract knowledge from event logs (e.g., transaction logs in an ERP system or audit trails in a WFM system), i.e., to do *process mining*. Unfortunately, tools use different formats for reading/storing log files and present their results in different ways. This makes it difficult to use different tools on the same data set and to compare the mining results. Furthermore, some of these tools implement concepts that can be very useful in the other tools but it is often difficult to combine tools. As a result, researchers working on new process mining techniques are forced to build a mining infrastructure from scratch or test their techniques in an isolated way, disconnected from any practical applications. To overcome these kind of problems, we have developed the ProM framework, i.e., an "pluggable" environment for process mining. The framework is flexible with respect to the input and output format, and is also open enough to allow for the easy reuse of code during the implementation of new process mining ideas. This paper introduces the ProM framework and gives an overview of the plug-ins that have been developed.

1   Introduction

The research domain *process mining* is relatively new. A complete overview of recent process mining research is beyond the scope of this paper. Therefore, we limit ourselves to a brief introduction to this topic and refer to [3, 4] and the http://www.processmining.org web page for a more complete overview.

The goal of process mining is to extract information about processes from transaction logs. It assumes that it is possible to record events such that (i) each event refers to an *activity* (i.e., a well-defined step in the process), (ii) each event refers to a *case* (i.e., a process instance), (iii) each event can have a *performer* also referred to as *originator* (the actor executing or initiating the activity), and (iv) events have a *timestamp* and are totally ordered. Table 1 shows an example

G. Ciardo and P. Darondeau (Eds.): ICATPN 2005, LNCS 3536, pp. 444–454, 2005.

Table 1. An event log (audit trail)

case id	activity id	originator	case id	activity id	originator
case 1	activity A	John	case 5	activity A	Sue
case 2	activity A	John	case 4	activity C	Carol
case 3	activity A	Sue	case 1	activity D	Pete
case 3	activity B	Carol	case 3	activity C	Sue
case 1	activity B	Mike	case 3	activity D	Pete
case 1	activity C	John	case 4	activity B	Sue
case 2	activity C	Mike	case 5	activity E	Clare
case 4	activity A	Sue	case 5	activity D	Clare
case 2	activity B	John	case 4	activity D	Pete
case 2	activity D	Pete			

of a log involving 19 events, 5 activities, and 6 originators. In addition to the information shown in this table, some event logs contain more information on the case itself, i.e., data elements referring to properties of the case. For example, the case handling system FLOWer logs every modification of some data element.

Event logs such as the one shown in Table 1 are used as the starting point for mining. We distinguish three different perspectives: (1) the process perspective, (2) the organizational perspective and (3) the case perspective. The *process perspective* focuses on the control-flow, i.e., the ordering of activities, as shown in Figure 1(a). The goal of mining this perspective is to find a good characterization of all possible paths, e.g., expressed in terms of a Petri net [15] or Event-driven Process Chain (EPC) [11, 12]. The *organizational perspective* focuses on the originator field, i.e., which performers are involved and how are they related. The goal is to either structure the organization by classifying people in terms of roles and organizational units (Figure 1(b)) or to show relation between individual performers (i.e., build a social network as described in [2] and references there, and as shown in Figure 1(c)). The *case perspective* focuses on properties of cases. Cases can be characterized by their path in the process or by the originators working on a case. However, cases can also be characterized by the values of the corresponding data elements. For example, if a case represents a replenishment order, it is interesting to know the supplier or the number of products ordered.

Orthogonal to the three perspectives (process, organization, and case), the result of a mining effort may refer to *logical* issues and/or *performance* issues. For example, process mining can focus on performance issues such as flow time, the utilization of performers or execution frequencies.

After developing *ad hoc* tools for the mining of the process perspective (e.g., EMiT [1] and Little Thumb [17]) and other *ad hoc* tools (e.g., MinSoN [2]) for the other mining perspectives we started the design of a flexible framework in which different algorithms for each of the perspectives can be plugged in.

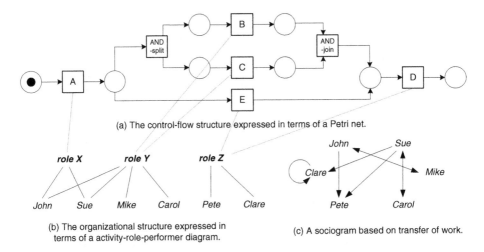

(a) The control-flow structure expressed in terms of a Petri net.

(b) The organizational structure expressed in terms of a activity-role-performer diagram.

(c) A sociogram based on transfer of work.

Fig. 1. Some mining results for the process perspective (a) and organizational (b and c) perspective based on the event log shown in Table 1

2   Architecture

As indicated in the introduction, the basis for all process mining techniques is a *process log*. Such a log is a file generated by some information system, with information about the execution of a process. Since each information system has its own format for storing log files, we have developed a generic XML format for the ProM framework to store a log in. This format was based on a thorough comparison of the input needs of various existing (ad-hoc) process mining tools and the information typically contained in an audit trail or transaction log of some complex information system (e.g., an ERP or a WFM system).

Another important feature of the ProM framework is that it allows for interaction between a large number of so-called plug-ins. A plug-in is basically the implementation of an algorithm that is of some use in the process mining area, where the implementation agrees with the framework. Such plug-ins can be added to the entire framework with relative ease: Once the plug-in is ready it can be added to the framework by adding its name to some *ini*-file. Note that there is no need to modify the ProM framework (e.g., recompiling the code) when adding new plug-ins, i.e., it is a truly "pluggable" environment. This in contradiction to open-source initiatives, such as the data mining software *Weka*[1].

In Figure 2, we show an overview of the framework that we developed. It explains the relations between the framework, the process log format, and the plug-ins. As Figure 2 shows, the ProM framework can read files in the XML format through the *Log filter* component. This component is able to deal with large data sets and sorts the events within a case on their timestamps before

---

[1] Weka is available from http://www.cs.waikato.ac.nz/~ml/weka/

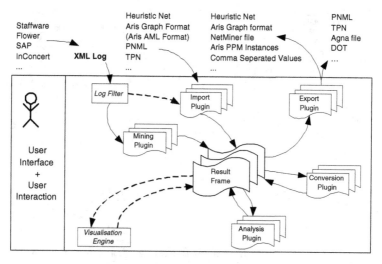

Fig. 2. Overview of the ProM framework

the actual mining starts. (If no timestamps are present, the order in the XML file is preserved.) Through the *Import plug-ins* a wide variety of models can be loaded ranging from a Petri net to logical formulas. The *Mining plug-ins* do the actual mining and the result is stored in memory, and in a window on the ProM desktop. The framework allows plug-ins to operate on each others results in a standardized way. Typically, the mining results contain some kind of visualization, e.g., displaying a Petri net [15], an EPC [12] or a Social network [2], or further analysis or conversion. The *Analysis plug-ins* take a mining result an analyze it, e.g., calculating a place invariant for a resulting Petri net. The *Conversion plug-ins* take a mining result and transform it into another format, e.g., transforming an EPC into a Petri net. In the remainder of this section, we describe both the process log format and the plug-ins.

2.1   Process Log Format

Figure 3(a) visualizes the XML schema that specifies the process log format. The root element is a *WorkflowLog* element. (The name "workflow log" is chosen for backwards compatibility and we prefer to talk about process log.) The *WorkflowLog* element contains (in the given order) an optional *Data* element, an optional *Source* element, and a number of *Process* elements. A *Data* element allows for storing arbitrary textual data, and contains a list of *Attribute* elements. A *Source* element can be used to store information about the information system this log originated from. A *Process* element refers to a specific process in an information system. Since most information systems typically control several processes, multiple *Process* elements may exist in a log file. A *ProcessInstance* is an instance of the process, i.e., a case. An *AuditTrailEntry* may refer to an activity (*WorkflowModelElement*), an eventtype (*Eventtype*), a timestamp (*Timestamp*), and a person that executed the activity (*Originator*).

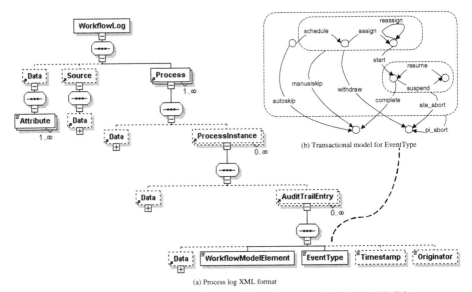

Fig. 3. Process log XML format (a) and transactional model (b)

As will be clear from what was mentioned earlier, a log file typically contains information about events that took place in a system. Such events typically refer to a case and a specific activity within that case. Examples of such events are:

- The activity *send message* is now ready to be executed.
- The activity *wait for incoming transmission* has not been started for three weeks.
- The case with ID 203453 was aborted.

In order to be able to talk about these events in a standard way, we developed a transactional model that shows the events that we assume can appear in a log. Again this model is based on analyzing the different types of logs in real-life systems (e.g., Staffware, SAP, FLOWer, etc.) Figure 3(b) shows the transactional model.

When an activity is created, it is either *schedule*d or skipped automatically (*autoskip*). Scheduling an activity means that the control over that activity is put into the information system. The information system can now *assign* this activity to a certain person or group of persons. It is possible to *reassign* an assigned activity to another person or group of persons. This can be done by the system, or by a user. A user can *start* working on an activity that was assigned to him, or some user can decide to *withdraw* the activity or skip it manually (*manualskip*), which can even happen before the activity was assigned. The main difference between a withdrawal and a manual skip is the fact that after the manual skip the activity has been executed correctly, while after a withdrawal it is not. The user that started an activity can *suspend* and *resume* the activity several times, but in the end he either has to *complete* or abort (*ate_abort*) it. Note the activity can get aborted (*pi_abort*) during its entire life cycle.

We do not claim that we have captured all possible behavior of all systems. However, we have verified our transactional model against several commercial systems and they all seem to fit nicely. Nonetheless, in the XML format, we allow for other event types to be defined on the fly.

2.2   Plug-ins

In this section, we provide an overview of the plug-ins as currently implemented in the context of the ProM framework. For more technical documentation and scientific publications, we refer to our website http://www.processmining.org. As shown in Figure 2 there are five kinds of plug-ins:

Mining plug-ins which implement some mining algorithm, e.g., mining algorithms that construct a Petri net based on some event log.

Export plug-ins which implement some "save as" functionality for some objects (such as graphs). For example, there are plug-ins to save EPCs, Petri nets (e.g., in PNML format [7]), spreadsheets, etc.

Import plug-ins which implement an "open" functionality for exported objects, e.g., load instance-EPCs from ARIS PPM.

Analysis plug-ins which typically implement some property analysis on some mining result. For example, for Petri nets there is a plug-in which constructs place invariants, transition invariants, and a coverability graph. However, there are also analysis plug-ins to compare a log and a model (i.e., conformance testing) or a log and an LTL formula.

Conversion plug-ins which implement conversions between different data formats, e.g., from EPCs to Petri nets.

The current version of the framework contains a large set of plug-ins. A detailed description of these plug-ins is beyond the scope of this paper. Currently, there are nine export plug-ins, four import plug-ins, seven analysis plug-ins, and three conversion plug-ins. Therefore, we only mention some of the available *mining plug-ins*. For each of the three perspectives which were mentioned in the introduction, there are different mining plug-ins.
For the process perspective, four plug-ins are available:

α-**algorithm** which implements the α-algorithm [5] and its extensions as developed by the authors. The α-algorithm constructs a Petri net which models the process recorded in the log.

Tshinghua-α algorithm which uses timestamps in the log files to construct a Petri net. It is related to the α algorithm, but uses a different approach. Details can be found in [18]. It is interesting to note that this mining plug-in was the first plug-in developed by researchers outside of our research group. Researchers from Tshinghua University in China (Jianmin Wang and Wen Lijie) were able to develop and integrate this plug-in without any help or changes to the framework.

Genetic algorithm which uses genetic algorithms to tackle possible noise in the log file as described in [13]. Its output format is a heuristics net (which can be converted into an EPC or a Petri net).

Multi-phase mining which implements a series of process mining algorithms that use instance graphs (comparable to runs) as an intermediate format. The two-phase approach resembles the aggregation process in Aris PPM.

For the organizational perspective, one plug-in is available:

Social network miner which uses the log file to determine a social network of people [2]. It requires the log file to contain the *Originator* element.

Finally, for the case perspective, also one plug-in is available:

Case data extraction which can be used for interfacing with a number of standard *knowledge discovering tools*, e.g., Viscovery and SPSS AnswerTree.

Sometimes a collection of plug-ins is needed to achieve the desired functionality. An example is the *LTL-checker* which checks whether logs satisfy some Linear Temporal Logic (LTL) formula. For example, the LTL-checker can be used to check the "four eyes" principle, i.e., two activities within the same case should not be executed by the same person to avoid possible fraud. The LTL-checker combines a mining plug-in (to get the log), an import plug-in (to load the file with predefined LTL formulas), and an analysis plug-in (to do the actual checking).

3   User Interface

Since the ProM framework contains a large number of plug-ins, it is impossible to discuss them all in detail. Therefore, we only present some screenshots of a few plug-ins that we applied to the example of Table 1. In Figure 4, we show the result of applying the α-mining plug-in to the example. The default settings of the plug-in were used, and the result is a Petri net that is behaviorally equivalent to the one presented in Figure 1. In Figure 5, we show the result of the social network mining plug-in. We used the *handover of work* setting, considering only direct succession, to generate this figure. Comparing it to Figure 1(c) shows that the result is an isomorphic graph (i.e. the result is the same).

Petri nets are not the only modelling language supported by the framework. Instead, we also have built-in support for EPCs (Event-driven Process Chains). In Figure 6, we show the result of the multi-phase mining plug-in. The result is an aggregated EPC describing the behavior of all cases. Note that it allows for more behavior than the Petri net, since the connectors are of the type *logical or*. In Figure 7 we show the user interface of the analysis plug-in that can be used for the verification of EPCs.

In this section, we have shown some screenshots to provide an overview of the framework. We would like to stress that we only showed a few plug-ins of the many that are available. We would also like to point out that most plug-ins allow for user interaction. The latter it important because process mining is often an interactive process where human interpretation is important and additional knowledge can be used to improve the mining result.

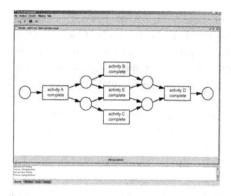

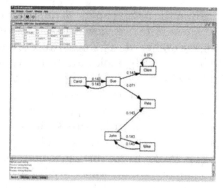

Fig. 4. The α-mining plug-in   **Fig. 5.** The social network mining plug-in

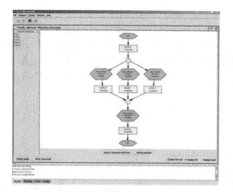

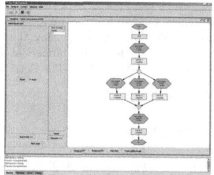

Fig. 6. The discovered EPC   **Fig. 7.** Analyzing the EPC for correctness

4   Related Work

Process mining can be seen as a tool in the context of Business Activity Monitoring (BAM) and Business (Process) Intelligence (BPI). In [9] a BPI toolset on top of HP's Process Manager is described. The BPI tools set includes a so-called "BPI Process Mining Engine". However, this engine does not provide any techniques as discussed before. Instead it uses generic mining tools such as SAS Enterprise Miner for the generation of decision trees relating attributes of cases to information about execution paths (e.g., duration). In [14] the PISA tool is described which can be used to extract performance metrics from workflow logs. Similar diagnostics are provided by the ARIS Process Performance Manager (PPM) [11]. The latter tool is commercially available and a customized version of PPM is the Staffware Process Monitor (SPM) [16] which is tailored towards mining Staffware logs.[2]

---

[2] Note that the ProM Framework interfaces with Staffware, SPM, ARIS Toolset, and ARIS PPM.

Given the many papers on mining the process perspective it is not possible to give a complete overview. Instead we refer to [3, 5]. Historically, Cook et al. [8] and Agrawal et al. [6] started to work on the problem addressed in this paper. Herbst et al. [10] took an alternative approach which allows for dealing with duplicate activities. The authors of this paper have been involved in different variants of the so-called α-algorithm [1, 5, 17]. Each of the approaches has its pros and its cons. Most approaches that are able to discover concurrency have problems dealing with issues such as duplicate activities, hidden activities, non-free-choice constructs, noise, and incompleteness.

The ProM framework subsumes process mining tools like EMiT [1], Little Thumb [17] and MinSon [2]. Most of these tools had their own format to store log files in, and had their own limitations. The tool EMiT for example was unable to deal with log files of more than 1000 cases. To be able to use all these tools together in an interactive way, we developed the ProM framework, which can be seen as a successor of all these tools. The framework allows researchers to seamlessly combine their own algorithms with algorithms from other people. Furthermore, using the framework allows you to interface with many existing tools, both commercial and public. These tools include: the Aris Toolset, Aris PPM, Woflan, The Petri net kernel, Netminer, Agna, Dot, Viscovery, etc.

5   Conclusion

The ProM framework integrates the functionality of several existing process mining tools and provides many additional process mining plug-ins. The ProM framework supports multiple formats and multiple languages, e.g., Petri nets, EPCs, Social Networks, etc. The plug-ins can be used in several ways and combined to be applied in real-life situations. We encourage developers and researchers to use the ProM framework for implementing new ideas. It is easy to add a new plug-in. For adding new plug-ins it suffices to add a few lines to the configuration files and no changes to the code are necessary, i.e., new mining plug-ins can be added without re-compiling the source code. Experiences with adding the *Thingua-α plug-in* and the *Social network miner* show that this is indeed rather straightforward.

Acknowledgements

The authors would like to thank all people that have been involved in the development and implementation of the ProM framework. In particular we would like to thank Minseok Song, Jianmin Wang and Wen Lijie for their contributions. Furthermore, we would like to thank IDS Scheer for providing us with Aris PPM and the Aris toolset. Last, but certainly not least, we would like to thank Peter van den Brand for doing the major part of the implementation work for us.

References

1. W.M.P. van der Aalst and B.F. van Dongen. Discovering Workflow Performance Models from Timed Logs. In Y. Han, S. Tai, and D. Wikarski, editors, *International Conference on Engineering and Deployment of Cooperative Information Systems (EDCIS 2002)*, volume 2480 of *Lecture Notes in Computer Science*, pages 45–63. Springer-Verlag, Berlin, 2002.

2. W.M.P. van der Aalst and M. Song. Mining Social Networks: Uncovering interaction patterns in business processes. In J. Desel, B. Pernici, and M. Weske, editors, *International Conference on Business Process Management (BPM 2004)*, volume 3080 of *Lecture Notes in Computer Science*, pages 244–260. Springer-Verlag, Berlin, 2004.

3. W.M.P. van der Aalst, B.F. van Dongen, J. Herbst, L. Maruster, G. Schimm, and A.J.M.M. Weijters. Workflow Mining: A Survey of Issues and Approaches. *Data and Knowledge Engineering*, 47(2):237–267, 2003.

4. W.M.P. van der Aalst and A.J.M.M. Weijters, editors. *Process Mining*, Special Issue of Computers in Industry, Volume 53, Number 3. Elsevier Science Publishers, Amsterdam, 2004.

5. W.M.P. van der Aalst, A.J.M.M. Weijters, and L. Maruster. Workflow Mining: Discovering Process Models from Event Logs. *IEEE Transactions on Knowledge and Data Engineering*, 16(9):1128–1142, 2004.

6. R. Agrawal, D. Gunopulos, and F. Leymann. Mining Process Models from Workflow Logs. In *Sixth International Conference on Extending Database Technology*, pages 469–483, 1998.

7. J. Billington and et. al. The Petri Net Markup Language: Concepts, Technology, and Tools. In W.M.P. van der Aalst and E. Best, editors, *Application and Theory of Petri Nets 2003*, volume 2679 of *Lecture Notes in Computer Science*, pages 483–506. Springer-Verlag, Berlin, 2003.

8. J.E. Cook and A.L. Wolf. Discovering Models of Software Processes from Event-Based Data. *ACM Transactions on Software Engineering and Methodology*, 7(3):215–249, 1998.

9. D. Grigori, F. Casati, U. Dayal, and M.C. Shan. Improving Business Process Quality through Exception Understanding, Prediction, and Prevention. In P. Apers, P. Atzeni, S. Ceri, S. Paraboschi, K. Ramamohanarao, and R. Snodgrass, editors, *Proceedings of 27th International Conference on Very Large Data Bases (VLDB'01)*, pages 159–168. Morgan Kaufmann, 2001.

10. J. Herbst. A Machine Learning Approach to Workflow Management. In *Proceedings 11th European Conference on Machine Learning*, volume 1810 of *Lecture Notes in Computer Science*, pages 183–194. Springer-Verlag, Berlin, 2000.

11. IDS Scheer. ARIS Process Performance Manager (ARIS PPM): Measure, Analyze and Optimize Your Business Process Performance (whitepaper). IDS Scheer, Saarbruecken, Gemany, http://www.ids-scheer.com, 2002.

12. G. Keller and T. Teufel. *SAP R/3 Process Oriented Implementation*. Addison-Wesley, Reading MA, 1998.

13. A.K.A. de Medeiros, A.J.M.M. Weijters, and W.M.P. van der Aalst. Using Genetic Algorithms to Mine Process Models: Representation, Operators and Results. BETA Working Paper Series, WP 124, Eindhoven University of Technology, Eindhoven, 2004.

14. M. zur Mühlen and M. Rosemann. Workflow-based Process Monitoring and Controlling - Technical and Organizational Issues. In R. Sprague, editor, *Proceedings of the 33rd Hawaii International Conference on System Science (HICSS-33)*, pages 1–10. IEEE Computer Society Press, Los Alamitos, California, 2000.
15. W. Reisig and G. Rozenberg, editors. *Lectures on Petri Nets I: Basic Models*, volume 1491 of *Lecture Notes in Computer Science*. Springer-Verlag, Berlin, 1998.
16. Staffware. Staffware Process Monitor (SPM). http://www.staffware.com, 2002.
17. A.J.M.M. Weijters and W.M.P. van der Aalst. Rediscovering Workflow Models from Event-Based Data using Little Thumb. *Integrated Computer-Aided Engineering*, 10(2):151–162, 2003.
18. L. Wen, J. Wang, W.M.P. van der Aalst, Z. Wang, and J. Sun. A Novel Approach for Process Mining Based on Event Types. BETA Working Paper Series, WP 118, Eindhoven University of Technology, Eindhoven, 2004.

High Level Petri Nets Analysis with Helena

Sami Evangelista

CEDRIC - CNAM Paris,
292, rue St Martin, 75003 Paris
evangeli@cnam.fr

Abstract. This paper presents the high level Petri nets analyzer Helena. Helena can be used for the on-the-fly verification of state properties, i.e., properties that must hold in all the reachable states of the system, and deadlock freeness. Some features of Helena make it particularly efficient in terms of memory management. Structural abstractions techniques, mainly transitions agglomerations, are used to tackle the state explosion problem. Benchmarks are presented which compare our tool to Maria.

Helena is developed in portable Ada and is freely available under the conditions of the GNU General Public License.

1 Introduction

Model checking is an automatic method for the verification of finite state systems. It consists of enumerating all the possible configurations or executions of the system to track the ones which do not match the specification.

Helena (a High LEvel Net Analyzer) is a model checker developed at the CNAM university in Paris. The formalism supported by Helena makes it suitable for the verification of concurrent software. Helena is part of the Quasar project (a tool for the analysis of concurrent Ada programs [1]), but it is also a fully autonomous tool which can be used independently. In its current version, Helena can be used for the verification of state properties, i.e., properties that must hold in all the reachable states of the system, and deadlock freeness. Helena is a command-line oriented tool without any graphical user interface though we consider to include a graphical interactive simulator.

This paper is structured as follows. Section 2 explains the reasons which motivated us to design and implement Helena. The main features of Helena are presented in Section 3. A set of benchmarks which compare our tool to Maria [2] are presented in Section 4. Finally we conclude in Section 5.

2 Motivations

The model checking of concurrent programs involves a translation task from the original programming language to a formalism suitable for the expression of concurrency, e.g., Promela, Petri nets. In order to limit the state explosion

G. Ciardo and P. Darondeau (Eds.): ICATPN 2005, LNCS 3536, pp. 455–464, 2005.

problem of a model checking procedure, the produced model has to remain as small as possible, but still must be an exact translation of the program to guarantee a correct verification process. Quasar is a tool which aims at the automatic verification of concurrent Ada programs based on colored Petri nets. It can not currently handle the whole language, though a large part of the language which is related to concurrency is supported. The main task of Quasar is to perform automatic abstraction (or slicing) of a program and to translate it into a colored Petri net. In its current state, Quasar does not perform reachability analysis but interfaces with another tool which realizes this task. The high level formalism supported by Maria [2] pushed us into choosing this tool, though Quasar can also interface with Prod [3].

After an intensive use of Quasar on various examples, we identified two main factors which limit the efficiency of our tool. Firstly, the translation from the source code to the colored Petri net is not straightforward. As the constructions of high level programming language do not have their exact counterpart in colored Petri nets, even in the Maria formalism, the translation step introduces additional transitions which generate intermediate states that are not relevant for the verification purpose. Secondly, the state vectors exhibited by models obtained from the translation of a program are usually quite large as software make heavy use of structured data types. Thus, though optimized to represent multiplicity in a compact way [4], the encoding scheme of Maria fails to represent efficiently state vectors with large color domains, and many tokens in places.

Helena has been designed from this previous experience to meet these two requirements : enable a straightforward and automatic translation of concurrent programs without resort to the introduction of many useless transitions and intermediate states, and handle state spaces with large state vectors.

3   Main Features of Helena

High Level Description Language. The class of high level Petri nets used in Helena was primarily designed to enable the simulation and verification of concurrent Ada programs. To achieve this, we naturally decided to include in Helena the possibility to define high level data types. There are four categories of data types : numerical type, enumerate type, structured type, and vector, i.e., array type. Another feature of Helena is the possibility for the user to define complex functions written in a pseudo-C syntax. These functions may then be used in arc expressions. Such a possibility allows to automatically translate sub programs and sequences of statements which do not include synchronizations into a single transition, provided that the input programming language can easily be mapped into Helena functions.

Compilation and Execution of the Model. A well known and efficient approach to reduce the execution time of a model checking procedure is to compile the model into a source code which will correspond to the actual reachability analyzer. Compiling and executing the model has numerous advantages over an interpretation of the model. Mainly, the evaluation of the expressions in arc mappings can be drastically fasten. Tools that use this technique include Prod, Spin [5], and Maria. It has been shown in [6] that this technique greatly reduces the execution time even for small models for which we may think that the compilation of the generated code is a too severe overhead (which does not exist if the model is interpreted).

Helena models are also translated into executable code. For performance and portability issues we chose the language C. In order to ease the readability of the generated code, and the debugging of the compilation process, this code is commented, nicely formatted, and divided into several libraries.

Enabling Test Algorithm. Verification and simulation tools based on high level Petri nets spend a significant amount of time in determining under which assignments (or bindings) a high level transition is firable. This non trivial problem is known under the term of enabling test. The enabling test algorithm implemented in Helena has been described in [7]. It basically includes two main components.

1. We exploit the locality principle of Petri nets which states that the firing of a transition only affects the status (enabled/disabled) of its neighbor transitions. Thus, a depth first search algorithm can benefit from this locality principle by
   (a) maintaining a set of enabled transitions
   (b) updating this set when a transition instance is fired by only inspecting the neighbor transitions of the fired instance.
   Our implementation of this locality principle is based on the definitions of structural conflict and causality [8]. The basic idea is to translate these relations into equivalent constraints systems [9] before the search algorithm. During the search, the systems built are solved in order to identify disabled and enabled bindings. We illustrate our purpose with an example. Let us consider the net of figure 1.
   From the structure of the net and the arc expressions we can deduce that the firing of an instance $(t, \langle X_t \rangle)$ disables the firing of an instance $(u, \langle X_y, Y_u \rangle)$ if the following constraint is verified : $[X_t = X_u \wedge Y_u = 0]$. During the search, at each firing of an instance of transition t we instantiate this system with the firing binding to identify the instances of u which are disabled. We also observe that the firing of an instance $(t, \langle X_t \rangle)$ enables the firing of an

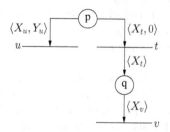

Fig. 1. Translating arc expressions into constraints system

instance $(v, \langle X_v \rangle)$ if the following holds : $[X_v = X_t]$. Consequently, the firing of an instance $(t, \langle X_t \rangle)$ causes the insertion of the instance $(v, \langle X_t \rangle)$ into the set of enabled transitions without any additional check.

2. We showed in [7] that this approach is unfortunately not always sufficient to determine valid transitions assignments at the new marking. In most cases, this must be followed by an unification algorithm. The algorithm we proposed is an improvement of Mäkelä's unification algorithm [10].

State Space Storage. Model checking tools usually represent states in two different forms :

- an expanded form which is convenient for the implementation of the transition relation
- a compressed form in which expanded states are encoded before their insertion into the state space in order to save memory

An interesting feature of Helena is the way it represents compressed states. The method, called Δ-markings method, has been described in [11]. The underlying idea is to store a large set of states in a non explicit way by only storing references on others states. Figure 2 illustrates the principle of the method. Markings met at a depth d such that $d \mod k_\delta = 0$ (with k_δ a user defined parameter) are stored explicitly whereas other markings are represented in the state space by a couple $(pred, (t, c_t))$, where $pred$ is the "address" in the hash table of one of the predecessors of the marking and (t, c_t) is the transition instance which firing leads from the predecessor to the marking. Retrieving the actual value of a marking from its "Δ encoding" can simply be done by following the links which point on the predecessors until a marking stored explicitly is reached. The marking is then obtained by firing the sequence of instances (t, c_t) which label the links followed to reach the explicit marking. For instance, the actual value of marking m can be retrieved by first backtracking to m' and then to m_0 and to apply on it the firing of sequence $(t, c_t).(t', c_{t'})$. For models exhibiting large state vectors, the compression ratios observed are quite impressive, whereas the method becomes less interesting for small models. The price to pay is an acceptable increase of the run time. Both the compression ratio and the run time increase can be influenced by parameter k_δ.

The state collapsing [12] method of Spin is also implemented in Helena. This method is based on the observation that even if the set of syntactical possible values for a token in a place is huge (let us denote it size by m), the set of values really met during the search (which size is denoted by n) is in practice usually much smaller. This is so because the types of the places and transitions of the net are usually over-approximations made by the user of the possible values really met during the search.

With the collapse method a token is represented with $log_2(n)$ bits. This "collapsed token" is in fact an index of a table of size n which stores all the "true tokens" already met during the search on $log_2(m)$ bits. This table is initially empty, and filled during the search by the token values met. Thus, we make use of the two following functions to query the tokens table:

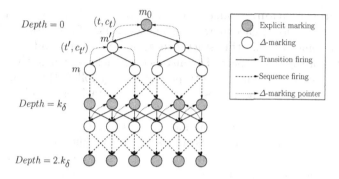

Fig. 2. Illustration of the Δ-markings method

- $index\_of(token\ t) \rightarrow int$ looks for token t into the tokens table and returns the index at which the token has been found. It may also add an entry in the table if the token is not already in it.
- $token\_at(int\ i) \rightarrow token$ returns the token at index i in the tokens table.

Since n can not be known a priori, Helena can detect overflows in this table, i.e., cases where the number of different tokens met is greater than n, and report it to the user who, in turn, can increase it and rerun the search. This restriction may seem quite bothering, but after intensive experiments, we observed that overflows are usually quickly detected during the search, causing only a small run time overhead.

Structural Abstraction Techniques. The efficiency of the explicit model checking approach is seriously limited by the state explosion problem inherent to the concurrent execution of several components. An efficient way to tackle this problem is to perform structural abstractions on the initial model in order to obtain a simpler model which is equivalent to the initial one for to the specified property. For Petri nets and Colored Petri nets, transitions agglomerations are surely the most efficient abstraction techniques. They have been defined by Berthelot in his doctoral thesis [13], and generalized by Haddad to Colored Petri nets in [14]. Transition agglomerations merge consecutive transitions into a virtual atomic one which effect is the composition of the effects of these transitions. It results in a drastic reduction of the combinatory explosion due to the elimination of some intermediate states. In addition, the complexity of these transformations is linear with respect to the size of the net, and their application is totally automatic.

These agglomerations are implemented in Helena. This implementation is based on the symbolic calculus of some structural relations which ensure correct agglomerations in the unfolded net. Though this symbolic computation is not always possible, it usually works fine. To our best knowledge, Helena is currently the only tool that implement structural agglomerations for high level Petri nets.

Stack Representation. Model checking algorithms based on a depth first search, e.g., LTL model checking, store the set of non fully processed states in a stack. For some models, this stack can grow very large, and include almost all the states of the state space. For instance, models for which the state space forms a single strongly connected component usually follow this behavior. Thus, an awkward choice for the representation of this stack can lead to important memory wastes.

In Helena, the stack is represented as a vector of bits, and it uses a nice property of Petri nets : the transition relation is a reversible mechanism, that is, given a marking m and a transition t, there is a single marking m' such that $m'[t\rangle m$. This marking m' is defined by $\forall p \in P, m'(p) = m(p) + W^-(p,t) - W^+(p,t)$. Instead of representing the search stack as a stack of states we therefore chose to represent it as the sequence of transition bindings which leads from the initial marking to the current marking. When a state has been fully expanded, the binding on top of the stack is "unfired", and the next enabled binding is processed. This representation can naturally be combined with the collapse method to store transition bindings of the stack more compactly.

Expressing State Properties. The query language of Helena is rich enough to express a wide range of state properties. It is based on four basic operations :

- The **count** operation allows to count the number of distinct items in a place which fulfill a given condition.
- The **mult** operation allows to count the cumulated multiplicities of distinct items in a place which fulfill a given condition.
- The **exists_token** operation allows to check that at least one token in a place fulfills a given condition.
- The **forall_token** operation allows to check that all the tokens present in a place fulfills a given condition.

For instance, the property

```
forall_token(p in P : exists_token(q in Q : q->1 = p->2))
```

could be stated in an informal way as : for each token p held in place P there is a token q in place Q which is such that the first component of q and the second component of p are equal.

4   Benchmarks

We have compared our tool with Maria to study the performances of both tools concerning time and memory consumption. We considered these six examples :

- The distributed database system
- The slotted ring protocol
- The dining philosophers

- The sieves of Eratosthene to find prime numbers
- The leader election protocol of Chang and Roberts
- The Peterson algorithm for the mutual exclusion problem

All these examples can be found in the Helena distribution (directory `samples`). Experiments were done on a Pentium 4 with 2.5Ghz and a main memory of 1Gb. To allow a fair comparison, Maria was invoked with option `--compile` which allows to compile arc expressions into a dynamically linkable library. In addition, no capacity constraint was indicated in the Maria model and structural agglomerations were disabled in Helena in order to obtain the same numbers of markings and arcs.

The results of our experiments are reported in table 1. For each model, row T reports the times observed for the exploration of the state space in seconds, row S reports the size of the state space in kilobytes, and row V reports the average size of the state vector in bytes. Compilation times of the models are included. Helena was invoked with several values of parameter k_δ within the set $\{1, 5, 10, 20, 50\}$. Let us recall that k_δ is the parameter of the compression method implemented in Helena. The higher this parameter is, the more compressed the state space

Table 1. Results of the Comparison of Helena with Maria

	Maria	Helena				
		$k_\delta = 1$	$k_\delta = 5$	$k_\delta = 10$	$k_\delta = 20$	$k_\delta = 50$
The distributed database system, N=12, 2 125 765 states						
T	900	452	492	533	778	782
S	303 409	336 330	80 111	52 496	16 493	15 997
V	146.15	162.01	38.59	25.29	7.95	7.71
The dining philosophers, N=12, 4 126 351 states						
T	360	306	425	474	549	647
S	57 992	67 947	35 679	31 587	29 527	28 290
V	14.39	16.86	8.85	7.84	7.33	7.02
The sieves of Eratosthene, N=40, 2 028 969 states						
T	116	90	121	151	206	368
S	82 056	92 868	35 274	28 046	24 432	22 294
V	41.41	46.87	17.80	14.15	12.33	11.25
The leader election protocol, N=14, 1 518 111 states						
T	155	103	141	181	264	354
S	29 367	33 258	15 930	13 667	12 786	11 616
V	19.81	22.43	10.75	9.22	8.62	7.84
The Peterson algorithm, N=4, 3 407 946 states						
T	136	91	133	154	188	315
S	35 899	43 366	28 287	26 511	25 630	25 094
V	10.78	13.03	8.50	7.97	7.70	7.54
The slotted ring protocol, N=8, 3 294 720 states						
T	214	164	245	276	325	401
S	37 622	42 760	26 370	23 456	23 342	22 730
V	11.39	13.29	8.20	7.57	7.25	7.06

will be, and the more slow the search will be. If $k_\delta = 1$ then no compression is performed. In addition, the state collapsing method was not helpful for these simple models and therefore disabled.

We observe that the results obtained by Helena without any compression technique are comparable to the ones obtained by Maria. The encoding scheme of Maria gives a slight advantage to Maria in terms of memory usage, while Helena usually explores the state space in less time. When using the Δ-markings method, Helena outperforms its competitor concerning the size of the state space, especially for models with large state vectors (the distributed database system, and the sieves of Eratosthene). The best compression ratio is obtained for the distributed database system, with $k_\delta = 50$. For this case, the size of the state space is divided by almost 19. The drawback is an increase of the execution time. This one is acceptable for low values of k_δ but tends to grow with it. For models with small vectors, e.g., the slotted ring protocol, our storage method is clearly less interesting though a reduction factor of 2, which is the average reduction factor observed for these models, can still be helpful for systems with large numbers of states. Finally, we observe that with our storage method, the average size of the state vector could be reduced to approximatively 10 bytes, whatever the model is.

5   Conclusions and Perspectives

The validation of software is a difficult problem. Few tools based on high level Petri nets have been designed to face this challenge as most of them focus on the representation of control, e.g. tools based on Well Formed Petri nets such as GreatSPN [15], over data. Thanks to multiple features, Helena can handle the validation of software systems.

The possibility to define high level data types and functions enables to model concurrent software written in high level programming languages such as Ada in a succinct way.

The Δ-markings method [11] is implemented in Helena. This one is particularly efficient when dealing with state spaces with large state vectors. This is why we believe that it should be adapted within the scope of the verification of software as these make heavy use of structured data types and usually exhibit large state vectors.

Lastly, structural abstractions help to tackle the state explosion problem.

We plan to extend Helena in the following ways :

- At the current implementation stage, Helena supports the verification of state properties and the deadlock freeness. We plan to extend the field of properties that Helena can verify by including a module for the verification of linear time temporal logic properties (LTL). An interface with the extensible library SPOT [16] is currently under study. The main interest of using SPOT is that it relies on transition-based generalized Büchi automata and allows translation of LTL formula to smaller automata and thus, smaller synchronized products.

- The current version of Helena tackles the state explosion problem by the use of structural abstraction techniques. We envisage to combine these with the stubborn set method of Valmari [17, 18, 19]. However, defining an algorithm which computes good stubborn sets for colored Petri nets is a difficult task. To our best knowledge, the only tool that compute stubborn sets for colored Petri nets (without unfolding) is CPN tools [20]. The algorithm has been described in [19].
- New transitions agglomeration rules have been recently defined for ordinary and colored Petri nets in [21, 22]. An implementation of these agglomerations in Helena is scheduled.

Availability. Helena is a free software available under the conditions of the GNU General Public License. It can be downloaded at the following URL : http://helena.cnam.fr.

References

1. Evangelista S., Kaiser C., Pradat-Peyre J.F., and Rousseau P. Quasar : a new tool for analyzing concurrent programs. In *Proceedings of the Ada-Europe International Conference on Reliable software technologies*, volume 2655 of *LNCS*, pages 166–181. Springer, 2003.
2. Mäkelä M. Maria : modular reachability analyser for algebraic system nets. In *Proceedings of the 23th International Conference on Application and Theory of Petri Nets*, volume 2360 of *LNCS*, pages 434–444. Springer, 2002.
3. Varpaaniemi K. PROD 3.4.00 — an advanced tool for efficient reachability analysis. Laboratory for Theoretical Computer Science, Helsinki University of Technology, Espoo, Finland, 2004. Software.
4. Marko Mäkelä. Condensed storage of multi-set sequences. In *Practical Use of High-Level Petri Nets*, number 547 in DAIMI report PB, pages 111–125. University of Århus, Denmark, 2000.
5. Holzmann G.J. The model checker spin. *IEEE Transactions on Software Engineering*, 23(5):279–295, 1997.
6. Mäkelä M. Applying compiler techniques to reachability analysis of high-level models. In *Workshop on Concurrency, Specification & Programming 2000*, number 140 in Informatik-Bericht, pages 129–142. Humboldt-Universität zu Berlin, Germany, 2000.
7. Evangelista S. and Pradat-Peyre J.F. An efficient algorithm for the enabling test of colored petri nets. In *Fifth Workshop and Tutorial on Practical Use of Coloured Petri Nets and the CPN Tools*, number 570 in DAIMI report PB, pages 137–156. University of Århus, Denmark, 2004.
8. Dutheillet C. and Haddad S. Conflict sets in colored petri nets. In *Proceedings of the 5th International Workshop on Petri Nets and Performance Models*, pages 76–85, 1993.
9. Evangelista S. Syntactical rules for colored petri nets manipulation. Technical Report 641, Cedric, CNAM, http://cedric.cnam.fr, 2004.
10. Mäkelä M. Optimising enabling tests and unfoldings of algebraic system nets. In *Proceedings of the 22th International Conference on Application and Theory of Petri Nets*, volume 2075 of *LNCS*, pages 283–302. Springer, 2001.

11. Evangelista S. and Pradat-Peyre J.F. Efficient state space storage in explicit model checking. Technical Report 682, Cedric, CNAM, http://cedric.cnam.fr/, 2004.

12. Visser W. Memory efficient state storage in spin. In *Proceedings of the Second Spin Workshop*, 1996.

13. Berthelot G. Transformations and decompositions of nets. In *Advances in Petri Nets*, volume 254 of *LNCS*, pages 359–376. Springer, 1986.

14. Haddad S. A reduction theory for colored nets. In *Advances in Petri Nets*, volume 424 of *LNCS*, pages 399–425. Springer, 1989.

15. Chiola G., Franceschinis G., Gaeta R., and Ribaudo M. Greatspn 1.7: graphical editor and analyzer for timed and stochastic petri nets. *Performance Evaluation*, 24(1-2):47–68, 1995.

16. Alexandre Duret-Lutz and Denis Poitrenaud. Spot: an extensible model checking library using transition-based generalized Büchi automata. In *Proceedings of the 12th IEEE/ACM International Symposium on Modeling, Analysis, and Simulation of Computer and Telecommunication Systems (MASCOTS'04)*, pages 76–83. IEEE Computer Society Press, 2004.

17. Valmari A. On-the-fly verification with stubborn sets. In *Proceedings of the 5th International Conference on Computer-Aided Verification*, volume 697 of *LNCS*, pages 397–408. Springer.

18. Valmari A. The state explosion problem. *Lectures on Petri Nets I : Basic Models*, 1491:429–528, 1998.

19. Lars Michael Kristensen and Antti Valmari. Finding stubborn sets of coloured petri nets without unfolding. In *Proceedings of the 19th International Conference on Application and Theory of Petri Nets*, volume 1420 of *LNCS*, pages 104–123. Springer, 1998.

20. Michel Beaudouin-Lafon, Wendy E. Mackay, Mads Jensen, Peter Andersen, Paul Janecek, Michael Lassen, Kasper Lund, Kjeld Mortensen, Stephanie Munck, Anne Ratzer, Katrine Ravn, Soren Christensen, and Kurt Jensen. CPN/tools: A tool for editing and simulating coloured petri nets. In *Proceedings of the 7th International Conference on Tools and Algorithms for the Construction and Analysis of Systems*, volume 2031 of *LNCS*, pages 574–pp. Springer, 2001.

21. Haddad S. and Pradat-Peyre J.F. New powerfull Petri nets reductions. Technical report, Cedric, CNAM, http://cedric.cnam.fr/, 2003.

22. Evangelista S., Haddad S., and Pradat-Peyre J.F. New coloured reductions for software validation. In *Proccedings of the 7th International workshop on discrete event systems*, pages 355–360, 2004.

Protos 7.0: Simulation Made Accessible

Eric Verbeek[1], Maarte van Hattem[2], Hajo Reijers[1], and Wendy de Munk[2]

[1] Department of Technology Management, Eindhoven University of Technology,
P.O. Box 513, NL-5600 MB, Eindhoven, The Netherlands
{h.m.w.verbeek, h.a.reijers}@tm.tue.nl
[2] Pallas Athena BV,
P.O. Box 747, NL-7300 AS Apeldoorn, The Netherlands
service@pallas-athena.com

Abstract. Many consider simulation to be a highly specialist activity: it is difficult to undertake and is even more difficult to understand its outcomes. The new version of the business process modeling tool Protos attempts to more closely integrate modeling and simulation facilities into one tool. The assumed benefit is that business professionals may more easily undertake simulation experiments when they are enabled with the same tool to extend their existing process models to carry out simulation experiments. This paper explains how the existing engine of the Petri-net based tool ExSpect is integrated into Protos 7.0. It also shows the extended user interface of Protos and the simulation reports it generates.

1   Introduction

In the early 90s, it was no less than a revolution when companies started focusing their attention on the performance of entire business processes. In the 21st century, "process thinking" has become mainstream organizational practice [11]. The pursuit of specifically supporting business processes with all kinds of methods, techniques, and systems has become known as Business Process Management (BPM) [2].

For any BPM effort, which includes designing, enacting, controlling, or analyzing a business process, it is important to have a conception of the business process in question. Various approaches exist to model such business processes, but particularly graphical approaches such as the UML and IDEF have become prevalent. Petri nets too have become quite popular in this respect, resulting in various proposed Petri net classes (see e.g. [8]). The influence of Petri nets on other modeling languages is substantial; consider for example how they inspired UML's Activity Diagrams. Next to widespread academic and industrial use of Petri nets for the purpose of business process modeling, various commercial vendors made Petri nets the backbone of their process-aware information systems. Consider, for example, the COSA Workflow Management System and the Dynamic Enterprise Modeler of the SSA ERP$_{LN}$ System (formerly known as BAAN).

G. Ciardo and P. Darondeau (Eds.): ICATPN 2005, LNCS 3536, pp. 465–474, 2005.

The BPM modeling and analysis tool that we will focus on in this paper is Protos from the company Pallas Athena. Current versions of Protos are in use by thousands of organizations in more than 25 countries. In The Netherlands alone, more than half of all municipalities use Protos for the specification of their in-house business processes. Protos is perfectly suitable to model well-defined Petri net structures. Nevertheless, it also permits freehand specifications of business processes without formal semantics, e.g. to support initial and conceptual modeling. In an earlier publication [1] we reported on the use of Protos in the context of the Petri-net based ExSpect tool. We showed how a specific set of Protos models could be automatically translated to be enacted and analyzed with the tool ExSpect (for an introduction, see [7]). Due to the formal Petri net semantics of Protos models, translations to various other process-based systems are feasible as well, e.g. to the workflow management system COSA and the workflow analyzer Woflan [12].

The main use of Protos is to define models of business processes as a step towards either the implementation of quality systems, the redesign of a business process, communication enhancement between process stakeholders, or the implementation of workflow management systems. Until now, Protos only allowed for basic analyses of the models. For example, it is possible to view a Protos process model from different perspectives, i.e. the data, user, or control logic perspective. Considering the main use of Protos, it can perhaps be imagined that a more quantitative analysis of a business process may be valuable. For example, ahead of the implementation of a workflow management system, a simulation of a process assuming such a system to be in place could give insight into the attractiveness of pursuing an implementation project. Similarly, alternative redesign scenarios of a business process could be compared in terms of their ability to meet due dates, execution cost, etc. In short, many of the advantages of simulation seem to be applicable in the realm of business processes, just as they are in the design of more tangible artifacts such as airplanes and cars.

Taking into account the popularity of process modeling, it may come as a surprise that simulation is hardly applied by the same business professionals that use tools such as Protos for modeling. Simulation is considered by many as a highly specialist activity, which is difficult to undertake, while it is even more difficult to understand its outcomes. Protos 7.0 is an attempt to more closely integrate modeling and simulation facilities into one tool. The assumed benefit is that business professionals may more easily undertake simulation experiments when they find out that they could use the same model and the same tool for this purpose. By restricting the simulation settings and the development of a simple user interface, it is expected that the parameterization of a business model is understandable and executable by most business professionals. By creating a standardized simulation report, it is expected that the analysis of a simulation becomes easier too. The main technological means to accomplish the integration is to incorporate the engine of the Petri-net based package ExSpect into the Protos software architecture.

This paper is structured as follows. In section 2, the main architecture of Protos 7.0 is explained. Section 3 presents a short tour through the user interface of the tool, illustrated with an example of the application of Protos 7.0. This paper ends with a conclusion that reflects on the envisioned use of Protos and future work.

2   Architecture

In a bird's eye view, Protos 7.0 embeds the ExSpect COM-server [1] (which we simply call the server from now on) and communicates with the server by means of two XML [4] formatted streams. The first formatted stream is used to communicate the process modeled in Protos to the server; the second stream is used to communicate the simulation report back to Protos. For ease of reference, we will call these streams the process stream and the report stream. Because the formats of both streams give a good overview of the simulation features offered by Protos 7.0, this section will go into details about them.

To map the process stream onto an ExSpect stream, an XSLT [5] file is used. One advantage of doing so is that by replacing this XSLT file we can also feed the process stream to a different simulation server. For example, we can use an XSLT file that maps the process stream onto a CPN Tools [10] stream, and use CPN Tools to obtain a simulation report.

2.1   Process Stream

The following figures are XMLSpy content model views [3], which visualize the XML schema [6] that underlies the process stream. The root element of the

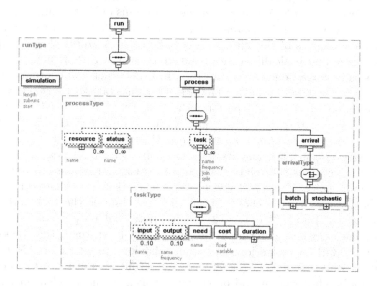

Fig. 1. Format of a process

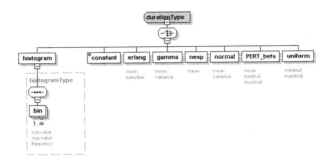

Fig. 2. Format of duration (and histogram)

schema is called *run*, for which Figure 1 shows a high level overview. From this figure, we learn for example, (i) that a *process* element contains (in the given order) a set of *resource* elements, a set of *status* elements, a set of *task* elements, and an *arrival* element, and (ii) that an *arrival* element contains either a *batch* element or a *stochastic* element. For some elements shown by Figure 1 the contained elements are not shown (which would take too much space), which is indicated by a plus sign underneath the element. For these elements, namely *resource*, *batch*, *duration*, and *stochastic*, Figure 2 shows the contained elements, where elements *resource* and *batch* have type *histogramType* and elements *duration* and *stochastic* have type *durationType*.

The *simulation* element prescribes the length of one subrun (*length*), the number of subruns used for measurements (*subruns*), and the number of subruns used to obtain a reasonable starting state (and for which the measurements are discarded) (*start*). Thus, the length of the entire simulation run equals (*start+subruns*)*length*. A *resource* element contains only the name of a particular resource class (or role) (*name*) and a histogram that describes the number of available resources of this class at a specific time. A *status* element contains only its name, and basically corresponds to a place in a Petri net. A *task* element basically corresponds to a transition in a Petri net and contains up to 10 names of input statuses (*input*), 10 names of output statuses (*output*), the name of the resource which should perform the task (*need*), the fixed and variable cost of performing this task (*cost*), and a function for determining the time it takes to perform it (*duration*). Furthermore, the task contains its name (*name*), its frequency (*frequency*), and its join and split behavior (*join*, *split*). A task frequency determines the relative weight of a task and is relevant only if it is involved in a choice. If two tasks with frequencies $f1$ and $f2$ share input places, and if both tasks can be chosen, then the probabilities the tasks are chosen are $f1/(f1 + f2)$ and $f2/(f1 + f2)$. Please note that output statuses contain also frequencies, which determine the chosen output status if that tasks split behavior is XOR. A task's join and split behavior can either be AND (all input/output statuses) or XOR (exactly one input/output status). A *batch* element results in a batch-driven arrival process, whereas a stochastic element results in a case-based arrival process with stochastic inter-arrival times.

2.2   Report Stream

Figure 3 visualizes the XML schema that underlies the report stream. The report contains the following items:

- the utilization rate of a resource class (*resource/utilization*), that is, the part of the time available resources of this class were actually working on some task;
- the wait time of a status (*status/waitTime*), that is, the average time a case has to wait for synchronization in this status;
- the combined wait and queue time of a status (*status/waitQueueTime*), that is, the average time a case spends in the corresponding status;
- the cost of a task (*task/cost*);
- the queue time of a task (*task/queueTime*), that is, the average time a case has to wait for a resource to become available for it;
- the work time of a task (*task/workTime*), that is, the average time it takes to perform this task for any case;
- the sojourn time of cases (*time*), that is, the average time a case spends in the entire process;
- the cost of cases (*cost*), that is, the average cost per case for the entire process;
- the work time of cases (*workTime*), that is, the average work time per case for the entire process.

As Figure 3 shows, the report contains for each item an estimated average (*mean*), a 90% confidence interval (*lo90* up to *hi90*) and a 99% confidence interval (*lo99* up to *hi99*).

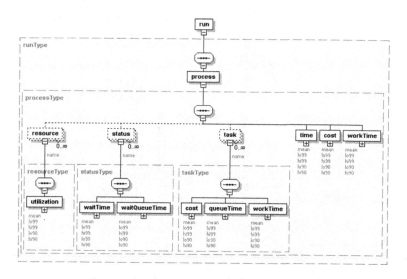

Fig. 3. Format of a report

3   User Interface

The User Interface for the simulation feature in Protos 7.0 includes native forms
to parameterize the process stream, and Microsoft Excel to inspect the (converted) report stream.

3.1   Parameterizing the Process Stream

Figure 4 shows a simplified complaint-handling process of an actual Dutch inspection agency, modeled using Protos. After a complaint has been received, we
first check whether the complaint is for the inspection agency or not. If not, the
complaint will be referred. The inspection agency first stores the relevant data
on both the complainant and the complaint, and then determines the procedure
to be followed etc.

If we take a closer look at the properties of one activity, for instance *Determine procedure*, we discover the role occupied with the execution of this ac-

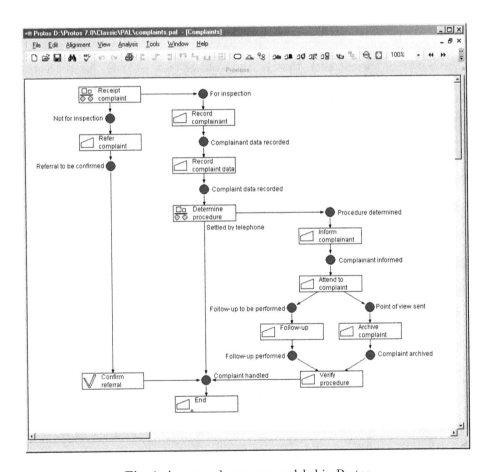

Fig. 4. An example process modeled in Protos

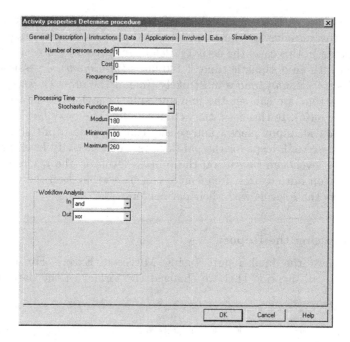

Fig. 5. Setting simulation parameters for activity determine procedure

Table 1. Simulation parameters on non-activity levels

Object	Parameters	
Role	Number of persons	
	Cost	
Connection	Frequency	
Trigger	Frequency	
Proces	Number of Sub-runs	
	Length of Sub-run	
	Number of Start-runs	
	Arrival Time of Cases	Stochastic Function

tivity. We also see the tab to set the simulation parameters; this is shown in Figure 5.

Figure 5 shows the simulation parameters we can set for the activity *Determine procedure*. An activity corresponds to a *task* element in the process stream, and every parameter corresponds to some element from that task element: a number of resources corresponds to a *resource* element (a one-bin histogram), a resource class corresponds to a *need* element, a fixed cost corresponds to the *fixed* attribute of a *cost* element, and a frequency corresponds to the *frequency* attribute.

At the top, we can set the number of resources needed to perform this activity (which resource class is needed is set on the *General* tab), the fixed cost of performing this activity, and its frequency.

At the middle, we can set a stochastic function to determine the time needed to perform this activity for the case at hand (*duration* element) together with its parameters. In this case, the activity *Determine procedure* uses a PERT Beta [9] function with an optimistic time of 100 units (*minimum*), a pessimistic time of 260 units (*maximum*), and a most likely time of 180 units (*modus*).

At the bottom, we can set the join and split behavior of the activity (*join* and *split* attribute). In this case, the activity *Determine procedure* synchronizes (*and*) cases on all input places, and selects exactly one output state (*xor*) to forward the case to. Except for the parameters on the activity level, parameters on four other levels can be set: on the process level, on the resource level, on the arc level (on output arcs, frequencies can be set), and on the trigger level. Table 1 shows these levels with their parameters.

3.2   Inspecting the Report

Figure 6 shows the final report (using Microsoft Excel). For the sake of completeness, we mention that we changed the format of the data cells (two

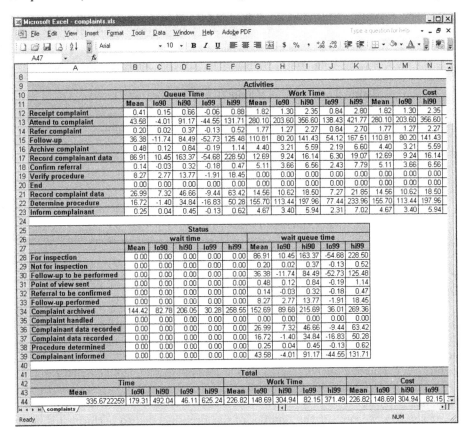

Fig. 6. Report for the example process

digits following the decimal point) and reduced the column width of the data columns.

At the moment, a report can only be inspected using this spreadsheet format. At a later stage, we plan to incorporate the possibility for inspection (and visualization) into the Protos tool.

A report like this reveals a lot of information. For example, from Figure 6 we can conclude that the wait time at the *Complaint archived* status is excessive and needs to be reduced. Furthermore, such a report could be used for validating the process modeled by comparing the results against measurements taken from the real-world situation.

4   Conclusion

In this paper, we have presented Protos 7.0, which extends the popular business process modeling tool Protos with simulation capabilities. We showed how the technical integration of the engine of the Petri-net based ExSpect tool into Protos 7.0 was established. Using a simplified case from practice, we illustrated how Protos models can be extended with various parameter settings so that the simulation of such models becomes feasible. Finally, we showed what the output of a simulation looks like in Protos 7.0 and how such output may be used to improve the design of the process.

As Protos 7.0 has been released only recently, we have no figures on industrial usage of the simulation facility. A beta version of the tool has been introduced recently in a course of the bachelor education program on industrial engineering at Eindhoven University of Technology. This has led to various improvements of the tool, but, more importantly, seems to confirm the relative ease of creating simulation models by non-simulation experts. Considering this and the widespread use of earlier Protos versions, we expect business professionals to get increasingly involved in simulation experiments. Nevertheless, the gathering of accurate data needed as input for a simulation still needs to be performed by professionals.

In upcoming versions of the tool, we plan to improve the tool in several ways. First, we plan to enable the process designer to specify histograms for certain simulation parameters, for example, for resource availability, or for interarrival times. The format of the process stream and the simulation engine already provide support for these histograms, but the user interface does not yet. Second, we plan to integrate the report into the tool. The current version of the tool relies on Microsoft Excel for displaying the report results, while we could display these results in the tool itself. For example, wait times could be displayed near the corresponding status and queue times near to the corresponding activity. Excessive queue and/or wait times and ditto utilization rates, which are likely to correspond to bottlenecks, can even be visualized using, for example, a coloring scheme.

References

1. W.M.P. van der Aalst, P. de Crom, R. Goverde, K.M. van Hee, W. Hofman, H. Reijers, and R.A. van der Toorn. ExSpect 6.4: An executable specification tool for hierarchical colored Petri nets. In M. Nielsen and D. Simpson, editors, *Application and Theory of Petri Nets 2000*, volume 1825 of *Lecture Notes in Computer Science*, pages 455–464. Springer, Berlin, Germany, 2000.

2. W.M.P. van der Aalst, A.H.M. ter Hofstede, and M. Weske. Business process management: A survey. In W.M.P. van der Aalst, A.H.M. ter Hofstede, and M. Weske, editors, *1st International Conference on Business Process Management (BPM 2003)*, volume 2678 of *Lecture Notes in Computer Science*, pages 1–12, Eindhoven, The Netherlands, June 2003. Springer, Berlin, Germany.

3. Altova. XMLSpy online manual: Content model view. http://link.xmlspy.com/manual2005/xmlspy/spyprofessional/contentmodelview.htm, last visited on November 2, 2004.

4. T. Bray, J. Paoli, C.M. Sperberg-McQueen, and E. Maler. eXtensible Markup Language (XML) 1.0 (second edition). http://www.w3.org/TR/REC-xml, 2000.

5. J. Clark. XSL Transformations (XSLT) version 1.0. http://www.w3.org/TR/1999/REC-xslt-19991116, 1999.

6. D.C. Fallside and P. Walmsley. XML Schema part 0, primer second edition. http://www.w3.org/TR/2004/REC-xmlschema-0-20041028, 2004.

7. K.M. van Hee, L.J. Somers, and M. Voorhoeve. Executable specifications for distributed information systems. In E.D. Falkenberg and P. Lindgreen, editors, *Proceedings of the IFIP TC 8 / WG 8.1 Working Conference on Information System Concepts: An In-depth Analysis*, pages 139–156, Namur, Belgium, 1989. Elsevier Science Publishers, Amsterdam.

8. G.K. Janssens, J. Verelst, and B. Weyn. Techniques for modeling workflows and their support of reuse. In W.M.P. van der Aalst, J. Desel, and A. Oberweis, editors, *Business Process Management: Models, Techniques, and Empirical Studies*, volume 1806 of *Lecture Notes in Computer Science*, pages 1–15. Springer, Berlin, Germany, 2000.

9. J.J. Moder and C.R. Phlips. *Project Management with CPM and PERT Second Edition*. Van Nostrand Reinhold, New York, 1970.

10. A.V. Ratzer, L. Wells, H.M. Lassen, M. Laursen, J.F. Qvortrup, M.S. Stissing, M. Westergaard, S. Christensen, and K. Jensen. CPN tools for editing, simulating, and analysing coloured Petri nets. In W.M.P. van der Aalst and E. Best, editors, *24th International Conference on Application and Theory of Petri Nets (ICATPN 2003)*, volume 2679 of *Lecture Notes in Computer Science*, pages 450–462, Eindhoven, The Netherlands, June 2003. Springer, Berlin, Germany.

11. A. Sharp and P. McDermott. *Workflow Modeling: Tools for Process Improvement and Application Development*. Artech House Publishers, Boston, 2001.

12. H.M.W. Verbeek, T. Basten, and W.M.P. van der Aalst. Diagnozing workflow processes using Woflan. *The Computer Journal*, 44(4):246–279, 2001.

Author Index

Lecture Notes in Computer Science

For information about Vols. 1–3444

please contact your bookseller or Springer